							VIII A
	III A	IV A	V A	VI A	VII A		2 $1s^2$ **He** 4.0026
	5 (He) $2s^2 2p^1$ **B** 10.81	6 (He) $2s^2 2p^2$ **C** 12.011	7 (He) $2s^2 2p^3$ **N** 14.007	8 (He) $2s^2 2p^4$ **O** 15.999	9 (He) $2s^2 2p^5$ **F** 18.998		10 (He) $2s^2 2p^6$ **Ne** 20.179
I B	II B	13 (Ne) $3s^2 3p^1$ **Al** 26.98	14 (Ne) $3s^2 3p^2$ **Si** 28.09	15 (Ne) $3s^2 3p^3$ **P** 30.974	16 (Ne) $3s^2 3p^4$ **S** 32.06	17 (Ne) $3s^2 3p^5$ **Cl** 35.453	18 (Ne) $3s^2 3p^6$ **Ar** 39.948

28 (Ar) $3d^8 4s^2$ **Ni** 58.71	29 (Ar) $3d^{10} 4s^1$ **Cu** 63.55	30 (Ar) $3d^{10} 4s^2$ **Zn** 65.37	31 (Ar) $3d^{10} 4s^2 4p^1$ **Ga** 69.72	32 (Ar) $3d^{10} 4s^2 4p^2$ **Ge** 72.59	33 (Ar) $3d^{10} 4s^2 4p^3$ **As** 74.92	34 (Ar) $3d^{10} 4s^2 4p^4$ **Se** 78.96	35 (Ar) $3d^{10} 4s^2 4p^5$ **Br** 79.909	36 (Ar) $3d^{10} 4s^2 4p^6$ **Kr** 83.80
46 (Kr) $4d^{10}$ **Pd** 106.4	47 (Kr) $4d^{10} 5s^1$ **Ag** 107.87	48 (Kr) $4d^{10} 5s^2$ **Cd** 112.40	49 (Kr) $4d^{10} 5s^2 5p^1$ **In** 114.82	50 (Kr) $4d^{10} 5s^2 5p^2$ **Sn** 118.69	51 (Kr) $4d^{10} 5s^2 5p^3$ **Sb** 121.75	52 (Kr) $4d^{10} 5s^2 5p^4$ **Te** 127.60	53 (Kr) $4d^{10} 5s^2 5p^5$ **I** 126.90	54 (Kr) $4d^{10} 5s^2 5p^6$ **Xe** 131.30
78 (Xe) $4f^{14} 5d^9 6s^1$ **Pt** 195.09	79 (Xe) $4f^{14} 5d^{10} 6s^1$ **Au** 196.97	80 (Xe) $4f^{14} 5d^{10} 6s^2$ **Hg** 200.59	81 (Xe) $4f^{14} 5d^{10} 6s^2 6p^1$ **Tl** 204.37	82 (Xe) $4f^{14} 5d^{10} 6s^2 6p^2$ **Pb** 207.19	83 (Xe) $4f^{14} 5d^{10} 6s^2 6p^3$ **Bi** 208.98	84 (Xe) $4f^{14} 5d^{10} 6s^2 6p^4$ **Po** (210)	85 (Xe) $4f^{14} 5d^{10} 6s^2 6p^5$ **At** (210)	86 (Xe) $4f^{14} 5d^{10} 6s^2 6p^6$ **Rn** (222)

In several cases, atomic weight is rounded to four or five significant figures. See inside back cover for listing of 1969 International Atomic Weights. Electron configurations taken from *Theoretical Inorganic Chemistry* by M. Clyde Day and Joel Selbin, Reinhold Publishing Corporation, except numbers 104 and 105 which are predicted by analogy.

[a] Value in parentheses denotes mass number of the most stable known isotope.

[b] Names and symbols for elements 104 and 105 have not been officially accepted. See page 283.

64 (Xe) $4f^7 5d^1 6s^2$ **Gd** 157.25	65 (Xe) $4f^9 6s^2$ **Tb** 158.92	66 (Xe) $4f^{10} 6s^2$ **Dy** 162.50	67 (Xe) $4f^{11} 6s^2$ **Ho** 164.93	68 (Xe) $4f^{12} 6s^2$ **Er** 167.26	69 (Xe) $4f^{13} 6s^2$ **Tm** 168.93	70 (Xe) $4f^{14} 6s^2$ **Yb** 173.04	71 (Xe) $4f^{14} 5d^1 6s^2$ **Lu** 174.97
96 (Rn) $5f^7 6d^1 7s^2$ **Cm** (245)	97 (Rn) $5f^8 6d^1 7s^2$ **Bk** (247)	98 (Rn) $5f^{10} 7s^2$ **Cf** (251)	99 (Rn) $5f^{11} 7s^2$ **Es** (254)	100 (Rn) $5f^{12} 7s^2$ **Fm** (253)	101 (Rn) $5f^{13} 7s^2$ **Md** (256)	102 (Rn) $5f^{14} 7s^2$ **No** (254)	103 (Rn) $5f^{14} 6d^1 7s^2$ **Lr** (257)

FUNDAMENTALS
 OF
COLLEGE
CHEMISTRY

FUNDAMENTALS OF COLLEGE CHEMISTRY

THIRD EDITION

JESSE H. WOOD
PROFESSOR EMERITUS OF CHEMISTRY

CHARLES W. KEENAN
PROFESSOR OF CHEMISTRY

WILLIAM E. BULL
PROFESSOR OF CHEMISTRY

THE UNIVERSITY OF TENNESSEE

HARPER & ROW, PUBLISHERS
NEW YORK, EVANSTON, SAN FRANCISCO, LONDON

The cover design shows the pattern etched in the surface of glass by a solution of HF with added ions. (Courtesy of R. M. Tichane and L. B. Wilson, *J. Appl. Physics*, **35**, 2538 [1964].)

FUNDAMENTALS OF COLLEGE CHEMISTRY, *Third Edition*
Copyright © 1963, 1968, 1972 by Harper & Row, Publishers, Inc.
Printed in the United States of America. All rights reserved. No part of this book may be used or reproduced in any manner whatsoever without written permission except in the case of brief quotations embodied in critical articles and reviews. For information address Harper & Row, Publishers, Inc., 49 East 33rd Street, New York, N.Y. 10016.

Standard Book Number: 06-047202-2

Library of Congress Catalog Card Number: 79-178102

CONTENTS

PREFACE vii
1. Introduction; Some Physical Measurements 1
2. Description of Matter; Periodic Classification 15
3. The Nature of Atoms 29
4. Electronic Arrangements in Atoms 41
5. Chemical Bonding 58
6. Formulas and Equations; Classes of Compounds 75
7. The Mole; Energy and Weight Relationships 94
8. Oxygen and Hydrogen 111
9. Kinetic Theory of Gases; Gas Laws 130
10. Gas Laws and Weight Relationships 146
11. Liquids and Solids 158
12. Water; Solutions 172
13. Concentrations of Solutions; Colligative Properties 191
14. The Colloidal State 207
15. Chemical Kinetics; Equilibria 228
16. Ionic Equilibria 251
17. Nuclear Chemistry 270
18. Inorganic Chemistry: The Alkali and Alkaline Earth Elements 299
19. Groups VIIA and VIIIA: The Halogens and the Noble Gases 314
20. Electrochemistry; Oxidation-Reduction 336

21. Group VIA: The Sulfur Family   363
22. Group VA: The Nitrogen Family   380
23. Carbon and Silicon   396
24. The Transition Metals and Their Neighbors   416
25. Metals: Their Production and Use   431
26. Organic Chemistry I: Hydrocarbons   445
27. Organic Chemistry II: Derivatives of the Hydrocarbons   465
28. Biochemistry   481
29. Applied Organic Chemistry   503
30. Chemical Nature of Our World   516

ANSWERS TO SOME EXERCISES   537

APPENDIX: TABLES   543

  1. Conversion Factors   545
  2. Vapor Pressure of Water at Different Temperatures   545
  3. Solubility Product Constants   546
  4. Solubilities of Common Metal Compounds in Water   547
  5. Four-place Logarithms   548

INDEX   551

PREFACE

This text is written for the student who chooses to take a course in chemistry as a part of his general cultural education in this technological age. The development of the sciences continues to have the most profound effect upon civilization, as indeed it has had for the past three centuries. Our very philosophy is affected, and the way in which we live is changed remarkably. Chemists have contributed to this development in many ways, particularly to our understanding of matter and the changes it undergoes—including the substances of our own bodies—and to the application of scientific knowledge in producing new and improved materials.

Today every citizen is confronted with chemically oriented problems: foods and nutrition, environmental pollution, use of synthetic substances, nuclear energy production, deceptive advertising of consumer products, and so on. A first year's course in chemistry is not meant to fit a person for solving such problems, but we hope that it enables one to ask the appropriate questions and to make an informed choice among the alternatives proposed as one discharges his obligations in a democratic society.

Although this text is not designed for the chemistry major, it does contain enough chemical facts, theories, and principles to provide for a subject-matter centered course. It is not a book about chemistry, but rather a book in chemistry which assumes that to understand the role of chemistry in our society one must understand certain fundamental parts of the science. To achieve this objective we have simplified and condensed many topics that are usually treated at length in a standard one-year course, for example, chemical bonding, thermochemistry, gas laws, and descriptive inorganic chemistry, but we have included more than the usual amount of material on nuclear chemistry, organic chemistry, biochemistry, and geochemistry. Because the treatment differs mainly in emphasis from a more rigorous text for a science major, our policy has been to permit students who thoroughly master the material in our course for nonscience majors to use that course as a prerequisite to further chemistry or other technical courses if they change their career objectives.

Based upon our own experience with the book in two courses—one for home economics majors and the other for nonscience majors—and on queries to about half of the more than 200 users of the first two editions, it seemed clear that the content of the third edition should not differ greatly in rigor or subject matter from that of the second edition, except, of course, for additions and deletions that are needed to keep abreast of the times. Significant changes have been made in the chapter dealing with colloids to bring it more in line with the thinking of modern colloid chemists. The chapter on chemical bonding has been expanded to include a simple discussion of MO theory, and orbital assignments for some homonuclear diatomic molecules have been included. At several places in the text, remarks have been added to relate a specific topic to environmental chemistry. Other changes include updating production figures, rewriting about eighty-five percent of the exercises at the end of chapters, adding material to Chapters 29 and 30 that deals with timely topics, and including recent literature references.

It is hoped that the student who studies this text, or any other chemistry text for that matter, will come to realize that chemistry is not the ogre that some would hold responsible for many present-day ills. One has only to look at nations that have not developed their chemical industries to realize that more of the world's population would be forced to accept hunger and disease as a way of life if the science of chemistry had not reached its present stage in the technically well-developed countries. Before condemning categorically even such a chemical as DDT, one must consider the millions of lives saved from malaria and other infectious diseases, as well as the millions who were better fed because of increased agricultural production. This is not to say that the harmful aspects of chemical products and processes should not be ferreted out and eliminated. But now that the bad effects of DDT have been recognized, it is very likely that chemists can produce a satisfactory substitute that will not have the unacceptable side effect of seriously polluting the environment. Or consider the case of nitrates and phosphates that are carried in the runoff from farm lands into lakes and rivers where they fertilize undesirable plant growth and ultimately make the water unsuitable for game fish, swimming, and boating. Yet to suddenly ban the use of these chemicals in fertilizers would have a devastating effect upon our food supply. These and many other problems of pollution associated with a large population and a high standard of living must be dealt with sanely and unemotionally, and with all the knowledge that chemistry and the other sciences can bring to bear.

Throughout we have endeavored to use systematic nomenclature and symbols, relying principally on the recommendations of *Chemical Abstracts* and the *Journal of the American Chemical Society*. We have provided lists of supplementary readings where appropriate. A systematic survey of pertinent articles was restricted to recent issues of the *Journal of Chemical Education* and to *Scientific American* on the ground that these are readily available to students.

We are greatly indebted to all the chemistry teachers who supplied us with constructive criticisms of the first two editions and to our colleagues at the University of Tennessee for their suggestions and technical advice regarding a variety of topics.

JESSE H. WOOD
CHARLES W. KEENAN
WILLIAM E. BULL

Knoxville, Tennessee

FUNDAMENTALS
  OF
COLLEGE
CHEMISTRY

1

INTRODUCTION; SOME PHYSICAL MEASUREMENTS

Someone has said that the abiding impulse in every human being is to seek order and harmony. This search for order is clearly visible in our study of nature. We see that many objects are rigid, but that others flow easily and do not have definite shapes. We want to distinguish these types and we even give them names—solids and liquids. When we try to describe substances such as tar, butter, and glass, which do not fit well into either the solid or the liquid class, we may be somewhat confused. We may decide arbitrarily to put them into one class or the other, or we may think of another class or a new subclass.

Man's historic attempt to understand his environment has been based in large part upon his success in arranging his growing collection of facts in what seems to him an orderly way. His search for order has been stimulated whenever lack of order has made him dissatisfied. This collection of facts has increased tremendously in recent centuries as a result of his ever-widening interest in careful observation and in purposeful, planned experiments.

SCIENCE

A field of study that deals with the observation and interpretation of reproducible facts is properly termed a **science.** A key word in this definition is reproducible. The facts studied must be such that they can be demonstrated to any interested person. This limits the scope of science a great deal, because if an observation of a fact is to be accepted, any interested person must be able to repeat it again and again. For example, one may doubt that water is always formed when hydrogen burns in air. But the combustion can be repeated until finally even the most skeptical observer must be convinced.

There are four words that are most useful in discussing scientific subjects: fact, hypothesis, law, and theory. We should have a definite understanding of the meaning of each of these words. A **fact** is an event or occurrence. A **hypothesis** is a guess, based on a few facts, which can be used as a basis for reasoning or for planning experiments to gather more information. A **law,** as the scientist defines it, is a concise statement that summarizes a large number of related facts. Sometimes such a statement is referred to as a law of nature. A **theory** is a statement, based on facts and reason, that is designed to explain the facts and laws of nature. A theory is similar to a hypothesis but usually is more formal and based on more facts.

MATTER AND ENERGY

Any study of science concerns itself with the things of nature and with what these things do or what can be done to them. Thus, we have the basic qualities of our environment: matter and energy.

Matter is anything that takes up space and has mass. The earth, the atmosphere, the heavenly bodies of outer space (and any parts of these) are examples of matter.

Energy is the ability to do work. Every action in nature involves energy. The energy of a body or system is that body's or system's capacity to do work. The flight of a bird, the breaking of the earth's crust by a new blade of grass, or, to become prosaic, the washing of dishes or the waxing of floors—all these actions require energy.

There are a number of varieties of energy. Heat energy, electrical energy, potential energy, kinetic energy, radiant energy, chemical energy, and atomic energy are common types.

The **heat energy** of a body is a measure of the internal energy of a substance that is due to its temperature. Heat energy can be transferred from one body to another either by radiation or by conduction (actual contact).

Potential energy may be thought of as the energy a body possesses because of its position, or because of its existence in a state other than its normal state. The water held in a reservoir behind a dam is in a position to do work by turning a turbine or water wheel and hence possesses potential energy. A tightly wound watch spring also possesses potential energy. As the water flows through turbines or the watch spring slowly uncoils, potential energy is transformed into **kinetic energy,** the energy a body possesses by virtue of its motion. A moving automobile, a pitched baseball, an airplane in flight all possess kinetic energy.

Kinetic energy depends on both mass and velocity. A change in the velocity of a moving body has a greater effect on its kinetic energy than has a proportionate change in its mass, because the kinetic energy is proportional to the square of the velocity. To illustrate, three identical automobiles, one moving at 20 mph, one at 40, and one at 60, have kinetic energies in the ratio of 1 : 4 : 9 (i.e., $1 : 2^2 : 3^2$). It would require nine times as much work to bring to a stop the one moving at 60 mph as is required for the one moving at 20 mph. But if the

Introduction; Some Physical Measurements

mass were tripled and the velocity not changed, the kinetic energy would be increased by just three times. That is, the kinetic energy is directly proportional to the mass. The dependence of kinetic energy upon mass and velocity is summed up in the expression:

$$K.E. = \tfrac{1}{2}mv^2$$

Radiant energy is the type of energy associated with electromagnetic radiations such as ordinary light, X rays, radio waves, etc. The velocity at which these radiations travel is 3×10^{10} cm (186,000 miles) per sec in a vacuum. They differ from each other in their wavelengths and frequencies (Fig. 1-1).

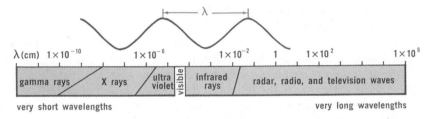

Fig. 1–1. The different types of radiant energy (λ is the wavelength).

Frequently the energy in coal, gasoline, dynamite, food, and an automobile battery is referred to as potential energy. This type of energy is more appropriately called **chemical energy**. Chemical energy is transformed into other kinds of energy when matter undergoes the proper kind of change. For example, when coal and gasoline burn, or when the food we eat is "burned" in our cells, chemical energy is converted into heat or light energy or both. Conversely, other kinds of energy can be transformed into chemical energy by the proper kind of change. Radiant energy from the sun is transformed in a growing corn plant into chemical energy, which becomes associated with the substances making up the stalk, corn grains, and other parts of the plant. It goes without saying that chemical energy is of great importance to man; we shall want to consider carefully its transformation into other kinds of energy as we study the science of chemistry.

A kind of energy called **nuclear energy** or **atomic energy** is associated with the manner in which atoms are constructed. Methods for transforming this type of energy into heat, light, and other kinds of energy have been developed in our generation. Nuclear energy is discussed in detail in Chap. 17.

CHEMISTRY

Chemistry is the branch of natural science that is concerned with the description and classification of matter, with the changes which matter undergoes, and with the energy associated with each of these changes.

The arbitrary divisions of natural studies into physics, astronomy, geology, botany, zoology, chemistry, and other sciences are not classifications imposed by nature, but by man. Chemistry draws heavily on all the sciences for helpful ideas, and it contributes to many areas in all other sciences. Chemistry has been called the servant science, because it supplies descriptions and understanding of the many kinds of matter that are studied in detail in other sciences. Whether a student is interested in medicine, meteorology, home economics, agriculture, or one of the fields of engineering, he will often find that chemical facts and theories are of great importance to him. For the nonscientist, a study of chemistry enriches his understanding of the world of nature and of the somewhat unnatural world of new materials that the chemical industry provides for him.

IMPORTANT MEASUREMENTS. THE METRIC SYSTEM

One of the hurdles which the American student of science must clear is the learning of a new system of weights and measures. In this country the units of measurement used in everyday life are the units of the English system. People in the countries of western Europe use the units of the metric system. The metric system is employed universally in scientific work. A young Frenchman, or Russian, or Italian uses the same system in his everyday life that he uses in studying science. He runs in the 100-meter dash, he drinks a liter of milk a day, and he goes with a girl who weighs about 50 kilograms. If he expected the temperature outside his home to be 35°, he would not bother to wear a sweater or a coat, for he would think of the temperature on the centigrade scale, where a temperature of 35° is equivalent to 95° Fahrenheit—a hot day!

The basic measurements of chemistry, or any science, includes these:

How long?	What is the distance between two points?
How much surface?	What is the area of a body?
How big?	How much space or volume does an object take up?
How heavy?	What is the mass (or weight) of a body?
How big and heavy?	What is the density of the body?
How long in time?	How is time measured?
How hot?	How is temperature measured?
How energetic?	How much work can be obtained from a process or an action? In what units is energy measured?

Length. The basic unit of length in the metric system is the **meter.** This length was defined by the French Academy of Science as one ten-millionth part of the distance from the North Pole to the equator. This distance was measured by the methods of astronomy; then the length corresponding to $\frac{1}{10,000,000}$ of this was carefully marked off on a bar of platinum. This bar, the standard meter, is kept at nearly constant temperature at the International Bureau of Weights and Measures just outside of Paris. In 1960 a new international standard of length was adopted; the meter was defined in terms of the wavelength of the orange-red light emitted by excited krypton. For all but the most precise measurements, however, the older standard meter may still be used.

The meter is equal to 39.37 in., or a little more than a yard. For measuring the small objects of the laboratory, it is usually more convenient to use $\frac{1}{100}$ meter, the **centimeter,** as the unit of length. One inch is equal to 2.54 cm. For measuring extremely small distances, such as the diameters of submicroscopic particles, the **angstrom** (abbreviated A or Å) is used.

$$1 \text{ A} = 1 \times 10^{-8} \text{ cm, or one hundred-millionth of a centimeter.}$$

Introduction; Some Physical Measurements

Area. In the English system a common unit of area for small surfaces is the square inch. In the metric system the **square centimeter** (cm^2) is used.

Volume. The size of an object can be expressed in terms of cubic centimeters (cc or cm^3). It can also be expressed in terms of liters. A liter is equal in volume to 1.06 quarts. As first defined, a liter is the volume of 1 kilogram (kg) of water at 4°C. By this definition, one cubic centimeter is a tiny bit smaller than 1 milliliter (ml), 1000.000 ml equaling 1000.028 cc. Recently, the definition was changed so that one **liter** equals exactly 1000 cc, and 1 ml equals 1 cc.

Volumes of materials used in the laboratory are usually expressed in terms of milliliters, although the cubic centimeter is often employed. These units are used interchangeably.

Mass and Weight. The standard unit of mass was originally defined as the mass of 1000 cc of water at 4°C. This mass was called the **kilogram,** and a piece of platinum of precisely this mass was placed in the International Bureau of Weights and Measures. It was found later, by more precise methods, that the

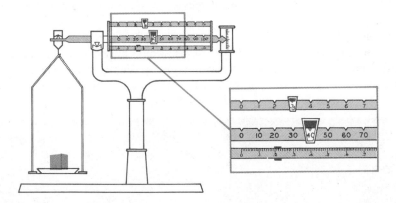

Fig. 1-2. A simple balance for determining masses. On the moon, this balance would show the mass of the object on the pan as being 43.22 g, the same as on earth, although the object's weight on the moon would be about one-sixth of its weight on earth.

piece of platinum so carefully made did not have the same mass as 1000 cc of water. The decision was made to keep the standard kilogram as it was to avoid having to change the standard masses the world over. Therefore, the gram is defined as the one-thousandth part of the standard kilogram. The U. S. Bureau of Standards is the custodian of the standard kilogram mass in our country.

Strictly speaking, mass and weight are different units. **Mass** refers to the quantity of matter in a body, whereas **weight** refers to the gravitational attraction on a body. (See Fig. 1-2.)

A coin that has a mass of 10.0 g also has a weight of 10.0 g when the coin is at sea level on earth. If the coin were on the surface of the moon, its mass still would be 10.0 g, but its weight would be only 1.67 g, because the gravitational attraction of the moon is about one-sixth that of the earth. In this book the terms mass and weight are used interchangeably, because it is assumed that any differences are negligible.

Density. Which is heavier, 20.0 g of lead or 20.0 g of cork? This is an old catch-question, which is answered by the single word—neither. But the question does bring to mind the fact that many people confuse the words "heavy" and "dense." "Heavy" refers to the *mass*, "dense" refers to the *mass per unit volume* of the substance. Consider the lead and cork just referred to. Which has the greater density?

Twenty grams of lead has a volume of only 1.77 ml. The mass of lead contained in 1 ml is its density.

$$\text{Density of lead} = \frac{\text{Mass}}{\text{Volume}} = \frac{20.0 \text{ g}}{1.77 \text{ ml}} = \frac{11.3 \text{ g}}{1.00 \text{ ml}} = 11.3 \text{ g per ml}$$

Twenty grams of cork has a volume of 95 ml. The mass of cork contained in 1 ml is:

$$\text{Density of cork} = \frac{\text{Mass}}{\text{Volume}} = \frac{20 \text{ g}}{95 \text{ ml}} = 0.21 \text{ g per ml}$$

In Fig. 1-3 the volumes of identical masses of some common substances are compared. See also Table 1-1.

From the definitions of the gram and the milliliter in the preceding paragraphs, we can see that 1 ml of water at 4°C would have a mass of exactly 1 g. The density of water, then, is 1 g per ml (varying slightly with changes in temperature). An object with a density greater than 1 will sink in water; an object with a density less than 1 will float.

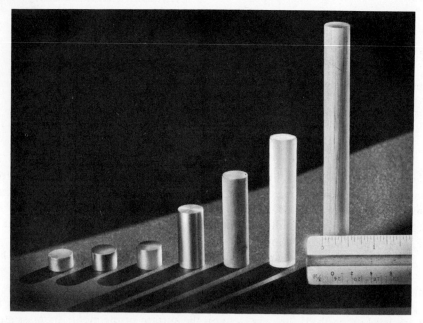

Fig. 1-3. Each rod here has the same diameter, 1.3 cm (0.5 in.), and the same mass, 10 g. Since volumes of the same masses vary inversely with the densities, the 10 g of lead has the smallest volume and the wood the largest. The densities in grams per milliliter are (from left to right): lead, 11.3; copper, 8.9; iron, 7.9; aluminum, 2.7; magnesium, 1.7; Plexiglas, 1.3; maple wood, 0.7.

Introduction; Some Physical Measurements

TABLE 1-1.

Densities of Common Substances in Grams per Milliliter (Room Temperature)

Hydrogen (gas)	0.00009	Table salt	2.16
Carbon dioxide (gas)	0.00198	Sand	2.32
Balsa wood	0.16	Aluminum	2.70
Cork wood	0.21	Iron	7.9
Oak wood	0.71	Silver	10.5
Water	1.00	Lead	11.3
Eucalyptus wood	1.06	Mercury	13.6
Magnesium	1.74	Gold	19.3

Problem 1: What volume of mercury has the same weight as 100 gallon (gal) of water?

Solution: From Table 1-1 we see that the weights of equal volumes of water and mercury are in the ratio of 1.00 wt unit of water to 13.6 wt units of mercury. The volumes of equal weights are in the inverse ratio: 1 volume of mercury to 13.6 volumes of water:

$$100 \text{ gal water} \times \frac{1.00 \text{ gal mercury}}{13.6 \text{ gal water}} = 7.35 \text{ gal of mercury}$$

Time. The amount of time a process takes, or the rate at which an event proceeds, is a fundamental kind of measurement in science. Here we are happy to find that the man in the laboratory uses the same units of time as the man on the street: seconds, hours, and years.

Temperature. In the United States, temperatures are commonly measured in degrees Fahrenheit, a scale named after the German scientist who devised it. On this scale, water freezes at 32°F and boils at 212°F. Scientists everywhere measure temperature in terms of the **centigrade** scale. The 0° and 100° points on this scale are defined in terms of the very common substance, water. The freezing point of water is called 0°C, the boiling point of water is called 100°C. In Fig. 1-4 the two common ways of measuring temperature are compared.

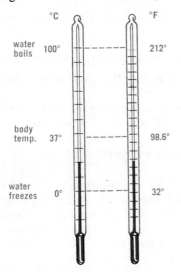

$100° - 0° = 100$ centigrade units
$212° - 32° = 180$ fahrenheit units

$$\frac{100}{180} = \frac{5}{9}$$

$$°C = (°F - 32°)\frac{5}{9}$$

$$°F = °C\left(\frac{9}{5}\right) + 32°$$

Fig. 1–4. A comparison of the centigrade and Fahrenheit temperature scales.

Heat. In what units do we express energy? Heat energy is measured so simply that it is common practice to convert other forms into the equivalent amount of heat energy and then measure the heat. The amount of heat necessary to increase the temperature of 1 g of water 1°C (from 15°C to 16°C) is called a **calorie** (cal). The calorie is the unit of energy measurement most often used by the chemist.

One common error the beginning student of science makes is failing to distinguish properly between a *temperature* change of an object and a change in its *heat* (more correctly, its heat content). The scientist often uses the temperature of a body to determine its energy content, but he must have other information besides. The following examples will help clarify this distinction. Suppose a glass of water and a pail of water are both at room temperature and we then raise their temperature to boiling. Which has more heat energy? Obviously the pail of water does. Even though both are at the same final temperature, 100°C, we had to apply more energy to the pail of water to cause the same temperature change, and consequently the heat content of the pail of water is larger.

As a second example, let us compare the effect of dropping a hot iron nail and a hot aluminum nail into the same quantity of water. If each is dropped into a pint of water at 25°C, we find that a hot iron nail weighing ¼ oz and heated to 650°C will heat the water to 26.4°C, whereas an aluminum nail of the same weight and at the same initial temperature will heat the water to 27.8°C. Additional experiments will show that the metal magnesium changes the temperature of the water more than aluminum does; gold changes it less than the iron does. In cooling approximately the same number of degrees, the aluminum loses about twice as much heat as the iron does. Therefore, it is clear that temperature alone is not a measure of the heat energy a body contains.

The amount of heat needed to raise the temperature of 1 g of a substance 1 centigrade degree is called its **specific heat.** This statement may be abbreviated thus: specific heat = cal/(g × deg C). The specific heats in cal/(g × deg C) of a few substances are: water, 1; aluminum, 0.214; iron, 0.107; alcohol, 0.581; and sand, 0.188.

To calculate the heat in calories gained or lost by a sample of a substance, a useful relationship is:

Heat in calories = (specific heat of substance) × (temp change) × (wt of substance)

Problem 2: Calculate the heat required to raise the temperature of 10 g of water from 10°C to 30°C.

Solution: Since the specific heat of water is 1 cal/(1 g × 1 deg C), the heat needed to raise the temperature of 10 g by 20° (from 10°C to 30°C) is:

$$\text{Heat needed} = \frac{1 \text{ cal}}{1 \text{ g} \times 1 \text{ deg C}} \times 10 \text{ g} \times 20 \text{ deg C} = 200 \text{ cal}$$

Problem 3: A piece of gold weighing 42.0 g was heated to 97.0°C and then quickly dropped into an insulated vessel equipped with thermometer and stirrer and containing 100.0 g of water at a temperature of 21.0°C. See Fig. 1-5. The water was stirred until its temperature ceased to rise. The final temperature was read as 21.9°C. The specific heat of water is known to be 1 cal/(1 g × 1 deg C);

Introduction; Some Physical Measurements

also, let it be given that this particular insulated vessel along with the thermometer and stirrer takes up 10.0 cal for each deg C rise in temperature. From these data calculate the specific heat of gold.

Solution: We can begin with the assumption that the heat lost by the gold is equal to the heat gained by the water and the vessel:

Heat lost by gold = Heat gained by water and vessel

$$\text{Heat gained by water} = \frac{1 \text{ cal}}{1 \text{ g} \times 1 \text{ deg C}} \times (21.9 - 21.0) \text{ deg C} \times 100 \text{ g} = 90 \text{ cal}$$

$$\text{Heat gained by vessel} = \frac{10 \text{ cal}}{1 \text{ deg C}} \times (21.9 - 21.0) \text{ deg C} = 9 \text{ cal}$$

Therefore,

$$\text{Heat lost by gold} = 90 + 9 = 99 \text{ cal}$$

Also,

99 cal = (specific heat gold) × (temp change of gold) × (wt gold)

99 cal = (specific heat gold) × (97.0 − 21.9) deg C × 42.0 g

$$\text{Specific heat of gold} = \frac{99 \text{ cal}}{42.0 \text{ g} \times 75.1 \text{ deg C}} = 0.031 \text{ cal/(g} \times \text{deg C)}$$

NUMERICAL METHODS

Expressing Numbers. Very large numbers and very small numbers are conveniently expressed as powers of 10. This method of expressing numbers is so simple that a few minutes' study will make you sufficiently expert for most purposes. Consider the following:

Ordinary Number	Exponential Form
1	1×10^0
10	1×10^1
100	1×10^2
1000	1×10^3
10,000	1×10^4
0.1	1×10^{-1}
0.01	1×10^{-2}
0.001	1×10^{-3}
0.0001	1×10^{-4}

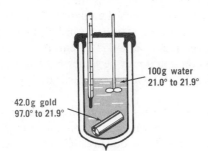

Fig. 1–5. In this insulated vessel, water gains the heat energy given up by the gold.

The number written to the right and above the figure 10 is called an **exponent**. A positive exponent tells how many times a number must be multiplied by 10 to give a certain number. Thus, 1×10^3 means to multiply 1 by 10 three times, so 1×10^3 equals $1 \times 10 \times 10 \times 10$ equals 1000. Conversely, 1×10^{-5} means to *divide* 1 by 10 five times. Hence, 1×10^{-5} equals $1 \div 10 \div 10 \div 10 \div 10 \div 10$ equals 0.00001.

Numbers which are not whole-number powers of 10 are written as follows:

Ordinary Number	Exponential Form
200	2×10^2
340,000	3.4×10^5
0.0000046	4.6×10^{-6}

Note that the number to be multiplied by 10 is always between 1 and 10. We write 2×10^2, not 20×10^1; 4.6×10^{-6}, not 0.46×10^{-5}.

In multiplying numbers expressed as powers of 10, exponents are added; in dividing, exponents are subtracted.

Usual method: $300 \times 4000 \quad\quad = 1,200,000$
Exponential method: $(3 \times 10^2)(4 \times 10^3) = 12 \times 10^5 = 1.2 \times 10^6$
Two other examples: $(7 \times 10^3)(5 \times 10^{-7}) = 35 \times 10^{-4} = 3.5 \times 10^{-3}$
                              $(6 \times 10^4) \div (3 \times 10^5) = 2 \times 10^{-1}$

Any considerable change in the exponent makes a fantastic change in the size of the number. It can be calculated that there are about 1×10^{23} atoms of copper in three pennies. How many atoms are there in the one and a half million tons of copper for electrical use produced annually in the United States? One times ten to the million? Far from it; there are about 1×10^{34} atoms in that huge amount of copper.

Taking Roots. The use of exponents is often of value in taking roots. Suppose $x^2 = 159,000$; to find x, we express the number exponentially, taking care to use an exponent that is divisible by two. To do this, we write, not 1.59×10^5, but 15.9×10^4:

$$x^2 = 15.9 \times 10^4$$
$$x = \sqrt{15.9} \times \sqrt{10^4}$$
$$x = 3.99 \times 10^2 = 399$$

Significant Figures. In recording a measured or calculated quantity, the number of figures used should indicate the precision with which the quantity is known. Measured quantities that have one uncertain figure may be recorded. Suppose we measure the length of a wooden stick and record it as 121.20 cm. It is understood that the last figure is significant even though it may be somewhat uncertain because of our inability to measure it exactly; we have tried to measure it to the nearest 0.01 cm.

In calculating quantities, the available data may vary in precision; i.e., different quantities may have different numbers of significant figures. As an example, consider calculating the volume of a rectangular wooden stick from these measurements: length 121.20 cm, width 3.31 cm, thickness 0.19 cm. In recording these measurements, we imply that in each case the last figure is significant but possibly uncertain. Although we have tried to measure to the nearest 0.01 cm, errors of ± 0.01 cm may have occurred. By simply multiplying length $\times$ width $\times$ thickness (L $\times$ W $\times$ T), we calculate the volume to be 76.222680 cm^3. But to record such a volume would be ridiculous, because it would indicate that we knew the volume precisely to eight significant figures (or to one part in over 76 million). Hence we round off the volume to 76 cm^3. This number has two significant figures, the same number as the least precise of the measurements (the thickness).

An answer to a multiplication or a division should have the same number of significant figures as there are in the least precise item of data.

In the volume calculation above, we could have saved time by rounding the quantities to two significant figures at the outset:

$$L \times W \times T = (1.2 \times 10^2 \text{ cm}) (3.3 \text{ cm}) (0.19 \text{ cm})$$
$$= 75.24 \text{ cm}^3 = 75 \text{ cm}^3$$

This answer is one part in 76 less than the first answer, but for the purpose of an answer with two significant figures it is fine. (A change of one part less in 19 in the thickness would result in an answer that would be even smaller than 75 cm^3.)

Quantities are often known with greater precision than is necessary for a particular problem. Atomic weights, specific heats, solubilities, etc., are usually tabulated very precisely. Although we sometimes need such precise figures, often we do not. The following problem illustrates this point.

Problem 4: Calculate the amount of heat in calories necessary to raise the temperature of 261 g of mercury from 25.0°C to 98.5°C.

Solution: From a handbook we find that the specific heat of mercury is 0.03294 cal/(g $\times$ deg C). The solution is given by:

$$\text{Heat in cal} = \text{cal}/(\text{g} \times \text{deg C}) \times \text{g} \times \text{deg C}.$$

Before multiplying, we round the handbook figure to three significant figures, 0.0330 cal/(g $\times$ deg C).

$$\text{Heat in cal} = 261 \text{ g} \times (98.5 - 25.0) \text{ deg C} \times 0.0330 \text{ cal}/(\text{g} \times \text{deg C})$$

$$\text{Heat in cal} = 633 \text{ cal}$$

A little practice will enable you to determine the number of significant figures in different quantities very quickly. The following examples illustrate some typical cases:

Quantity	Significant figures
1.062 grams	4
751 students	3
0.006110 centimeter	4
1.2×10^8 feet	2
7,685,000 people	4
$683,462.02	8

Factor-Units Method. To help master units of measurement, it is desirable to set down the units along with the numbers that they qualify, and then multiply and divide these units as indicated in solving the problem. The following problems illustrate the handling of units in calculations.

Problem 5: Calculate the area of a table that measures 30 in. by 40 in.
Solution: Area = 30 in. × 40 in.
= (30 × 40)(in. × in.)
= 1200 in.^{2} (i.e., inches multiplied by inches, or 1200 sq in.)

Problem 6: Convert 40 in. to yards. Given: 12 in. = 1 ft; 3 ft = 1 yd.

At this point, it is important to fit the term *per* into our mathematical language by noting that *this term implies division*. Thus, 12 in. = 1 ft may be read 12 inches per foot and written $\frac{12 \text{ in.}}{1 \text{ ft}}$. Similarly, the relationship between feet and yards may be written $\frac{3 \text{ ft}}{1 \text{ yd}}$. In arithmetical operations, we may wish to use the reciprocals of these ratios, i.e., $\frac{1 \text{ ft}}{12 \text{ in.}}$ and $\frac{1 \text{ yd}}{3 \text{ ft}}$, respectively.

Now we set up the ratios or factors so that units cancel to give the conversion of 40 in. to yards.

Solution: Length = $40 \text{ in.} \times \frac{1 \text{ ft}}{12 \text{ in.}} \times \frac{1 \text{ yd}}{3 \text{ ft}} = \frac{40 \text{ yd}}{12 \times 3} = 1.1$ yd

Problem 7: Calculate in square meters the area of a table that measures 30 in. by 40 in. Given: 36 in. = 1 yd; 1 m = 1.09 yd

Solution: Sq m = $\left(30 \text{ in.} \times \frac{1 \text{ yd}}{36 \text{ in.}} \times \frac{1 \text{ m}}{1.09 \text{ yd}}\right) \times \left(40 \text{ in.} \times \frac{1 \text{ yd}}{36 \text{ in.}} \times \frac{1 \text{ m}}{1.09 \text{ yd}}\right)$

$= \frac{30 \times 1 \times 1 \text{ m}}{36 \times 1.09} \times \frac{40 \times 1 \times 1 \text{ m}}{36 \times 1.09}$

= 0.78 m^2 or sq m

Solving problems by setting up factors so that units cancel to give an answer containing the desired unit can be quite helpful. However, this method can also become entirely too mechanical. The problems in this text are useful not as exercises in arithmetic, but as a means of developing your understanding of chemical methods and principles. Consequently, problem solving should begin with an analysis of the terms and language used in the problem. This analysis should lead to a definite prediction of what the answer will be before any precise calculations are made. In general, then, solving a problem should involve:

1. Analyzing and predicting the approximate magnitude and units of the answer.
2. Setting up mathematical equations and carrying out calculations.
3. Checking work by canceling units and inspecting the magnitude of the answer.

The next two problems and their solutions illustrate this three-step attack.

Problem 8: When 12 tons of coke are burned to carbon dioxide, 32 tons of oxygen are required. How much oxygen is needed to burn 100 tons of coke?

Analysis: Because 100 is just over 8 times 12, the amount of oxygen used will be between 8 and 9 times 32 tons, i.e., between 256 and 288 tons of oxygen.

Solution: For purposes of calculation we set up our equation so that the units cancel to give an answer in tons of oxygen.

$$32 \text{ tons of oxygen} \times \frac{100 \text{ tons of coke}}{12 \text{ tons of coke}} = 267 \text{ tons of oxygen}$$

Since we made a thorough analysis of the problem, we have only to compare this answer to our prediction to see that the answer is reasonable. Such an approach should help in detecting careless mistakes, especially decimal point errors, which might lead to such incorrect answers as 26.7 or 2670 tons of oxygen. Note that two of the weights used in the calculation have only two significant figures. Therefore, the answer should be rounded to 270 tons.

Problem 9: A quantity of gas measures 100 ml at a pressure of 15 psi (lb per sq in.). What is its volume at 25 psi? Given: As the pressure on a gas increases, the volume decreases proportionately.

Analysis: The volume will be considerably smaller; i.e., the gas will be compressed. However, the volume will not be reduced to one-half its original amount (i.e., to 50 ml), because the pressure has not been doubled.

Solution:

$$100 \text{ ml} \times \frac{15 \text{ psi}}{25 \text{ psi}} = 60 \text{ ml}$$

THE INTERNATIONAL METRIC SYSTEM

The metric system had its beginning in a committee of the French Academy in 1791. In 1875, the United States signed the Treaty of the Meter, an agreement between 17 nations that provided for the establishment of the International Bureau of Standards (to be situated near Paris, France) and for international conferences on weights and measures. From these conferences, the international metric system has gradually evolved. This system defines six fundamental units: kilogram (mass), meter (length), degree Celsius or Kelvin (temperature), second (time), ampere (electricity), and candela (luminosity). All other units of physical quantities are defined in terms of these six in a manner so as to be consistent with accepted equations of physics.

In 1948, the centigrade scale was designated the Celsius scale, after Andrew Celsius who developed the scale in 1742. There is no practical difference between the old centigrade and the new Celsius scales.

EXERCISES

The questions at the end of each chapter include both numerical and verbal exercises which require the application of principles discussed in the chapter. For problems, your attention is called to the Appendix which contains various data collected in table form as well as a table of logarithms.

The answers to many problems are given in the Appendix. They should be used to check your work only after you have obtained the solution.

1. Choose two of the following occupations and cite specific examples of how a knowledge of chemistry would be of value to a person working in the following

areas: farmer, chef, civil engineer, nurse, weather forecaster, executive of clothing manufacturing plant, druggist, politician, automobile mechanic.
2. Distinguish between hypothesis and theory, between fact and law.
3. Argue both sides of the statement: "The game of football has become a science."
4. Distinguish between weight and mass.
5. Some of the English-speaking countries are in the process of changing from English units of measure to metric units. What are some of the advantages of this change? What are some of the disadvantages?
6. Cite specific examples of the conversion of (a) kinetic energy to electrical energy; (b) chemical energy to electrical energy; (c) heat energy to kinetic energy; (d) radiant energy to chemical energy; (e) potential energy to kinetic energy.
7. Calculate the kinetic energy of a 250-g ball thrown at a velocity of 100 km/hr.
8. Calculate the height of each of the rods shown in Fig. 1–3.
9. Carry out the following; when the numbers are not in exponential form, first convert them to this form:
    (a) $(2.4 \times 10^2) + (8.35 \times 10^3) - (4.78 \times 10^2) =$
    (b) $(6.245 \times 10^3) \times (2.5 \times 10^{-2}) =$
    (c) $187{,}600 \times 5000 \times (2 \times 10^{-2}) =$
    (d) $(6.34 \times 10^1) \div (8.6 \times 10^{-2}) =$
    (e) $1000 \times 2.78 =$
10. A quantity of lead shot weighing 29.4 g was immersed in water in a graduated cylinder. The volume of water in the cylinder was originally 12.6 ml, and the volume of the water and the lead shot was 15.2 ml. Calculate the density of lead from these data.
11. Convert each of the following to °C:
    (a) 0°F, the temperature of a cold day;
    (b) 350°F, the temperature of a moderate oven;
    (c) -175°F, the temperature of Dry Ice;
    (d) 3075°F, the melting point of iron.
12. The distance to the sun is 9.3×10^7 miles. What is this distance in kilometers?
13. The velocity of light is 3.0×10^{10} cm/sec. Calculate the time required for light to travel from the sun to the earth (see Exercise 12).
14. (a) The radius and weight of an atom of hydrogen are 5.3×10^{-9} cm and 1.8×10^{-24} g, respectively. Assuming a spherical atom, calculate its density.
    (b) The radius of the earth is approximately 3960 miles and the earth is estimated to weigh 6.58×10^{21} tons. Calculate the average density of the earth in grams per milliliter.
15. What quantity of heat energy is required to raise the temperature of 600 g of water (a) from 20°C to 35°C and (b) from 40°F to 55°F?
16. A 23.0-g sample of metallic copper at 90°C was placed in 100 g of water at 22.0°C. The temperature of the water increased to 23.4°C. Assuming no loss of heat to the surroundings, calculate the specific heat of the copper.
17. (a) Calculate the weight of 1 liter of magnesium in grams. (b) Twenty-four grams of magnesium contain 6.0×10^{23} atoms of magnesium. How many magnesium atoms are in 1 liter? (c) What is the weight in grams of one magnesium atom?
18. Using conversion factors located in the Appendix, calculate (a) your weight in grams; (b) your height in centimeters; (c) your height in meters; (d) the velocity in miles per hour of a 3000-lb automobile that has a kinetic energy of 5.00×10^{10} g cm^2 sec^{-2}.

2

DESCRIPTION OF MATTER; PERIODIC CLASSIFICATION

PROPERTIES

Every substance—for example, water, sugar, salt, silver, or copper—has a set of characteristics or properties that distinguish it from all other substances and give it a unique identity. Both sugar and salt are white, solid, crystalline, soluble in water, and odorless. But the former has a sweet taste, melts and turns brown on heating in a saucepan, and burns in air. The latter tastes salty, does not melt till heated above red heat, does not turn brown no matter how high it is heated, and does not burn in air, although it gives off a bright yellow light when heated in a flame. We have described both of these substances by listing some of their respective intrinsic properties.

Intrinsic properties are qualities that are characteristic of any sample of a substance, regardless of the shape or size of the sample. *Extrinsic properties* are qualities that are not characteristic of the substance itself. Size, shape, length, weight, and temperature are extrinsic properties.

Physical Properties. A number of intrinsic properties are especially useful in describing materials. One group of these includes taste, odor, color, and transparency. The qualities in this group are common enough, but it is difficult to assign definite values to them expressed in numbers. A second group of intrinsic properties includes melting point, boiling point, density, viscosity, refractive index, and hardness. The qualities in this group can be measured easily and expressed in definite numbers. The substance that we call grain alcohol melts at

−114.6°C (−174.3°F), boils at 78.3°C (173°F), has a density of 0.785 g/ml, and a refractive index of 1.3624. No other substance has precisely this set of unique properties; such a substance is grain alcohol and nothing else.

Intrinsic properties of the type that we have been discussing (color, odor, taste, melting point, density, etc.) are commonly referred to as *physical properties*.

Chemical Properties. The tendency of a substance to change, either alone or by interaction with other substances, and in so doing to form different materials involves a second class of intrinsic properties. For example, it is characteristic of alcohol to burn, of iron to rust, of wood to decay. These characteristics are referred to as *chemical properties*.

CHANGES IN MATTER AND ENERGY

Changes in Matter. The materials around us are subject to constant change. Plant and animal materials decay, metals corrode, water changes to ice when the temperature drops sufficiently and changes back to the liquid form when the temperatures rises, land areas erode, and lakes and seas evaporate. When we study these changes, we find we can classify them under two headings: chemical change and physical change. **Chemical changes** are those which result in the disappearance of substances and the formation of new ones. For example, when pieces of magnesium metal burn in a photoflash bulb, the magnesium and some oxygen from the bulb disappear. In their place, we find a powdery, incombustible solid, magnesium oxide, that has its own unique set of properties. Or, as another example, consider some of the changes in matter that occur as a stalk of corn matures. In this process, carbon dioxide and water disappear in the sense that they are converted to glucose sugar in the growing plant. Much of this sugar accumulates in the ear of the corn, and as the ear matures, the sugar is converted to starch. The glucose sugar that appears has its own set of identifying properties that are completely different from the carbon dioxide and water from which it was made. The starch, in turn, has different properties from the sugar. Chemical changes are also referred to as *chemical reactions*.

The second type of change, **physical change,** is that which does not result in the formation of new substances. For example, when ice melts to water, or when sand is ground to a fine powder, no new substance is formed. However, it should be noted that in physical changes some properties do change and energy transformations do occur.

Exothermic and Endothermic Changes. As a result of every physical or chemical change, there is a change in energy. If substances, or a single substance, change in such a way that energy is given to the surroundings, the change is said to be **exothermic** (heat comes out). The combining form *-thermic* originally referred to heat energy, but exothermic now refers to a change in which any type of energy is given off.

For example, when magnesium burns in oxygen to make magnesium oxide, chemical energy is converted to heat and light energy that is emitted to the surroundings. This is an exothermic chemical change. Or when a hot bowl loses

Description of Matter; Periodic Classification

heat to the table on which it has been placed, the cooling of the bowl is an exothermic physical change.

When carbon dioxide and water are changed to glucose in a growing plant, radiant energy from the sun is converted to chemical energy. Such a change in which materials take up energy from the surroundings is said to be **endothermic.** This formation of glucose is an endothermic chemical reaction. In the example of the hot bowl on the table, mentioned previously, the taking up of heat by the table is an endothermic physical change. For the bowl the change is exothermic, but for the table the change is endothermic.

A process that is exothermic in one direction is always endothermic in the opposite direction, and vice versa. The reaction of magnesium and oxygen to form magnesium oxide is exothermic, whereas the breaking down of magnesium oxide to yield magnesium and oxygen is an endothermic change. In fact, the amount of energy required to break down a given amount of magnesium oxide is equal to the amount of energy given off when that much magnesium oxide is formed. See Fig. 2-1.

Energy is neither created nor destroyed in any transformation of matter. This statement, the **law of conservation of energy,** seems to describe all natural phenomena accurately, whether in the field of chemistry, physics, biology, geology, astronomy, or any other. Consider the following example:

$$\text{magnesium} + \text{oxygen} \rightarrow \text{magnesium oxide} + \text{heat and light}$$

chemical energy   =   chemical energy   +   energy emitted
of the reactants   of the product

If no energy is created or destroyed, the total energy of the system before the reaction equals the energy after the reaction. In this reaction, therefore, the chemical energy of the product must be less than the chemical energy of the reactants, because some chemical energy is transformed to heat and light during the reaction. In any endothermic reaction, just the reverse would be true: the chemical energy of the products would be greater than the chemical energy of the reactants.

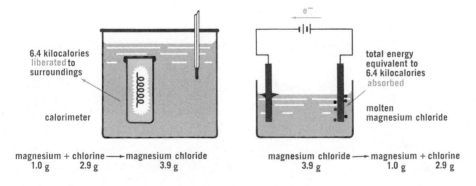

Fig. 2-1. The energy liberated when magnesium unites with chlorine can be measured in a calorimeter; the energy required for decomposing a sample of magnesium chloride can be obtained by measuring the electrical energy required for the decomposition.

EARLY IDEAS ABOUT CHEMICAL REACTIONS

In the period from about 1600 to 1800 in Western Europe, great advances were made in the study of chemical changes, and chemists stated the first laws describing these changes. Although modern research has brought about modification of these early ideas, an understanding of them is fundamental to the study of chemistry.

Classes of Matter. Single pure substances were early classified as either elements or compounds. **Elements** were described as substances that could not be decomposed by simple chemical change into two or more different substances. Some elements familiar to the early chemist were copper, silver, gold, sulfur, carbon, and phosphorus. **Compounds** were described as substances of definite composition that could be decomposed by simple chemical change into two or more different substances. Common salt is an example of a compound. It can be decomposed into a shiny, active metal (sodium) and a poisonous, greenish-yellow gas (chlorine). The properties of the substances obtained by the decomposition of a compound, of course, are completely unrelated to the properties of the compound. Today, just over 100 elements are known, but there are over 1 million compounds; some familiar compounds are water, sugar, alcohol, carbon dioxide, and ammonia.

Most materials cannot be classified as single pure substances, because they contain a number of different substances more or less intimately jumbled together. Such materials are called **mixtures**. A mixture has no unique set of properties; rather, it possesses the properties of the substances of which it is composed. Air is an example of a gaseous mixture; it is composed principally of nitrogen, oxygen, argon, water vapor, and carbon dioxide, and each of these substances displays its own unique properties in the mixture. Further, it is usually possible to separate the components of a mixture by physical rather than chemical changes. For example, when the temperature of air is lowered, water vapor tends to separate as liquid or solid water, that is, as dew or frost. On extreme cooling,

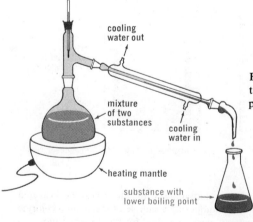

Fig. 2–2. Apparatus for distilling mixtures with widely separated boiling points.

the carbon dioxide solidifies and then the remainder of the air liquefies. If the liquid air is carefully boiled, the mixture can be separated, because each component tends to boil away at a particular temperature range, depending on its own boiling point. This method of separating the substances in a mixture is called distillation. Figure 2-2 shows a typical simple apparatus that is used in laboratories to carry out distillations of common liquid mixtures that contain components whose boiling points are widely separated.

In describing the appearance of materials, the chemist often finds two terms useful: *homogeneous* refers to material in which no differing parts can be distinguished even with a microscope; *heterogeneous* refers to material in which there are visible differing parts.

Law of Conservation of Mass. In the eighteenth century, experimental methods were developed for measuring volumes of gases; weighing gases, liquids, and solids; and conducting chemical reactions in a manner so that the weights of the reactants and the products of a reaction could be precisely measured (Fig. 2-3). Such experiments provided investigators with many facts and led to the

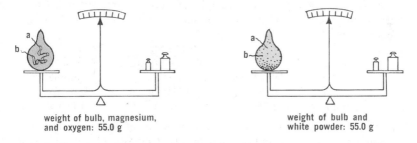

weight of bulb, magnesium, and oxygen: 55.0 g

weight of bulb and white powder: 55.0 g

Fig. 2-3. Magnesium ribbon and oxygen are sealed in a glass bulb and weighed (left). Electrical contact is then made across *a b* to ignite the magnesium. The magnesium disappears completely, leaving a white powder (magnesium oxide) on the walls of the bulb. However, the weight remains unchanged (right).

discovery of a number of fundamental laws describing chemical behavior. According to one of these laws, *mass is neither created nor destroyed in any transformation of matter.* This statement, which sums up the results of thousands of painstaking experiments, is the **law of conservation of mass.** The facts necessary to support this statement were correlated by M. V. Lomonosov, a Russian, in 1756. Perhaps translation difficulties kept his work from becoming widely known in Western Europe. Antoine Lavoisier, a Frenchman, formulated the law independently in 1783. As an illustration of the law, consider the complete combustion of gasoline. The following relationship is true within the limits of our ability to determine the weights[1] of the reacting substances and the products of the reaction:

$$\underset{\text{weight of reactants}}{\text{gasoline} + \text{oxygen}} \quad \underset{=}{\rightarrow} \quad \underset{\text{weight of products}}{\text{carbon dioxide} + \text{water vapor}}$$

---

[1] The interconversion of matter and energy is discussed in Chap. 17.

Law of Definite Composition. To determine the composition of a compound, one can decompose a weighed sample of the compound into its constituent elements and determine their individual weights, or one can determine the weight of a compound formed by the chemical union of known weights of the elements. (Other methods are also available; some of them are discussed in Chap. 7.) The study of the composition of many compounds led to the discovery of the following law: *A pure compound is always composed of the same elements combined in a definite proportion by weight.* This is the **law of definite composition** (also called the law of definite proportions).

By way of illustration consider water, whose composition has been determined by experiment many times. The same answer is always obtained. Water as it occurs in nature is composed of hydrogen and oxygen only, and these elements are always in the proportion of 11.19 percent hydrogen and 88.81 percent oxygen by weight. Or consider table sugar. The composition of sugar (no matter whether it comes from sugar cane, sugar beet, or maple syrup) as determined by analysis is carbon, 42.1 percent; hydrogen, 6.5 percent; oxygen, 51.4 percent.

Problem 1: Analysis of two samples of pure salt, one obtained from a salt deposit in Louisiana and the other from a deposit in New York State, gave the following results:

	Wt of salt	Wt of sodium obtained from sample	Wt of chlorine obtained from sample
Sample 1:	0.2925 g	0.1150 g	0.1775 g
Sample 2:	1.755 g	0.690 g	1.065 g

Show that these data are in accord with the law of definite composition.

Solution: We note that both samples are composed of the same elements, sodium and chlorine (in chemical union). The following calculations show that these elements are present in the same percentages by weight:

$$\text{Percent sodium in sample 1} = \frac{0.1150 \text{ g}}{0.2925 \text{ g}} \times 100 = 39.3 \text{ percent}$$

$$\text{Percent sodium in sample 2} = \frac{0.690 \text{ g}}{1.775 \text{ g}} \times 100 = 39.3 \text{ percent}$$

$$\text{Percent chlorine in sample 1} = \frac{0.1775 \text{ g}}{0.2925 \text{ g}} \times 100 = 60.7 \text{ percent}$$

$$\text{Percent chlorine in sample 2} = \frac{1.065 \text{ g}}{1.755 \text{ g}} \times 100 = 60.7 \text{ percent}$$

Law of Multiple Proportions. Often, two elements unite to form more than one compound. Analyses show that, although the compounds have different compositions by weight, these compositions are related in a simple way. If a given weight of the first element is considered, say 1.00 g, it is found that the weights of the second element that will combine with this 1.00 g are related to each other in the ratio of small whole numbers. To take a specific example, let us consider the two common combinations of carbon with oxygen. With good conditions for burning, carbon burns in air to form a dense, nonpoisonous, noncombustible gas; however, if not enough oxygen is present during the burning, a poisonous, combustible gas is also formed. An analysis of these compounds reveals that each gas has its own definite composition. In the noncombustible gas, 1.00 g of carbon is always combined with 2.67 g of oxygen; whereas in the gas that will burn, 1.00 g of carbon is always combined with only 1.33 g of oxygen. We see that the ratio of the weights of oxygen that combine with the same weight of carbon is 2.67 : 1.33, or 2 : 1.

Iron and chlorine also form two compounds, iron(II) chloride and iron(III) chloride. The amount of chlorine that combines with 1.00 g of iron in iron(II) chloride is 1.26 g; the amount of chlorine combining with 1.00 g of iron in iron (III) chloride is 1.89 g. The weights of chlorine that combine with the same weight of iron are in the ratio of 1.26 : 1.89, or 2 : 3. A formal statement of these facts is known as the **law of multiple proportions:** *When two elements combine to form more than one compound, the different weights of one that combine with a fixed weight of the other are in the ratio of small whole numbers.*

EARLY ATOMIC THEORY

Simple chemical and physical changes have been studied since the dawn of history. Some of the earliest questions that occurred to the student of nature were: What is matter made of? Is sand made of the same stuff that wood is made of? Is a piece of gold made of tiny pieces of gold, or is it made of tiny pieces of several substances mixed together to make gold?

Questions such as these were pondered by the ancient Greeks, and the rational conclusion that some of them came to was this. If a large piece of gold is cut (*tomos*) in half, both pieces are still gold. If these halves could be divided again and then again and again, one would finally get down to the smallest possible piece of gold. This tiny particle could not be cut (*a-tomos* is Greek for not cut), for it would be the unit particle of gold, a gold atom.

For more than 2,000 years after this early rational beginning, the concept of small particles called atoms influenced man's thinking very little. It was not till the latter part of the seventeenth century that the birth of modern chemistry focused attention on the investigation of material things and led to the recognition of basic differences between elementary and complex substances. Indeed, it was not till early in the nineteenth century, about 1803, that the English chemist John Dalton stated his famous atomic theory of matter. Dalton's atomic theory was based directly on the ideas of elements and compounds and on the three empirical laws of chemical combination just described.

Dalton's Atomic Theory. John Dalton was an English schoolteacher who developed the first modern theory of *atoms* as the smallest particles of elements, and *molecules* as the smallest particles of compounds. To explain the properties of elements, he developed the idea that an element contained only one kind of atom and that an atom was a simple, indestructible particle of matter. Elements, he said, could not be changed to simpler substances, because their atoms could not be broken down.

Dalton explained the constant composition of compounds by the theory that atoms of elements were joined to make more complex particles called molecules, which were the simplest units of compounds. According to Dalton, the favored atomic combination for just two elements was probably 1 : 1. Because all the molecules were identical, the compound would have a constant composition, having a greater percentage by weight of the element that had the heavier atom.

The law of conservation of mass was easily explained also. The theory held that in any chemical reaction atoms could change their partners in composition, or molecules could be broken down into atoms; but the total number of atoms in the reactants and the products would be the same. If atoms were indeed indestructible, no mass could be gained or lost in a chemical reaction.

The law of multiple proportions is very nicely accounted for if one assumes that under some conditions atoms of two types combine in a 1 : 1 combination and under other conditions they combine in a 1 : 2 or 1 : 3 or 2 : 3 or some other combination. If we go back to the example that we considered earlier of the two oxides of carbon, we will recall that the ratio of weights of oxygen that combined with a given weight of carbon under two different conditions was 2 : 1. If we assume a 1 : 1 atomic combination in one case and a 1 : 2 in the other, as shown schematically in Fig. 2-4, it is apparent that the weight of oxygen must

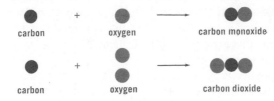

Fig. 2–4. Dalton postulated that in one oxide of carbon one atom of oxygen is combined with one atom of carbon, and that in the other oxide two atoms of oxygen are combined with one atom of carbon.

be twice as great in one as in the other. In modern symbols, a molecule of the first oxide is given the formula CO and is named carbon monoxide. The formula for a molecule of the second is CO_2 and is named carbon dioxide.

Dalton's atomic theory can be summarized by listing the following assumptions:

1. All matter is made up of tiny, indestructible unit particles called atoms.
2. The atoms of a given element are all alike.
3. During chemical reactions atoms may combine, or combinations of atoms may break down; but the atoms themselves are unchanged.
4. When atoms form molecules, they unite in small whole-numbered ratios, such as 1 : 1, 1 : 2, 1 : 3, 2 : 3.

Although some of these assumptions have been shown to be incorrect by later work, Dalton's theory was a guiding principle for a century of brilliant chemical discoveries.

Relative Atomic Weights. Dalton's atomic theory, coupled with the determination of the composition of many compounds, led to the development of a scale of relative weights of atoms. Consider the example of carbon monoxide. We saw earlier that, to form this compound, 1.00 g of carbon combined with 1.33 g of oxygen. If we assume that the carbon monoxide is a collection of billions upon billions of molecules all with the formula CO, it is clear that each oxygen atom must be one-third heavier than each carbon atom. If we could determine the weight of the oxygen atom, the weight of the carbon atom could be calculated.

Although Dalton thought water should have the formula HO, later workers showed that H_2O explained the known facts better. In the case of water, 1 g of hydrogen combines with 8 g of oxygen. If we assume that the compound is a collection of billions upon billions of molecules all with the formula H_2O, it is clear that each oxygen atom must be sixteen times heavier than each hydrogen atom.

It was not possible for Dalton and his contemporaries to determine the weight of a single atom or even to show for sure that atoms existed at all. But they could assume that atoms had definite weights and assign relative atomic weights to them that agreed with the known compositions of compounds.

A few years after Dalton's initial work, oxygen atoms were arbitrarily assigned the relative weight of 16. The weights of other atoms were compared with oxygen by analyzing as many compounds as possible and working out the most likely formulas for these compounds. In the case of the three elements that we have mentioned, the relative weights of H : C : O are 1 : 12 : 16.

In Chap. 7, we say more about an early chemical method of determining relative atomic weights. The chemical method, however, is rarely used today, because of the discovery of a more precise, modern method that we describe in the next chapter.

Problem 2: If we choose to represent the weight of a sodium atom with the number 100, what number must we use to represent the weight of a chlorine atom, based on the data in Problem 1? Repeat the calculation, using 23.0 to represent the weight of a sodium atom.

Solution: We assume that the atoms are combined in a 1 : 1 ratio (this is now known to be true). Since the total weight of the chlorine atoms in sample 1 is 0.1775/0.1150 times heavier than the total weight of the sodium atoms, a single chlorine atom must be 0.1775/0.1150 times heavier than a single sodium atom:

$$\text{Wt of Cl atom} = 100 \times \frac{0.1775}{0.1150} = 154, \text{ based on Na} = 100$$

$$\text{Wt of Cl atom} = 23.0 \times \frac{0.1775}{0.1150} = 35.5, \text{ based on Na} = 23.0$$

EARLY CLASSIFICATION OF ELEMENTS

The concept of an atom as a simple, indestructible particle was generally accepted by scientists during the nineteenth century. One reason for this acceptance was that no methods were available for studying particles as small as atoms. In this chapter we trace the development of an early scheme of classifying elements, and in the next two chapters we take up some of the marvelous discoveries of the present century that revealed details of atomic structure and provided a basis for our modern scheme of classifying elements.

Symbols. The medieval alchemists used symbols to stand for elements, such as a crescent ☽ for silver, symbolic of the silvery color of the moon, or a circle ○ for gold, symbolic of the golden sun and of perfection. Our present system of using letters as symbols was begun by a contemporary of Dalton's, the Swedish chemist J. J. Berzelius (1779–1848). He began by using the first letter of the name of the element as a symbol. Examples include those which we have already used in this chapter: H is the symbol for hydrogen, O for oxygen, and C for carbon. Because the names of several elements begin with the same letter, Berzelius found it convenient to use two letters in some symbols. Thus, carbon, calcium, chlorine, and cobalt are designated by the symbols C, Ca, Cl, and Co, respectively. Note that the first letter of the symbol is capitalized, the second is not. In some cases, the symbols we use today are related to the Latin names that were common centuries ago. For example, the symbols for silver (Ag), copper (Cu), and iron (Fe) are derived from the Latin names *argentum*, *cuprum*, and *ferrum*, respectively. The symbols for the known elements are listed with their names inside the back cover.

Periodic Behavior of Elements. One of the first attempts at grouping similar elements was made by Johann Döbereiner from 1817 to 1830, when he grouped similar elements in threes, or *triads*. He noted that iron, cobalt, and nickel were alike in many ways, as were also chlorine, bromine, and iodine.

Between 1864 and 1869 a number of men, including John Newlands in England, Lothar Meyer in Germany, and Dmitri Mendeleev in Russia, came to realize that an exciting unifying principle applied to the elements that they knew. These early chemists found that, if they listed the elements in the order of increasing atomic weight, elements with similar properties appeared at fairly regular intervals in the list. Properties such as melting points, boiling points, and chemical activities were found to vary in a roughly periodic way, rising, falling, rising again, falling again as the atomic weight increased.

Mendeleev's Periodic Table. Dmitri Mendeleev was the first person to arrange the elements clearly and concisely according to their periodic similarities. In his first periodic table he listed the elements in order of increasing atomic weight unless this order conflicted with his desire to group similar elements. As we shall see in a later chapter, the *atomic number*, a feature unknown to Mendeleev, is a more fundamental guide in correlating properties than is atomic weight. In modern periodic tables the elements are listed strictly in order of increasing atomic number. Among the naturally occurring elements, there are four instances in which an increase in atomic number is not accompanied by an increase in atomic weight: for numbers 18 and 19, 27 and 28, 52 and 53, and 90 and 91. A modern statement of the **periodic law** is: *The physical and chemical characteristics of the elements are periodic functions of their atomic numbers.*

The significance of this law is illustrated by the plot of boiling points versus atomic numbers in Fig. 2-5. We see from this plot that, as the atomic number increases, the boiling points go through a cyclic or periodic type of change. In this case, the change involves the irregular rise to a maximum boiling point and then a decline to a minimum for a period or cycle of elements. Modern data are used in this plot; also, a number of elements not known to the chemists of a century ago are included. It is of interest to find that the colorless, gaseous elements He, Ne, Ar, Kr, and Xe not only have very low boiling points but are the least active chemically of the elements. Because of their resistance to chemical reactions they are called the *noble gas family*. Another group of similar elements, each of which follows one of the noble gases, is Li, Na, K, Rb, and Cs. Not only do these elements have similar boiling points, but each is a soft, malleable

Description of Matter; Periodic Classification

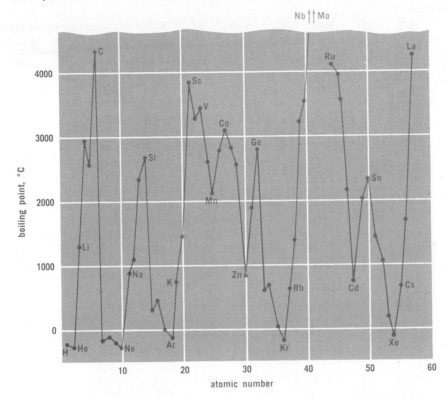

Fig. 2–5. As the atomic numbers increase, the boiling points of the elements vary in a roughly periodic way. Several elements are identified by symbols on the graph; others may be identified by checking their atomic numbers and symbols in Fig. 2–6. The boiling points of niobium and molybdenum are so high that they are off the graph.

metal; and as a group they are the most active chemically of all the metals. Observations such as these indicate that the periodic variation of physical and chemical properties is an extremely important characteristic of the elements.

Figure 2-6 shows a modern periodic table that is arranged on a basis similar to that used in Mendeleev's original table. Above the symbol for each element is its atomic number, and below the symbol is the atomic weight as precisely determined by modern methods.

In 1869 the atomic numbers were unknown, and some of the elements had not been discovered. It was part of Mendeleev's outstanding contribution to chemistry that he recognized a more fundamental way of arranging the elements than a rigid adherence to a regular increase in atomic weight. His arrangement was based largely on periodic physical similarities such as melting points, boiling points, and densities, and on such periodic chemical similarities as the tendencies to combine with other elements and the types of compounds formed. Knowing that there were groups of similar elements, Mendeleev arranged most of the known elements in a table so that similar elements appeared in the same vertical column. This and other features of periodic tables will be taken up in detail later.

Fig. 2-6. A modern periodic table based on Mendeleev's original arrangement. Elements in the darker boxes were not known in 1869. Elements 104 and 105 have been reported by scientists in the U.S.S.R. and the U.S.A., as pointed out in Chap. 17.

Description of Matter; Periodic Classification

EXERCISES

1. Make a list of five compounds and five elements found in your home. For each describe the intrinsic properties that make it suitable for a particular use.
2. Describe how physical changes differ from chemical changes.
3. For each of the following state whether a chemical change or a physical change is occurring: (a) souring of milk; (b) mixing a cake; (c) baking a cake; (d) grinding wheat to flour; (e) melting of butter; (f) bleaching of hair; (g) brewing of coffee.
4. Classify each of the following as either an element, a compound, or a mixture: (a) salt; (b) a fruit cake; (c) sugar; (d) sugar solution; (e) gold; (f) a penny; (g) soil.
5. Suggest experimental procedures that would allow you to determine if brass is a pure substance or a mixture. Do the same for water.
6. Ammonia is a compound of hydrogen and nitrogen. It has a very characteristic odor and a boiling point of $-33°C$, and it is very soluble in water. How might the properties of a mixture of nitrogen and hydrogen differ from this compound?
7. Explain carefully why a piece of aluminum appears to gain weight while a piece of wood appears to lose weight upon burning in air.
8. Upon passing an electric current through water, hydrogen and oxygen are produced. Is this a chemical or physical change? Would this be an exothermic or endothermic change? Explain.
9. If a tennis ball is dropped from a height of 3 ft, its original potential energy is changed to kinetic energy as it falls. After striking the floor, it bounces and regains potential energy as it rises. However, it does not bounce to a height of 3 ft, and if allowed to continue, the ball will come to rest finally on the floor. Is this an exception to the law of conservation of energy? Explain.
10. If 20.0 g of calcium combines with 8.00 g of oxygen, what weight of calcium oxide will be formed when 75.0 g of calcium combines with oxygen?
11. Analysis of a 2.016-g sample of pure magnesium oxide gave 1.216 g of magnesium and 0.800 g of oxygen. A second sample weighing 4.479 g was found to contain 2.701 g of magnesium and 1.778 g of oxygen. Show that these data are in accord with the law of definite composition.
12. A metal forms two oxides, A and B. One gram of A contains 0.881 g of metal, while one gram of B contains 0.788 g of metal. Show that these data are in agreement with the law of multiple proportions.
13. If compound A of Exercise 12 has the formula M_2O, what is the formula of compound B?
14. Discuss how Dalton's atomic theory accounts for the law of conservation of mass during chemical changes.
15. Criticize the statement: "An element is a substance made up of atoms all of which are alike."
16. Dalton used relative atomic weights of 6 for carbon and 8 for oxygen. A few years later these weights were changed to 12 for carbon and 16 for oxygen. Suggest a reason for making this change.
17. Can the periodicity of the properties of the elements be explained in terms of Dalton's atomic theory? How (if answer is yes)? Why not (if answer is no)?
18. Using the data recorded on the graph shown in Fig. 2–5, prepare a list of elements that have boiling points below $0°C$. What group of the periodic table is most highly represented in your list? Would you predict the boiling point of the noble gas radon to be less than $0°C$? Why?

19. Photosynthesis, the process in plants that converts carbon dioxide and water into sugar and oxygen, requires radiant energy from the sun. What happens to this energy?
20. The reverse of any exothermic process must be endothermic. Is this a satisfactory statement of the law of conservation of energy? Explain.
21. About 100 years ago, Mendeleev predicted the existence of several elements not yet discovered. How was he able to do this?
22. Calculate an approximate atomic weight in grams for an atom of kurchatovium (rutherfordium), mass number 260, relative to an atom of hydrogen having a weight of 1.6×10^{-24} g. The atomic number is 104.

SUGGESTED READING

Astin, A. V., "Standards of Measurements," *Sci. Amer.*, **218** (6), 50 (1968).

Coward, H. F., "John Dalton (b. 1766, d. 1844)," *J. Chem. Educ.*, **4**, 22 (1927).

Hammond, C. R., "Collecting the Chemical Elements," *J. Chem. Educ.*, **41**, 401 (1964).

Pinkerton, R. C., and C. E. Gleit, "The Significance of Significant Figures," *J. Chem. Educ.*, **41**, A27 (1964).

Steere, N. V., "Lab Safety," *J. Chem. Educ.*, **41**, A27 (1964).

Wilson, M., "Count Rumford," *Sci. Amer.*, **203** (4), 158 (1960).

3

THE NATURE OF ATOMS

Before continuing our study of chemistry, we will find it most helpful to study in considerable detail the structure of the atom. We will find that present ideas about atoms are quite unlike those of Dalton, which were described over a century and a half ago. In fact, as is true of many theories,[1] man's concept of the atom has undergone almost continual change since Dalton's time and will probably change in the future. The tiny atoms are still the subject of the most serious and careful study. Our knowledge of them increases day by day, and it still appears that we have much to learn.

An important postulate of Dalton's atomic theory dealt with the indivisibility of the atom. To Dalton, the atom could not be subdivided and was, therefore, the ultimate or smallest division of matter. Today it is well established that atoms are complex organizations of matter and energy. Many particles smaller than atoms have been described by physicists. These subatomic particles include the proton, neutron, electron, positron, neutrino, and several types of mesons and hyperons. However, so far as the structure of atoms is related to their chemical behavior, we can confine our attention to just three of these particles: the proton, the neutron, and the electron. Atoms of various kinds differ from one another in the number and arrangement of these three subatomic particles.

---

[1] Today the arguments in favor of atoms are so numerous and convincing that the concept is nearly universally accepted as an established fact rather than a theory.

CHARGED BODIES

Since earliest recorded history men have known about the effect of rubbing two dissimilar objects together; each of the objects, if handled carefully, has a strange effect on certain other objects it touches. A comb rubbed against dry hair attracts strands of the hair or even picks up pieces of tissue paper; if two balls of pith are touched with a glass rod that has been rubbed with a silk cloth, the two balls repel each other. These and many similar observations have led to the conclusion that when two dissimilar objects are rubbed together briskly, one of the objects may take on a positive charge, the other a negative charge. Two important rules resulting from these studies are: (1) an object with one charge attracts an object with a charge of opposite sign; (2) two objects having charges of the same sign repel each other.

Experience with high-voltage discharge tubes has shown that these same rules describe the behavior of particles far too small to be seen even with the most powerful of miscroscopes. Figure 3-1 shows schematically the movement of

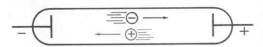

Fig. 3–1. A charged particle is attracted to an electrode of opposite charge and repelled from an electrode of the same charge.

two charged particles, positive (+) and negative (−), caused by the presence of charged wires or plates called electrodes. The positively charged electrode in a discharge tube is called the **anode** and the negatively charged electrode is called the **cathode**. The greater the charges on the electrodes, the greater the attraction they have for the charged particles.

The direction of travel of charged particles can be influenced by bringing electrically charged plates (**electrostatic forces**) or magnets (**magnetic forces**) close to their paths. If the charged plates are not too highly charged, the speeding particles are not drawn to the plates but are only deflected from their original paths (see Fig. 3-2). In the case of a magnet, the charged particles are not

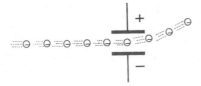

Fig. 3–2. Negatively charged particles in passing between charged plates are deflected toward the positive plate and away from the negative plate.

attracted to either pole of the magnet; instead their path of travel is in a plane that is at right angles to the magnetic field, with positive particles being forced in one direction, negative particles in the opposite direction.

The velocities of particles, their charges and masses, and the magnitudes of the electrical and magnetic forces required to deflect the particles by certain amounts—all these quantities are interrelated by well-established physical laws.

The Nature of Atoms

Knowing the direction in which particles moving in an electrical or magnetic field are deflected and the amount of deflection under carefully controlled conditions, physicists can calculate the charges and masses of particles that are too small to be seen directly.

Finally, we should note two experimental techniques of great practical importance. First, most of the experiments with tiny moving charged particles are carried out in evacuated containers of glass or metal, because particles of air interfere with the free movement of tiny particles. Second, because the moving particles are too small to be seen, special methods are required to follow their motion. These methods depend on the effects that the particles have when they collide with certain substances. Some substances emit visible light (**fluoresce**) when struck by fast-moving particles. Another simple detector is photographic film, which is affected by tiny fast-moving particles in much the same way as it is by being exposed to light. If we put a piece of film in the path of a beam of particles, we can detect the areas in which a number of particles strike the film. The developed film will be darkened in the spots struck by particles; the area and degree of darkening are related to the number of bombarding particles.

The Electron. If two wires are subjected to a high electrical potential and then brought close together, a spark or arc jumps from one wire to the other. If the ends of the two wires are sealed in a glass tube, which is then highly evacuated, the discharge from one wire to the other is more gentle. The greenish glow or luminescence that is emitted from the negative wire, or cathode, is caused by the **cathode ray.**

First studied intensively in 1858 by J. Plücker, cathode rays have been shown to have the following properties:

1. They travel in straight lines away from the cathode.
2. They are negatively charged. This is evident from the fact that they are attracted by a positively charged plate. Also, the rays' path is bent by a magnetic field in the same way as the path of particles known to be negative (Fig. 3-3).
3. The cathode rays consist of particles with definite mass. From the speed of the particles (about one-tenth that of light), their charge, and their deflection in a magnetic field of known strength, the mass of individual particles has been

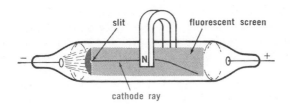

Fig. 3-3. When the direction of travel of charged particles is in a plane at right angles to a magnetic field, the path of the particles is bent but remains in the plane at right angles to the field. The direction of bending depends on the type of charge; the amount of bending depends on the velocity, the mass, and the amount of charge of the particle as well as on the strength of the magnetic field.

calculated as 1/1838 that of the lightest known atom, the hydrogen atom. These particles are called **electrons**.

4. The nature of the cathode rays (electrons) is the same irrespective of (a) the material of which the cathode is made, (b) the type of residual gas present in the evacuated tube, (c) the kind of metal wires used to conduct the current to the cathode, and (d) the materials used to produce the current.

All this evidence, especially the last item, indicates that electrons are fundamental particles found in all matter. No charge smaller than that on one electron has ever been found. For convenience, this charge has been assigned a value of $1-$.

The Proton. Even before the electron was identified, E. Goldstein, in 1886, noticed that a fluorescence appeared on the inner surface of a cathode-ray tube behind a cathode perforated with holes. (See Fig. 3-4.) This finding indicated that positive rays were moving in such a tube, some of which sped through the holes in the cathode and struck the end of the tube.

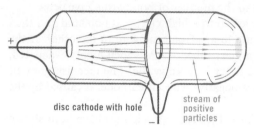

Fig. 3–4. Schematic diagram showing the formation of positive rays (protons) in a discharge tube.

After the discovery of the electron, physicists devoted a great deal of effort to seeking a fundamental particle with a positive charge. Studies of cathode-ray tubes indicated that many different types of positive particles could exist, depending on the gas used to flush out the tube prior to evacuation. In the course of these studies a positive particle was found, which had a *charge* equal in size but opposite in sign to that of the electron and which had a *mass* almost equal to that of a hydrogen atom. This subatomic positive particle, the **proton**, is formed in cathode-ray tubes that contain a little hydrogen gas. It is assumed that the high-speed electrons knock electrons off the neutral hydrogen atoms, leaving the positive nuclei behind. These positive particles travel in a direction opposite to that of the negative cathode rays and can be shown to be particles of unit positive charge with a mass equal to 1837/1838 that of the hydrogen atom.

THE NUCLEAR ATOM

Though the nature of the proton was well established by 1900, the role of this positive particle in the structure of matter was uncertain. A clue to the arrangement of the positive and negative particles within atoms was finally provided by the famous experiments of Lord Rutherford.

The Nature of Atoms

Rutherford's Experiment. One type of particle emitted by radioactive materials is the **alpha particle,** a unit that has a mass about four times that of a hydrogen atom and a charge opposite in sign and twice the magnitude of the charge of the electron, i.e., 2+. Because they have velocities of about 10,000 miles/sec, alpha particles can be thought of as tiny high-speed projectiles.

In 1908–1909 H. Geiger and E. Marsden, working under the great English physicist Ernest Rutherford, reported some of the most meaningful experiments of modern times. By a fluorescent method, they were able to show that alpha particles of low penetrating power could go right through a thin sheet of solid gold. However, the most important facts that they observed were that a few of the alpha particles were deflected from their straight path and that some of these even bounced back from the gold foil. As Rutherford remarked later, "It was almost as incredible as if you had fired a 15-inch shell at a piece of tissue paper and it came back and hit you."

Gold was chosen as the target because it is a very malleable metal that can be beaten into extremely thin sheets, possibly only 100 atoms thick. Like other solids, however, gold can hardly be compressed at all, so one assumes that its atoms are packed tightly together. As shown in Fig. 3-5, the experimenters

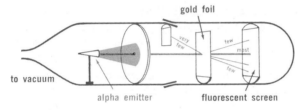

Fig. 3–5. A representation of several experiments in which gold foil was bombarded with a beam of alpha particles.

observed three things, principally: most of the alpha particles went straight through the supposedly closely packed gold atoms; a small percentage of the particles was apparently deflected by something; and a very few of the alphas were bounced back from the gold.

Even for the fertile mind of a genius, it takes time to correlate new facts and to frame theories to explain them. According to Geiger, Rutherford walked into the laboratory on a day in 1911, saying, "I've got it." What he had was a clear mental picture of a relationship between the bombarding alpha particles and the gold atoms. He had worked out an explanation of all the observed facts in terms of a new concept of atomic structure. He pictured a model similar to that shown in Fig. 3-6. The nucleus of each atom is a tiny, massive, positive unit. The nucleus is so small that only rarely does an alpha particle chance to pass near. The nucleus evidently repels alphas, so the nucleus must be positive. The fact that an alpha particle occasionally bounces back shows that the nucleus of a gold atom must be considerably heavier than the bombarding alpha particle. The outer electrons are far away from the nucleus; they form the outside surface of the atom. Between the nucleus and the outer electrons is empty space except for other electrons.

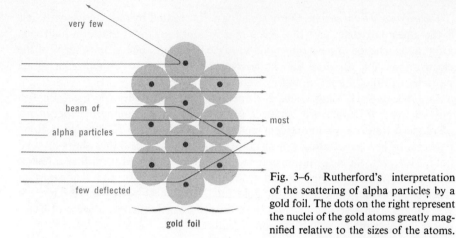

Fig. 3-6. Rutherford's interpretation of the scattering of alpha particles by a gold foil. The dots on the right represent the nuclei of the gold atoms greatly magnified relative to the sizes of the atoms.

Our present information, gleaned largely from X-ray investigations, indicates that the diameter of an atom is about 10^{-8} cm. It has been estimated that a nucleus has a diameter of only about 1/50,000 that of the atom.

Atomic Number. The fact that the proton was found to be about 1837 times as heavy as the electron supported Rutherford's theory of the nuclear atom very well. The light electrons could, at relatively great distances, surround a nucleus that might be made of the heavy, positive protons collected together. Additional support for this theory was given by the experiments of H. G. J. Moseley who discovered that the various elements emitted characteristic X rays when bombarded with fast-moving electrons. (See Fig. 3-7.) When Moseley tabulated his

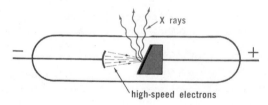

Fig. 3-7. An X-ray tube.

X-ray results, he found that the wavelengths of the X rays emitted decreased regularly as the atomic weights of the elements increased. After studying his data, Moseley was able to state that the number of positive charges on the nucleus "increases from atom to atom by a single electronic unit...." This number of positive charges is now called the **atomic number** and is 1 for hydrogen, 2 for helium, 3 for lithium, etc. We can define an element in terms of modern atomic theory: an **element** is a substance all of whose atoms have the same atomic number.

Of great importance is the fact that Moseley's determination of atomic numbers agreed with Mendeleev's periodic arrangement. It further justified Mendeleev's decision to rely on properties even when the order of increase in atomic weights was misleading. For example, after Moseley's discovery, the atomic numbers of tellurium and iodine were found to be 52 and 53, respectively; this is consistent with Mendeleev's placement of Te before I.

ISOTOPES AND ATOMIC WEIGHT

The Mass Spectrograph. In order to determine if atoms could be thought of as being made solely of protons and electrons, it was necessary to determine the weights of specific atoms and to compare these weights with the total weights of the protons and electrons thought to be involved per atom. Seeking to compare the weights of atomic-sized particles directly, J. J. Thomson and F. W. Aston, just prior to World War I, experimented with the deflection of streams of charged gaseous particles by means of known magnetic and electrostatic forces. After the war, Aston in England and other workers around the world perfected an apparatus called the mass spectrograph with which they compared the weights of atoms precisely.

A schematic diagram of a simple mass spectrograph is shown in Fig. 3-8.

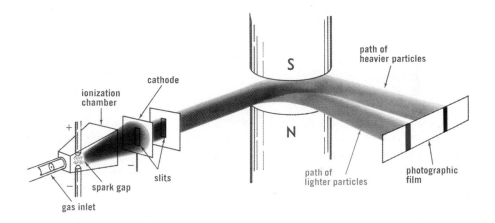

Fig. 3-8. Schematic diagram of a mass spectrograph.

Atoms of some element, say chlorine, are subjected to a high-voltage electric discharge in a spark gap. A fast-moving electron in the electric discharge may hit one of the electrons of a chlorine atom so hard that it knocks an electron off this atom and changes the atom into a particle called an ion. An **ion** is a charged particle that is formed when an atom, or a group of atoms, gains or loses electrons. The ion formed in this case is a positive chlorine ion, Cl^+.

In studying Fig. 3-8, imagine that there are two species of chlorine atoms present, so that ions of two different masses are produced within the ionization chamber in the spark gap and attracted to the slit cathode. The ions passing through both the cathode and the second slit emerge in a thin beam with constant and equal velocities. Upon entering the magnetic field, the ions move in a curved path; the path of the lighter ions is curved more than that of the heavier ions. A separation into two distinct beams results, and these are detected on the photographic film.

In the case of the element chlorine, the photographic film from a mass spectrograph is exposed in two places. This is interpreted to mean that there are heavier

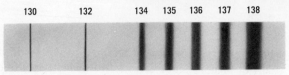

Fig. 3-9. Artist's copy of a mass spectrum of barium isotopes (A. J. Dempster, *Phys. Rev.*, **49**, 947 [1936]).

and lighter atoms of chlorine. An element that has several isotopes has a more complicated mass spectrum (see Fig. 3-9).

Atoms of the same element that have different weights are called **isotopes.** The first isotopes discovered were those of neon by Thomson and Aston in 1912–1913. Since that time isotopes for practically all the elements have been discovered. The mass spectrograph has been developed into a very precise instrument that can determine atomic masses to 1 part in 10,000.

Precise mass spectrographic measurements provide the best means of determining atomic weights of elements. The standard for atomic weights has recently (1961) been established as the most common isotope of carbon. The weight of this isotope is arbitrarily defined as exactly 12 atomic weight units, and all other atoms are compared with it.[2] One **atomic weight unit** (awu) is defined as one-twelfth the weight of one carbon-12 atom.

For chlorine, measurements show that about 75.4 percent of the atoms weigh 35.0 awu and 24.6 percent of the atoms weigh 37.0 awu. The average weight of chlorine atoms can be calculated. Average atomic weight = $0.754 \times 35.0 + 0.246 \times 37.0 = 35.5$. Using more precise measurements, the atomic weight of chlorine has been determined to be 35.453 awu.

In Table 3-1, the weights of the isotopes and the average atomic weights of several common elements are listed.

The Neutron. By means of the mass spectrograph it has been determined that the proton has a weight of 1.0073 awu; the weight of the electron is 0.00055 awu. We see from Table 3-1 that the mass of the lighter hydrogen isotope can be accounted for by picturing a nucleus of 1 proton and 1 electron outside the nucleus. But none of the other cases is so simple. Consider the isotopes of nitrogen. This element has an atomic number of 7, which indicates that nitrogen atoms contain 7 protons and 7 electrons. But such atoms would weigh only a little more than 7 awu, whereas nitrogen atoms weigh about 14 and 15 awu, respectively. All known atoms, except the lighter hydrogen isotope described in Table 3-1, have weights that are much greater than the sum of the weights of their protons and electrons.

The fact that the weight of the protons and electrons does not account for the total masses of atoms led scientists to search for an uncharged particle whose presence would explain the additional weight of an atom but not upset the balance of charges between protons and electrons. Because it had no charge, this elusive particle escaped detection for years. But in 1932 neutrons were

---

[2] Prior to 1961, the standard for chemical atomic weights was natural oxygen, which was given the weight of 16. The adoption of carbon-12 as a standard changed all atomic weights by 43 parts in 1 million: (old atomic weight) = (1.000043) (new atomic weight). Carbon-12 was chosen to replace oxygen, because it is more convenient to use in the mass spectrograph.

The Nature of Atoms

TABLE 3-1.

Isotopic Weights and Average Atomic Weights

ELEMENT	MASS NUMBER (NONRADIOACTIVE ISOTOPES)	ISOTOPIC WEIGHT, AWU	ISOTOPIC ABUNDANCE, PERCENT	AVERAGE ATOMIC WEIGHT, AWU
Hydrogen	1	1.0078	99.98	1.008
	2	2.0141	0.02	
Carbon	12	12.0000	98.9	12.011
	13	13.0033	1.1	
Nitrogen	14	14.0031	99.63	14.007
	15	15.0001	0.37	
Oxygen	16	15.9949	99.76	15.9994
	17	16.9991	0.04	
	18	17.9992	0.20	
Sulfur	32	31.972	95.06	32.064
	33	32.971	0.74	
	34	33.968	4.18	
	36	35.967	0.02	
Chlorine	35	34.969	75.53	35.453
	37	36.966	24.47	

identified by J. Chadwick.[3] A **neutron** has no charge and has a mass of 1.0087 awu, i.e., about the same as that of a proton.

With the discovery of the neutron, chemists were able to account for the main features of atoms in terms of three fundamental particles (see Table 3-2). An atom is defined as an extremely small, electrically neutral particle that has a tiny but massive positive core or nucleus and one or more electrons relatively far outside its nucleus. The number of protons plus neutrons in the nucleus is the

TABLE 3-2.

Fundamental Particles of Matter

NAME	SYMBOL	MASS, AWU	RELATIVE CHARGE[a]
Proton	p	1.00728	+1
Neutron	n	1.00867	0
Electron	e^-	0.00055	−1

[a] The charge of an electron is defined as the elementary unit of charge. Its absolute value is 4.80298×10^{-10} electrostatic unit (esu).

---

[3] The experimental details of the identification of the neutron are presented in Chap. 17.

mass number of the atom.[4] It follows that the mass number minus the number of protons equals the number of neutrons. Consider the heaviest isotope of sulfur listed in Table 3-1:

$$\text{mass number} - \text{atomic number} = \text{number of neutrons}$$
$$36 \qquad\qquad 16 \qquad\qquad 20$$

To indicate a specific isotope in a concise way, the symbol for the element is used with the mass number at the upper left and the atomic number at the lower left. For example, the isotope of sulfur just mentioned is designated as $^{36}_{16}S$, and the two isotopes of chlorine discussed earlier as $^{35}_{17}Cl$ and $^{37}_{17}Cl$.

SUMMARY. Atoms are small particles of matter with tiny, but massive, positive nuclei, which are surrounded by electrons. The total space occupied by the electrons in an atom is quite large relative to the volume of the nucleus. The following rules serve to help predict the composition of individual atoms:

1. The atomic number is the number of protons.
2. The number of electrons equals the number of protons.
3. The mass number is the total number of neutrons and protons.
4. The isotopes of an element differ from each other only in the numbers of neutrons in their nuclei.

Let us apply these rules to working out the composition of neon atoms of mass numbers 20 and 21. Reference to the periodic table shows the atomic number to be 10. Hence, each atom contains 10 protons as a part of the nucleus and 10 electrons surrounding the nucleus. The number of neutrons is 10 and 11, respectively (see rule 3).

Neon atom of mass number 20: 10 protons, 10 neutrons, 10 electrons
Neon atom of mass number 21: 10 protons, 11 neutrons, 10 electrons

The organization of the electrons about the nucleus will be discussed in the next chapter.

WEIGHTS AND SIZES OF ATOMS

Atoms and subatomic particles are so minute that it is difficult for us to compare them with any familiar object. A proton weighs

0.00000000000000000000000000368 lb (i.e., 3.68×10^{-27} lb)

or

0.000000000000000000000000167 g (i.e., 1.67×10^{-24} g)

and an electron weighs only 1/1837 of this, or 9.11×10^{-28} g. Because we are usually interested in comparing atoms and subatomic particles with one another, we find it convenient to speak of their weights in atomic weight units. It requires about 602,300,000,000,000,000,000,000, or 6.023×10^{23} awu, to equal 1 g.

Just how big are particles of atomic dimensions? No atom has ever been seen, even through the most powerful optical microscopes. Physicists have been able,

---

[4] The alert student will find that the weight of a particular isotope of an element is not precisely equal to the sum of the weights of the protons, neutrons, and electrons that it contains. A detailed discussion of this interesting point is delayed till Chap. 17.

TABLE 3-3.

Measurements, Great and Small

Distance to farthest observed galaxy	5.0×10^9 light-years	4.5×10^{27} cm
Distance to nearest star	4.3 light-years	4×10^{18} cm
Diameter of earth	8000 miles	1.3×10^9 cm
Height of average man	68 in.	1.7×10^2 cm
Diameter of a penny	0.75 in.	1.9 cm
Diameter of red blood cell	3×10^{-4} in.	7.6×10^{-4} cm
Diameter of smallest virus	4×10^{-7} in.	1×10^{-6} cm
Diameter of uranium atom	1×10^{-8} in.	2.8×10^{-8} cm
Diameter of hydrogen atom (smallest atom)	4×10^{-9} in.	1×10^{-8} cm
Diameter of proton	4.7×10^{-14} in.	1.2×10^{-13} cm

however, to measure the volume occupied by a known number of atoms and thus calculate the volume of the individual particle. If the atom is assumed to be spherical, its diameter can be calculated. Table 3-3 lists a number of common measurements of large and small objects. From the value of 2.5×10^{-8} cm given for the diameter of the uranium atom, we can calculate that it would take a line of 7.6×10^7, or 76 million, uranium atoms placed side by side to span the diameter of a penny, a distance of 1.9 cm. Man's successes in measuring the vastness of the universe and the minuteness of the atom stand as two of his great experimental achievements.

Tiny fractions of centimeters are as inconvenient for expressing lengths of atomic size as are tiny fractions of grams for expressing weights. A common unit for describing the sizes of atoms is the **angstrom** (A), defined in Chap. 1. Atoms range in size from the hydrogen atom, with a diameter of about 1 A, to the cesium atom, with a diameter of about 5 A.

EXERCISES

1. How do X rays differ from the light produced by an ordinary light bulb? From cathode rays?
2. Describe an experiment to show that cathode rays consist of particles which have mass.
3. How are the atomic numbers of elements determined?
4. If the elements are arranged in order of increasing atomic number, how do the successive elements differ in (a) number of protons; (b) number of electrons; (c) number of neutrons; (d) nature of X rays produced from anodes made of the elements; (e) atomic weight?
5. What evidence indicates that electrons are fundamental particles of *all* substances?
6. Is it probable that a new element with an atomic number less than 80 will be discovered as man expands his exploration of space and spatial bodies? Explain.

7. What led Lord Rutherford to postulate that an atom is largely empty space with most of its mass concentrated in a small positively charged core?
8. Zinc has an atomic number of 30 and an atomic weight of 65.37 awu.
   (a) An atom of zinc contains how many protons? How many electrons?
   (b) Do all zinc atoms contain the same number of neutrons? Explain.
9. Describe two ways by which one might decide experimentally if a beam of fast-moving particles consists of (a) protons or electrons; (b) neutrons or protons.
10. Distinguish between the mass number of an atom and the mass of an atom.
11. Why did Lord Rutherford choose gold for the target material in his experiments? Could some other material have been used? If so, what material?
12. Calculate the number of atoms in (a) 1.0 g of hydrogen; (b) 1.0 g of carbon; (c) 1.0 g of uranium.
13. Magnesium has three naturally occurring isotopes of masses 23.9924, 24.9938, and 25.9898 awu. These have abundances of 78.6, 10.1, and 11.3 percent, respectively. Calculate the average atomic weight of magnesium.
14. The radius of an atom of copper is 1.29 A. How many copper atoms are necessary to span a distance of 1 ft?
15. There are three isotopes of hydrogen called protium, deuterium, and tritium, with mass numbers of 1, 2, and 3, respectively. How do these isotopes differ? How are they alike?
16. What is meant by the statement: "The average weight of boron atoms is 10.81 awu"?
17. Answer the following questions about the mass spectrograph:
    (a) Why is it necessary to produce ions in the mass spectrograph?
    (b) The ions are produced, travel, and are finally detected in an evacuated chamber. How?
    (c) What is the purpose of the magnet?
    (d) Why must the ions emerge from the slits with constant and equal velocities?
18. Prior to 1961, the standard for chemical atomic weights was natural oxygen, which was given a weight of exactly 16. Today, with carbon-12 as the standard, natural oxygen has an average weight of 15.9994 awu. Calculate the pre-1961 atomic weight of chlorine from the presently accepted value (listed inside the back cover).
19. A sample of metal M was found to react with 0.3475 g of oxygen and to produce 1.4750 g of metal oxide. If the ratio of metal atoms to oxygen atoms in the oxide is 1:1, calculate the atomic weight of M.

SUGGESTED READING

Davies, M., "The Electromagnetic Spectrum in Chemistry," *J. Chem. Educ.*, **31**, 89 (1954).

Duveen, D. I., and H. S. Klickstein, "John Dalton's 'Autobiography,'" *J. Chem. Educ.*, **32**, 333 (1955).

Garrett, A. B., "The Nuclear Atom: Sir Ernest Rutherford," *J. Chem. Educ.*, **39**, 287 (1962).

Garrett, A. B., "X-Rays, W. C. Roentgen," *J. Chem. Educ.*, **39**, 360 (1962).

Labbauf, A., "The Carbon-12 Scale of Atomic Masses," *J. Chem. Educ.*, **39**, 282 (1962).

Redfern, J. P., and J. E. Salmon, "Periodic Classification of the Elements," *J. Chem. Educ.*, **39**, 41 (1962).

Standen, A., "The Fairy Story of Atomic Weights," *J. Chem. Educ.*, **24**, 143, 453 (1947).

4

ELECTRONIC ARRANGEMENTS IN ATOMS

As discussed in the preceding chapter, interesting and important discoveries relating to the nucleus of the atom were made at the beginning of the twentieth century. At that time the puzzle of electron arrangement was also being intensively investigated by many persons. One of the properties of elements that had been studied carefully for many years, since the invention of the spectroscope in 1859, was the radiation emitted by excited elements. It was in 1913 that Niels Bohr showed how the fascinating wavelength patterns revealed by the spectroscope could be related to electronic structures of atoms, an achievement for which he received the Nobel Prize in 1922. Today the Bohr concept of electrons traveling in circular and elliptical orbits about the positive nucleus, like planets around the sun, is known to be an oversimplified picture. A precise physical model is still not available, although great strides toward a mathematical description have been made.

ATOMIC SPECTRA

Emission Spectra. When an element absorbs sufficient energy, for example, from a flame or an electric arc, it emits radiant energy. The radiation emitted may fall within the range of visible light, but not necessarily so. When this radiation is passed through a prism in a spectroscope, it is separated into component wavelengths to form an image called an emission spectrum. Emission spectra are of two types: *continuous* and *discontinuous*. For the latter, the image (spectrum) consists of a characteristic pattern of bright lines on a dark field.

Although any element can be heated to incandescence, some elements (or their compounds) have only to be heated in a bunsen flame to make them emit a characteristic colored light. Such elements include lithium, sodium, potassium, calcium, and strontium. A convenient method of testing a substance for the presence of these elements is to dissolve a bit of the material in water and then dip a loop of platinum wire into the solution. If a droplet of the solution is carefully evaporated on the wire and then heated in the hot flame of a laboratory burner, the flame has a color characteristic of the elements present.

An element can often be identified by visual observation of the flame and reference to a list such as this one:

Element	Flame color	Element	Flame color
Lithium	Red	Cesium	Blue
Sodium	Yellow	Calcium	Orange-red
Potassium	Violet	Strontium	Brick red
Rubidium	Red	Barium	Green

For truly positive identification visual observation is insufficient; for example, many persons would have difficulty in distinguishing between a lithium and a strontium flame.

Precise analysis of the color of a flame can be made with a relatively simple prism spectroscope, such as the one shown schematically in Fig. 4-1. The heart

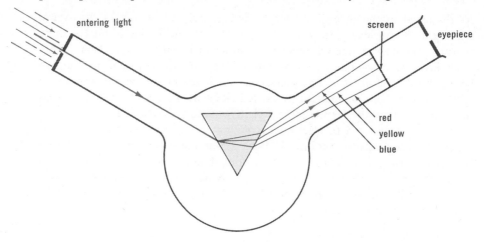

Fig. 4–1. Representation of a simple prism spectroscope. Light of various wavelengths (colors) are bent varying amounts upon passing through the prism. (See also Fig. 4–3.)

of the instrument is a glass prism that bends the path of any light going through it and bends the paths of different colors to different degrees. (See also Fig. 4-3.) By looking at a flame through the eyepiece and rotating the prism slowly, one can determine the component colors of the spectrum of a flame or an arc. This type spectrum consists of a series of bright lines on a dark field and is known as a **bright-line emission spectrum.**

Electronic Arrangements in Atoms

In very careful work, a photographic record of the spectrum may be made, and wavelengths of radiation invisible to the human eye may be recorded on the photographic plate. (In Fig. 4-4, the spectra of several elements are shown.)

Emission spectra have played important roles in scientific investigations, for the spectrum of a given element is as individual as a fingerprint. The elements rubidium, cesium, thallium, indium, gallium, and scandium were discovered (between 1860 and 1879) as a result of spectroscopic examination of minerals that revealed spectral lines unlike those of any previously known elements.

The element helium was discovered in the sun, 93,000,000 miles away, before it was known to exist on earth. In 1868 European astronomers traveled to India to be in a good location to make observations during an eclipse of the sun. A French astronomer, Pierre Janssen, noted some unidentifiable lines in the spectrum of the sun's corona, whereupon one of his English contemporaries, J. N. Lockyer, suggested that these lines might belong to an element that existed in the sun but had never been found on earth. The unknown element was named *helium* from the Greek word *helios* (sun). It was not until 27 years later, in 1895, that Sir William Ramsay discovered that helium did exist on the earth in association with certain minerals. On heating these minerals he found that gases were evolved, one of which had a spectral pattern that matched the pattern discovered by Janssen and Lockyer.

Spectra and Electron Energies. The present viewpoint holds that the electrons outside the nucleus are normally in places of relatively low energy known as

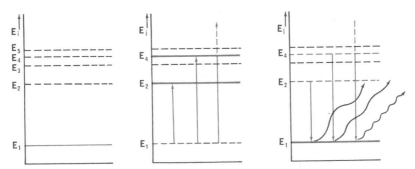

Fig. 4–2. Excitation of electrons from low energy levels to higher energy levels is a process that requires energy. When electrons fall from the excited levels, e.g., E_2, E_4, or E_i, radiant energy is given off, possibly as visible light. The greater the energy emitted, the shorter the wavelength of the radiant energy. (E_i is the energy necessary for ionization, i.e., to remove an electron completely from the atom.)

ground states. When the atoms are subjected to high temperatures or to bombardment by other electrons (e.g., in an electric arc), the electrons, especially the outer ones, absorb energy and are forced out to places of higher energy, or **excited states** (Fig. 4-2). When these excited electrons fall back to lower energy levels a certain amount of energy is given off, sometimes as visible light.

The facts are that (1) samples of the same element always emit the same wavelengths of radiation (emission spectrum), and (2) under the right conditions only certain wavelengths are emitted by any one element. This leads to the belief that electrons are arranged around the nucleus in *definite* ground-state **energy levels,** E_1, and that when they are excited they go to *definite* excited levels, E_2. This means that the difference in the energy of the electrons, $E_2 - E_1$, is the same for a given transition and thus helps explain the fact that the energies emitted by a given excited element always have the same wavelengths.

Absorption Spectra. At very high temperatures most solids become "white-hot," emitting radiation of all visible wavelengths. Such a radiation is said to give a *continuous emission spectrum* because no absence of color (dark spaces) can be detected when the light is passed through a prism in a spectroscope. See Figs. 4-3 and 4-4. Elements and compounds with high melting points can be used as convenient sources of continuous spectra. Tungsten, commonly used as a filament material in incandescent lights, is an example of such an element.

When continuous electromagnetic radiation (e.g., white light) passes through a substance, certain wavelengths of radiation may be absorbed. These wavelengths are characteristic of the substance that absorbs the radiation, and the pattern of these lines is referred to as an **absorption spectrum.**

Part of the continuous spectrum emitted by our sun is absorbed by the gases in the sun's atmosphere. Because of this, there are narrow dark lines (called Fraunhofer lines after their discoverer) in the otherwise continuous spectrum that reaches us, and the position of these dark lines enables us to identify the gases that absorbed this light. See Fig. 4-4.

Study of the absorption spectra of gases has led to the development of methods for testing unknown substances, regardless of whether they are gaseous, liquid, or solid. Transparent colored materials absorb certain wavelengths of visible light. Materials that do not absorb visible light may absorb characteristic wavelengths of ultraviolet or infrared radiation. Modern electronic devices are used to record absorption spectra automatically. The pattern of absorption of radiant energy by a material gives an almost certain indication of the substances, elements or compounds, in the material.

Fig. 4-3. When a beam of white light (from the projector at the left) passes through a glass prism, the path of the light is bent from the original path. Because light of short wavelength (blue) is bent more than light of long wavelength (red), the beam of white light is spread out as a spectrum. In this case the spectrum is continuous (there are no dark areas), showing that this white light is made up of all visible wavelengths (colors).

Fig. 4-4. (1) A continuous spectrum from an incandescent solid. (2) The sun's spectrum, showing several dark (Fraunhofer) lines. Also shown are the discontinuous emission spectra of sodium, hydrogen, calcium, mercury, and neon.

Figures follow page 52.

Electronic Arrangements in Atoms

IONIZATION ENERGIES OF ATOMS

Suppose that a sample of a gaseous element is collected at very low pressure in a cathode-ray tube. The energy of the cathode ray, measured in electron volts,[1] required to knock the most loosely bound electron off a gaseous atom is called the first ionization energy of the element:

$$\text{atom} + \text{energy} \rightarrow \text{ion} + \text{electron}$$

For atoms with enough electrons, a second electron can be knocked off at a higher potential, a third at a still higher potential, and so on.

In Table 4-1 some of the ionization energies of the first 22 elements, expressed in electron volts, are listed. Careful study of these data has helped the physicist to

TABLE 4-1.

Ionization Energies, ev

ATOMIC NUMBER	SYMBOL	1st e^-	2nd e^-	3rd e^-	4th e^-	5th e^-
1	H	13.595				
2	He	24.580	54.40			
3	Li	5.390	75.6193	122.420		
4	Be	9.320	18.206	153.850	217.657	
5	B	8.296	25.149	37.920	259.298	340.127
6	C	11.264	24.376	47.864	64.476	391.986
7	N	14.54	29.605	47.426	77.450	97.863
8	O	13.614	35.146	54.934	77.394	113.873
9	F	17.42	34.98	62.646	87.23	114.214
10	Ne	21.559	41.07	64	97.16	126.4
11	Na	5.138	47.29	71.65	98.88	138.60
12	Mg	7.644	15.03	80.12	109.29	141.23
13	Al	5.984	18.823	28.44	119.96	153.77
14	Si	8.149	16.34	33.46	45.13	166.73
15	P	11.0	19.65	30.156	51.354	65.007
16	S	10.357	23.4	35.0	47.29	72.5
17	Cl	13.01	23.80	39.90	53.5	67.80
18	Ar	15.755	27.62	40.90	59.79	75.0
19	K	4.339	31.81	46	60.90	
20	Ca	6.111	11.87	51.21	67	84.39
21	Sc	6.56	12.89	24.75	73.9	92
22	Ti	6.83	13.63	28.14	43.24	99.8

SOURCE: Reprinted with permission from Therald Moeller, *Inorganic Chemistry*, John Wiley & Sons, Inc., New York, 1952.

---

[1] One electron volt (ev) is the energy acquired by a singly charged particle when it falls through a potential of one volt.

form his ideas about the arrangement of the electrons outside the atomic nucleus. For example:

1. The elements potassium, K, sodium, Na, and lithium, Li, have low first ionization energies; this indicates that each has 1 electron that is lost easily. That the second ionization energy for these elements is much higher indicates that their other electrons are more strongly held.

These facts indicate that potassium, sodium, and lithium atoms each have 1 electron that is far from the positive nucleus and is loosely held. This outer electron is in a high energy level; the other electrons are in one or more energy levels nearer the nucleus.

2. Magnesium, Mg, and calcium, Ca, have low first and second ionization energies; this indicates that each has 2 electrons that are knocked off easily. Note that these elements have two of the lowest second ionization energies listed.

These data indicate that magnesium and calcium atoms each have 2 electrons that are far from the positive nucleus and are loosely held. Their remaining electrons are in energy levels nearer the nucleus and are more tightly held.

3. The elements helium, He, neon, Ne, and argon, Ar, have very large first ionization energies. This means that atoms of these elements hold on to all their electrons very tightly. (These so-called noble gases are found to be chemically quite inactive; i.e., they hardly ever interact with other substances in chemical reactions.)

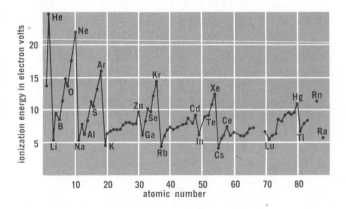

Fig. 4-5. As the atomic numbers increase, the first ionization energies of the elements vary in a periodic way.

From the total number of electrons in these noble-gas atoms it has been calculated that helium has a single energy level containing 2 electrons and that neon and argon each have 8 electrons in their outside energy levels.

When the first ionization energies are plotted against atomic numbers, as in Fig. 4-5, the energy required to knock an electron off an atom is seen to vary in a periodic way as the atomic numbers increase. Note that the noble gases are at the top of each of the periodic peaks of the curve.

Electronic Arrangements in Atoms

PERIODICITY OF PROPERTIES

One of the most important indications that there are differing numbers of major energy levels in atoms is the fact that elements with similar properties appear at intervals in the periodic classification. We saw in Chap. 2 that Mendeleev associated the cyclic change in properties with an increase in the atomic weights. Since Mendeleev's time, it has become quite clear that the periodic change in properties is associated with a periodic change in the electron arrangement in atoms as the atomic numbers increase, rather than the atomic weights. Today, we take the atomic number as equal to the number of electrons in an atom, which is a concept discovered since Mendeleev's death.

As we see from Mendeleev's table in Chap. 2, or from the long-form of the table to be discussed in the next section, there are seven periods of elements, a fact which indicates seven major repetitions in electron arrangement. Before discussing the electron arrangements, let us look at a modern version of the periodic table.

The Long-Form Periodic Table. Since Mendeleev's time, a number of different arrangements of the elements have been proposed—pyramids, spirals, and three-dimensional models of various shapes. Many of these have some unique merit, but the only one that has received wide acceptance is the **long-form periodic table**, a copy of which appears inside the front cover.

The long form of the periodic table accents a little more clearly than Mendeleev's table that changes in properties depend upon the atomic number.

There are sixteen vertical divisions into **groups** or **families**, since the A and B families of groups I through VIII are listed separately.

There are seven horizontal rows or **periods**, with each of the first six periods ending with a noble gas.

For each element there are given (1) the symbol, (2) the atomic number, (3) the atomic weight, and (4) the number of electrons in each of the principal energy levels of the atom (this will be discussed in the next section). There are six complete periods; the seventh is incomplete. The first has only two members; the other complete ones have 8, 18, or 32 members. There are eight groups numbered I through VIII. In each group, the A family always includes two elements from periods 2 and 3, whereas the B family has no members in these short periods. The following families illustrate these points:

IIA family: Be, Mg, Ca, Sr, Ba, Ra
IIB family: Zn, Cd, Hg
VIIA family: F, Cl, Br, I, At
VIIB family: Mn, Tc, Re

The table is arranged so that similar elements are in the same family. For instance, family IB, the copper family, is made up of the metals copper, Cu, silver, Ag, and gold, Au. An element is usually more like a member of its own family than like any element in another family.

The elements in the A family of a group are usually not physically similar to the elements in the B family of the same group. In most areas of the table the

A and B families of a group are also chemically dissimilar. For example, the A family of group I is made up of very active metals (Na, K, etc.), whereas the B family is made up of inactive metals (Cu, Ag, Au). Perhaps the most important similarity between A and B elements is in the formulas of some of their compounds. This is illustrated by the following lists of oxide compounds:

Group	A Family	B Family
I	K_2O	Cu_2O
II	BaO	HgO
III	Al_2O_3	Sc_2O_3
IV	CO_2	TiO_2
V	N_2O_5	V_2O_5
VI	SO_3	MoO_3
VII	Cl_2O_7	Re_2O_7

SUMMARY. On the basis of the study of atomic spectra, ionization energies, and the periodic table, physicists have come to think of the electrons in unexcited atoms as being arranged in one to seven major energy levels, depending on the

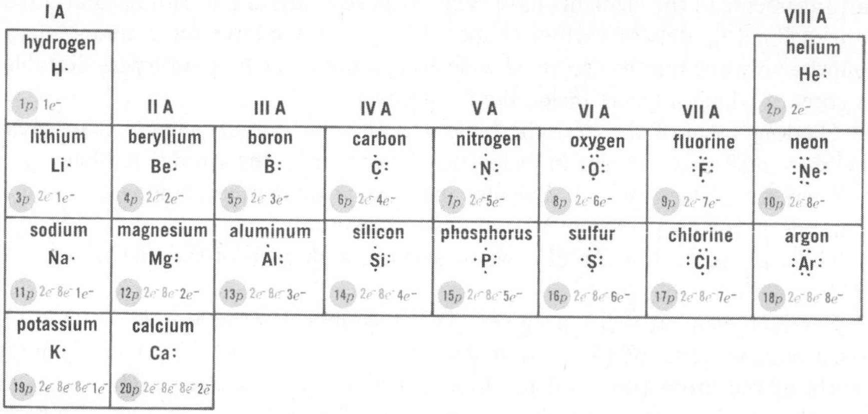

I A	II A	III A	IV A	V A	VI A	VII A	VIII A
hydrogen H· $1p$ $1e^-$							helium He: $2p$ $2e^-$
lithium Li· $3p$ $2e^- 1e^-$	beryllium Be: $4p$ $2e^- 2e^-$	boron B: $5p$ $2e^- 3e^-$	carbon C: $6p$ $2e^- 4e^-$	nitrogen ·N: $7p$ $2e^- 5e^-$	oxygen ·O: $8p$ $2e^- 6e^-$	fluorine :F: $9p$ $2e^- 7e^-$	neon :Ne: $10p$ $2e^- 8e^-$
sodium Na· $11p$ $2e^- 8e^- 1e^-$	magnesium Mg: $12p$ $2e^- 8e^- 2e^-$	aluminum Al: $13p$ $2e^- 8e^- 3e^-$	silicon Si: $14p$ $2e^- 8e^- 4e^-$	phosphorus ·P: $15p$ $2e^- 8e^- 5e^-$	sulfur ·S: $16p$ $2e^- 8e^- 6e^-$	chlorine :Cl: $17p$ $2e^- 8e^- 7e^-$	argon :Ar: $18p$ $2e^- 8e^- 8e^-$
potassium K· $19p$ $2e^- 8e^- 8e^- 1e^-$	calcium Ca: $20p$ $2e^- 8e^- 8e^- 2e^-$						

Fig. 4–6. Schematic representation of atoms of the first 20 elements, showing the periodic arrangements of electrons in the outer energy levels. In the abbreviated version, using dots around the symbol, each dot represents an electron in the outermost energy level.

complexity of the atom. The first energy level, which can hold only 2 electrons, is thought to be closest to the nucleus, because these electrons are held very strongly. In the energy levels at greater distances from the nucleus, the number of electrons that each level can hold increases, and their attraction for the nucleus decreases. The main energy levels are designated as 1, 2, 3, 4, etc., or K, L, M, N, etc.

The arrangement of electrons for the first 20 elements into main energy levels is thought to be as shown in Fig. 4-6. Note that the maximum numbers of electrons in the first and second main energy levels are 2 and 8, respectively. The third main energy level does not fill beyond 8 until element 21 is reached in the fourth period.

ENERGY SUBLEVELS

More intensive studies, especially more precise and sensitive photographic measurements of the radiations emitted by excited atoms, reveal that the energies of electrons *within* a given main energy level differ from one another. It is necessary to postulate that within a main energy level there must be **energy sublevels** to account for the large number of wavelengths of radiant energy emitted by excited atoms.

Recognition of the various groups of spectral lines corresponding to sublevels did not occur all at once, but rather it comprised a series of discoveries in the early years of this century. The sublevels were given names suggested as each new series of lines in the spectra was discovered: *sharp, principal, diffuse,* and *fundamental*. Today we speak of these energy sublevels as the s, p, d, and f sublevels, respectively.

The number of sublevels in any main energy level is apparently equal to the number of that level. That is, in the first main level there is only one energy level, in the second main level there can be two energy sublevels, in the third main level there can be three energy sublevels, etc. To say it another way, in the first main level there is only the s sublevel, in the second there can be s and p sublevels, in the third there can be $s, p,$ and d sublevels, etc.

Figure 4-7 is a schematic representation of the relative energies of the main energy levels and the energy sublevels for an atom. Such a representation is called an energy-level diagram.

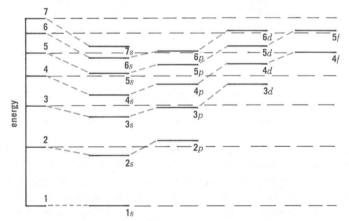

Fig. 4–7. An energy-level diagram showing the relative energies of the main energy levels and the energy sublevels. Note that the energies of the lowest levels differ considerably from one another, but at higher levels the differences become smaller. In some cases, a sublevel of a lower number may be of higher energy than a sublevel of a higher number, for example the $4d$ sublevel is of higher energy than the $5s$ sublevel.

WAVE MECHANICS AND ORBITALS

Although a tremendous amount of spectroscopic data indicate that electrons in atoms have various energies and that these energies may be described as main energy levels and sublevels, these data do not give information about the movement of electrons in atoms. Actually, according to the *Heisenberg uncertainty principle*, it is not possible to measure precisely the position and the velocity of an electron at the same time. Moreover, an electron is such a small particle that it is disturbed or set off in some unpredictable motion by an attempt to examine it, say by irradiating it with light or X rays. Although we may attempt to determine where the electron is, we cannot determine its position exactly, and we cannot tell where it is going after we observe it.

Through the use of mathematical methods known as wave mechanics, Erwin Schrödinger, starting in 1926, was able to calculate the probability of locating the electron in a region of space about the nucleus. His mathematical procedure leads to descriptions of regions of space about a nucleus in which an electron is most likely to be located. These regions of space are called **orbitals.** An electron may be in any place within an orbital at a given time, although it tends to occupy certain portions of its orbital to a greater extent than other portions. No more than two electrons may occupy each orbital.[2] Magnetic studies indicate that electrons act as if they were spinning and that two electrons in the same orbital must be spinning in opposite directions.

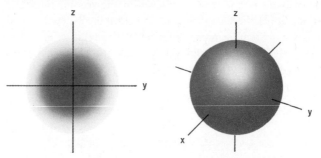

Fig. 4-8. The 1s orbital. On the left is a cross section showing the one region (darker region) of high electron density. The drawing on the right shows the spherical shape of the 1s orbital.

The first energy level (the K or 1 level) contains only one orbital; therefore, this main level can contain no more than two electrons. (Also, because this simplest main level is limited to two electrons, it is not, strictly speaking, divided into sublevels; "the 1s sublevel" and the "first energy level" are merely different ways of referring to the same energy level.) The shape of the orbital occupied by these two electrons is believed to be spherical with the nucleus at the center (Fig. 4-8). The orbitals of s sublevels have spherical shapes and are called s orbitals.

The second main energy level, with a maximum of eight electrons, consists of four orbitals. One of these is an s orbital, i.e., spherical (Fig. 4-9). The other three orbitals are dumbbell-shaped (Fig. 4-10) and are called p orbitals. Thus, the

---

[2] Even when there is only one electron in an orbital, this electron acts as if it occupied the whole region available to the orbital. Just one electron can be thought of as an electron "cloud" that has a high probability of occupying some region of space about a nucleus.

Electronic Arrangements in Atoms

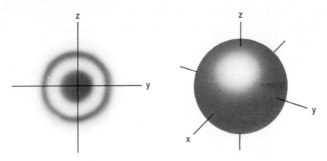

Fig. 4–9. The 2s orbital. On the left is a cross section showing the two regions of high electron density. The drawing on the right shows the spherical shape of the 2s orbital.

second main level consists of two sublevels, the 2s sublevel and the 2p sublevel; the 2s sublevel consists of a single s orbital, and the 2p sublevel consists of three p orbitals.

The third main energy level, with a maximum of 18 electrons, contains three sublevels consisting of nine orbitals: one s orbital, three p orbitals, and five d orbitals. The fourth main energy level, consisting of four sublevels, contains 16 orbitals: one s, three p, five d, and seven f orbitals.[3] The d and f orbitals are thought to have more complicated shapes than the s and p orbitals.

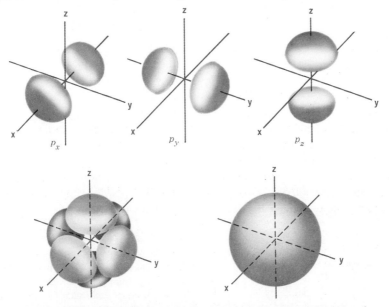

Fig. 4–10. Each of the three 2p orbitals is represented, as is a composite of the three orbitals. On the lower right is shown the spherical electron distribution which results when each of the three orbitals contains one or two electrons.

---

[3] There are also g, h, i, etc., sublevels, which may be occupied by electrons that have been excited or activated; however, the outermost *ground-state sublevels* that are occupied in the most complicated atoms known today are the 7s, 6d, and 5f sublevels.

To summarize:
1. The main energy levels have number designations of 1, 2, 3, 4, 5, 6, 7.
2. The number of sublevels within a main level is numerically equal to the number designation of that level.
3. The square of the number designation of a main energy level gives the number of orbitals in that level.
4. The number of orbitals multiplied by 2 gives the maximum number of electrons in a main energy level.

Table 4-2 shows how these statements apply to the first four main levels.

TABLE 4-2.

Subdivision of Main Energy Levels

Main energy level	1	2		3			4			
Number of sublevels (n)	1	2		3			4			
Number of orbitals (n^2)	1	4		9			16			
Kind and number of orbitals	s 1	s 1	p 3	s 1	p 3	d 5	s 1	p 3	d 5	f 7
Maximum number of electrons	2	2	6	2	6	10	2	6	10	14
Total maximum number of electrons ($2n^2$)	2	8		18			32			

Order of Filling Orbitals. The first two main energy levels are widely separated in energy value, but for the third, fourth, fifth, and higher main levels there can be overlapping of energies. For example, as the atomic number increases, the $5s$ sublevel is added to before the $4d$ and $4f$ sublevels.

The sublevels may be listed in order of increasing energy as follows: $1s$, $2s$, $2p$, $3s$, $3p$, $4s$, $3d$, $4p$, $5s$, $4d$, $5p$, $6s$, $4f$, $5d$, $6p$, $7s$, $5f$, $6d$, $7p$. Figure 4-11 is useful in helping one remember the order of increasing energies.

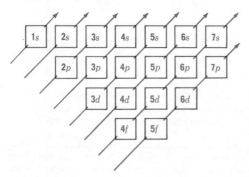

Fig. 4–11. The approximate order in which sublevels are filled with increasing numbers of electrons. Follow through each slanting arrow in turn, starting at the left and continuing with the next one to the right. For example, after $2s$ is filled, there follow $2p$, $3s$, $3p$, $4s$, $3d$, etc. (Adapted with permission from Therald Moeller, *Inorganic Chemistry*, John Wiley and Sons, Inc., New York, 1952.)

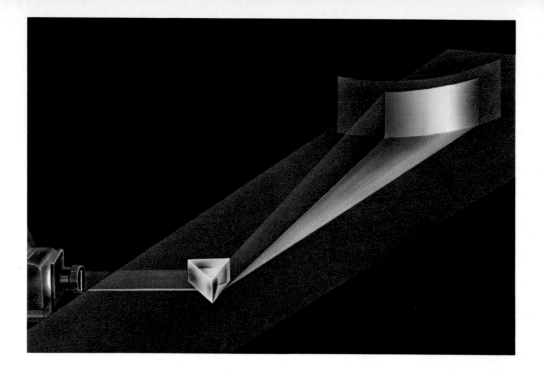

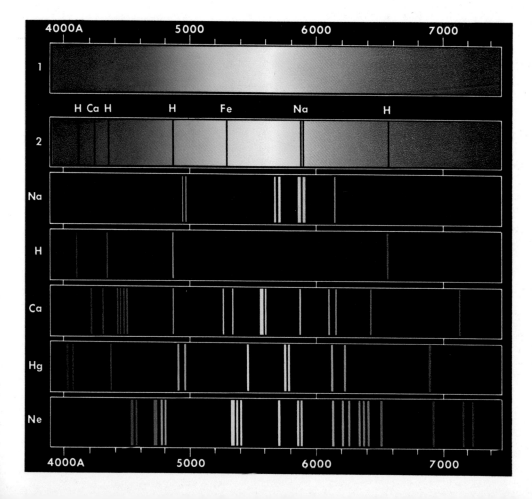

Electronic Arrangements in Atoms

As the number of electrons increases from atom to atom in the periodic table, two general principles govern the order in which the sublevels and orbitals in the different energy levels are filled. (1) The order in which sublevels are filled usually follows the energy ranking, as shown in Fig. 4-7, starting with the sublevel lowest in energy and working up. (2) Within a given sublevel, each orbital is usually occupied by a single electron before any orbital has two electrons.[4] Although there are exceptions to the order in which orbitals are filled, these two principles apply in enough cases to make them reliable guides. The application of these principles to twelve elements is shown in Table 4-3; each box represents

TABLE 4-3.

Electron Arrangements

MAIN LEVELS	1	2		3	SUMMARY
SUBLEVELS	s	s	p	s	
H	↓				$1s^1$
He	↓↑				$1s^2$
Li	↓↑	↓			$1s^2\ 2s^1$
Be	↓↑	↓↑			$1s^2\ 2s^2$
B	↓↑	↓↑	↓ □ □		$1s^2\ 2s^2\ 2p^1$
C	↓↑	↓↑	↓ ↓ □		$1s^2\ 2s^2\ 2p^2$
N	↓↑	↓↑	↓ ↓ ↓		$1s^2\ 2s^2\ 2p^3$
O	↓↑	↓↑	↓↑ ↓ ↓		$1s^2\ 2s^2\ 2p^4$
F	↓↑	↓↑	↓↑ ↓↑ ↓		$1s^2\ 2s^2\ 2p^5$
Ne	↓↑	↓↑	↓↑ ↓↑ ↓↑		$1s^2\ 2s^2\ 2p^6$
Na	↓↑	↓↑	↓↑ ↓↑ ↓↑	↓	$1s^2\ 2s^2\ 2p^6\ 3s^1$
Mg	↓↑	↓↑	↓↑ ↓↑ ↓↑	↓↑	$1s^2\ 2s^2\ 2p^6\ 3s^2$

an orbital and each arrow represents an electron. The following convention is used to summarize the number and location of electrons in atoms:

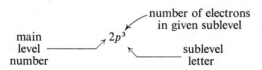

---

[4] This rule for the filling of orbitals is known as Hund's *principle of maximum multiplicity*.

To illustrate a rather complicated case, we can apply these principles to element 73, tantalum. There are 73 electrons to be accounted for. Following the scheme used in Fig. 4-11, we find the following electronic configuration for tantalum:

$$1s^2 2s^2 2p^6 3s^2 3p^6 4s^2 3d^{10} 4p^6 5s^2 4d^{10} 5p^6 6s^2 4f^{14} 5d^3$$

Or it can be abbreviated as:

$$(\text{xenon core})\ 6s^2 4f^{14} 5d^3$$

This agrees with the configuration given for element 73 in the periodic table inside the front cover. The term "xenon core" stands for the electron arrangement of xenon, element 54; it indicates the filling of all sublevels through $5p^6$. Summing up, the number of electrons per main energy level for tantalum is 2, 8, 18, 32, 11, 2.

The electronic structure of an atom as determined experimentally does not always agree with that predicted by the use of the scheme pictured in Fig. 4-11. Chromium, Cr, for example, according to its spectrum has the structure (argon core) $4s^1 3d^5$; however, according to the regular prediction, its expected structure would be (argon core) $4s^2 3d^4$.

Filling Orbitals and the Periodic Table. The order in which electrons are added in building up atomic structures is clearly related to the periodic table, as summarized in Fig. 4-12. One of the main features to note is that with the first element of each period, a member of group IA, a new main level begins to fill with the addition of an electron to an s sublevel. A second important feature is that each period contains the number of elements corresponding to the filling of

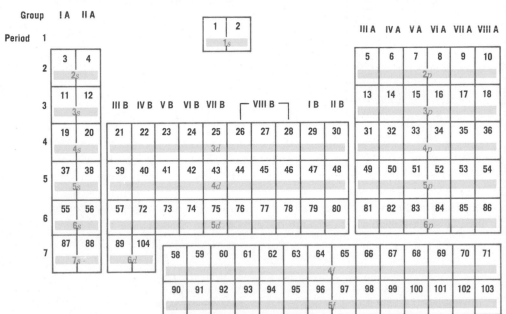

Fig. 4-12. The order in which orbitals are filled is related to the organization of the periodic table.

certain types of sublevels: period 1, involving only an s sublevel, contains just 2 elements; periods 2 and 3, involving s and p sublevels, contain 8 elements each; periods 4 and 5, involving the filling of s, p, and d sublevels, contain 18 elements each; period 6, with s, p, d, and f sublevels to fill, contains 32 elements; and period 7 contains the remaining known elements. Presumably, in the event enough elements are ever found or synthesized, period 7 will contain 32 elements, also.

A third important feature to note is that each period, except the first, ends with the filling of a p sublevel. The elements in group VIIIA are called the noble gases because they are gaseous at room temperature and because they have little tendency to react chemically with other elements. Only a few compounds containing them have been made. Their extreme inactivity is evidently related to their electronic structures, a topic discussed more fully in Chap. 5.

Orbitals and Ionization Energies. In Table 4-1 and Fig. 4-5 we noted the periodic character of changes in the first ionization energies of elements. We can understand some of the variations from ideal periodic behavior within each major period by examining the changes in energy as related to the arrangement of electrons in orbitals and sublevels.

In Fig. 4-5 it is seen that the first electron is knocked off a boron atom (B), atomic number 5, more easily than off the previous atom (beryllium); the same behavior is shown by aluminum (number 13) following magnesium. A glance at Fig. 4-12 shows that both elements 5 and 13 are in family IIIA, in which the first of the six p electrons in a sublevel is added. This first p electron is in a new sublevel; it is somewhat less strongly held than the previous electron. Examination of Fig. 4-5 indicates that this same behavior is shown by elements 31, 49, and 81 (gallium, indium, and thallium), which are also in family IIIA and which have but one p electron in a new sublevel.

Dips in the graph in Fig. 4-5 show that a somewhat lower ionization energy than expected is found for oxygen (8), sulfur (16), selenium (34), and tellurium (52). This is apparently associated with the pairing of p electrons in orbitals. As shown in Table 4-3, when electrons are added within a sublevel, one electron is added to each orbital before a second is added to any orbital. In each element in family VIA, as for oxygen in Table 4-3, a fourth p electron is added to make a pair of electrons in one orbital. Probably owing to the repulsion between two electrons close to one another, this second electron in an orbital is less strongly held, so that the ionization energy needed to remove it is less than would be predicted. In Fig. 4-5 the elements boron, carbon, and nitrogen lie on one smooth line, and the elements oxygen, fluorine, and neon on another. The members of the second set, in which p orbital electrons are being paired, have lower ionization energies than predicted on the basis of the first set.

To summarize:

1. When a new main outer level is begun with the addition of the first s electron, this electron is relatively loosely held, and the ionization energy is low.

2. Within a period there is a general increase in ionization energy, as no new main levels are added and the attractive positive charge on the nucleus increases steadily.

3. The completion of *p* sublevels leads to especially high ionization energies (VIIIA family elements); the completion of *s* and *d* sublevels (IIB family elements) is also marked by peaks in ionization energy, as shown in Fig. 4-5.

EXERCISES

1. How does the emission spectrum of calcium differ from its absorption spectrum?
2. How do the ground states of atoms differ from their excited states? Suggest two ways of producing neon atoms in excited states.
3. As iron is heated, its temperature rises, and it progressively changes color from dull red to bright red to white. Account for these color changes in the appearance of the iron in terms of the electrons.
4. Hydrogen and helium are believed to be in the atmosphere of most stars. On what is this belief based?
5. The emission spectrum of neon consists of many more lines than the emission spectrum of helium. Why?
6. For each of the following elements listed predict the number of electrons in the outside energy level. Check your answers by referring to Table 4-1, then to Fig. 4–6.

Ionization Energies, ev

	1st e^-	2nd e^-	3rd e^-	4th e^-
Element X	8.30	25.1	37.9	259.3
Element Y	5.1	47.3	71.6	98.9
Element Z	9.3	18.2	153.8	217.7

7. While electrons are emitted by gaseous cesium when subjected to radiant energy of wavelength of 3200 A or less, electrons are not emitted by gaseous rubidium unless the radiation has a wavelength of 2900 A or less. What do these data suggest about the relative magnitudes of the ionization energies of these two metals? Explain.
8. Draw diagrams to show the distribution of electrons, protons, and neutrons in atoms of helium, carbon, sodium, silicon, chlorine, and potassium.
9. Criticize the following statement: "An electron moves about the nucleus of an atom in a manner analogous to the movement of a planet about the sun."
10. What is the Heisenberg uncertainty principle? Why is it applicable to electrons in atoms but not to moving automobiles?
11. With the aid of a periodic table predict each of the following:
    (a) an element in which the atoms have no unpaired electrons;
    (b) the atomic number of the yet to be discovered element that has only one $7p$ electron;
    (c) the element in family VIIA that has the lowest ionization energy;
    (d) the element in the same period as krypton but in group IIIA;
    (e) at least ten elements whose atoms contain two unpaired electrons.

12. Without referring to the periodic table, show the placement of electrons in orbitals for each of the following: (a) calcium, atomic number 20; (b) technetium, atomic number 43; (c) europium, atomic number 63; (d) radon, atomic number 86.
13. What information about electrons in atoms is obtained from the fact that the emission spectra of the elements are frequently discontinuous?
14. Indicate the maximum number of electrons per atom that can be accommodated in (a) all the $3d$ orbitals; (b) one $3d$ orbital; (c) all the $5f$ orbitals; (d) one $5f$ orbital; (e) all orbitals of the fourth main level.
15. Sketch the shapes of the $2s$ and $2p$ orbitals.
16. In Fig. 4-5, note that the first ionization energy of oxygen is less than the first ionization energies of its two neighbors, nitrogen and fluorine. Account for this in terms of the electronic structures of the atoms.
17. Why is the first ionization energy of fluorine greater than that of chlorine?
18. Explain how the long form of the periodic table is an aid in keeping track of the order of filling electrons in orbitals.
19. On the basis of your knowledge of atomic structure, predict which is the larger: (a) an atom of sodium or an atom of potassium; (b) an atom of sodium or an atom of magnesium; (c) an atom of sodium or a positive sodium ion; (d) an atom of sodium or an atom of neon. Give reasons for your prediction in each case, then check your answer with Fig. 18-1.
20. We usually describe the light coming from a neon sign as red or orange-red. Yet examination of the neon emission spectrum (Fig. 4-4) shows lines in the blue, green, and yellow portions of the spectrum as well as in the red. Why do we not see these colors?

SUGGESTED READING

Adamson, A. W., "Domain Representations of Orbitals," *J. Chem. Educ.*, **42**, 140 (1965).

Becker, C., "Geometry of the f Orbitals," *J. Chem. Educ.*, **41**, 358 (1964).

Cohen, I., "The Shape of the $2p$ and Related Orbitals," *J. Chem. Educ.*, **38**, 20 (1961).

Cohen, I., and T. Bustard, "Atomic Orbitals, Limitations and Variations," *J. Chem. Educ.*, **43**, 187 (1966).

Feinberg, G., "Light," *Sci. Amer.*, **219** (3), 50 (1968).

Garrett, A. B., "The Bohr Atomic Model: Niels Bohr," *J. Chem. Educ.*, **39**, 534 (1962).

Garrett, A. B., "Quantum Theory: Max Planck," *J. Chem. Educ.*, **40**, 262 (1963).

Keller, R. N., "Energy Level Diagrams and Extranuclear Building of the Elements," *J. Chem. Educ.*, **39**, 289 (1962).

Lambert, F. L., "Atomic Orbitals from Wave Patterns," *Chemistry*, **41** (2), 10 (1968); **41** (3), 8 (1968).

Weisskopf, V. F., "How Light Interacts with Matter," *Sci. Amer.*, **219** (3), 60 (1968).

5

CHEMICAL BONDING

Over a hundred essentially different atoms are now known, corresponding to the number of known elements. We would not have to rise from our chair to observe many hundreds of different kinds of matter. The paper of this page, the ink, the glue, and thread that go into this book; the hand with which we turn the page; the skin, bones, blood, fingernails, and tendons; each of these in turn (examined under a microscope) appears to be made of a number of things. Before we had catalogued the materials in ourselves, our clothes, and our immediate possessions, we would have a list with many more than 100 entries. Most of these items would not be elements at all. Indeed, we find that most elements do not exist in nature in the form of their simple atoms. Rather, we find their atoms combined with atoms of other elements to form the myriad substances with which we come in contact. These compound substances have little or no resemblance to the elements from which they are formed; hence, they must be considered as fundamentally different types of matter.

HOW ATOMS COMBINE

When a piece of coal burns, the carbon (the main element in coal) *combines* with the element oxygen in the air. The result of this combination is the formation of a gas called carbon dioxide. We can understand this and other such changes better if we focus our attention on the individual particles of the elements that are involved, the atoms.

Atomic Structure and Chemical Reactions. The most important structural feature of atoms in determining chemical behavior is the number of electrons in their outermost energy levels. When atoms of one element combine with those of another, there is always some change in the distribution of electrons in the outermost energy levels. The study of many elements and compounds has led to the idea that, in compound formation, atoms of certain elements tend to gain electrons, and others tend to lose electrons. As a result of these tendencies, two atoms may *transfer* or *share* electrons, and either process may provide for a stable arrangement of electrons such that the atoms are held together as a unit of a compound. The electrons that are involved in a chemical reaction are called the **valence electrons** of an atom.

The elements are roughly divided into four classes. Elements whose atoms usually lose electrons in compound formation are *metals*; those whose atoms often gain electrons are *nonmetals*. A class between these is called the *metalloids*, or *borderline elements*. The fourth class consists of the *noble gases*, six elements that do not combine with other substances readily. No compounds of the lighter noble gases are known, and only a few compounds have been prepared of the heavier ones, notably xenon.

These four classes of elements are indicated in the periodic table inside the front cover. The colored vertical line at the far right of the table marks off the six noble gases in group VIIIA. In the middle of the right-hand page there is a colored zigzag line that separates the metals (to the left) from the nonmetals (to the right). This division is not at all sharp. Near this line are the metalloids, such as germanium, Ge, arsenic, As, and antimony, Sb.

Inferences from Properties of Noble Gases. The mere fact that there is a family of elements whose members form compounds rarely, if at all, indicates that atoms of these elements have very stable electron arrangements. Evidently atoms of the noble gases have little tendency to gain, lose, or share electrons with other atoms; their electron levels are of relatively low energy. Prior to 1962, the elements in group VIIIA were called *inert*, because they were thought to be completely unreactive.

The arrangement of electrons in the noble gases is summarized in Table 5-1.

TABLE 5-1.

Arrangement of Electrons in Noble Gases

NOBLE GAS	SYMBOL	ATOMIC NUMBER	NUMBER OF ELECTRONS BY MAIN ENERGY LEVELS[a]					
			1	2	3	4	5	6
Helium	He	2	2					
Neon	Ne	10	2	8				
Argon	Ar	18	2	8	8			
Krypton	Kr	36	2	8	18	8		
Xenon	Xe	54	2	8	18	18	8	
Radon	Rn	86	2	8	18	32	18	8

[a] The numbers 1 to 6 are the principal quantum numbers that denote the different main levels. See Table 4-2.

The table shows that, except in the first main level, there can be no more than eight electrons in any level when it is the outer one. In Table 4-3 the electron arrangement of neon is given as

$$1s \quad 2s \quad 2p$$

[↓↑] [↓↑]   [↓↑][↓↑][↓↑]   or $1s^2 2s^2 2p^6$

For each of the larger noble-gas atoms a similar outer electron structure is formed in which a *p* sublevel is completed. The eight electrons in the outer main level of an argon atom are designated as $3s^2 3p^6$, in krypton as $4s^2 4p^6$, in xenon as $5s^2 5p^6$, and in radon as $6s^2 6p^6$.

The arrangement of electrons in these six atoms that do not combine or rarely combine chemically with others gives us a clue to how atoms interact with one another in general. It is thought that, by combining with one another, atoms of many of the other elements tend to achieve electron stability comparable with that of the noble gases. Such stability is attained in one of two ways: (1) by the *transfer* of outer-shell electrons from the atoms of one element to those of another, or (2) by the *sharing* of electrons by two or more atoms.

The new substances that result when two or more elements combine by sharing or transferring electrons are called **compounds.**

TRANSFER OF ELECTRONS

In general, when a metallic element combines with a nonmetallic element, electrons are lost by atoms of the metal and gained by atoms of the nonmetal.

Atoms of lithium, sodium, and potassium lose one electron easily; beryllium, magnesium, and calcium lose two. These six elements are metals. Fluorine and chlorine atoms each gain one electron; oxygen and sulfur atoms each gain two. These four elements are nonmetals. When sodium combines with chlorine to make

(11p) 2e⁻ 8e⁻ 1e⁻ + (17p) 2e⁻ 8e⁻ 7e⁻ ⟶ (11p) 2e⁻ 8e⁻ + (17p) 2e⁻ 8e⁻ 8e⁻
sodium   chlorine   sodium   chlorine
atom   atom   ion   ion

Fig. 5–1. Sodium and chlorine atoms react by the transfer of an electron to yield ions.

sodium chloride, NaCl (ordinary table salt), the sodium atoms each lose an electron and the chlorine atoms each gain one. The reaction between the sodium and chlorine atoms can be indicated as in Fig. 5-1, or it can be written thus:

$$\text{Na} + \text{Cl} \rightarrow \text{Na}^+ + \text{Cl}^-$$

The above representation of this reaction is called a **chemical equation.** In an equation all the substances are represented by symbols, the reactants being shown on the left of the arrow and the products on the right.

The symbol Na refers to the sodium atom, which has a net charge of zero; the symbol Na$^+$ refers to the sodium *ion*, which has a net charge of $1+$ ($11p$ in the nucleus, $10e^-$ outside). The symbol Cl$^-$ refers to an *ion* of chlorine, which has a net charge of $1-$.

Chemical Bonding

When electrons are transferred from one atom to another, ions are formed. Some of the ions will be positive, others negative. Let us show how magnesium atoms combine with fluorine atoms. Magnesium tends to lose two electrons and have an outside shell with eight electrons (Fig. 5-2). Fluorine tends to gain one electron in order to complete its outside shell:

$$Mg + 2F \rightarrow Mg^{2+} + 2F^-$$

12p 2e⁻ 8e⁻ 2e⁻	+2 9p 2e⁻ 7e⁻	⟶	12p 2e⁻ 8e⁻	+2 9p 2e⁻ 8e⁻
magnesium atom	two fluorine atoms		magnesium ion	two fluorine ions

Fig. 5-2. A magnesium atom gives up two electrons to form a stable Mg^{2+} ion; a fluorine atom gains one electron to form a stable F^- ion. Therefore, magnesium and fluorine react in a 1:2 ratio by atoms.

The magnesium atom loses two electrons to form a magnesium ion with a charge of $2+$. The particle formed when the fluorine atom gains one electron is an ion of fluorine, which has a charge of $1-$. The formation of *one* magnesium ion must result in the formation of *two* ions of fluorine.

Electronic Structure of Ions. The remarkable stability of an outside shell that has eight electrons is evident not only in the noble-gas atoms but in ions of other atoms. Thus, magnesium (atomic number 12) forms a dipositive ion, whereas chlorine (atomic number 17) forms a uninegative ion. The arrangement of electrons of these atoms and ions is as follows:

Atoms				Ions			
Mg	2e⁻	8e⁻	2e⁻	Mg^{2+}	2e⁻	8e⁻	
Cl	2e⁻	8e⁻	7e⁻	Cl^-	2e⁻	8e⁻	8e⁻

The magnesium ion has an electronic structure similar to that of the neon atom, $1s^2 2s^2 2p^6$; the chloride ion is electronically similar to the argon atom, $1s^2 2s^2 2p^6 3s^2 3p^6$. There is a tendency for atoms with atomic numbers within about three units of that of a noble gas to gain or lose electrons so as to form ions with "noble-gas structures" having eight electrons outside. This tendency is called the *rule of eight*.[1]

Limitation of Rule of Eight. The idea that ions are formed when atoms gain or lose electrons to attain the structure of a noble gas is a rule that is sometimes of value, sometimes not. In the examples that we have given in this chapter, the rule of eight enables us to predict the ionic structures of Na^+, Cl^-, Mg^{2+}, F^-, Y^{3+}, and As^{3-} correctly. But the structures of As^{3+}, Fe^{2+}, and Fe^{3+} are not similar to any of the noble gases, so that the rule of eight is of no use in these cases. In beginning our study of ions we shall limit our predictions to those ions which do achieve a structure with eight electrons in their outer main levels. However, we shall also encounter examples of other types of ions formed by many common elements, such as the ions of iron and arsenic just mentioned.

---

[1] Elements with atomic numbers 1, 3, 4, and 5 tend to form ions with electron structures similar to that of helium atoms, i.e., a $1s^2$ configuration.

We shall memorize these ionic charges as we need them; as we develop more skill in using sublevel structures, we shall be able to predict charges on some ions that do not follow the simple rule of eight.

Ions Differ from Atoms. When the atoms of two elements combine with each other by transferring electrons, the substance formed does not resemble either of the original materials. The new substance is composed, not of atoms, but of ions. These ions, some positive, some negative, are strongly attracted to one another by the electrostatic attraction that exists between unlike charges. The modern theory regarding the structure of a substance such as sodium chloride is that it consists of positive and negative ions arranged in well-ordered fashion in a crystal. Each positive ion is surrounded by negative ions, each negative ion by positive ones, as shown in Fig. 5-3. The attraction that binds unlike ions together is termed an **electrovalent bond.** Both sodium chloride and magnesium fluoride are called **electrovalent compounds.** Their chemical formulas are written $NaCl$ and MgF_2, respectively.

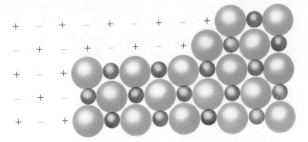

Fig. 5-3. A representation of one layer of particles in one type of ionic crystal. Except at the surface, each positive ion has six negative ions as its closest neighbors.

SHARING OF ELECTRONS

Two atoms, both of which tend to gain electrons, may combine with each other by sharing one or more pairs of electrons. For example, an atom of fluorine and an atom of chlorine, each having seven electrons in its outer shell, unite by sharing two of these fourteen electrons between them. This is shown diagrammatically as follows, using dots to indicate only the electrons in the outside main levels:

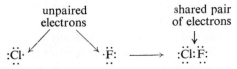

As a result of sharing a pair of electrons, each atom has a complete shell of eight electrons. By "complete" is meant an outside shell similar to a noble gas. For many simple compounds, the rule of eight is a satisfactory guide for predicting the number of electrons to be shared in building up the outside shell of electrons.

The strong force that binds the chlorine atom to the fluorine is the attraction of each for the electrons that are held jointly. A shared pair of electrons is called a

Chemical Bonding

covalent bond. Compounds whose atoms are joined by covalent bonds are called **covalent compounds.**

The *single particle* formed by the union of one chorine atom with one fluorine atom is uncharged, because it contains the same number of protons as it does electrons, namely 26 of each. An uncharged particle resulting from the union of two or more atoms is called a **molecule.**[2] The molecule ClF is the smallest particle of chlorine fluoride.

Molecular Orbitals. In Chap. 4, an atomic orbital was defined as a region of space about the nucleus of an atom in which an electron is likely to be located. For molecules the regions of space in which an electron (or pairs of electrons) has a probability of being located is called a **molecular orbital.**[3] As in the case of the electron orbitals in atoms, electrons in molecular orbitals are best described in mathematical terms. However, a schematic approximation of a molecular orbital is obtained by sketching the overlapping atomic orbitals as shown in Fig. 5-4. A molecular orbital is not simply the region of space where two atomic

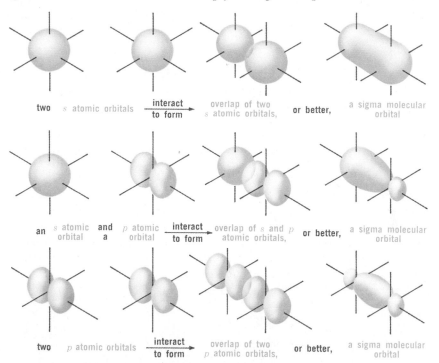

Fig. 5-4. A representation of molecular orbitals resulting from the interaction of two s atomic orbitals, an s and a p atomic orbital, and two p atomic orbitals.

---

[2] The term molecule refers to the smallest particle of a substance that has the characteristics of the substance In the case of a compound the molecule must consist of two or more atoms, but in the case of elements the single atoms may be characteristic of a quantity of the element. (See section on Individual Particles in Gases, Chap. 9.)

[3] The term molecular orbital, as used in this text, simply refers to a bonding orbital in a covalent compound. The student is referred to advanced texts for specific descriptions of the valence bond (VB) or molecular orbital (MO) theories, and other theories of chemical bonding.

64   FUNDAMENTALS OF COLLEGE CHEMISTRY

orbitals overlap. The molecular orbital has a shape all its own that is determined by the types of atoms and energy sublevels involved in the combination.

A molecular orbital that is symmetrical about the axis connecting two atoms is called a *sigma orbital* (σ). Sigma orbitals from the interaction of two *s* atomic orbitals, as in the hydrogen molecule, H : H, or two *p* orbitals as in the fluorine molecule, :F̈:F̈: , or an *s* and a *p* orbital, as in the hydrogen fluoride molecule, H:F̈: , are most common. (See Fig. 5-4.)

The sigma orbital is the simplest type of molecular orbital. Later we shall discuss the *pi* (π) type of orbital, which arises from the interaction of *p* orbitals that are not directed along the axis joining the centers of the two bonded atoms.

The hydrogen atom and the carbon atom are well known for their tendencies to form covalent bonds with other atoms. The electron-dot formulas for water (H_2O), ammonia (NH_3), carbon tetrachloride (CCl_4), and methane (CH_4) can be written on the basis of the rule of eight (rule of two for hydrogen):

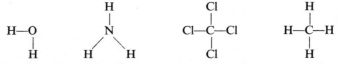

    water   ammonia   carbon   methane
                                                    tetrachloride

It is also common, in indicating the structure of molecules, to use a dash for a pair of shared electrons when only the electrons that are shared are indicated:

$$\begin{array}{cccc}
& H & Cl & H \\
& | & | & | \\
H\!-\!O & N & Cl\!-\!C\!-\!Cl & H\!-\!C\!-\!H \\
| & /\ \backslash & | & | \\
H & H\ \ \ H & Cl & H
\end{array}$$

The molecular orbitals in these molecules can be diagrammed by considering the atomic orbitals available for bonding and by using information about the molecular structure that has been derived from X-ray studies and other methods. The angle between the two H—O bonds in water is known to be about 104°; the angle between two H—N bonds in ammonia is 107°; and the angle between Cl—C bonds or H—C bonds in the two carbon compounds is 109°28′, the well-known tetrahedral angle. The diagrams in Fig. 5-5 are drawn on the basis

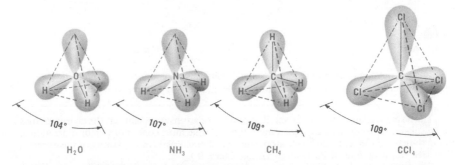

Fig. 5–5. Diagrams showing the sigma bonds and bond angles in molecules of water, ammonia, methane, and carbon tetrachloride.

Chemical Bonding

of these data. Each of the four molecules has some relation to the regular four-sided figure, the tetrahedron. CH_4 and CCl_4 have perfect tetrahedral symmetry so far as is known. NH_3 has its three hydrogens directed almost toward three corners of an imaginary tetrahedron and its pair of nonbonding electrons toward the fourth corner. In H_2O, the two hydrogen atoms are directed roughly toward two corners of a tetrahedron and the two nonbonding orbitals toward the other two.

Hybridized Orbitals. As pointed out previously, the shape and direction of a molecular orbital differ from the atomic orbitals that form it. The case of carbon and its compounds is of great importance, for carbon forms such a wide variety of compounds. In an isolated carbon atom, the most stable state is thought to be that in which two of the four outside electrons occupy the $2s$ orbital as paired electrons and each of the remaining two occupies separate $2p$ orbitals, i.e., $1s^2 2s^2 2p^2$. On the basis of this atomic orbital structure and assuming the use of only the two p orbitals, we might predict that a carbon atom would react with two hydrogen atoms to form CH_2 and that the two C—H bonds would be at right angles. However, it is found experimentally that a carbon atom reacts with four hydrogen atoms to form CH_4 and that the molecule has the symmetrical shape indicated in Fig. 5-5. Because the molecule is symmetrical, the four bonds are assumed to be alike.

To account for the formation of four identical C—H bonds, we assume that all four valence electrons of carbon participate in bond formation. Although the elevation of one of the $2s$ electrons into a vacant higher-energy $2p$ orbital would increase the energy of the isolated atom, this condition would provide four separate orbitals (one $2s$ and three $2p$) with four unpaired electrons. Such an atom could then form four energy-lowering bonds with other atoms to produce a stable molecule (see Fig. 5-6). The energy required to elevate the s electron (an endothermic process) is more than compensated for by the energy given off in the formation of more bonds with hydrogen (an exothermic process).

Carbon, of course, does form four covalent bonds with other atoms. A striking characteristic of these bonds in the simplest carbon compounds is their equivalence. For example, the four carbon-hydrogen bonds in methane are identical in stability and reactivity. To account for the four identical bonds, we assume that

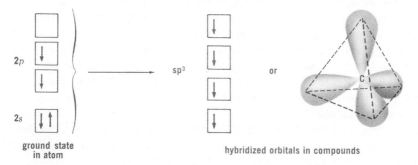

Fig. 5-6. A schematic representation of the four sp^3 bonding orbitals of a carbon atom.

the elevation of a 2s electron to a 2p orbital is accompanied by an interaction or **hybridization** of the four resulting orbitals to form orbitals that are identical in all respects. Because these hybrid orbitals result from the blending of one s and three p orbitals, they are referred to as sp^3 (pronounced s-p-three) orbitals.

The most important characteristics of the carbon sp^3 orbitals are:
1. There are four such orbitals.
2. They are identical.
3. They are directed in space from a carbon nucleus at the center toward the four corners of an imaginary tetrahedon.

The hybridization of orbitals is also characteristic of atoms of other elements. For example, in boron atoms containing two 2s and one 2p electrons, a 2s electron during compound formation is elevated to a vacant 2p orbital with the resultant hybridization of the three orbitals (sp^2 orbitals). In boron trichloride (BCl_3), the B—Cl bonds are identical and all atoms are in a single plane, the bond angles being 120°.

Pi Bonds and Pi Orbitals. A pair of atoms in a molecule may be joined by two or even three bonds, that is, double or triple bonds. For example, ethylene,

$$C_2H_4 \text{ or } \begin{array}{c} H-C=C-H \\ | \quad | \\ H \quad H \end{array}$$

has a double bond and nitrogen, N_2, or N≡N, has a triple bond. In multiple bonds like these, one bond is a sigma bond and the remaining bonds are of a type now to be discussed, the pi bond. For this discussion we will consider a molecule of nitrogen, :N:::N: being its electron-dot formula. The electronic configuration for a nitrogen atom is $1s^2 2s^2 2p_x^1 2p_y^1 2p_z^1$. Since each p orbital of a nitrogen atom contains only one electron, it is postulated that all three p orbitals of one atom overlap the corresponding p orbitals of a second nitrogen atom to form three bonds. How this may take place is shown schematically in Fig. 5-7. For clarity, only the p_x and p_y orbitals of the N atoms are shown in the top part of the figure. As the two nitrogen atoms approach each other, the p_x orbitals overlap end on end to form a sigma bond. At the same time, the p_y orbitals overlap along their edges to form an electron cloud that lies above and below the nitrogen nuclei. The bond and orbital resulting from this lateral overlap are called a **pi (π) bond** and a **pi orbital**, respectively. Now consider the p_z orbitals that are shown at the bottom part of Fig. 5-7. Looking at the left part that shows the p_z orbitals, we can visualize the p_z axis as passing through the paper at right angles to the axes of the p_x and p_y orbitals, with the lobes of the p_z orbitals lying in front and behind the plane of the paper. As the two nitrogen atoms approach each other closely, the p_z orbitals overlap laterally to form electron clouds lying in front of and behind the nitrogen nuclei. This overlap is similar to the overlap of the p_y atomic orbitals and results in the formation of a second pi bond. The triple bond, N≡N, is therefore composed of one sigma and two pi bonds; further, the electron clouds of the two pi bonds are thought to merge to form a hollow, cylindrical double-molecular orbital. While pi bonding is not

Chemical Bonding

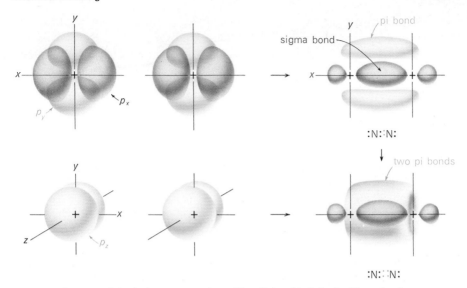

Fig. 5–7. Schematic representation of bonding orbitals in the N_2 molecule.

at all unusual among inorganic substances, we shall largely limit its discussion to the structures of organic molecules (Chap. 26).

MOLECULAR ORBITAL (MO) THEORY

Our discussion of covalent bonding thus far has been limited to molecular orbitals that result when pairs of electrons are shared between two atoms. The remaining electrons in each atom have not been considered, so that one may assume that they occupy atomic orbitals associated with a single atom. Thus in a molecule of fluorine, $:\!\ddot{F}\!:\!\ddot{F}\!:$, only one pair of electrons is considered to be in an orbital that encompasses both fluorine nuclei. While this simple picture of covalent bonding is adequate for explaining much of the physical and chemical behavior of molecules, a more complete theory, such as the MO theory, is often needed.

The significant feature of MO theory is that all electrons in outer main-energy levels are assigned to molecular orbitals. In doing this, some of the electrons have to be promoted to higher-energy orbitals than the constituent atomic orbitals. These electrons with higher energies act as a disruptive force in the molecule; they are referred to in MO language as **antibonding electrons**. The remaining electrons are placed in lower-energy orbitals and are referred to as **bonding electrons**. The electron density in the internuclear region is high for bonding electrons and low for antibonding electrons. See Fig. 5-8 for the approximate shapes of σp and $\sigma^* p$ molecular orbitals, the asterisk denoting an antibonding orbital.

In describing the structure of a molecule, the electrons are placed in molecular orbitals according to a plan very similar to that for filling atomic orbitals. That

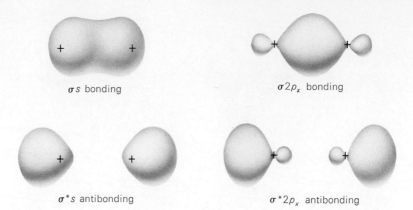

Fig. 5–8. Approximate shapes of sigma bonding and antibonding molecular orbitals in an A_2 molecule.

is, electrons are added to the orbitals in the order of increasing energy and in the order demanded by Hund's rule and the Pauli exclusion principle. Figure 5-9 gives a comparison of energies of atomic and molecular orbitals for homonuclear diatomic molecules. Table 5-2 shows the assignment of electrons to molecular orbitals for five such molecules: H_2, He_2, N_2, O_2, and F_2.

TABLE 5-2.

Filling of Molecular Orbitals

MOLECULE	NO. OF e^-	MO CONFIGURATION[a]	NET NO. OF BONDS
H_2	2	$(\sigma 1s)^2$	1
He_2	4	$(\sigma 1s)^2(\sigma^*1s)^2$	0
N_2	14	$KK(\sigma 2s)^2(\sigma^*2s)^2(\pi 2p_y)^2(\pi 2p_z)^2(\sigma 2p_x)^2$	3
O_2	16	$KK(\sigma 2s)^2(\sigma^*2s)^2(\sigma 2p_x)^2(\pi 2p_y)^2(\pi 2p_z)^2(\pi^*2p_y)(\pi^*2p_z)$	2
F_2	18	$KK(\sigma 2s)^2(\sigma^*2s)^2(\sigma 2p_x)^2(\pi 2p_y)^2(\pi 2p_z)^2(\pi^*2p_y)^2(\pi^*2p_z)^2$	1

[a] K refers to main level 1; KK represents the electrons in both atoms which occupy the two $1s$ shells, a total of four electrons. These are treated as nonbonding electrons and are assigned to atomic orbitals associated with their respective nuclei.

Table 5-2 also lists the net number of bonds per molecule; for homonuclear diatomic molecules this number is calculated as:

$$\tfrac{1}{2}(\text{no. of bonding } e^- - \text{no. of antibonding } e^-)$$

For O_2, the net number of bonds calculated from this relationship is 2:

$$\tfrac{1}{2}(8 - 4) = 2$$

The MO theory predicts that an He_2 molecule does not exist, because for a hypothetical helium molecule the net number of bonds is zero:

$$\tfrac{1}{2}(\sigma 1s^2 - \sigma^*1s^2) \text{ or } \tfrac{1}{2}(2 - 2) = 0$$

Chemical Bonding

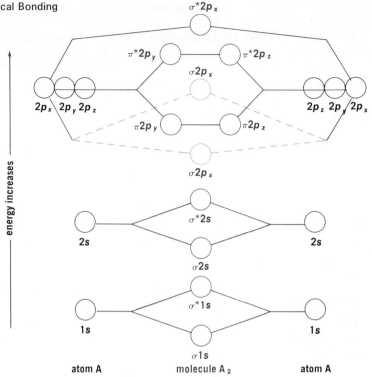

Fig. 5-9. Energy order for filling molecular orbitals for diatomic homonuclear molecules. For some molecules $\sigma 2p$ is above $\pi 2p$, for others it is below.

To illustrate the application of MO procedures, let us see how the assignment of electrons in orbitals as shown in Table 5-2 is achieved. With an atomic number of 8, the outermost shell of an oxygen atom contains 6 electrons; so, 12 electrons are to be assigned to molecular orbitals for the O_2 molecule (the 2 electrons of the first main level of each atom are left in atomic orbitals). In making the assignment, we begin with the lowest energy orbital of the second main level, the $2s$ orbital (see Fig. 5-9), and add 2 electrons to each orbital, following the order of increasing energy, until the pi antibonding orbitals are reached. At this point, 10 of the 12 electrons have been assigned, 2 each for the $\sigma 2s$, $\sigma^* 2s$, $\pi 2p_x$, $\pi 2p_y$, and $\pi 2p_z$ orbitals. The 2 remaining electrons are assigned 1 to each of the next two orbitals, $\pi^* 2p_y$ and $\pi^* 2p_z$, since these orbitals are of equal energy (Hund's rule). Now compare the assignment we have made with that in Table 5-2.

ELECTROVALENT AND COVALENT BONDS

It is a great temptation to try to classify things according to hard and fast rules. We must beware of generalizing too quickly and too definitely in our study of chemistry. Although electrovalent compounds and covalent compounds are the two principal types of chemical substances, there are many compounds that do not fit well into either class.

There are compounds that are held together by both electrovalent and covalent bonds. In the compound ammonium chloride, NH_4Cl, the four hydrogens are held to the nitrogen by covalent bonds to form the ion particle NH_4^+. This positive ammonium ion is attracted as a unit to neighboring Cl^- ions. The attraction of such unlike charges on the ions are made clear by the following formulas, which indicate the number of outer electrons:

$$\begin{bmatrix} \text{H} \\ \text{H:N:H} \\ \text{H} \end{bmatrix}^+ \quad :\!\ddot{\text{Cl}}\!:^-$$

A nitrogen atom has 5 outer electrons; a hydrogen atom has 1. One nitrogen plus 4 hydrogens involves 9 outer electrons; because there are only 8 electrons in the NH_4^+ particle, this ion has a net charge of $1+$. The electron missing from the NH_4^+ ion is held by the chlorine in its complete outer level (forming a Cl^- ion).

Another example is sodium sulfate, Na_2SO_4. A sodium ion, Na^+, results when a sodium atom loses 1 electron (the outer level of the ion then has 8 electrons). For 1 sulfur atom (6 outer electrons) and 4 oxygen atoms (24 outer electrons), there are a total of 30 outer electrons. But the SO_4^{2-} ion, as indicated by the dot formula, includes 32 outer electrons:

$$:\!\ddot{\text{Na}}\!:^+, :\!\ddot{\text{Na}}\!:^+ \quad \begin{bmatrix} :\!\ddot{\text{O}}\!: \\ :\!\ddot{\text{O}}\!:\!\text{S}\!:\!\ddot{\text{O}}\!: \\ :\!\ddot{\text{O}}\!: \end{bmatrix}^{2-}$$

This excess of 2 electrons is in agreement with its charge of $2-$. This 2-electron excess also accounts for the electrons lost by 2 sodium atoms in forming ions.

Compounds such as NH_4Cl and Na_2SO_4 are classed as electrovalent compounds, because they contain ions held together by electrovalent bonds. However, within the NH_4^+ ion and the SO_4^{2-} ion there are covalent bonds between the various parts. These covalent bonds are sufficiently strong so that the whole ion acts in many ways as a unit particle. Thus ammonium ions and sulfate ions maintain their identity in combining with other ions to form, for example, NH_4Br, $(NH_4)_2S$, $CaSO_4$, K_2SO_4, and $(NH_4)_2SO_4$.

Ions made from two or more atoms are called **polyatomic ions.** Other examples include the hydroxide, OH^-, the phosphate, PO_4^{3-}, and the nitrate, NO_3^-, ions.

Valence. In the compounds discussed above we saw that different atoms have different capacities for transferring or sharing electrons. The number of electrons given up or received in forming ions is called the **electrovalence** of the atom. In the examples of the transfer of electrons earlier in this chapter, sodium, magnesium, fluorine, and chlorine had electrovalences of $1+$, $2+$, $1-$, and $1-$, respectively. The number of electron pairs shared in forming molecules is called the **covalence** of the atom. In the examples in Fig. 5-5, hydrogen, oxygen, nitrogen, carbon, and chlorine have covalences of 1, 2, 3, 4, and 1, respectively. Both electrovalent and covalent bonds are called **valence bonds.**[4]

Polar Covalent Bonds. A simple covalent bond is defined as a pair of electrons shared between two atoms. However, electrons are shared equally only if the atoms are the same, as is the case with diatomic molecules of the elements shown in Fig. 5-10. In Figs. 5-10 and 5-11 diagrams of molecular models are shown. It is frequently helpful to visualize molecular structure with the use of models. In such models the nonbonding electrons are not indicated.

Fig. 5–10. Molecular models of diatomic molecules.

If the atoms are different, the shared electrons are attracted more toward one of the atoms. Such a pair of electrons constitutes a **polar bond,** making one part of a molecule partially negative and another partially positive. In the case of a diatomic molecule, this results in a **polar molecule** or a **dipole,** which has one end relatively positive and the other relatively negative. Examples are HCl and ClF, as shown in Fig. 5-11. In these diagrams the $\delta+$ or $\delta-$ signifies a partial charge that is less in magnitude than the charge on a proton or electron. We will see later that the polarities of molecules greatly influence the physical properties (melting point, solubility, etc.) and chemical properties of compounds.

Electronegativity. The power of an atom in a molecule to attract electrons is called the **electronegativity** of the atom. Studies have been made showing that, in general, the higher the ionization potential of an element, the greater its electronegativity. Also, electronegativities are calculated on the basis of the strengths of the bonds formed between atoms; the stronger the bond, the greater the dif-

Fig. 5–11. Polar diatomic molecules. Note that the more negative end of the molecules is the end at which the more electronegative atom is located.

---

[4] The term valence comes from the Latin *valentia* meaning capacity. At one time the valence of an element was defined as the number of hydrogen atoms an element combines with or displaces in a chemical reaction. However, this definition is not completely satisfactory. For example, in compounds such as acetylene, C_2H_2, and benzene, C_6H_6, the valence of carbon appears to be 1, although it is not. Many chemists prefer to use the term "oxidation numbers" rather than "valence numbers" (see Chap. 6).

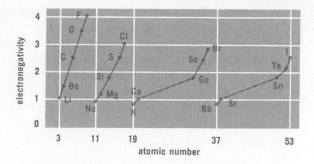

Fig. 5-12. As the atomic numbers increase, the electronegativities of the elements vary in a periodic way, increasing from Group I to Group VII in each period.

ference in electronegativities. Linus Pauling has derived a scale of electronegativity values based on bond strengths. On Pauling's scale, the element fluorine, F, whose atoms have the greatest attraction for electrons, is given an arbitrary electronegativity of 4.0. All other elements are compared with this standard; for example, potassium, K, an atom with a very slight affinity for electrons, has a relative electronegativity of only 0.8, whereas silicon, Si, whose electron affinity is moderate, has a value of 1.8. The higher this value, the greater the atom's attraction for an electron pair. In Fig. 5-12 the periodic variation in electronegativities with change in atomic number is shown for four periods. As the atomic number increases from 3 to 9, the electronegativity rises; but at 11, Na, it is low once more, then starts to rise till 17, Cl, is reached. (See also Table 5-3.)

When two atoms of different electronegativities are joined by a covalent bond, the pair of electrons is drawn toward the atom of higher electronegativity. This

TABLE 5-3.

Electronegativities

IA	IIA	IVA	VIA	VIIA
Li	Be	C	O	F
1.0	1.5	2.5	3.5	4.0
Na	Mg	Si	S	Cl
0.9	1.2	1.8	2.5	3.0
K	Ca	Ge	Se	Br
0.8	1.0	1.8	2.4	2.8
Rb	Sr	Sn	Te	I
0.8	1.0	1.8	2.1	2.5
Cs	Ba	Pb	Po	At
0.7	0.9	1.8	2.0	2.2

results in one of the atoms being more negative than the other and causes the bond to be polar or somewhat ionic. Pauling states that, if atoms differ sufficiently (by about 2 units) in electronegativity, they form bonds that are mainly ionic, that is, ionic-covalent bonds. If the electronegativity difference is less than this, the bonds are mainly covalent, that is, covalent-ionic. Bonds may be covalent, covelant-ionic, ionic-covalent, or ionic.

The atoms of lowest electronegativity are those in which a new main energy level has just been started with one or two s electrons. The atoms of highest electronegativity are those in which a p sublevel is almost complete. Atoms that have partially filled d and f sublevels are intermediate in electronegativity.

EXERCISES

1. List several different facts which indicate that the noble gases have very stable electron structures. What are these structures?
2. Draw diagrams similar to Fig. 5-1 and Fig. 5-2 to illustrate the formation of ionic compounds when the following elements combine: (a) sodium and oxygen; (b) beryllium and sulfur; (c) calcium and nitrogen; (d) magnesium and bromine; (e) lithium and fluorine.
3. Show diagrammatically, using dots to represent electrons in the outside main energy levels, the covalent compounds formed between the following pairs of elements: (a) sulfur and hydrogen; (b) fluorine and bromine; (c) silicon and hydrogen; (d) phosphorus and hydrogen; (e) sulfur and oxygen.
4. In which of the following molecules are the bonds mainly ionic and in which are they mainly covalent: $LiBr$, CH_4, K_2O, Al_2O_3, PCl_3, CaF_2, CO, CO_2, NF_3, AlF_3?
5. In which of the following compounds is the "rule of eight" violated: NO_2, NF_3, CaO, NH_4F, Na_2SO_4, SF_6, CuF_2?
6. Describe the manner in which sodium ions and chloride ions differ from sodium atoms and chlorine atoms. With respect to the ultimate particles that compose each, how does sodium chloride differ from metallic sodium and gaseous chlorine?
7. Using a periodic table, place each of the following elements into one of the four classes of elements: potassium, bromine, krypton, germanium, hafnium, sulfur, scandium, boron.
8. Describe how the electronegativity scale can be used to predict the nature of the bond formed between two elements.
9. Consult Table 5-3 and predict the electronegativity value of each member of group VA. Based on the predicted values, which elements are metals and which are nonmetals?
10. What is meant by sp^3 hybridized orbitals? How do they differ from s and p orbitals?
11. What determines whether a given covalent bond is termed a sigma bond or a pi bond?
12. What is the MO configuration of a molecule of neon with formula Ne_2?
13. What is the net number of bonds in a molecule of Ne_2 (see Exercise 12)? Based on this information, would such a molecule exist? Explain.
14. Magnesium and sulfur react to form magnesium sulfide. Discuss qualitatively the composition of the mixture that results when 10 g of Mg and 50 g of S are heated in a closed container until the reaction is complete.

15. How many electrons are present in a phosphate ion, PO_4^{3-}?
16. What is a polar covalent bond? Do all polar molecules contain polar covalent bonds? Are polar covalent bonds found only in polar molecules? Support your answers with some specific examples.
17. In general, ionic compounds have higher melting points than covalent compounds. Why?
18. How many atoms of hydrogen will combine with (a) 500 atoms of oxygen; (b) 200 atoms of chlorine; (c) 6.02×10^{23} atoms of carbon; (d) 34 atoms of sodium?
19. Many triatomic (three-atom) molecules are bent with an angle approaching 90°. Suggest a reason for this observation.

SUGGESTED READING

Frey, J. E., "Discovery of the Noble Gases and Foundation of the Theory of Atomic Structure," *J. Chem. Educ.*, **43**, 371 (1966).

Friedman, H. G., G. R. Choppin, and D. G. Feuerbacher, "The Shapes of the *f* Orbitals," *J. Chem. Educ.*, **41**, 354 (1964).

Little, E. J., Jr., and M. M. Jones, "A Complete Table of Electronegativities," *J. Chem. Educ.*, **37**, 231 (1960).

Neville, R. G., "The Sceptical Chemist, 1661," *J. Chem. Educ.*, **38**, 106 (1961).

Ogryzlo, E. A., "On the Shapes of *f* Orbitals," *J. Chem. Educ.*, **42**, 150 (1965).

Ogryzlo, E. A., and G. B. Porter, "Contour Surfaces for Atomic and Molecular Orbitals," *J. Chem. Educ.*, **40**, 256 (1963).

Sanderson, R. T., "Principles of Chemical Bonding," *J. Chem. Educ.*, **38**, 382 (1961).

6

FORMULAS AND EQUATIONS; CLASSES OF COMPOUNDS

Ultimate Particles in Matter. As was pointed out in Chap. 2, there are two major classes of matter: elements and compounds. A single element or a single compound is spoken of as a pure substance. Most materials cannot be classified as pure, because they may contain a number of different elements or compounds more or less jumbled together. In elements, the individual particles are usually the atoms of the element. In this case a single atom may also be called a molecule of the element. In a few very common elements, the individual particle that is normally present is not the single atom, but a molecule composed of two or more atoms. Oxygen gas consists of molecules made up of pairs of oxygen atoms. The symbol for oxygen gas is written O_2, indicating that each molecule contains two atoms. The elements hydrogen, H_2, nitrogen, N_2, fluorine, F_2, chlorine, Cl_2, bromine, Br_2, and iodine, I_2, also consists of *diatomic* (two-atom) molecules.

In compounds, the individual particles are either molecules (covalent compounds) or positive and negative ions (electrovalent compounds).

In mixtures, the individual particles that are present depend upon what elements and compounds are collected together to make the mixture. A mixture such as soil might contain all three of the tiny particles that characterize different substances—atoms, ions, and molecules.

FORMULAS AND EQUATIONS

CHEMICAL FORMULAS

A **formula** is a representation of the composition of a compound. For example, the formula for water, H_2O, shows that water contains the combined elements hydrogen, H, and oxygen, O. It also shows that in one unit of water there are two hydrogen atoms and one oxygen atom. The formula does not indicate whether the atoms are held together by the sharing of electrons (covalent bonds), or by the electrostatic attraction of oppositely charged ions (electrovalent bonds).

Oxidation Numbers. A system of small whole numbers is useful in helping us remember formulas for compounds and in correlating certain chemical properties. These numbers are called **oxidation numbers** or **oxidation-state numbers**.

The system of oxidation numbers has been developed during a period of years from a consideration of (1) the composition of compounds, (2) the relative electronegativities of the elements comprising the compounds, and (3) arbitrary rules and guides.

Three of the arbitrary rules are:

1. The oxidation state of an uncombined element is zero.
2. In a compound the more electronegative elements are assigned negative oxidation states, and the less electronegative elements are assigned positive states.
3. In a formula for a compound the sum of the positive oxidation states equals the sum of the negative oxidation states.

Examples of assigned oxidation numbers are:

$$\overset{+1\ -1}{NaCl},\quad \overset{(+1)_2+6(-2)_4}{Na_2SO_4},\quad \overset{+4(-2)_2}{CO_2},\quad \overset{(+7)_2(-2)_7}{Cl_2O_7}\quad \overset{+2(-1)_2}{OF_2}$$

THE PERIODIC TABLE AND OXIDATION NUMBERS. We can formulate some general guides based on the periodic table that will be helpful in predicting oxidation numbers of some of the common elements when in chemical union. Note that in electrovalent compounds the oxidation number of an ion is the same as the charge of the ion. However, in covalent compounds the oxidation number of an atom does not necessarily correspond with the number of covalent bonds joining the atom to other atoms.

1. The metals in the A family of group I have oxidation states of $+1$ in compounds. For example, NaCl and KCl are compounds in which sodium and potassium have oxidation numbers of $+1$. Hydrogen has an oxidation state of $+1$ when combined with nonmetals and -1 when combined with metals.

The members of the B family of group I have several oxidation states. Among these, we should remember that silver has a $+1$ oxidation state and that copper is usually present in compounds with a $+2$ oxidation state, although compounds in which copper exhibits a $+1$ oxidation state are known.

2. The members of both the A and B families of group II commonly exhibit a $+2$ oxidation state. Thus, in $MgCl_2$, $CaBr_2$, $ZnSO_4$, and $Hg(NO_3)_2$, the oxi-

dation state of the metal is $+2$. Because these all represent electrovalent compounds, the metallic ions have 2 plus charges, for example, Mg^{2+}, Ca^{2+}. Mercury, in the B family, also forms compounds in which its oxidation state is $+1$, for example, Hg_2Cl_2.

3. Aluminum, in group IIIA, has an oxidation state of $+3$. Examples: $AlCl_3$, $Al(NO_3)_3$.

4. Tin and lead, in group IVA, have oxidation states of $+2$ or $+4$. Examples: $SnCl_2$, $SnCl_4$, $PbCl_2$, $PbCl_4$.

Carbon, in group IVA, commonly has oxidation states of $+4$ or -4. Examples: $\overset{+4(-2)2}{CO_2}$, $\overset{-4(+1)4}{CH_4}$. However, the assignment of oxidation states to carbon can be very involved.

5. The nonmetals of the A family of group V often have oxidation states of -3 or $+5$. Example of -3: NH_3. Example of $+5$: N_2O_5. (Many other oxidation states are commonly displayed by these elements.)

6. The nonmetals of the A family of group VI often have oxidation states of -2 and $+6$; in rare cases, -1. Examples of -2: H_2O, H_2S, CaO, ZnS. Example of $+6$: SO_3. Group VIA elements also form compounds with a $+4$ oxidation, for example, SO_2. Because oxygen has the second highest electronegativity (3.5), it almost always has an oxidation number of -2; in rare cases, -1, for example, H_2O_2; only in fluorine-containing compounds is oxygen's oxidation number positive.

7. The nonmetals of the A family of group VII have oxidation states of -1 when combined with hydrogen or metals. Examples: HCl, $NaBr$, ZnI_2. (Some of them also display oxidation states of $+7$, $+5$, $+3$, and $+1$.)

SUMMARY. Common oxidation states for a number of elements are shown in Fig. 6-1. Oxidation numbers that we can predict from the periodic table at this stage, noting that we are on safer ground with A families, are:

Group	I	II	III	IV	V	VI	VII
Oxidation number	$+1$	$+2$	$+3$	$+2$, $+4$ -4	$+5$ -3	$+6$ -2	$+7$ -1

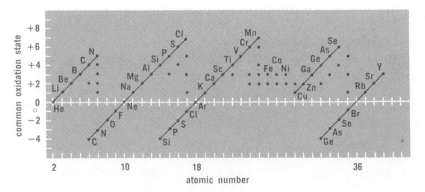

Fig. 6–1. From the periodic law we would expect oxidation numbers to be a periodic function of the atomic numbers. Study of this graph reveals that this is indeed the case for the maximum and minimum states (through which the lines have been drawn for emphasis). It is important to note that certain elements have a number of oxidation states that are not evident from the position of the element in the periodic table.

Because the sum of the oxidation states for all the components of a compound is zero, we can calculate the oxidation state of an unfamiliar element in a compound, provided we know the formula for the compound and the oxidation states of all the other elements in it.

Problem 1: What is the oxidation state of gold in gold chloride, $AuCl_3$?
Solution: Because the chloride ion has an oxidation state of -1, the oxidation state of gold must be $+3$: $\overset{+3(-1)_3}{AuCl_3}$, $(+3)+(-3)=0$.

Problem 2: What is the oxidation state of selenium in selenous acid, H_2SeO_3?
Solution: If we assume the oxidation states of hydrogen and oxygen are $+1$ and -2, respectively, the oxidation state of selenium is $+4$: $\overset{(+1)_2 + 4(-2)_3}{H_2SeO_3}$, $(+2)+(+4)+(-6)=0$.

It should be borne in mind that the oxidation-state numbers do not imply isolated charges. For example, in the compound NO the oxidation state of oxygen is -2, although oxygen is covalently bonded to nitrogen and no ions are present; in the compound CaO the oxidation state of oxygen is also -2, but the oxygen in this compound is in the form of ions, each with a doubly negative charge, O^{2-}. Whether a specific compound, such as NO or CaO, is actually covalent or ionic we can recognize only with experience, as we learn more and more about individual compounds and classes of compounds. We shall follow the plan of designating the oxidation numbers by placing a $-$ or $+$ in front of the number; to show the magnitude of the charge on an ion, the number precedes the $+$ or $-$ sign, except that the number 1 is not written. Examples: Cl^-, O^{2-}, Al^{3+}.

Writing Formulas. In writing the formula of an electrovalent compound, the symbol for the component that has a positive oxidation number is usually written first. In the case of covalent compounds, this simple rule for the order of components may or may not be followed. For example, the formulas for carbon dioxide, methane, hydrogen sulfide, and ammonia are written $\overset{+4}{C}O_2$, $\overset{-4}{C}H_4$, $H_2\overset{+1}{S}$, and $\overset{-3}{N}H_3$, respectively.

If the oxidation numbers are known, we can readily deduce the ratio of atoms in the compound and hence the formula. A somewhat mechanical method of doing this is to place the oxidation number over the respective symbol of the element or ion and then transfer each number (omitting the sign) as a subscript to the other symbol. The subscript 1 is never shown in a formula; it is implied if there is no subscript. This "crisscross" method is illustrated in the accompanying examples.

Name of Compound	Oxidation Number over Symbol		Formula
Aluminum oxide	$\overset{+3}{Al}$	$\overset{-2}{O}$	Al_2O_3
Calcium chloride	$\overset{+2}{Ca}$	$\overset{-1}{Cl}$	$CaCl_2$
Barium nitrate	$\overset{+2}{Ba}$	$\overset{-1}{(NO_3)}$	$Ba(NO_3)_2$
Calcium sulfate	$\overset{+2}{Ca}$	$\overset{-2}{(SO_4)}$	$CaSO_4$

Note that on crisscrossing the numbers in the last example, we first obtain $Ca_2(SO_4)_2$. The ratio 2 : 2 is then simplified to the ratio 1 : 1, and the simplest formula for the electrovalent compound calcium sulfate is $CaSO_4$.

The formulas for electrovalent compounds ordinarily indicate the simplest ratio of ions, whereas the formulas for covalent molecules show the actual number of atoms which have combined to form a molecule. That the molecule may or may not be formed by the union of the smallest possible number of atoms is shown in the following examples: NH_3, ammonia; CO_2, carbon dioxide; H_2O_2, hydrogen peroxide; C_6H_6, benzene; P_4O_{10}, phosphorus pentoxide; C_2H_2, acetylene. It is usually not possible to predict the formula of a covalent compound on the basis of oxidation states.

INTRODUCTION TO THE SYSTEMATIC NAMING OF COMPOUNDS

More than a million compounds are described in the chemical literature. Most of these were named systematically according to rules set up by international committees. The systematic names are based partly upon the composition of the compound and partly upon the class or type. The naming of compounds composed of large molecules can be very complicated, hence abbreviations[1] are fairly common. Furthermore, nonsystematic names are used for certain compounds that were well known before systematic naming was initiated by Antoine Lavoisier and his contemporaries in the eighteenth century. Examples of nonsystematic names are water, ammonia, and acetic acid.

The naming of simple inorganic salts and acids is described in this chapter; the naming of more complicated compounds will be discussed briefly in later chapters.

Compounds Containing Two Elements (Binary Compounds). The name shows the two elements present and ends in *-ide*. The more metallic element is mentioned first. Examples:

H_2O	Hydrogen oxide
HCl	Hydrogen chloride
K_2S	Potassium sulfide
$AlBr_3$	Aluminum bromide

Although the names of binary compounds end in *-ide*, many compounds that contain more than two elements are also named in this manner. Metallic hydroxides and ammonium compounds of the type NH_4X, where X is a simple negative ion, are two cases in which the compounds are so named.

$Ca(OH)_2$	Calcium hydroxide (not a binary compound)
NH_4Cl	Ammonium chloride (not a binary compound)

---

[1] Examples of some abbreviations are TNT for 2,4,6-trinitrotoluene and DDT for 1,1-bis(*p*-chlorophenyl)-2,2,2-trichloroethane. The DDT is derived from the partial abbreviation: *d*ichloro*d*iphenyl*t*richloroethane.

WATER SOLUTIONS OF BINARY ACIDS. The first term in the name is built around the name of the element other than hydrogen with the prefix *hydro-* and the suffix *-ic*. The name is completed by adding the word *acid* to the first term. Examples:

$$\begin{array}{lll} HCl & \text{(in water)} & \text{Hydrochloric acid} \\ HBr & \text{(in water)} & \text{Hydrobromic acid} \\ H_2S & \text{(in water)} & \text{Hydrosulfuric acid} \end{array}$$

Compounds Containing Three Elements, One of Which Is Oxygen. There are several classes of compounds that contain oxygen and two other elements. A very common one is the *-ate* class. The two elements other than oxygen are indicated in the name, and the name ends in *-ate*. (This rule will be extended later to include various other types.) Examples:

$$\begin{array}{ll} H_2SO_4 & \text{Hydrogen sulfate} \\ HNO_3 & \text{Hydrogen nitrate} \\ Na_2SO_4 & \text{Sodium sulfate} \\ NaClO_3 & \text{Sodium chlorate} \\ Ca_3(PO_4)_2 & \text{Calcium phosphate} \end{array}$$

WATER SOLUTIONS OF -ATE ACIDS. The distinguishing part of the name indicates the element other than oxygen and hydrogen; the name ends in *-ic acid*. Examples:

$$\begin{array}{lll} H_2SO_4 & \text{(in water)} & \text{Sulfuric acid} \\ HNO_3 & \text{(in water)} & \text{Nitric acid} \\ H_2CO_3 & \text{(in water)} & \text{Carbonic acid} \end{array}$$

Two or More Oxidation States. Certain elements share or transfer their electrons in more ways than one; hence, these elements have more than one oxidation number. For example, iron forms two series of compounds, one in which its oxidation number is $+2$, and one in which it is $+3$. To give these compounds unambiguous names, the oxidation number is indicated with Roman numerals, thus: $FeCl_2$, iron(II) chloride; $FeCl_3$, iron(III) chloride; $FeSO_4$, iron(II) sulfate; $Fe_2(SO_4)_3$, iron(III) sulfate; etc. Copper also has two oxidation states in compounds, namely, $+1$ and $+2$. For example, $CuCl$, copper(I) chloride; $CuCl_2$, copper(II) chloride; etc. Because copper(I) compounds are not common, it is customary to omit the oxidation number in the name of copper(II) compounds. Therefore, if no oxidation state is shown, it is assumed that the copper(II) compound is meant.

Unfortunately for the student of chemistry, the method of naming just described is of recent origin, and many common compounds were already named in accordance with other systems. Thus $CuCl$, $FeCl_2$, $CuCl_2$, and $FeCl_3$ are called cuprous chloride, ferrous chloride, cupric chloride, and ferric chloride, respectively. These older names persist, which means that the student is sometimes faced with the necessity of knowing both names.

Formulas and Equations; Classes of Compounds

CHEMICAL EQUATIONS

Equations are used to represent chemical reactions. The equation

$$2H_2 + O_2 \rightarrow 2H_2O$$

states that hydrogen and oxygen react to form water. It further states that for each molecule of oxygen, 2 molecules of hydrogen are required and 2 molecules of water result. The formulas show the composition of the reacting molecules and of the resulting molecules. A chemical equation does not necessarily indicate the temperature, concentration, etc., that are required for the reaction to take place, nor does it tell how rapidly the reaction will occur.

Finally, an equation is never correct unless the reaction actually occurs as indicated. For example,

$$Cu + 2NaCl \rightarrow 2Na + CuCl_2$$

is an incorrect equation because it descibes an reaction that does not take place. Copper and sodium chloride do not react to produce sodium and copper(II) chloride. It is not always easy to figure out what the products of a chemical reaction will be. The only way to be sure is to determine experimentally what they are. This has already been done for thousands of reactions, so for any of these we can turn to the chemical literature to find out what products are formed.

There are three distinct steps in writing an equation. They are illustrated here by two examples, the first showing the reaction of aluminum and oxygen, the second showing the reaction of iron(III) sulfate and calcium chloride.

Step 1. Show in words on the left side of the arrow the substances that are brought together and on the right those which are formed. Use a plus sign to separate the different substances:

$$\text{aluminum} + \text{oxygen} \rightarrow \text{aluminum oxide}$$

Step 2. Write the correct formula for each of the substances mentioned in the word equation.

a. Elemental substances are usually shown in equations as monatomic molecules. (Their symbols require no subscript.) Some notable exceptions are the diatomic molecules N_2, H_2, O_2, F_2, Cl_2, Br_2, and I_2.

b. Compound substances have formulas that can often be obtained from the oxidation numbers of the elements in the compounds.

In the equation that we have begun, the molecules of aluminum and oxygen must be represented, respectively, as monatomic and diatomic (see [*a*] above), and the formula of aluminum oxide is shown as Al_2O_3 (see [*b*] above). Consequently,

$$Al + O_2 \rightarrow Al_2O_3 \quad \text{(incomplete)}$$

Step 3. Balance the incomplete equation. This is done by putting numbers *in front* of the formulas so that the same number and kinds of atoms will be represented on the right of the arrow as on the left.

A most important thing to remember is that none of the subscript numbers, shown correctly in step 2, can be changed in order to balance the equation:

$$?Al + ?O_2 \rightarrow ?Al_2O_3$$

It will be noted that the number of oxygen atoms must be divisible by both 2 and 3, because each O_2 molecule contains 2 atoms of oxygen and each Al_2O_3 unit contains 3 atoms of oxygen. The smallest number divisible by both 2 and 3 is 6. In our equation we show 3 oxygen molecules (6 atoms of oxygen) and 2 aluminum oxide units (6 atoms of oxygen, because each unit contains 3):

$$?Al + 3O_2 \rightarrow 2Al_2O_3$$

As the equation now stands, 4 atoms of Al are represented on the right and only 1 on the left. We must also show 4 on the left. The finished equation is

$$4Al + 3O_2 \rightarrow 2Al_2O_3 \quad \text{(complete)}$$

The second example follows.
Step 1.

calcium chloride + iron(III) sulfate → calcium sulfate + iron(III) chloride

Step 2.

$$?CaCl_2 + ?Fe_2(SO_4)_3 \rightarrow ?CaSO_4 + ?FeCl_3$$

Step 3.

$$3CaCl_2 + Fe_2(SO_4)_3 \rightarrow 3CaSO_4 + 2FeCl_3$$

Ionic Equations. In many chemical reactions involving ionic compounds in water solutions, only two species of ions participate in a particular reaction. This point is illustrated by the following equations, which represent three seemingly different reactions that are carried out in water solutions. The downward arrow indicates that a compound is not soluble and separates or precipitates from the solution.

$$BaCl_2 + Na_2SO_4 \rightarrow BaSO_4\downarrow + 2NaCl$$
$$Ba(NO_3)_2 + MgSO_4 \rightarrow BaSO_4\downarrow + Mg(NO_3)_2$$
$$BaBr_2 + (NH_4)_2SO_4 \rightarrow BaSO_4\downarrow + 2NH_4Br$$

As far as the formation of insoluble barium sulfate is concerned, these three reactions all require that barium ions and sulfate ions come together to form insoluble $BaSO_4$. Hence, the single observable chemical change for the reactions above could be represented as

$$Ba^{2+} + SO_4^{2-} \rightarrow BaSO_4\downarrow$$

This type of equation is called an *ionic equation*. In such an equation, the sum of the charges on the left of the arrow must equal the sum on the right.

CLASSIFICATION OF COMPOUNDS

ELECTROVALENT AND COVALENT COMPOUNDS

An important classification of compounds is based on the kind of bonds that hold the atoms in the compound together (Chap. 5). If all the bonds are pairs of shared electrons, the compound is said to be **covalent**. If one or more bonds are primarily due to the attraction between ions of unlike charge, the compound is said to be **electrovalent**.

ORGANIC AND INORGANIC COMPOUNDS

One of the oldest methods of classifying compounds is based on whether the compound contains carbon as one of the combined elements. Compounds that contain carbon are called **organic compounds;** those which do not are called **inorganic compounds**.

Organic Compounds. The organic compounds that occur in nature are found mostly in plants and animals, or in materials of plant and animal origin, such as coal, natural gas, and petroleum. Carbohydrates, fats, proteins, and alcohols represent familiar classes of organic compounds; these and other types are discussed in later chapters. Thousands of organic compounds that do not occur in nature have been synthesized in chemical laboratories. Organic compounds make up possibly more than 90 percent of the number of known compounds. Organic compounds are usually covalent, although many electrovalent ones exist.

Inorganic Compounds. Inorganic compounds constitute rocks, clay, sand, and other earthy materials. Most of the ionic substances that we have mentioned are inorganic, for example, $Ca(OH)_2$, K_2S, and Na_2SO_4. Actually, a few compounds that contain carbon may be classified as inorganic rather than organic because they are earthy or rocklike and are quite similar to other mineral substances. Calcium carbonate, $CaCO_3$, magnesium carbonate, $MgCO_3$, and sodium cyanide, $NaCN$, are examples of such inorganic compounds.

ELECTROLYTES AND NONELECTROLYTES

Another method of classifying compounds in two broad divisions is based on whether the molten compound or a solution of it will conduct a current of electricity. If a molten compound or its aqueous solution is a conductor of an electric current, the compound is an **electrolyte;** if not, the compound is a **nonelectrolyte**. The test is made quite easily. In the apparatus shown in Fig. 6-2, the electrical cord is the usual type of light cord with a socket for a light bulb and a plug for insertion into a regular 110-volt outlet. However, one of the wires is cut in two so that the current cannot flow. When the two ends of the cut wire are immersed in pure water, the light will not burn; but when they are dipped into a water solution of common salt (sodium chloride) or hydrochloric acid or sodium hydroxide, the electric current is conducted by the solution, and the light burns. Sodium chloride, hydrochloric acid, and sodium hydroxide are therefore classed as electrolytes. On the other hand, when the two ends of the cut wire are immersed in a water solution of sugar or alcohol or glycerin, the light does not burn. Sugar, alcohol, and glycerin are therefore classed as nonelectrolytes.[2]

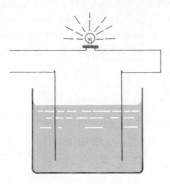

Fig. 6–2. Apparatus for testing a solution for electrical conductivity.

It is important to note at this point that all electrovalent compounds that dissolve in water make an electrolyte solution, whereas some covalent compounds that dissolve in water are electrolytes and some are not. The case of molten substances is more clear-cut: molten electrovalent compounds conduct the electric current, but covalent ones do not.

How Current Flows through an Electrolyte. Suppose that we arrange our "cut wire" apparatus as shown in Fig. 6–3. We shall use a direct current[3] from a battery of dry cells, instead of the 110-volt alternating current, and our electrolyte will be copper chloride, $CuCl_2$, dissolved in water. When the current is turned on, we are sure that it is flowing through the circuit, because the light bulb burns. We also notice that chlorine, a greenish-yellow gas, bubbles from the solution at the positive electrode (the **anode**[4]) and that metallic copper begins to plate out on the negative electrode (the **cathode**[4]). When the battery is disconnected, the light goes out and the chemical changes cease. When the battery is connected again, the light begins to burn, additional copper plates out, and more

---

[2] There are substances that act as electrolytes in some solutions and nonelectrolytes in others. For example, hydrogen chloride is a nonelectrolyte when dissolved in benzene.

[3] There are two kinds of current: **alternating** and **direct**. Alternating current, the type commonly used in homes and industry for lighting, heating, and operating electric motors, etc., is produced by rotating coils of copper wire in a magnetic field. In such a current, the electrons oscillate first in one direction and then in the opposite direction. In direct current, produced by batteries and by the **rectification** of alternating current, the electrons flow only in one direction.

[4] Whether the anode is positive or negative depends upon definition. The definition we will follow is: *The anode is the electrode at which oxidation occurs; the cathode, the electrode at which reduction occurs.* This means that the anode is positive and the cathode negative in electrolytic cells. However, in voltaic cells, a flashlight battery, for example, the anode is negative and the cathode is positive. See Chap. 20.

Formulas and Equations; Classes of Compounds

chlorine bubbles appear. If the battery is left connected for a long period of time, used cells being replaced as necessary, the light dims and eventually goes out. If we now examine the water solution, we find that no copper chloride remains. It has all been changed into elemental copper, which plated out, and into elemental chlorine, which escaped as a gas:

$$CuCl_2 \xrightarrow{\text{electric current}} Cu + Cl_2$$

The chemical reaction that occurs when an electric current flows through an electrolyte—$CuCl_2$ in the example above—is called **electrolysis**. The process is of great importance in the production of many elements and compounds and in electroplating.

These observations naturally raise several questions:
1. How did the electric current pass through the solution?
2. What caused the copper to be formed at the cathode?
3. What caused the chlorine to be formed at the anode?
4. What caused the current to stop flowing eventually, although the worn-out batteries were replaced by fresh ones?

The answers to these questions begin with our recognition of the fact that copper chloride, $CuCl_2$, is composed of ions. A more precise way of representing

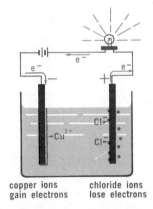

Fig. 6–3. Electrolysis of copper chloride.

a unit quantity of copper chloride is as follows: Cu^{2+}, Cl^-, Cl^-. When copper chloride is dissolved in water, the ions become separated and mix with the water, every ion being free to move around at random among the water molecules. When the current is turned on, there is a mass movement of Cu^{2+} ions toward the negative pole or cathode; there is also a mass movement of Cl^- ions toward the positive pole or anode. At the cathode, electrons are picked up by the copper ions. In this manner, the copper ions are changed to copper atoms:

$$Cu^{2+} + 2e^- \rightarrow Cu$$

At the anode, electrons are given up by the chloride ions. The chlorine atoms that are formed then join in pairs to make chlorine molecules.

$$Cl^- \rightarrow Cl + e^-$$
$$Cl + Cl \rightarrow Cl_2$$

The equation representing the over-all reaction may be written this way:

$$CuCl_2 \rightarrow Cu + Cl_2$$

We can summarize as follows:
1. A solution of an electrolyte contains ions.
2. The current (a flow of electrons) enters the solution at the cathode. The entering electrons are taken up by the positive ions (**cations**).
3. Electrons leave the solution at the anode. Negative ions (**anions**) give up these electrons.
4. When all the ions originally present have been changed to neutral particles, the current can no longer flow.

Sources of Ions.

ELECTROVALENT COMPOUNDS. Electrovalent compounds are composed of ions even when in dry, solid form. It is only when such substances are melted or dissolved in a solvent that the ions are free to migrate to an anode or cathode. All electrovalent compounds are electrolytes.

It should be emphasized that in the case of electrovalent compounds such as NaCl and $CuCl_2$, the water plays no special part in conducting the current beyond furnishing a medium in which the ions can move about. When pure sodium chloride is melted, at about 800°C, this pure liquid is an excellent conductor of an electric current. X-Ray studies show that sodium and chloride ions are present even in solid salt.

POLAR COVALENT COMPOUNDS. It will be recalled that a polar covalent molecule as a whole is an electrically neutral particle (Chap. 5). Pure liquid hydrogen chloride, HCl, pure liquid ammonia, NH_3, pure liquid hydrogen acetate, $HC_2H_3O_2$, and pure water, H_2O, are very poor conductors of electric current. But if we make a water solution of hydrogen chloride and test it with our "cut wire" apparatus, we find that the light burns brilliantly; i.e., the water solution of hydrogen chloride, HCl, is a conductor of electricity. On the other hand, if hydrogen chloride is dissolved in benzene, the solution does not conduct electricity. What is the reason for this difference? In order to account for the conductivity, we must assume that the covalent hydrogen chloride molecules are able to form ions in the water solution but are unable to do so in the benzene solution. (See Fig. 6-4.) Actually the ions result from a chemical reaction between hydrogen chloride molecules and water molecules, as shown in the following equation:

$$\underset{\text{hydrogen chloride}}{HCl} + \underset{\text{water}}{H_2O} \rightleftharpoons \underset{\text{hydronium ion}}{H_3O^+} + \underset{\text{chloride ion}}{Cl^-} \qquad (1)$$

This water solution is hydrochloric acid.

Acetic acid, when dissolved in water, also forms a solution that is a conductor of electricity. Because pure acetic acid (hydrogen acetate) is a covalent compound, we assume that the ions result from a reaction similar to the one above:

$$\underset{\text{hydrogen acetate}}{HC_2H_3O_2} + \underset{\text{water}}{H_2O} \rightleftharpoons \underset{\text{hydronium ion}}{H_3O^+} + \underset{\text{acetate ion}}{C_2H_3O_2^-} \qquad (2)$$

Ammonia, a covalent compound, also reacts with water to produce ions:

$$\underset{\text{ammonia}}{NH_3} + \underset{\text{water}}{HOH} \rightleftarrows \underset{\substack{\text{ammonium} \\ \text{ion}}}{NH_4^+} + \underset{\substack{\text{hydroxide} \\ \text{ion}}}{OH^-} \quad (3)$$

In these reactions, the covalent compound is said to *ionize*, and the process is referred to as *ionization*. The water solution of ammonia is ammonium hydroxide and is often represented by the formula NH_4OH.

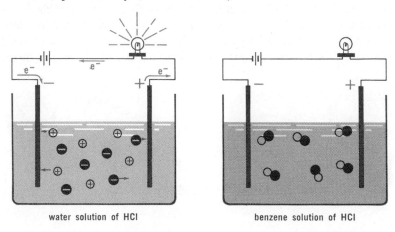

Fig. 6–4. Hydrogen chloride ionizes in water but not in benzene. The benzene solution does not conduct the electric current.

Chemical Equilibria. Note that double arrows are used in Eqs. (1) to (3). As in the case with a great number of chemical reactions, the conditions required for a reaction to occur also allow the products to react to form the original substances; that is, the reaction is **reversible**. Consider what happens when ammonia is added to water—Eq. (3). At the instant that the two are brought together, only NH_3 and H_2O molecules can react, because only these two are present. A very short time later some NH_4^+ and OH^- ions will have formed, and the reverse reaction starts, slowly at first, because very few of these ions are present to react. Both reactions continue to take place simultaneously. In time the concentrations of NH_3, HOH, NH_4^+, and OH^- become adjusted so that both reactions are occurring at the same speed. This condition is described as a **chemical equilibrium** and is indicated in equations by double arrows. In an equilibrium system, the rates of the competing reactions are equal; the longer arrow used in some equations points toward the reactants whose concentrations are greater at equilibrium.

The factors that determine the relative amounts of reactants and products at equilibrium are discussed in detail in Chap. 15. Let us note here that when equilibrium is reached for reaction (1), very little HCl is left in the mixture; its concentration is low, and the concentrations of H_3O^+ and Cl^- are relatively high. The opposite is true for reactions (2) and (3); in these solutions relatively few ions are formed, and the solutions are rather poor conductors of the electric current.

An interesting and important question is: Why do hydrogen chloride and other similar compounds form ions when dissolved in solvents like water but do not when dissolved in certain other solvents, such as benzene? This question can be answered in terms of the structures of the molecules involved. As was pointed out in Chap. 5, a polar bond results when the electron pair constituting a covalent bond is shifted toward the atom of higher electronegativity. This unequal attraction for the electron pair causes the oxygen side of the H_2O molecule and the chlorine side of the HCl molecule to be relatively negative, and the hydrogen side of each molecule to be relatively positive. We account for the formation of H_3O^+ and Cl^- ions by noting the attraction of the oxygen side of the H_2O molecule (see Fig. 6-5) for the proton (hydrogen nucleus) of the HCl molecule;

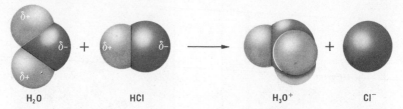

Fig. 6–5. A hydrogen chloride molecule reacts with a water molecule to form a pair of ions.

we also note the strong attraction of the highly electronegative chlorine atom for the electron pair of the H–Cl bond; these two effects result in the transfer of the proton to the water molecule. The additional electron left with the chlorine atom produces a negative ion (Cl^-), and the positive proton gained by the H_2O molecule produces a positive ion (H_3O^+).[5]

On the other hand, when hydrogen chloride is dissolved in the nonpolar liquid benzene (C_6H_6), protons are not pulled away from HCl molecules to form ions, because the localized negative charges in benzene molecules are not as large as in water molecules.

Strong and Weak Electrolytes. It is a fact that water solutions of sodium chloride and other electrovalent compounds as well as water solutions of certain covalent compounds are excellent conductors of electricity. Such compounds are said to be *strong electrolytes*. Strong electrolytes exist in solution completely or almost completely in the form of ions.

On the other hand, water solutions of many covalent compounds are poor conductors of electricity. Water solutions of ammonia (ammonium hydroxide) and of hydrogen acetate (acetic acid) are examples. Such compounds are said to be *weak electrolytes*. We assume in these cases that only a small percentage of the dissolved molecules have reacted with water to form ions. The great bulk of the dissolved substance is present as covalent molecules.

---

[5] Attempts to show conclusively that H_3O^+ is a separate ion in water have failed. Instead, evidence from spectroscopic studies indicate that the particle is more complex, perhaps being $H_9O_4^+$ (or $4H_2O \cdot H^+$). However, the symbol H_3O^+ (or $H_2O \cdot H^+$) is so widely used to refer to the hydrogen ion in water solution that we shall use it throughout the text.

Formulas and Equations; Classes of Compounds

Actually, the terms strong and weak electrolytes cannot be considered clear-cut classifications, for strong electrolytes may be weakly strong, moderately strong, strong, very strong, etc., and weak electrolytes may be subclassified in the same manner. That is, there are all degrees of weak and strong, so that the dividing line between the two is not always clear. This topic is discussed in greater detail in the chapter on ionic equilibrium.

In addition to being classified as to strength, electrolytes may be classified as to type. The three common types to be considered in this chapter are *acids*, *bases*, and *salts*. There are many examples of strong and weak electrolytes in the first two classes, but practically all salts are strong electrolytes.

ACIDS AND BASES

Acids and bases were defined by chemists centuries ago in terms of the properties of their water solutions. In these terms, an acid is a substance whose water solution has a sour taste, turns blue litmus red, reacts with active metals to form hydrogen, and neutralizes bases. Following a similar pattern, a base is defined as a substance whose water solution has a bitter taste, turns red litmus blue, feels soapy, and neutralizes acids.

Some common examples of acids and bases include the following. *Acids:* hydrochloric, HCl; nitric, HNO_3; sulfuric, H_2SO_4; and acetic, $HC_2H_3O_2$. *Bases*: sodium hydroxide, $NaOH$; potassium hydroxide, KOH; calcium hydroxide, $Ca(OH)_2$; and ammonium hydroxide, NH_4OH. The first three in each group are highly or completely ionized in water solutions. These are classed as *strong acids* or *strong bases*, respectively. Acetic acid and ammonium hydroxide are sparingly ionized in water solutions and are accordingly classed as a *weak acid* and a *weak base*, respectively.

Although the definition of acids and bases in terms of properties of their water solutions is of practical value, it greatly limits the scope of this field of chemistry. In 1923, J. N. Brønsted (Denmark) and J. M. Lowry (England) independently suggested that acids be defined as **proton donors** and bases as **proton acceptors.** By this definition, a great variety of closely related chemical properties and chemical reactions can be correlated. In this chapter we are chiefly concerned with some of the more common acid-base relationships as they apply to water systems; in Chap. 16 other aspects of the Brønsted–Lowry theories are taken up.

Reactions of Acids in Water Solutions. Let us consider from the Brønsted point of view what the proton-donating species is in water solutions of acids like HCl, HNO_3, and H_2SO_4. These are strong electrolytes; their water solutions, if not too concentrated, consist almost wholly of ions. The following equations show the reactions which occur when the covalent compounds HCl, HNO_3, and H_2SO_4 are dissolved in water:

$$HCl + H_2O \rightleftharpoons H_3O^+ + Cl^-$$
$$HNO_3 + H_2O \rightleftharpoons H_3O^+ + NO_3^-$$
$$H_2SO_4 + H_2O \rightleftharpoons H_3O^+ + HSO_4^-$$
$$HSO_4^- + H_2O \rightleftharpoons H_3O^+ + SO_4^{2-}$$

Quite often these equations are written in a simplified form in which the water as a reactant is omitted:

$$HCl \rightleftarrows H^+ + Cl^-$$
$$HNO_3 \rightleftarrows H^+ + NO_3^-$$
$$H_2SO_4 \rightleftarrows H^+ + HSO_4^-$$
$$HSO_4^- \rightleftarrows H^+ + SO_4^{2-}$$

It is apparent from these equations that in water solutions of hydrochloric, nitric, and sulfuric acids there is only one acid, the hydronium ion, H_3O^+.

Even for weak acids such as acetic acid, the main proton donating species is the hydronium ion:

$$HC_2H_3O_2 + H_2O \rightleftarrows H_3O^+ + C_2H_3O_2^-$$

or

$$HC_2H_3O_2 \rightleftarrows H^+ + C_2H_3O_2^-$$

However, because the equilibrium mixture contains a low concentration of H_3O^+ ions, acid reactions that depend on this ion proceed slowly in acetic acid

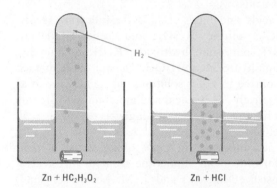

Fig. 6-6. The volumes of hydrogen liberated in an equal period of time give an approximate comparison of the amount of ionization of the two acids.

as compared to the speed in a strong acid such as hydrochloric acid. A good example of such a reaction is the action of zinc with acetic acid as compared with hydrochloric acid (each acid solution made by dissolving the same number of covalent HCl and $HC_2H_3O_2$ molecules in equal amounts of water). The reaction proceeds approximately fifty to one hundred times faster in the hydrochloric acid (Fig. 6-6). However, the reaction that occurs is the same in both cases:

$$Zn + 2H_3O^+ \to Zn^{2+} + H_2 + 2H_2O$$

It has long been customary to represent such reactions by over-all equations that imply that the proton comes directly from the HCl or the $HC_2H_3O_2$:

$$Zn + 2HCl \to ZnCl_2 + H_2$$
$$Zn + 2HC_2H_3O_2 \to Zn(C_2H_3O_2)_2 + H_2$$

These equations have the advantage of showing what specific acid was originally dissolved in water, but it must be borne in mind that they summarize a stepwise process that includes ionization of covalent molecules and reaction of hydronium ions with metal atoms.

Neutralization. When acids and bases are brought together in a water solution, the hydrogen ions of the acid and the hydroxide ions of the base combine to form water. The equation is

$$\underset{\text{acid}}{H^+} + \underset{\text{base}}{OH^-} \rightarrow \underset{\text{water}}{HOH}$$

This is the fundamental reaction that occurs when acids and bases neutralize each other in water solutions. The characteristic properties of both acid and base disappear when the ions responsible for these properties react to form water.

SALTS

Ionic compounds are commonly called salts. Some examples are

NaCl or Na$^+$, Cl$^-$	sodium chloride
KCl or K$^+$, Cl$^-$	potassium chloride
Ca(NO$_3$)$_2$ or Ca^{2+}, NO$_3^-$, NO$_3^-$	calcium nitrate
Na$_2$SO$_4$ or Na$^+$, Na$^+$, SO$_4^{2-}$	sodium sulfate

Because salts are electrovalent compounds, they exist in water solutions entirely as ions. Strictly speaking, ionic compounds such as NaOH and Ca(OH)$_2$ are salts. Because the anion OH$^-$ is a very strong base, metallic hydroxides are more often called bases to emphasize this property. We shall see in Chap. 16 that the anions in all common salts, for example, NaCl, NaNO$_3$, Na$_2$SO$_4$, Na$_2$CO$_3$, can be proton acceptors and are therefore bases.

Salt Formation During Neutralization. The formation of a salt solution in conjunction with an acid-base reaction is an important type of reaction. In terms of the Brønsted–Lowry definitions, the neutralization reaction in the case of water solutions of HCl and NaOH can be described as

$$H_3O^+ + OH^- \rightarrow H_2O + H_2O$$

After this reaction is complete, there remains a solution of Na$^+$ and Cl$^-$ ions. Although these two ions are not involved in the neutralization, we can say that the NaCl solution is formed as a result of the acid-base reaction. An older method of representing the neutralization reaction uses an over-all equation to show the initial acid and base that are brought together and what substances are present when the reaction is over, without regard to the solvents used, if any. The reaction between HNO$_3$ and KOH, either in the pure form or in water solution, is written as:

$$\underset{\text{acid}}{HNO_3} + \underset{\text{base}}{KOH} \rightarrow \underset{\text{salt}}{KNO_3} + \underset{\text{water}}{HOH}$$

Over-all neutralization equations are still commonly written in this fashion. Other examples are:

$$\underset{\substack{\text{hydrochloric}\\\text{acid}}}{HCl} + \underset{\substack{\text{sodium}\\\text{hydroxide}}}{NaOH} \rightarrow \underset{\substack{\text{sodium}\\\text{chloride}}}{NaCl} + \underset{\text{water}}{HOH}$$

$$\underset{\substack{\text{sulfuric}\\\text{acid}}}{H_2SO_4} + \underset{\substack{\text{calcium}\\\text{hydroxide}}}{Ca(OH)_2} \rightarrow \underset{\substack{\text{calcium}\\\text{sulfate}}}{CaSO_4} + \underset{\text{water}}{2HOH}$$

Each of the following four equations may be used to represent the neutralization of hydrochloric acid and sodium hydroxide in water solutions:

$$HCl + NaOH \rightarrow NaCl + HOH$$

$$H^+, Cl^- + Na^+, OH^- \rightarrow Na^+, Cl^- + HOH$$

$$H_3O^+ + OH^- \rightarrow 2H_2O$$

$$H^+ + OH^- \rightarrow HOH$$

The first equation merely shows what substances are brought together and what substances are present when the reaction is completed. The second, in addition to showing this, indicates that hydrochloric acid, sodium hydroxide, and sodium chloride are composed of ions and that water is made up of covalent molecules. The last two equations are ionic equations and show as reactants or products only those particles which took part in the *chemical reaction*. These two emphasize that the Na^+ ion and the Cl^- ion are merely "spectator ions," which do not change chemically in any way. The third equation is a satisfactory way of emphasizing the donation and acceptance of the proton, but because the hydronium ion may be $H_9O_4^+$ or something else, the fourth equation may be just as satisfactory.

EXERCISES

1. Using the rules stated in this chapter, assign oxidation states to each element in the following: $KClO_4$, CF_4, CH_4, Na_2O, $NaNO_3$, $NaNO_2$, P_4O_{10}, $BaSO_4$, NO_2, $XeOF_4$, $KMnO_4$, $(NH_4)_3AsO_4$.
2. Name the following binary compounds: SO_2, FeF_3, FeF_2, Rb_2O, ZnS, Mg_3N_2, HBr, NH_3, PCl_3.
3. Name the following compounds using the *-ate* convention: $KClO_3$, $BaSO_4$, $CaCO_3$, HNO_3, $(NH_4)_3PO_4$, $CuSO_4$, $Al_2(CrO_4)_3$.
4. Name the following as acids: H_2S, H_2SO_4, H_3PO_4, HBr, HNO_3, H_2CrO_4, $HC_2H_3O_2$.
5. It was found experimentally that 45 ml of gas A reacted with 15 ml of gas B to produce 45 ml of a gaseous compound of A and B. (The volumes were measured at the same temperature and pressure.) If it is assumed that equal volumes of gases contain the same number of molecules and that A and B are elemental, what theoretical explanation can be given to account for the volumes?
6. Write formulas for the following compounds: copper(I) chloride, beryllium hydroxide, sulfur trioxide, mercury(II) sulfide, aluminum sulfate, barium peroxide, phosphorus pentachloride, lithium nitride, hydroselenic acid.
7. Balance the following equations:
   (a) $H_2 + N_2 \rightarrow NH_3$
   (b) $H_2 + O_2 \rightarrow H_2O$
   (c) $CH_3OH + O_2 \rightarrow CO_2 + H_2O$
   (d) $Pb(NO_3)_2 + NaCl \rightarrow PbCl_2 + NaNO_3$
   (e) $H_2O + CO_2 \rightarrow C_6H_{12}O_6 + O_2$

Formulas and Equations; Classes of Compounds

8. Write balanced equations for the following reactions:
   (a) hydrogen + bromine → hydrogen bromide
   (b) barium + water → hydrogen + barium hydroxide
   (c) calcium oxide + phosphoric acid → calcium phosphate + water
   (d) ammonium carbonate + tin(II) nitrate → ammonium nitrate + tin(II) carbonate
9. Write ionic equations for the reactions that occur at the anode and the cathode when a direct current is passed through each of the following:
   (a) molten magnesium chloride;
   (b) a water solution of hydrogen chloride.
10. Explain how a current of electricity is conducted through a water solution of HI.
11. Distinguish between electrolyte and nonelectrolyte and give three examples of each.
12. Distinguish between strong and weak electrolytes and give three examples of each.
13. In the most common chemical test for the presence of Ag^+ ions, hydrochloric acid is added to the solution being tested. The formation of a precipitate of AgCl signifies the presence of Ag^+. Write an ionic equation that illustrates this reaction. List several other compounds that could be substituted for the hydrochloric acid (see Table 4 of the Appendix for water solubilities).
14. Identify the acid in each of the following and give a reason for your choice:
    (a) $H_2O + HCl \rightarrow H_3O^+ + Cl^-$
    (b) $HCO_3^- + NH_3 \rightarrow NH_4^+ + CO_3^{2-}$
    (c) $CaO + H_2O \rightarrow Ca^{2+} + 2OH^-$
    (d) $2H_3O^+ + CO_3^{2-} \rightarrow CO_2 + 3H_2O$
    (e) $NaCl + H_2SO_4 \rightarrow HCl + NaHSO_4$
15. For each of the reactions indicated in Exercise 14, what acted as the base? Justify your choice.
16. List four characteristics common to a water solution of HCl and a water solution of H_2SO_4. How do we account for these common properties?
17. If three separate water solutions, each containing 40.0 g of NaOH, are neutralized, one with HCl, one with HNO_3, and one with HBr, about 13,700 cal of heat are liberated in each case. Why?
18. Oxygen reacts with hydrogen chloride at high temperature to produce chlorine and water.
    (a) Write a balanced equation for the reaction.
    (b) How many chlorine molecules will be produced from 800 molecules of HCl?
    (c) How many HCl molecules will react with 6.02×10^{23} oxygen molecules?
19. Account for the fact that a water solution of HBr contains ions, while a solution of HBr in the nonpolar liquid carbon tetrachloride does not contain ions.
20. Identify the "spectator ions" in a reaction between water solutions of silver nitrate and sodium chloride. (An insoluble silver salt is formed.)

SUGGESTED READING

Clever, H. L., "The Hydrated Hydronium Ion," *J. Chem. Educ.*, **40**, 37 (1963).

7

THE MOLE; ENERGY AND WEIGHT RELATIONSHIPS

In this chapter we take up the solution of problems that involve simple energy and weight relationships. In every case the solution is based on one or more of the three most fundamental laws of chemistry:

1. The law of conservation of energy.[1]
2. The law of conservation of mass.
3. The law of definite composition.

These laws were discussed in Chap. 1. It might be mentioned here that the equivalence of matter and energy as defined by Einstein's relationship, $E = mc^2$, is not a matter of concern in ordinary chemical reactions. We will have more to say of this equivalence in Chap. 17.

THE MOLE

In many kinds of chemical problems, it is necessary to consider quantities of substances in terms of the number of atoms, ions, or molecules present. For example, if we compare the amounts of heat liberated by different chemical reactions, our comparisons have more meaning when the heats are measured for reactions in which the same numbers of atoms, molecules, or ions react.

The unit devised by chemists to express numbers of atoms, ions, or molecules is called the mole. A **mole** is defined as that quantity of a substance that contains

---

[1] This law is often referred to as the *first law of thermodynamics*.

the same number of ultimate particles (atoms, molecules, ions, or units of ions) as are contained in 12 g of carbon-12. The purpose of relating our unit number of particles to the standard of atomic weights is to provide a ready method of calculating the weight of a mole of any substance (**molar weight**).

Consider a mole of carbon-12 atoms and a mole of magnesium atoms. By definition, a mole of ^{12}C is 12 g of this substance. What weight of magnesium atoms contains the same number of atoms as are present in the 12 g of carbon-12? We see from a table of atomic weights that the average weight of magnesium atoms is 24.31 awu. Because each magnesium atom is slightly more than twice as heavy as a single carbon atom, it follows that a mole of magnesium weighs slightly more than twice the weight of a mole of ^{12}C, or 24.31 g, precisely. This quantity, 24.31 g of magnesium, contains the same number of atoms as are contained in 12 g of ^{12}C and is referred to as *mole of magnesium*.

Let us consider additional examples. The weights of hydrogen, oxygen, and sulfur atoms are 1.008 awu, 15.999 awu, and 32.06 awu, respectively, on the scale $^{12}C = 12$ awu. It follows that 1.008 g of hydrogen, 15.999 g of oxygen, or 32.06 g of sulfur each contains the same number of atoms as are contained in 12 g of ^{12}C. These are the weights of a mole of hydrogen atoms, oxygen atoms, or sulfur atoms, respectively.

The mole concept is not limited to atoms. We may also apply it to matter in which the ultimate particles are molecules, ions, or units of ions. For example, a mole of oxygen molecules (O_2) weighs 31.998 g, and a mole of ozone molecules (O_3) weighs 47.997 g.[2] Each contains the same number of molecules as there are atoms in a mole of carbon-12.

Our guide to the weight of a mole is the formula for the substance. Suppose we wish to find the weight of a mole of aluminum fluoride, AlF_3. The simplest unit of this compound is one aluminum ion (Al^{3+}) and three fluoride ions (F^-, F^-, F^-). The weight of this unit is 26.98 + 3(18.998) or 83.974 awu. The weight of AlF_3, which contains as many AlF_3 units as there are atoms in 12 g of ^{12}C, is 83.974 g. This quantity is a mole of AlF_3. See Table 7-1 and Fig. 7-1 for additional examples.

Modern experimental methods for counting atoms, ions, and molecules show that in one mole of a substance there are 6.0228×10^{23} particles or formula units. This huge number is called **Avogadro's number,** named in honor of the brilliant Italian contemporary of Dalton who recognized that the simplest molecules of compounds might very well be triatomic, tetratomic, or more complex, in contrast to Dalton's emphasis on diatomic molecules.

On page 77 it was stated that an equation indicates the ratio of molecules that participate in a chemical reaction. An equation may also be interpreted in terms of moles of reactants and products. For the reaction of hydrogen and oxygen to form water, the equation may be read: 2 moles of hydrogen react with 1 mole of oxygen to form 2 moles of water. And because the weights of a

---

[2] The formula weight expressed in pounds is a pound mole; that is, a pound mole of ^{12}C is 12 lb, of CO_2, 44.009 lb. In this country, quantities are often expressed in terms of the pound mole in the chemical industry. Because 1 lb = 454 g, a pound mole contains 454 times as many particles as a mole.

TABLE 7-1.

Mole Relationships

NAME	FORMULA	WEIGHT OF UNIT INDICATED BY FORMULA, AWU[a]	WEIGHT OF 1 MOLE, G	NUMBER AND KIND OF PARTICLES IN 1 MOLE[a]
Nitrogen	N_2	28.0	28.0	6.02×10^{23} molecules
Atomic nitrogen	N	14.0	14.0	6.02×10^{23} atoms
Silver	Ag	108	108	6.02×10^{23} atoms
Silver ions	Ag^+	108	108	6.02×10^{23} Ag^+ ions
Methanol	CH_3OH	32.0	32.0	6.02×10^{23} molecules
Sodium chloride	NaCl	58.5	58.5	6.02×10^{23} NaCl units, or 6.02×10^{23} Na^+ ions 6.02×10^{23} Cl^- ions
Barium chloride	$BaCl_2$	208	208	6.02×10^{23} $BaCl_2$ units, or 6.02×10^{23} Ba^{2+} ions 12.0×10^{23} Cl^- ions
Ammonium sulfate	$(NH_4)_2SO_4$	132	132	6.02×10^{23} $(NH_4)_2SO_4$ units 12.0×10^{23} NH_4^+ ions 6.02×10^{23} SO_4^{2-} ions

[a] Weights and numbers of particles have been rounded to three significant figures.

Fig. 7-1. Arranged from left to right around 1 mole of sugar (342 g) are 1 mole of aluminium (27 g), 1 mole of water (18 g), 1 mole of copper (64 g), and 1 mole of iron (56 g). Each sample contains 6×10^{23} atoms if an element, or that number of molecules if a compound.

mole of hydrogen and a mole of oxygen can be readily inferred from the atomic weights, the weight relationship between reactants and products can be calculated:

$$2H_2 + O_2 \rightarrow 2H_2O$$
$$\text{2 moles} \quad \text{1 mole} \quad \text{2 moles}$$
$$\text{4.02 g} \quad \text{32.00 g} \quad \text{36.02 g}$$

When an equation is being interpreted in terms of moles, the equation may be written to indicate fractions of moles. For example, the above equation may be written

$$H_2 + \tfrac{1}{2}O_2 \rightarrow H_2O$$
$$\text{1 mole} \quad \tfrac{1}{2}\text{ mole} \quad \text{1 mole}$$
$$\text{2.01 g} \quad \text{16.00 g} \quad \text{18.01 g}$$

THERMOCHEMISTRY

The study of the changes in energy that accompany chemical reactions is called *thermochemistry*. The subject is of fundamental importance, not only because we often want to know how much energy can be obtained from reactions, but also because we can gain much information about the structures of molecules and about other chemical problems from the energy changes that occur when substances are heated, or when they react to liberate or absorb heat.

ENTHALPY

Ordinary chemical equations, such as we have been discussing, show only the substances that react and the substances that are formed. They do not show the energy changes that take place. As a rule, knowledge of the energy changes is important.

The heat given up during an exothermic chemical reaction is usually indicated in an equation, called a *thermochemical equation*, as follows. The letters in parentheses, s, l, and g, stand for solid, liquid, and gas, respectively.

$$CH_4(g) + 2O_2(g) \rightarrow CO_2(g) + 2H_2O(l) + 212{,}800 \text{ cal} \quad (1)$$

$$SO_2(g) \rightarrow S(s) + O_2(g) - 70{,}960 \text{ cal} \quad (2)$$

$$H_2(g) + \tfrac{1}{2}O_2(g) \rightarrow H_2O(l) + 68{,}320 \text{ cal} \quad (3)$$

The first two equations are read thus: (1) When 1 mole (16 g) of gaseous methane unites with 2 moles (64 g) of gaseous oxygen to form 1 mole (44 g) of gaseous carbon dioxide and 2 moles (36 g) of liquid water, 212,800 cal of heat is liberated to the surroundings. (2) When 1 mole (64 g) of gaseous sulfur dioxide is decomposed into 1 mole (32 g) of solid sulfur and 1 mole (32 g) of gaseous oxygen, 70,960 cal of heat is absorbed from the surroundings.

In discussing energy changes during chemical reactions, the chemist finds it convenient to think of each substance as having a certain heat content or **enthalpy** (Greek *enthalpein*, to warm in). The symbol for enthalpy is H; a heat change in a chemical reaction is termed a change in enthalpy or ΔH. The sign Δ means "the difference in." Strictly, the term "change in enthalpy" refers to the heat change during a process carried out at constant pressure. If the energy is to be precisely specified, the initial and final conditions of pressure and temperature must be known. In this text, most values for ΔH are for processes at 25°C and 1 atm.

For an exothermic reaction ΔH is negative, so that reaction (1) can be written

$$CH_4(g) + 2O_2(g) \rightarrow CO_2(g) + 2H_2O(l) \qquad \Delta H = -212{,}800 \text{ cal}$$

The heat energy that is transformed to chemical energy during an endothermic chemical reaction may be indicated as follows:

$$N_2(g) + O_2(g) \rightarrow 2NO(g) - 21{,}600 \text{ cal}$$

or

$$N_2(g) + O_2(g) \rightarrow 2NO(g) \qquad \Delta H = 21{,}600 \text{ cal}$$

Either equation is read thus: When 1 mole (28 g) of gaseous nitrogen unites with 1 mole (32 g) of gaseous oxygen to form 2 moles (60 g) of gaseous nitric oxide, 21,600 cal of heat energy is absorbed from the surroundings.

It is important to note again that enthalpy changes are measured at constant pressure. If there is no significant difference between the volume of the reactants and products, the energy liberated as heat for an exothermic reaction will not differ appreciably when carried out in an open vessel (constant pressure) or in a closed container (constant volume). However, if the volume of the products is significantly greater than the volume of the reactants, which is often the case when gases are involved, work must be done in pushing the atmosphere back to make room for the extra volume, and less heat will be liberated when the reaction is carried out in an open vessel. The opposite is true for an exothermic reaction, which gives rise to a smaller volume. When a reaction is carried out in a sealed container (constant volume), the pressure increases if the products have a larger volume, but no work is done because no expansion occurs. Problem 1 calls attention to some of the difficulties encountered in measuring enthalpy changes.

Problem 1: A sample of acetylene, C_2H_2, weighing 0.418 g was placed in a steel bomb. Enough oxygen was then introduced to insure complete combustion of the acetylene to carbon dioxide and water. The bomb was immersed in 1020 g of water in a calorimeter (Fig. 7-2), and the assembly was allowed to stand until the temperature became constant at 25.00°C. The acetylene was then ignited electrically. The maximum temperature attained by the water was 29.20°C. The steel vessel, calorimeter walls, and stirrer were known to absorb 158 cal per degree rise in temperature. From these data, calculate the heat taken up by the water and the calorimeter; describe the additional calculations necessary to arrive at the enthalpy of combustion of one mole of acetylene at 25°C and 1 atm.

The Mole; Energy and Weight Relationships

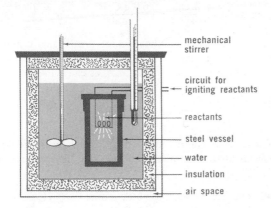

Fig. 7-2. A bomb calorimeter.

Solution:
Rise in temperature = 29.20°C − 25.00°C = 4.20 deg C

Heat taken up by water = 1020 g × $\dfrac{1 \text{ cal}}{1 \text{ g} \times 1 \text{ deg C}}$ × 4.20 deg C = 4280 cal

Heat taken up by calorimeter = $\dfrac{158 \text{ cal}}{\text{deg C}}$ × 4.20 deg C = 664 cal

Total = 4280 + 664 = 4940 cal

Additional Calculations. The heat taken up by the water and calorimeter does not represent all of the energy liberated by the combustion of 0.418 g of acetylene. Since the products of the combustion, H_2O and CO_2, are at a higher temperature than the original reactants, a small amount of heat is retained by them. The actual amount is calculated as follows:

(Wt of H_2O × specific heat × 4.20 deg)
+ (wt of CO_2 × specific heat × 4.20 deg)

Also, a small amount of water is present as a vapor. When this water is condensed to a liquid, 540 cal/g is released. The amount of water present as a vapor at 29.20°C can be calculated from water vapor pressures (see Table 1, Appendix). The heat retained by the water in the vapor form is calculated as follows:

g in vapor form × $\dfrac{540 \text{ cal}}{\text{g}}$ = heat retained by water vapor

Finally, the energy change involving the expansion or contraction of reactants and products can be calculated from the pressure and volume changes. Data needed for these calculations include the pressures and volumes of reactants and products. We will not attempt to describe these calculations. (In this reaction there is a decrease in the pressure; the products have a smaller volume than the reactants.)

By adding the heat liberated to the water and calorimeter, the heat retained by the products at a higher temperature, the heat retained by water vapor, and the energy involved in carrying out the reaction at constant volume rather than at a constant pressure of one atmosphere, we would obtain the enthalpy change as X calories for the combustion of 0.418 g of acetylene at 1 atm and 25°C:

$$C_2H_2(g) + 2\tfrac{1}{2}O_2(g) \rightarrow 2CO_2(g) + H_2O(l)$$

We could then calculate the enthalpy change per mole:

$$\Delta H \text{ per mole} = \frac{X \text{ cal}}{0.418 \text{ g}} \times \frac{26.0 \text{ g C}_2\text{H}_2}{1 \text{ mole}}$$

Note that the heat liberated per mole when the reactants are at 25.00°C and the products are at 29.20°C with part of the water in the vapor form and with the pressure of the products less than that of the reactants, as calculated from our data, is:

$$\frac{4940 \text{ cal}}{0.418 \text{ g}} \times \frac{26.0 \text{ g}}{1 \text{ mole}} = 307,000 \text{ cal} \quad \text{or} \quad 307 \text{ kcal}$$

The accepted value of ΔH, at 25°C and 1 atm, and all the water in the liquid form, is -310.61 kcal. The negative sign for ΔH indicates that the combustion is exothermic.

Hess's Law. It was in 1840 that the Swiss-Russian chemist G. H. Hess stated one of the most useful generalizations of thermochemistry. A modern version of **Hess's law** is, *for a given over-all reaction, the change in enthalpy is always the same, whether the reaction is performed directly or whether it takes place indirectly and in different steps.*

As an example of Hess's law, consider the exothermic reaction between sulfur and oxygen to produce sulfur dioxide, followed by the exothermic reaction between sulfur dioxide and more oxygen to produce sulfur trioxide:

$$S(s) + O_2(g) \rightarrow SO_2(g) \qquad \Delta H = -70.96 \text{ kcal/mole of product}$$

$$SO_2(g) + \tfrac{1}{2}O_2(g) \rightarrow SO_3(g) \qquad \Delta H = -23.49 \text{ kcal/mole of product}$$

If these two steps are considered to take place as a simple one-step over-all reaction, the heat evolved is the sum of the two steps:

$$S(s) + 1\tfrac{1}{2}O_2(g) \rightarrow SO_3(g) \qquad \Delta H = -94.45 \text{ kcal/mole of product}$$

It will be recalled that the enthalpy change during a reaction, ΔH, refers to a change in heat content of a system of chemicals due to a chemical reaction. In each of the three reactions just described, the product has a smaller heat content than the reactants; hence, the decrease in heat content is shown by the minus sign. The heat lost by the chemicals is given off to the surroundings in the exothermic reaction.

One useful consequence of Hess's law is that thermochemical equations can be added or subtracted to produce data that are difficult to determine experimentally. For example, carbon and carbon monoxide are important commercial fuels; therefore, it is of interest to compare the amount of heat liberated when carbon is burned to carbon dioxide with the amount of heat liberated when carbon is burned to carbon monoxide. The latter enthalpy change or heat of reaction is difficult to determine, because carbon monoxide is burned more readily than carbon. Consequently, a reaction in which carbon is burned in the theoretical amount of oxygen needed to form carbon monoxide produces a mixture of carbon dioxide, carbon monoxide, and unburned carbon. However, we can calculate the heat of combustion of carbon to carbon monoxide by determining the heat of combustion (1) of carbon to carbon dioxide and (2) of carbon monoxide to carbon dioxide. Subtracting the thermochemical equation for the latter reaction from that of the former gives the desired information:

$$\begin{array}{ll} & \Delta H \text{ values} \\ 2C(s) + 2O_2(g) \rightarrow 2CO_2(g) & -188.1 \text{ kcal} \\ \text{(minus)} \ 2CO(g) + O_2(g) \rightarrow 2CO_2(g) & -135.3 \\ \hline 2C(s) - 2CO(g) + O_2(g) \rightarrow & -52.8 \\ \text{or} \ \ \ \ 2C(s) + O_2(g) \rightarrow 2CO(g) & -52.8 \end{array}$$

The Law of Dulong and Petit. In 1819, two French chemists, P. L. Dulong and A. T. Petit, carefully measured the specific heats of a number of elements (the amount of heat required to raise the temperature of 1 g 1 deg C, see Chap. 1). From their work they came to the conclusion that *atoms of all elements have the same capacity for heat*. This statement is now known as the **law of Dulong and Petit**.

As an illustration of this law, consider a tiny piece of silver that contains 1000 billion silver atoms and a tiny piece of gold that contains 1000 billion gold atoms. The piece of gold, of course, weighs almost twice as much as the piece of silver, because gold atoms weigh almost twice as much as silver atoms. However, since the two pieces contain the same number of atoms, the law of Dulong and Petit predicts that the same amount of heat energy is required to raise the temperature of each 1 deg C.

As a check on the law, let us calculate from experimentally determined specific heats the amount of heat required to raise the temperature of 6.02×10^{23} atoms of several metals by 1 deg C, say from 25°C to 26°C. We should come out with the same amount of energy for each.

The metals chosen for the calculation are gold, silver, copper, and iron; the specific heats in cal/(g × deg C) are 0.0298, 0.0557, 0.0949, and 0.1100, respectively. Since 6.02×10^{23} atoms is a mole of atoms, the weights of gold, silver, copper, and iron are 197 g, 108 g, 63.5 g, and 55.8 g, respectively. Our problem then is to calculate the heat necessary to increase the temperature of each of these quantities 1 deg C.

Gold: $\dfrac{197 \text{ g}}{\text{mole}} \times \dfrac{0.0298 \text{ cal}}{1 \text{ g} \times 1 \text{ deg C}} = 5.9 \text{ cal/(mole} \times \text{deg C)}$

Silver: $\dfrac{108 \text{ g}}{\text{mole}} \times \dfrac{0.0557 \text{ cal}}{1 \text{ g} \times 1 \text{ deg C}} = 6.0 \text{ cal/(mole} \times \text{deg C)}$

Copper: $\dfrac{63.5 \text{ g}}{\text{mole}} \times \dfrac{0.0949 \text{ cal}}{1 \text{ g} \times 1 \text{ deg C}} = 6.0 \text{ cal/(mole} \times \text{deg C)}$

Iron: $\dfrac{55.8 \text{ g}}{\text{mole}} \times \dfrac{0.1100 \text{ cal}}{1 \text{ g} \times 1 \text{ deg C}} = 6.1 \text{ cal/(mole} \times \text{deg C)}$

It is apparent from these four calculations that the law of Dulong and Petit is not an exact one. However, there was a period in the development of chemistry when the law was quite important, because it provided a method of determining approximate atomic weights. That is,

$$\text{mole wt} \times \text{specific heat} \simeq 6.0 \text{ cal/(mole} \times \text{deg C)}$$

or

$$\text{mole wt} \simeq \frac{6.0 \text{ cal/(mole} \times \text{deg C)}}{\text{specific heat}}$$

Problem 2: The specific heat of bismuth was experimentally determined as 0.0288 cal/(g × deg C). Calculate the approximate atomic weight of bismuth.
Solution:

$$\text{Mole wt} \simeq \frac{6.0 \text{ cal/(mole} \times \text{deg C)}}{0.0288 \text{ cal/(g} \times \text{deg C)}}$$

$$\simeq 209 \text{ g/mole}$$

$$\text{Atomic wt} \simeq 209 \text{ awu}$$

As was pointed out in Chap. 3, the mass spectrograph is now used for precise determinations of atomic weights.

WEIGHT RELATIONSHIPS

PERCENTAGE COMPOSITION OF COMPOUNDS

The weights of the elements comprising a mole of a compound are readily deduced from the formula of the compound and the atomic weights of the elements. The percentage composition by weight is then calculated from the weights of the elements and the weight of a mole of the compound.

Problem 3: Calculate the percent weight of each element in magnesium chloride, $MgCl_2$.
Solution: In 1 mole of $MgCl_2$ there are 1 mole of Mg^{2+} ions and 2 moles of Cl^- ions. The weights of these ions and the weight of 1 mole of the compound are:

$$\begin{aligned}\text{Wt of } Mg^{2+} &= 1 \text{ mole} \times 24.31 \text{ g/mole} = 24.3 \text{ g}\\ \text{Wt of } Cl^- &= 2 \text{ moles} \times 35.45 \text{ g/mole} = 70.9 \text{ g}\\ \text{Wt of 1 mole of } MgCl_2: &\phantom{= 2 \text{ moles} \times 35.45 \text{ g/mole} =} 95.2 \text{ g}\end{aligned}$$

Note that the calculated weights have been rounded to three significant figures.

The percentages by weight are then calculated as follows:

$$\text{Magnesium:} \frac{24.3 \text{ g}}{95.2 \text{ g}} \times 100 = 25.5 \text{ percent}$$

$$\text{Chlorine:} \frac{70.9 \text{ g}}{95.2 \text{ g}} \times 100 = 74.5 \text{ percent}$$

Check: Chlorine atoms are about $1\frac{1}{2}$ times as heavy as magnesium atoms, and there are 2 times as many of the former in a sample of $MgCl_2$. Hence, we would

The Mole; Energy and Weight Relationships

expect the weight due to chlorine to be about 3 times the weight due to magnesium, that is, 3 parts out of 4, or about 75 percent.

It is interesting to note that the percentages that we have calculated are not those of a single $MgCl_2$ unit, because no single ion of Mg^{2+} has a mass of 24.31 awu and no single ion of Cl^- has a mass of 35.45 awu. However, any natural sample of $MgCl_2$ that we could weigh, even with the most sensitive analytical balance, would have this composition, because it would be composed of atoms of all the natural isotopes.

Problem 4: Calculate the percentage composition by weight of ether. $(C_2H_5)_2O$.

Solution:

Wt due to C = 4 moles × 12.0 g/mole = 48.0 g
Wt due to H = 10 moles × 1.01 g/mole = 10.1 g
Wt due to O = 1 mole × 16.0 g/mole = 16.0 g
Wt of 1 mole of ether: 74.1 g

Percent by wt of C = $\frac{48.0 \text{ g}}{74.1 \text{ g}} \times 100$ = 64.8 percent

Percent by wt of H = $\frac{10.1 \text{ g}}{74.1 \text{ g}} \times 100$ = 13.6 percent

Percent by wt of O = $\frac{16.0 \text{ g}}{74.1 \text{ g}} \times 100$ = 21.6 percent

Total   100.0 percent

WEIGHT CHANGES IN CHEMICAL REACTIONS

Consider a chemical reaction such as the combustion of methane, the main component of natural gas:

$$\underset{1 \text{ mole}}{CH_4} + \underset{2 \text{ moles}}{2O_2} \rightarrow \underset{1 \text{ mole}}{CO_2} + \underset{2 \text{ moles}}{2H_2O}$$

From the balanced equation it is seen that two moles of oxygen is required to burn one mole of methane and that one mole of carbon dioxide and two moles of water are formed. Because the molar weights are obtainable from the atomic weights, the weight relationship between reactants and products is easily established.

Problem 5: Calculate the weight of oxygen needed to burn 20.0 g of methane. What weights of carbon dioxide and water are formed? (See equation above.)

Solution: The first step in the solution of this and similar problems is to calculate the number of moles of the reactant available for the reaction. Only 20.0 g of CH_4 is to react. By adding the atomic weights (rounded to three significant figures), we find the weight of one molecule to be 16.0 awu. The weight of one mole of CH_4 is therefore 16.0 g. The number of moles in 20.0 g is:

$$20.0 \text{ g } CH_4 \times \frac{1 \text{ mole}}{16.0 \text{ g } CH_4} = 1.25 \text{ moles}$$

We are now ready to calculate the number of moles of oxygen required to react with 1.25 moles of methane and the number of moles of carbon dioxide and water that will be formed. The equation states that 2 moles of O_2 is required for each mole of CH_4 and that 1 mole of CO_2 and 2 moles of H_2O will be formed:

$$\text{Moles } O_2 = 1.25 \text{ moles } CH_4 \times \frac{2 \text{ moles } O_2}{1 \text{ mole } CH_4} = 2.50 \text{ moles}$$

$$\text{Moles } CO_2 = 1.25 \text{ moles } CH_4 \times \frac{1 \text{ mole } CO_2}{1 \text{ mole } CH_4} = 1.25 \text{ moles}$$

$$\text{Moles } H_2O = 1.25 \text{ moles } CH_4 \times \frac{2 \text{ moles } H_2O}{1 \text{ mole } CH_4} = 2.50 \text{ moles}$$

We have calculated that when 1.25 moles (20.0 g) of methane burns, 2.50 moles of O_2 is consumed, and 1.25 moles of CO_2 and 2.50 moles of H_2O are formed. To find the weights of these quantities we make use of the weight of a mole of each compound:

$$\text{Wt of } O_2 = 2.50 \text{ moles} \times 32.0 \text{ g/mole} = 80.0 \text{ g}$$

$$\text{Wt of } CO_2 = 1.25 \text{ moles} \times 44.0 \text{ g/mole} = 55.0 \text{ g}$$

$$\text{Wt of } H_2O = 2.50 \text{ moles} \times 18.0 \text{ g/mole} = 45.0 \text{ g}$$

Check: Wt of reactants = wt of products

20.0 g of CH_4 + 80.0 g of O_2 = 55.0 g CO_2 + 45.0 g H_2O

100 g reactants = 100 g of products

Problem 6: The equation for the reaction between elemental magnesium and a solution of gold(III) chloride is

$$3Mg + 2AuCl_3 \rightarrow 3MgCl_2 + 2Au$$

If a piece of magnesium weighing 1.00 g is placed in a solution that contains 3.00 g of gold(III) chloride dissolved in water, which of the reactants is present in excess? What weight of gold (Au) is formed? How much of the excess reactant is left when the reaction is over?

Solution: From the balanced equation we get the mole ratios of reactants and products:

$$\underset{3 \text{ moles}}{3Mg} + \underset{2 \text{ moles}}{2AuCl_3} \rightarrow \underset{3 \text{ moles}}{3MgCl_2} + \underset{2 \text{ moles}}{2Au}$$

Calculation of the number of moles of each reactant available:

$$\text{Moles of Mg} = 1.00 \text{ g of Mg} \times \frac{1 \text{ mole of Mg}}{24.3 \text{ g of Mg}} = 0.0412 \text{ mole of Mg}$$

$$\text{Moles of AuCl}_3 = 3.00 \text{ g of AuCl}_3 \times \frac{1 \text{ mole}}{303 \text{ g of AuCl}_3} = 0.00990 \text{ mole of Au}$$

It is seen from the equation that, if 0.0412 mole of Mg reacts, $0.0412 \times 2/3$ mole or 0.0274 mole of $AuCl_3$ must be present to react with it. However, there is only 0.00990 mole of $AuCl_3$, so that not all the magnesium can react. That is, the magnesium is present in excess.

Because all the $AuCl_3$ can react, we base our calculations of actual reactants and products on the amount of gold(III) chloride present and not on the amount of magnesium:

$$\text{Moles of Au formed} = 0.00990 \text{ mole of AuCl}_3 \times \frac{2 \text{ moles of Au}}{2 \text{ moles of AuCl}_3}$$

$$= 0.00990 \text{ mole of Au}$$

The Mole; Energy and Weight Relationships

$$\text{Wt of Au formed} = 0.00990 \text{ mole of Au} \times \frac{197 \text{ g of Au}}{1 \text{ mole of Au}}$$
$$= 1.92 \text{ g of Au}$$

$$\text{Moles of Mg that react} = 0.00990 \text{ mole of AuCl}_3 \times \frac{3 \text{ moles of Mg}}{2 \text{ moles of AuCl}_3}$$
$$= 0.0148 \text{ mole of Mg}$$

$$\text{Moles of Mg left in excess} = 0.0412 \text{ mole of Mg} - 0.0148 \text{ mole of Mg}$$
$$= 0.0264 \text{ mole of Mg}$$

$$\text{Wt of Mg left in excess} = 0.0264 \text{ mole} \times \frac{24.3 \text{ g of Mg}}{1 \text{ mole of Mg}}$$
$$= 0.642 \text{ g of Mg}$$

EMPIRICAL FORMULAS

Even though the number of known compounds may be almost two million, it is estimated that 10,000 new compounds are synthesized each year in the university and industrial research laboratories of the world. As these new compounds are obtained, their compositions must be experimentally determined. This is the first step in arriving at their formulas.

There are many methods for determining the percentage by weight of the different elements in a compound. These methods vary, depending on the nature of the compound and the elements in it. For instance, if a substance contains carbon and hydrogen, a weighed sample of the compound can be burned in a closed tube in a stream of oxygen to form carbon dioxide and water. The combustion products are swept from the tube by the stream of oxygen into two absorbing chemicals, one of which absorbs water vapor and the other carbon dioxide. See Fig. 7-3.

The gain in weight of each of the absorbers gives the weight of water and carbon dioxide, respectively. Because water is known to be $\frac{2}{18}$ hydrogen by weight, $\frac{2}{18}$ of the weight of the water is equal to the amount of hydrogen originally present in the compound. Similarly, CO_2 is $\frac{12}{44}$ carbon, and therefore $\frac{12}{44}$ of the weight gained by the CO_2 absorber is the weight of the carbon originally present in the sample. Then:

$$\text{Percent H} = \frac{\text{wt of hydrogen}}{\text{wt of sample}} \times 100$$

$$\text{Percent C} = \frac{\text{wt of carbon}}{\text{wt of sample}} \times 100$$

If such an experiment is carried out with benzene, a colorless liquid obtainable from coal, the results of the experimental determination show that benzene is composed of 7.7 percent hydrogen and 92.3 percent carbon.

Knowing the percentage composition and the atomic weights of the elements involved, we can calculate the simplest or empirical formula for a compound. In this type of problem it is convenient to consider 100 g of the compound as the basis for calculation.

One hundred grams of benzene contains 7.7 g of hydrogen and 92.3 g of carbon in chemical union. We now calculate the number of moles of carbon and of hydrogen atoms in these weights. Since 1 mole of hydrogen atoms weighs 1.0 g, 7.7 g contains

$$7.7 \text{ g} \times 1 \text{ mole}/1.0 \text{ g} = 7.7 \text{ moles of H atoms}$$

One mole of carbon weighs 12.0 g. The number of moles in 92.3 g of carbon is

$$92.3 \text{ g} \times 1 \text{ mole}/12.0 \text{ g} = 7.7 \text{ moles of C atoms}$$

This calculation brings us to an interesting and informative conclusion. In 100 g of benzene there is an equal number of carbon and hydrogen atoms. The number is $7.7 \times (6 \times 10^{23})$ carbon atoms and $7.7 \times (6 \times 10^{23})$ hydrogen atoms. Now we are in a position to write possible formulas for the compound. We assume that because a large sample of benzene contains an equal number of atoms of carbon and hydrogen, the tiniest sample, a molecule, also contains an equal number. Therefore, the formula for a benzene molecule could be any one that shows this 1 : 1 relationship, e.g., CH, C_2H_2, C_3H_3, C_4H_4, C_5H_5, C_6H_6, etc. Until we have additional experimental information about the size of the benzene molecule, we have no way of choosing among these formulas because all of them agree with the information that benzene is 7.7 percent hydrogen and 92.3 percent carbon. Of all the various formulas that show this 1 : 1 ratio, the simplest is CH. This formula, the simplest one, is called the **empirical formula**; it indicates the ratio of the atoms in a compound.

MOLECULAR FORMULAS

The formula we commonly use for a covalent compound shows not only the composition but the actual number and kind of atoms in a molecule. Although this formula, called the **molecular formula,** may be the same as the simplest or empirical formula—for example, methane, CH_4—frequently it is a multiple of the empirical formula—for example, ethane, C_2H_6, and glucose, $C_6H_{12}O_6$.

In an ionic compound, however, there are no molecules. The formula merely indicates the ratio of the ions present; it is not a molecular formula. Consequently, the formula that is normally written is the empirical or simplest formula. But in some cases the formula for an ionic compound is not the simplest one. For example, the formula for the ionic compound sodium dithionate is written $Na_2S_2O_6$ and not $NaSO_3$, because it has been shown that the dithionate ion contains two sulfur and six oxygen atoms held together by covalent bonds to form the polyatomic ion $S_2O_6^{2-}$.

The molecular formula for a compound is either the same as the empirical formula or a whole-number multiple of it. In order to calculate this whole number, the molecular weight must be known. Methods for experimentally determining molecular formulas will be discussed in Chaps. 10 and 13. Let us assume here that the determined molecular weight of benzene is 77 awu, and then show how this weight can be used, along with the empirical formula, to deduce the molecular formula. In all the formulas in the following list there is a 1 : 1 ratio of carbon to hydrogen, but only one of these formulas, C_6H_6, has a *calculated* molecular weight that agrees well with the *experimentally determined* molecular weight.

Possible Formula for Benzene	Calculated MW = Sum of Atomic Weights (in awu)
CH	13.0
C_2H_2	26.0
C_3H_3	39.0
C_4H_4	52.0
C_5H_5	65.0
C_6H_6	78.0
C_7H_7	91.0

At first it might seem that C_6H_6 is not necessarily the correct formula, because previously the experimental molecular weight was determined as 77, whereas the calculated value is 78.0. Actually most experimental methods give only *approximate* molecular weights; for benzene, values between 76 and 80 might be obtained in the laboratory. This does not complicate the problem, however. Since the empirical formula is *exact*, only the molecular weights in the right-hand column are possible, and the true molecular weight must be one of these.

In summary, the molecular formula of a compound is either the same as the empirical formula or a whole-number multiple of it. In order to arrive at the molecular formula, two separate determinations must be made in the laboratory. (1) The percentage composition must be found; from this the empirical formula is calculated. (2) The molecular weight must be determined, whereupon the true molecular formula may then be calculated.

Problem 7: A 0.1800-g sample of an organic compound, containing only C, H, and O, was burned in apparatus similar to that shown in Fig. 7-3. The increases

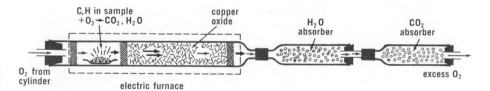

Fig. 7-3. Apparatus for determining percent of C and H in a compound. The copper oxide serves as an additional oxidizing agent for converting C and CO to CO_2.

in weight of the CO_2 and H_2O absorbers were 0.2640 g and 0.1081 g, respectively. The molecular weight of the compound was determined as 179 awu. Calculate the percentage composition, the empirical formula, the molecular formula, and the precise molecular weight.

Solution:
Calculation of percentage composition:

$$0.2640 \text{ g of } CO_2 \times \frac{12.01 \text{ g C}}{44.01 \text{ g } CO_2} = 0.07204 \text{ g C in sample}$$

$$\frac{0.07204 \text{ g}}{0.1800 \text{ g}} \times 100 = 40.02\% \text{ C in sample}$$

$$0.1081 \text{ g of } H_2O \times \frac{2.016 \text{ g } H_2}{18.016 \text{ g } H_2O} = 0.01210 \text{ g H in sample}$$

$$\frac{0.01210 \text{ g}}{0.1800 \text{ g}} \times 100 = 6.72\% \text{ H in sample}$$

Percent oxygen $= 100\% - (40.02\% + 6.72\%) = 53.26\%$.

Calculation of empirical formula:

$$40.02 \text{ g C} \times \frac{1 \text{ mole C}}{12.01 \text{ g C}} = 3.33 \text{ moles C in 100 g of compound}$$

$$6.72 \text{ g H} \times \frac{1 \text{ mole H}}{1.008 \text{ g H}} = 6.67 \text{ moles H in 100 g of compound}$$

$$53.26 \text{ g O} \times \frac{1 \text{ mole O}}{16.00 \text{ g O}} = 3.33 \text{ moles O in 100 g of compound}$$

Ratio of atoms $= 3.33 : 6.67 : 3.33 = 1 : 2 : 1$.

Empirical formula $= CH_2O$

Calculation of molecular formula:
A molecule of CH_2O would have a weight of $12 + 2 + 16 = 30$ awu. However, the experimentally determined molecular weight is 179. So, the actual molecule is 179/30 or six times heavier than the one having the formula CH_2O.

Molecular formula $= C_6H_{12}O_6$

Calculation of precise molecular weight:

Once the molecular formula is known, we abandon the experimental molecular weight in favor of the precise molecular weight obtained by adding atomic weights (four significant figures are used here):

Wt of 6 C atoms $= 6 \times 12.01$	$=$	72.06 awu
Wt of 12 H atoms $= 12 \times 1.008$	$=$	12.10
Wt of 6 O atoms $= 6 \times 16.00$	$=$	96.00
Wt of $C_6H_{12}O_6$ molecule		$=$ 180.2 awu

EXERCISES

1. Which is longer, the distance to the nearest star, 2.5×10^{13} miles, or the distance spanned by 1 mole of hydrogen atoms (radius 1×10^{-8} cm) aligned in a single straight row?
2. Calculate the weight of 1 mole of each of the following: sulfur atoms, S; sulfide

ions, S^{2-}; calcium carbonate, $CaCO_3$; oxygen molecules, O_2. What do these quantities of the different substances have in common?

3. Calculate the number of oxygen molecules in a pound mole of oxygen. Show by calculation that a pound mole of hydrogen contains an equal number of molecules.
4. Calculate the number of (a) molecules in 10 g of SO_3; (b) atoms in 2.0 g of Cu; (c) calcium ions in 20 g of $Ca_3(PO_4)_2$.
5. Calculate the percentage composition of (a) calcium oxide; (b) sodium hydroxide; (c) ethanol, C_2H_5OH; (d) ammonium sulfate.
6. One method of producing acetic acid is by the oxidation of ethanol according to the equation

$$C_2H_5OH(l) + O_2(g) \rightarrow HC_2H_3O_2(l) + H_2O(l)$$

Calculate ΔH per mole for this synthesis, given the following:

$$C_2H_5OH(l) + 3O_2(g) \rightarrow 2CO_2(g) + 3H_2O(l) \qquad \Delta H = -327.6 \text{ kcal}$$
$$HC_2H_3O_2(l) + 2O_2(g) \rightarrow 2CO_2(g) + 2H_2O(l) \qquad \Delta H = -209.4 \text{ kcal}$$

7. A sample of propane, C_3H_8, weighing 1.034 g was placed in a steel bomb along with an excess of oxygen to insure complete combustion. The bomb was immersed in a calorimeter containing 1350 g of water and allowed to stand until the temperature became constant at 22.5°C (see Fig. 7-2). The propane was ignited and a maximum temperature of 30.3°C was attained by the water. If the calorimeter, steel bomb, and stirrer were known to absorb 225 cal per deg C rise in temperature, calculate the enthalpy change per mole of propane, neglecting the small amount of heat retained by the products and the excess oxygen.
8. Which produces the more heat energy, the combustion of 10.0 g of methane, CH_4, or 10.0 g of ethylene, C_2H_4?

$$CH_4(g) + 2O_2(g) \rightarrow CO_2(g) + 2H_2O(l) \qquad \Delta H = -212.8 \text{ kcal}$$
$$C_2H_4(g) + 3O_2(g) \rightarrow 2CO_2(g) + 2H_2O(l) \qquad \Delta H = -331.6 \text{ kcal}$$

9. Calculate the approximate specific heat of nickel, using the law of Dulong and Petit.
10. Calculate the weight of ammonium nitrate needed to produce 10.0 g of nitrous oxide according to the reaction

$$NH_4NO_3 \rightarrow 2H_2O + N_2O$$

11. Water reacts with magnesium nitride as follows:

$$Mg_3N_2 + 6H_2O \rightarrow 3Mg(OH)_2 + 2NH_3$$

What weight of magnesium nitride reacts with 1.0 mole of water? What weight of ammonia is produced?

12. What weight of hydrogen is needed to produce 10 lb of ammonia?
13. A 3.000-g sample of a copper iodide compound on analysis gave 1.001 g of copper and 1.999 g of iodine. Calculate the empirical formula of the compound.
14. (a) An organic compound was found to have the following composition: C, 53.3%; H, 11.1%; O, 35.6%. From these data calculate the empirical formula of the compound. (b) The molecular weight was determined experimentally to be 89 awu. Calculate the molecular formula.

15. A 0.500-g sample of a salt known to contain chloride ions was dissolved in nitric acid. The addition of silver nitrate to this solution gave a precipitate of AgCl that weighed 1.675 g. Calculate the percentage chlorine in the salt.
16. (a) A 1.200-g sample of an organic compound known to contain C, H, and O was burned in an apparatus similar to that shown in Fig. 7-3. The increases in weight of the CO_2 and H_2O absorbers were 2.640 g and 1.440 g, respectively. Calculate the percent composition of the compound. (b) What is the empirical formula of the compound?
17. Starting with 100 lb of a sulfur ore that contains 95 percent sulfur, what weight of sulfuric acid can be obtained if the process of manufacture is 80 percent efficient? The reactions may be represented by

$$S + O_2 \rightarrow SO_2$$
$$2SO_2 + O_2 \rightarrow 2SO_3$$
$$H_2O + SO_3 \rightarrow H_2SO_4$$

18. Why is it necessary to specify conditions of pressure and temperature in describing the energy changes that accompany chemical reactions?
19. Why do we write H_2 instead of H for the gas hydrogen? Why do we write CO_2 instead of C_2O_4 for carbon dioxide while we write C_2H_4 instead of CH_2 for ethylene?
20. A compound of xenon chloride was found to be 55.2 percent Xe and 44.8 percent Cl by weight. Calculate the empirical formula.
21. Calculate the enthalpy change when a solution of nitric acid containing 20 g of HNO_3 in a liter of solution is neutralized by dilute $Ca(OH)_2$.

SUGGESTED READING

Armstrong, G. T., "The Calorimeter and Its Influence on the Development of Chemistry," *J. Chem. Educ.*, **41**, 297 (1964).

Benson, S. W., "Bond Energies," *J. Chem. Educ.*, **42**, 502 (1965).

Feifer, N., "The Relationship Between Avogadro's Principle and the Law of Gay-Lussac, *J. Chem. Educ.*, **43**, 411 (1966).

Fitzgerel, R. K., and F. H. Verhoek, "The Law of Dulong and Petit," *J. Chem. Educ.*, **37**, 545 (1960).

Guggenheim, E. A., "The Mole and Related Quantities," *J. Chem. Educ.*, **38**, 86 (1961).

Nash, L. K., "Elementary Chemical Thermodynamics," *J. Chem. Educ.*, **42**, 64 (1965).

Slabaugh, W. H., "Avogadro's Number by Four Methods," *J. Chem. Educ.*, **46**, 40 (1969).

Szabadváry, F. (trans. R. E. Oesper), "The Birth of Stoichiometry," *J. Chem. Educ.*, **39**, 267 (1962).

8

OXYGEN AND HYDROGEN

In general, our approach to the study of the more than 100 elements will be based on the periodic table. However, before starting this systematic plan, we need to be better acquainted with certain chemical principles. In this chapter we consider two elements more or less independently of the periodic system in order to apply some of the principles we have already learned, and to study factual and illustrative material on which to base additional principles.

HISTORICAL BACKGROUND

About the time of the American Revolution, chemists in Europe were beginning to solve the mysteries concerning gases. Up to this time the chemical nature of such gases as oxygen, hydrogen, nitrogen, and carbon dioxide was unknown; indeed, these names had not yet appeared in any language. All gases were referred to simply as "airs." The few types known were designated by appropriate modifiers such as atmospheric air, dephlogisticated air (oxygen), "inflammable air" (hydrogen), noxious or burnt air (nitrogen), and foul air (gases arising from decaying substances).

The problem of identifying these "airs" was unduly complicated by the phlogiston theory. This theory had been advanced about 100 years earlier to explain combustion, and most of the learned men of the period still believed in it. According to it, combustible substances such as wood and coal contained a material called phlogiston; this escaped when substances burned, and an ash remained. This theory was plausible up to a point, for it explained why common fuels appear to lose weight upon burning. It could also be used to explain why a burning candle is soon extinguished if a jar is placed over it, the assumption being that the air in the jar became saturated with phlogiston so that no more phlogiston could be given off by the burning candle.

In 1774, Joseph Priestley, a British clergyman, heated mercuric oxide sufficiently to bring about its decomposition into mercury and a gas (Fig. 8-1). Because a candle burned brilliantly in the gas and for a longer time than in an equal volume of atmospheric air, Priestley argued that he had prepared new or fresh air, air that contained no phlogiston to impede the escape of phlogiston from the burning candle. He called the gas "dephlogisticated air."

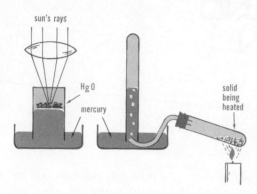

Fig. 8-1. Priestley's preparation of oxygen. He frequently improvised his apparatus from household articles, using tall beer glasses for the collection of gases over mercury or water, or even a gun barrel for heating solids in his fireplace.

During this same period, another Englishman, Henry Cavendish, was experimenting with a combustible gas. He found that he could prepare the gas by the action of metals (zinc, iron, and tin) with dilute acids (HCl and H_2SO_4) and collect it over water. He also found that equal weights of a given metal released the same amount of the gas from either hydrochloric acid or sulfuric acid. This probably caused him to conclude that the gas came from the metal. Cavendish performed numerous experiments involving burning his "inflammable air" (hydrogen) in atmospheric air and in Priestley's dephlogisticated air (oxygen). The mixtures of hydrogen and oxygen (or air) were ignited electrically in a glass globe and in a manner that prevented the products from being lost. In these experiments, Cavendish determined that inflammable air and dephlogisticated air burned in a 2 : 1 ratio by volume, and that water resulted. Thus, like Priestley who had done some work of this kind, Cavendish was on the threshold of discovering the two substances, hydrogen and oxygen, and the true nature of combustion. However, he concluded first that hydrogen (his "inflammable air") was phlogiston, and later that it was a compound of phlogiston and water. Also, like Priestley, he held to the phlogiston theory until his death.

At this time also, a French scientist, Antoine Lavoisier, was in touch with both Cavendish and Priestley. He devised additional experiments on combustion, in which phosphorus, sulfur, and mercury were burned in air. These experiments, extending from about 1772 to 1783, enabled Lavoisier to show that a substance gained weight when it burned, instead of losing weight. For example, he proved that when phosphorus burns, it combines with a substance in air to form a compound (phosphorus oxide) and that the compound weighs more than the phosphorus because it also contains the substance it extracted from the air (oxygen); i.e., the weight of the compound equals the weight of the phosphorus burned plus the weight of the substance removed from the air. Lavoisier was the first to recognize that this substance was an element. He named it *oxygen*. He also demonstrated that Cavendish's inflammable air was a gas that combined with oxygen to form water. He named this gas *hydrogen*, meaning "water former."

Although other investigators had prepared both hydrogen and oxygen prior to the time of Cavendish and Priestley, historians usually credit Cavendish with the discovery of hydrogen and Priestley with the discovery of oxygen.

Today the burning of hydrogen in air or in oxygen is represented by the following equation:

$$2H_2 + O_2 \rightarrow 2H_2O$$
$$\text{hydrogen} \quad \text{oxygen} \quad \text{water}$$

OCCURRENCE

Oxygen and hydrogen rank first and ninth, respectively, in abundance by weight in the earth's crust (see Fig. 8-2). Both occur abundantly in compounds (Table

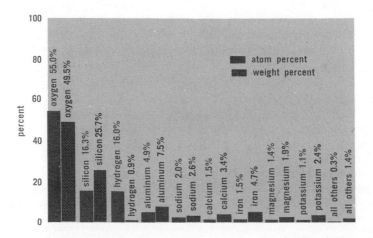

Fig. 8-2. Percentage distribution of the elements in the earth's crust, by atoms, and by weight.

8-1), but only oxygen occurs appreciably in the elemental form. It is present in air to the extent of about 21 percent by volume. However, if we consider the universe as a whole, hydrogen is by far the most abundant of all elements. For example, it comprises about 75 percent of the sun.

TABLE 8-1.

Natural Compounds of Oxygen and Hydrogen

COMPOUNDS CONTAINING BOTH	COMPOUNDS CONTAINING ONLY ONE
Water, H_2O	Oxygen
Sugars, $C_6H_{12}O_6$, $C_{12}H_{22}O_{11}$	Limestone or marble, $CaCO_3$
Starch and cellulose, $(C_6H_{10}O_5)_x$	Sand or quartz, SiO_2
Fats	Many silicate minerals
Proteins	Hydrogen
Clays and certain other minerals	Natural gas, CH_4, C_2H_6 ...
	Gasoline, C_6H_{10}, C_7H_{16}, ...
	Other petroleum products
	Natural rubber

PHYSICAL PROPERTIES

Both oxygen and hydrogen are colorless, odorless, tasteless gases that are only slightly soluble in water. Some physical properties are listed in Table 8-2. It is interesting to observe that hydrogen has the lowest density of any known substance.

TABLE 8-2.

Physical Properties of Oxygen and Hydrogen

	OXYGEN	HYDROGEN
Density	1.429 g/liter	0.08987 g/liter
Freezing point	−218°C	−259°C
Boiling point	−183°C	−253°C
Critical temperature[a]	−119°C	−240°C
Critical pressure[a]	49.7 atm	12.8 atm
Solubility in water	5 ml/100 g at 0°C	2 ml/100 g at 0°C

[a] The **critical temperature** of a gas is the highest temperature at which the gas can be liquefied. The pressure required to bring about liquefaction at this temperature is called the **critical pressure** (see Chap. 10).

LABORATORY PREPARATION

Both oxygen and hydrogen are easily prepared in the laboratory in small quantities with common laboratory apparatus and chemicals.

Oxygen. Oxygen is readily prepared by decomposing potassium chlorate by heating (Fig. 8-3). Other unstable compounds of oxygen may also be used. The equations below show some typical reactions.

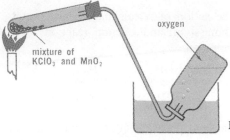

Fig. 8-3. Laboratory preparation of oxygen.

$$2KClO_3 \xrightarrow{heat} 2KCl + 3O_2$$
potassium chlorate → potassium chloride + oxygen

$$2HgO \xrightarrow{heat} 2Hg + O_2$$
mercury(II) oxide → mercury + oxygen

$$2KNO_3 \xrightarrow{heat} 2KNO_2 + O_2$$
potassium nitrate → potassium nitrite + oxygen

Table 8-3 shows that the oxides of inactive metals, such as mercury oxide, are easily decomposed.

Pure potassium chlorate decomposes slowly when it is heated moderately. If a small amount of either manganese dioxide, MnO_2, or iron(III) oxide, Fe_2O_3, is first added to the potassium chlorate, the decomposition proceeds rapidly with moderate heating. However, under these conditions neither the manganese dioxide nor the iron(III) oxide is changed chemically. Substances that influence the rate of a reaction without themselves being changed chemically are called **catalysts**. A substance that is a catalyst for a certain reaction may or may not catalyze some other reaction. No substance serves as a catalyst for reactions in general (see Chap. 15).

Hydrogen. Hydrogen is most readily prepared in the laboratory by the action of moderately active metals (see Table 8-3) with dilute acids (Fig. 8-4):

$$Zn + H_2SO_4 \rightarrow ZnSO_4 + H_2$$
zinc + sulfuric acid → zinc sulfate + hydrogen

$$2Al + 6HCl \rightarrow 2AlCl_3 + 3H_2$$
aluminum + hydrochloric acid → aluminum chloride + hydrogen

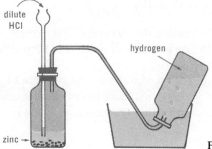

Fig. 8-4. Laboratory preparation of hydrogen.

Hydrogen may also be prepared by the action of certain metals with water. Very active metals such as sodium, potassium, and calcium react at room temperature to form hydroxides and hydrogen.

$$2\underset{\text{sodium}}{Na} + 2\underset{\text{water}}{HOH} \rightarrow 2\underset{\text{sodium hydroxide}}{NaOH} + \underset{\text{hydrogen}}{H_2}$$

$$\underset{\text{calcium}}{Ca} + 2\underset{\text{water}}{HOH} \rightarrow \underset{\text{calcium hydroxide}}{Ca(OH)_2} + \underset{\text{hydrogen}}{H_2}$$

Moderately active metals such as magnesium react at high temperatures with steam to form the oxide and hydrogen.

$$Mg + H_2O \xrightarrow{\text{heat}} MgO + H_2$$

Inactive metals, e.g., gold and silver, do not react with water or with acids to yield hydrogen.

COMMERCIAL PREPARATION

Commercial methods of preparation are based upon inexpensive, abundant raw materials. However, unlike laboratory methods of preparation, commercial methods may entail a high initial cost for apparatus and machinery required for the process; this is justified on the grounds that such equipment will probably be used continuously for many years.

Oxygen. Most commercial oxygen is prepared by separating it from air. The oxygen in the air (21 percent by volume) is in the elemental form, and hence no chemical reaction is necessary. The separation involves (1) liquefaction of the air by applying pressure and lowering the temperature, and (2) distillation of the liquid air. (Nitrogen, which constitutes 78 percent of air, and the inert elements, amounting to 1 percent, distill first, leaving liquid oxygen.) In practice, both these processes are combined into a single operation, which means that the heat liberated by the liquefaction can be used in the distillation.

Hydrogen. Large amounts of commercial hydrogen are prepared by the steam-hydrocarbon method, in which mixtures of hydrocarbons and steam are passed over a nickel catalyst at high temperatures. (Hydrocarbons are compounds that contain only carbon and hydrogen. They are the chief components of natural gas and petroleum.) The equation for the reaction involving the simplest hydrocarbon, methane, is

$$CH_4 + 2H_2O \xrightarrow[\text{hot}]{\text{Ni}} CO_2 + 4H_2$$

The carbon dioxide is readily removed by passing the mixture of the two gases through water under pressure. Most of the carbon dioxide dissolves; very little of the hydrogen does.

From Water by Electrolysis. The decomposition of water is brought about by passing a direct current through water to which a small amount of sulfuric acid has been added (Fig. 8-5). The over-all decomposition reaction is as follows:

$$2H_2O \xrightarrow[\text{current}]{\text{direct}} 2H_2 + O_2$$

Oxygen and Hydrogen

Hydrogen is evolved at the cathode:

$$2H^+ + 2e^- \rightarrow H_2$$

Oxygen is evolved at the anode:

$$2H_2O \rightarrow O_2 + 4H^+ + 4e^-$$

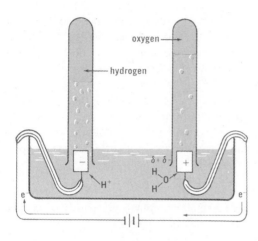

Fig. 8-5. The electrolysis of water.

In this endothermic reaction electrical energy must be continuously supplied or the reaction will stop. The reverse process (the uniting of oxygen and hydrogen to form water) is an exothermic reaction that produces 136.64 kcal of heat when 36 g of water is formed:

$$2H_2 + O_2 \rightarrow 2H_2O \qquad \Delta H = -136.64 \text{ kcal}$$

Conversely, when 36 g of water is decomposed by an electric current into hydrogen and oxygen, electrical energy equivalent to 136.64 kcal is used. Because of the high energy requirement, the preparation of oxygen and hydrogen by the electrolysis of water is too expensive for most commercial uses. However, the process is important in obtaining high-purity hydrogen and oxygen.

CHEMICAL PROPERTIES

Bond Energies. At normal temperatures both hydrogen and oxygen are rather inactive. They may remain in contact with each other or with many other substances at room temperature without a perceptible reaction occurring. Apparently the oxygen or hydrogen atoms are joined in O_2 or H_2 molecules by strong covalent bonds. We may take as a measure of the strength of the bonds the enthalpy change that occurs when a mole of molecules is broken up into atoms (note that ΔH is positive, i.e., energy must be added to bring about the dissociation):

$$O_2 \rightarrow 2O \qquad \Delta H = 118.3 \text{ kcal/mole}$$
$$H_2 \rightarrow 2H \qquad \Delta H = 104.2 \text{ kcal/mole}$$

The energy that must be added to break up a diatomic molecule is called the *bond energy* of the molecule. The bond energies of H_2 and O_2 are higher than for most diatomic molecules and are sufficiently high so that when H_2 and O_2 molecules collide with each other at room temperature, the molecules rebound from the collision without change. No reaction occurs. As the temperature goes up, the molecules move about more rapidly. Consequently, they collide with a greater force, and existing bonds are more apt to be broken so that new bonds may be formed. In later chapters, we will take up in more detail the influence of temperature changes on the motions of molecules and the speeds of chemical reactions. However, let us note here that at elevated temperatures both hydrogen and oxygen combine with a great number of other elements and compounds. In a relatively few cases, the reaction of hydrogen or of oxygen with another substance may occur even at room temperature at an appreciable speed.

Reactions of Oxygen. Under the proper conditions, oxygen combines directly with all the elements except the noble gases, the halogens, and two or three of the very inactive metals. For example, calcium burns brilliantly in pure oxygen, forming calcium oxide (lime). The equation is

$$2Ca + O_2 \rightarrow 2CaO$$

Magnesium in the form of a ribbon or a small wire burns vigorously in oxygen if heated to a high temperature.

$$Mg + O_2 \rightarrow 2MgO$$

Phosphorus burns with a dazzling flame and forms the dense white smoke of phosphorus pentoxide.

$$4P + 5O_2 \rightarrow P_4O_{10}$$

Carbon in the form of coke or charcoal burns to carbon dioxide.

$$C + O_2 \rightarrow CO_2$$

Sulfur burns to sulfur dioxide.

$$S + O_2 \rightarrow SO_2$$

Many compounds unite rapidly with oxygen with the liberation of heat and light. For almost all such reactions to start, the temperature must be raised. The burning of wood, gasoline, or natural gas are examples. Note from the following equations that these common fuels contain carbon and hydrogen and that the products of complete combustion are carbon dioxide and water.

$$C_6H_{10}O_5 + 6O_2 \rightarrow 6CO_2 + 5H_2O$$
cellulose, a component of wood

$$CH_4 + 2O_2 \rightarrow CO_2 + 2H_2O$$
methane, a component of natural gas

$$C_7H_{16} + 11O_2 \rightarrow 7CO_2 + 8H_2O$$
heptane, a component of gasoline

Oxygen and Hydrogen

A reaction in which a substance combines with oxygen is called **oxidation**. If oxidation proceeds so rapidly that light is emitted, it is also called **combustion**. For combustion to occur, a minimum temperature, called the **kindling temperature**, is required. For example, aluminum reacts slowly with oxygen at room temperature (corrodes), forming a thin coat of aluminum oxide on the surface. At an elevated temperature, aluminum burns brilliantly in oxygen and forms aluminum oxide. In both cases, the reaction is an **oxidation reaction**; only in the latter case is it called combustion.

$$4Al + 3O_2 \rightarrow 2Al_2O_3$$

Some substances—e.g., new cut hay, drying oils in paints—undergo slow oxidation at room temperature with the liberation of heat. If the heat is not dissipated to the surroundings, the temperature may gradually rise to the kindling temperature, and combustion will start. This process is known as **spontaneous combustion**.

Reactions of Hydrogen. At elevated temperatures, hydrogen combines directly with most nonmetallic elements. Examples are:

$$H_2 + Br_2 \rightarrow 2HBr \text{ (hydrogen bromide)}$$
$$H_2 + Cl_2 \rightarrow 2HCl \text{ (hydrogen chloride)}$$
$$H_2 + S \rightarrow H_2S \text{ (hydrogen sulfide)}$$
$$3H_2 + N_2 \rightarrow 2NH_3 \text{ (ammonia)}$$

Hydrogen does not have the same tendency to combine with metallic elements as with nonmetallic ones. However, under proper conditions very reactive metals do form compounds with hydrogen. Sodium hydride, NaH, lithium hydride, LiH, and calcium hydride, CaH_2, are examples. In these compounds of

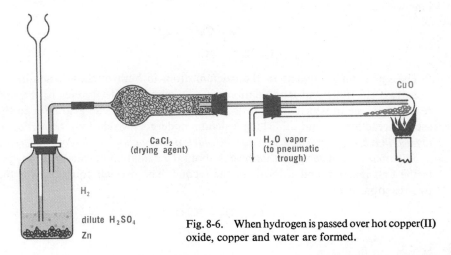

Fig. 8-6. When hydrogen is passed over hot copper(II) oxide, copper and water are formed.

metals of very low electronegativities, hydrogen has its unusual oxidation state of -1 and exists as the H^- ion.

Hydrogen is an excellent agent for removing oxygen from many metallic oxides. This process is called **reduction;** it is the opposite of oxidation.

$$2Cu + O_2 \rightarrow 2CuO \qquad \text{(copper is oxidized)}$$

$$CuO + H_2 \rightarrow Cu + H_2O \qquad \text{(copper oxide is reduced)}$$

The reduction of copper oxide with hydrogen as the reducing agent is easily carried out in the laboratory (Fig. 8-6).

A GENERALIZED CONCEPT OF OXIDATION AND REDUCTION

The development of a better understanding of the nature of chemical reactions has made it desirable to extend the definition of oxidation and reduction to include certain reactions in which oxygen plays no part. At first glance it may seem illogical to speak of a chemical reaction as an oxidation reaction when oxygen is not even present, but careful consideration of the following examples will show the desirability of thus extending the definition of oxidation.

First, let us consider what happens to calcium atoms when they combine with atoms of oxygen. A calcium atom has two electrons in its outermost shell. During the oxidation of calcium, these electrons are transferred from the calcium to the oxygen atoms to form calcium oxide, CaO. Because of the loss of electrons, calcium atoms are changed to calcium ions with an oxidation state of $+2$.

$$Ca \rightarrow Ca^{2+} + 2e^-$$

$$O + 2e^- \rightarrow O^{2-}$$

Now consider the reaction of calcium with chlorine. Here each atom of calcium loses two electrons from its outer shell as it combines with chlorine atoms to form calcium chloride. Again calcium atoms are changed to calcium ions with an oxidation state of $+2$.

$$Ca \rightarrow Ca^{2+} + 2e^-$$

$$Cl_2 + 2e^- \rightarrow 2Cl^-$$

The same thing happens to the calcium atom in both of these reactions: it loses two electrons and its oxidation state is increased. If we refer to the first reaction as one in which calcium undergoes oxidation, we can also refer to the second reaction as one in which calcium undergoes oxidation. *Oxidation is defined in a broad sense as a reaction in which atoms or ions undergo an increase in oxidation state.* The agent that brings about the oxidation of calcium is oxygen in the first reaction and chlorine in the second. The over-all equations for the two reactions are:

$$2Ca + O_2 \rightarrow 2CaO$$

$$Ca + Cl_2 \rightarrow CaCl_2$$

Similarly, *reduction is defined in a broad sense as a reaction in which atoms or ions undergo a decrease in oxidation state.* Actually, oxidation and reduction occur simultaneously, because the increase in oxidation state by one atom or ion is always accompanied by a decrease in oxidation state by another atom or ion. In the above reactions in which calcium is oxidized by chlorine and by oxygen, each of these substances is reduced.

As another example, consider the reaction of sulfur and oxygen to form the covalent compound sulfur dioxide:

$$S + O_2 \rightarrow SO_2$$

In this case no electrons are transferred. However, the shared electron-pairs are held more closely by the element with the higher electronegativity, oxygen. Hence, oxygen in SO_2 has a negative oxidation number, -2, and sulfur a positive one, $+4$. Since sulfur undergoes an increase in its oxidation state from 0 to $+4$, it is oxidized, and, since oxygen undergoes a decrease in its oxidation state, 0 to -2, it is reduced. Oxidation-reduction reactions are among the most common chemical reactions; many additional examples will be presented in this and later chapters.

THE ACTIVITY SERIES

In discussing the preparation and properties of oxygen and hydrogen, several differences in the chemical activity of the metals were mentioned. Some metals burn brilliantly in oxygen, others do not; certain metals displace hydrogen from water at room temperature; and a number of metals displace hydrogen from acids. These differences in activity can be correlated by arranging the metals in the order of decreasing tendency to undergo oxidation, i.e., the decreasing tendency of their atoms to give up electrons to other substances. Table 8-3, a portion of the **activity series** or **electrochemical series,** shows such an arrangement of several common metals and one nonmetal, hydrogen. (In a later chapter several additional nonmetals and metals will be added to the list.) In this series, the element at the top has the greatest tendency to undergo oxidation and the element at the bottom has the least. Conversely, binary electrovalent compounds of the metal at the top (RbCl, Rb_2O) are the most difficult to decompose (its metallic ions have the least tendency to gain electrons), and those of the metal at the bottom ($AuCl_3$, Au_2O_3) are the easiest.

Let us now consider the possibility of preparing oxygen by heating the two metallic oxides, calcium oxide and silver oxide.

$$2CaO \xrightarrow{heat} 2Ca + O_2 \quad \text{(will this reaction occur?)}$$
$$2Ca^{2+} + 4e^- \longrightarrow 2Ca \quad \text{(reduction)}$$
$$2O^{2-} \longrightarrow O_2 + 4e^- \quad \text{(oxidation)}$$
$$2Ag_2O \xrightarrow{heat} 4Ag + O_2 \quad \text{(will this reaction occur?)}$$
$$4Ag^+ + 4e^- \longrightarrow 4Ag \quad \text{(reduction)}$$
$$2O^{2-} \longrightarrow O_2 + 4e^- \quad \text{(oxidation)}$$

TABLE 8-3.
Activity Series of Metals

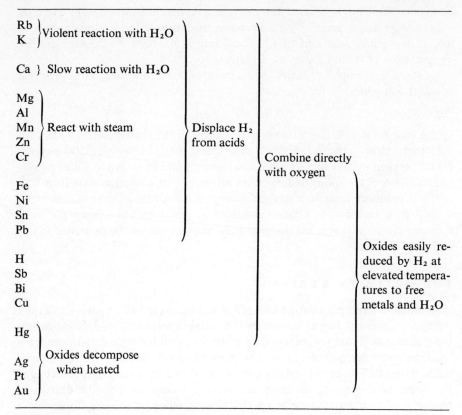

Calcium is toward the top of the activity series. This means that calcium ions are difficult to reduce. We would expect that calcium would react readily with oxygen to form calcium oxide (lime) and that this compound would be difficult to decompose by heating. Actually, calcium oxide cannot be decomposed by temperatures obtainable by ordinary methods. On the other hand, we would predict from its position in the activity series that silver would probably not combine readily with oxygen, and that silver oxide would decompose upon heating.

Next we consider the possible reactions of zinc and copper with dilute hydrochloric acid:

$$Zn + 2HCl \rightarrow ZnCl_2 + H_2 \quad \text{(will this reaction occur?)}$$
$$Zn \rightarrow Zn^{2+} + 2e^- \quad \text{(oxidation)}$$
$$2H^+ + 2e^- \rightarrow H_2 \quad \text{(reduction)}$$
$$Cu + 2HCl \rightarrow CuCl_2 + H_2 \quad \text{(will this reaction occur?)}$$
$$Cu \rightarrow Cu^{2+} + 2e^- \quad \text{(oxidation)}$$
$$2H^+ + 2e^- \rightarrow H_2 \quad \text{(reduction)}$$

Since zinc is above hydrogen in the activity series, we would expect zinc atoms to lose electrons readily to hydrogen ions; i.e., zinc atoms have a greater tendency to give up electrons than hydrogen atoms do, and hydrogen ions have a greater attraction for electrons than zinc ions do. On the other hand, copper is below hydrogen in activity; copper atoms have a greater tendency to hold on to their electrons than hydrogen atoms do, so we would not expect copper to react with hydrochloric acid.

These considerations lead to the following general conclusion about the reaction of metals with dilute acids: any metal above hydrogen in the activity series will displace hydrogen from dilute acids; metals below hydrogen will not. (Nitric acid dissolves inactive metals like copper and silver, but hydrogen is not liberated.)

Only very active metals react with water at room temperature. The reduction of HOH molecules is more difficult than the reduction of the H^+ ions of acid solutions; therefore a very active reducing agent is required.

$$Ca + 2HOH \rightarrow Ca(OH)_2 + H_2$$

$$Ca \rightarrow Ca^{2+} + 2e^- \quad \text{(oxidation)}$$

$$2HOH + 2e^- \rightarrow H_2 + 2OH^{-1} \quad \text{(reduction)}$$

NATURAL ISOTOPES OF OXYGEN AND HYDROGEN

Three isotopes of oxygen have been identified in nature. These are diagrammed in Fig. 8-7. About 99.8 percent of all the oxygen atoms existing in nature are of

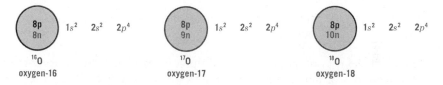

Fig. 8-7. The natural isotopes of oxygen.

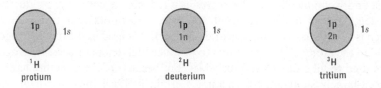

Fig. 8-8. The natural isotopes of hydrogen.

the mass 16 variety, about 0.16 percent of mass 18, and 0.04 percent of mass 17.

There are three known isotopes of hydrogen. These are shown diagrammatically in Fig. 8-8. Hydrogen as it occurs in nature is composed of about 7000 atoms of protium to 1 of deuterium. There is only an infinitesimal trace of tritium.

The properties shown by an element as it exists in nature are the average properties of the several isotopes of the element, rather than the properties of an individual isotope. However, the isotopes of an element differ very little in their chemical properties, and they differ only slightly in certain physical properties such as melting point, boiling point, and density. These differences are more pronounced in the isotopes of hydrogen than in any other element.

USES OF OXYGEN AND HYDROGEN

Most of the uses of commercial oxygen are based upon the property that the element acts as an oxidizing agent. A heavy consumer is the steel industry, oxygen being used in some of the more recently built blast furnaces to burn coke (air was long used for this purpose). The chemical industry uses oxygen to make such products as nitric acid, methanol (wood alcohol), and ethylene oxide.[1] Large quantities of liquid oxygen are consumed in the propulsion of certain types of rockets (see Fig. 8-9). A few of the commonplace uses of oxygen include skin-diving, special tents for respiratory troubles, high-altitude flying, and oxyacetylene torches for the high-temperature cutting of metals.

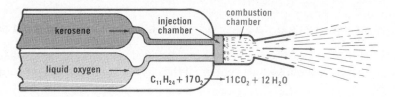

Fig. 8-9. The expanding gas from the combustion of the fuel gives the thrust for rocket propulsion.

Some of the uses of hydrogen depend upon its combustibility. For example, large quantities of liquid hydrogen are now being used as the fuel in certain types of rocket engines (Fig. 8-9); liquid oxygen is the oxidizer.

Hydrogen is used in the synthesis of a number of useful compounds such as ammonia and wood alcohol. It also serves as a reducing agent in the production of tungsten, and as a cooling agent for large generators and transformers (hydrogen is a better conductor of heat than is air); it is used in the oxy-hydrogen torch and in the atomic hydrogen torch to produce high temperatures. Hydrogen was once used to fill balloons and dirigibles, but because of the great fire hazard it has been largely supplanted by helium. Certain fuel cells now use hydrogen to produce electrical energy (see Chap. 20).

Large amounts of hydrogen are used to convert liquid fats into solid fats. Fats obtained from plants are generally liquids at room temperature. Cottonseed oil, coconut oil, soybean oil, and peanut oil are examples. At high temperatures in the presence of a nickel catalyst, these liquid fats combine with hydrogen

---

[1] Ethylene oxide is used to synthesize ethylene glycol, a common antifreeze.

and in so doing produce fats that are solids. Such fats are used for foods and for making soap. The solid fats such as Crisco are produced by the hydrogenation of liquid fats.

OZONE

The element oxygen can exist as triatomic molecules, O_3. This form of oxygen is called **ozone**. There is evidence that considerable ozone is formed in an upper layer of the earth's atmosphere as the result of the action of ultraviolet light on atmospheric oxygen. Since most of the ultraviolet light from the sun is absorbed by the air at higher altitudes, there is usually very little ozone in the air at low altitudes.

Preparation. Ozone is commonly prepared by allowing a cool electric discharge to pass through oxygen. A simple *ozonizing* apparatus is shown in Fig. 8-10. It consists of a glass tube with two large metal electrodes: one, a copper foil

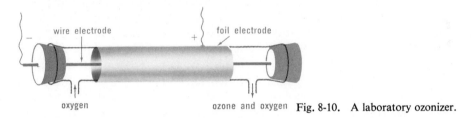

Fig. 8-10. A laboratory ozonizer.

wrapped around the tube; the other, a metal rod extending through the tube. The electrodes are connected to a source of high-voltage current. When the current is on, the two electrodes become highly charged so that there is a "silent" electrical discharge from one to the other. Some of the oxygen passing through the tube during the discharge is converted to ozone. This discharge, observable in a dark room as a blue glow, gives a greater yield of ozone than does a spark discharge, which produces a high temperature in the vicinity of the spark (or arc). When pure dry oxygen at 20°C is passed through the apparatus, about 19 percent is converted into ozone; at the temperature of liquid air about 90 percent is converted. The equation for the reaction is:

$$3O_2 \xrightarrow{\text{silent electrical discharge}} 2O_3 \qquad \Delta H = 68 \text{ kcal}$$
$$\text{oxygen} \qquad\qquad \text{ozone}$$

Properties. Ozone is a pale-blue gas with a characteristic odor. Its name comes from the Greek word meaning "to smell." The odor of ozone may be detected in the vicinity of electrical apparatus where considerable arcing occurs. In general, its chemical characteristics are like those of ordinary oxygen, except that ozone is more reactive. The oxidation of mercury and silver illustrates this point. At room temperature, these two metals are oxidized by ordinary oxygen at an imperceptible rate, if at all. When placed in ozone, they rapidly become coated with an oxide film (become tarnished), because of the oxidizing action of ozone.

Combustion reactions in ozone release a greater amount of heat than they do in oxygen. For example, when 12 g of carbon is burned in oxygen, 94,000 cal of heat energy is evolved; when this same weight of carbon is burned in ozone, 116,700 cal of heat energy is evolved. In either case the same weight of oxygen or ozone is required.

$$3C + 3O_2 \rightarrow 3CO_2 \quad \Delta H = 3(-94) \text{ kcal}$$
$$3C + 2O_3 \rightarrow 3CO_2 \quad \Delta H = 3(-116.7) \text{ kcal}$$

Recent research on rats and other experimental animals indicates that ozone is a highly poisonous gas.

Uses. All the commercial uses of ozone are based on its activity as an oxidizing agent. It is used to bleach oils, flour, and delicate fabrics, to sterilize water, and to destroy odors; it is also used as an oxidizing agent in the preparation of a number of useful compounds.

Many natural materials are slightly colored because of the presence of small amounts of complex organic compounds. For example, flour, cellulose materials, and vegetable oils are usually slightly colored when first obtained from their plant sources. To obtain a colorless product, the colored compounds are oxidized to colorless compounds. The process is referred to as **bleaching,** and the oxidizing agent employed is called a **bleaching agent.** Ordinary oxygen is not a suitable bleaching agent because a rather high temperature is required to start the oxidation. Furthermore, once started, both the pigment and the product oxidize, and combustion may even occur. On the other hand, ozone reacts with many plant pigments at room temperature and converts them into colorless substances. The less active starch, cellulose, and fat molecules are not affected. Impurities that give rise to objectionable odors and tastes are also removed from certain foods by mild oxidation in which ozone is the oxidizing agent.

Bond Energies in Ozone Molecules. The dissociation energy of ozone may be calculated by means of Hess's law from the thermochemical data we have already considered:

$$O_2 \rightarrow 2O \quad \Delta H = 118.3 \text{ kcal} \quad (1)$$

$$3O_2 \rightarrow 2O_3 \quad \Delta H = 68 \text{ kcal} \quad (2)$$

$$O_3 \rightarrow 3O \quad \Delta H = ?$$

By rewriting Eqs. (1) and (2) as shown below and then adding, we have the dissociation energy per mole of ozone:

$$1\tfrac{1}{2}O_2 \rightarrow 3O \quad \Delta H = 1\tfrac{1}{2} \times 118.3 \text{ kcal}$$
$$O_3 \rightarrow 1\tfrac{1}{2}O_2 \quad \Delta H = -(\tfrac{1}{2} \times 68) \text{ kcal}$$
$$\overline{O_3 \rightarrow 3O \quad \Delta H = 177 + (-34) = 143 \text{ kcal/mole}}$$

Since in an ozone molecule there are two sets of bonds joining the three atoms, the average energy of the bond system that joins two atoms is $\tfrac{1}{2}(143 \text{ kcal})/\text{mole}$ or 71.5 kcal/mole of bonds. We see from these calculations that the bond energies are not as great in O_3 molecules as in O_2 molecules—71.5 kcal versus

Oxygen and Hydrogen

118 kcal per mole of bonds.[2] Hence, we would expect O_3 molecules to enter into oxidation reactions at lower temperatures at which O_2 molecules might be relatively unreactive.

MAGNETIC PROPERTIES AND THE STRUCTURE OF MOLECULES

Substances that are slightly repelled by a magnetic field are said to be **diamagnetic**; those that are slightly attracted are termed **paramagnetic**. (See Fig. 8-11 for a method of measuring.) It is rather well established that in a diamagnetic substance all electrons are paired, but that in molecules of a paramagnetic substance there are unpaired electrons.

Based on the usual Lewis formulas for H_2, O_2, and N_2 molecules, H:H, :Ö::Ö:, and :N:::N:, we would expect these substances to be diamagnetic. However, such is not the case for oxygen; it is paramagnetic. The solving of the O_2 puzzle was one of the first cases in which the MO theory proved superior to other bonding theories. In filling molecular orbitals in the order of increasing energy according to the MO scheme in Fig. 5-9, 1 electron is placed in each of the two π orbitals of equal energy before either of the orbitals is filled with a pair of electrons (Hund's rule, see Table 5-2). The application of these procedures to O_2 results in unpaired electrons in the π^*2p_y and π^*2p_z orbitals and accounts for the paramagnetism of O_2. For diamagnetic substances such as H_2 or N_2, the MO theory shows all electrons paired (see Table 5-2).

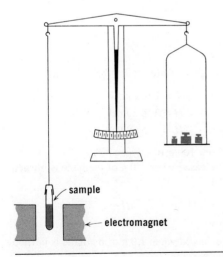

Fig. 8-11. The Gouy magnetic balance. When the electromagnet is turned on, a paramagnetic substance is attracted by the field and appears to weigh more; a diamagnetic substance appears to weigh less.

---

[2] Note that our calculations give an average energy for the two bonds O—O—O. The energy for splitting off one oxygen atom, $O_3 \to O_2 + O$, is given by:

$O_3 \to 3O$	$\Delta H =$	143 kcal
$2O \to O_2$	$\Delta H =$	−118.3 kcal
$O_3 \to O_2 + O$	$\Delta H =$	24.7 kcal

EXERCISES

1. When 1.0 g of wood is burned in air, the ash is found to weigh 0.1 g. How might this result be explained in terms of the phlogiston theory? How is it explained today?
2. When 1.0 g of aluminum is burned, the ash has a weight of 1.9 g. How could this observation be explained in terms of the phlogiston theory? By present day theories?
3. Show by writing balanced equations:
   (a) three reactions for producing hydrogen;
   (b) three reactions for producing oxygen.
4. How could you prove that ordinary commercial oxygen contains a small amount of a chemically inert gas?
5. What weight of water must be electrolyzed to produce 50 g of hydrogen? What weight of oxygen is produced during this electrolysis?
6. Suggest a method for determining if a bottle containing a colorless, odorless gas is filled with hydrogen.
7. Write balanced equations for each of the following:
   (a) $C_4H_{10} + O_2 \rightarrow$
   (b) $Ca + H_2 \rightarrow$
   (c) $NaH + H_2O \rightarrow$
   (d) $H_2 + N_2 \rightarrow$
   (e) $NiO + H_2 \rightarrow$
8. Magnesium reacts with fluorine to form magnesium fluoride. Using this reaction, illustrate with equations why oxidation must always be accompanied by reduction.
9. On the basis of the activity series, predict whether or not a reaction will occur in the following cases. Write an equation for those which will react:
   (a) $K + H_2O \rightarrow$
   (b) $MgO + H_2 \rightarrow$
   (c) $Zn + CuSO_4 \rightarrow$
   (d) $Ni + ZnCl_2 \rightarrow$
   (e) $Mg + AuCl_3 \rightarrow$
   (f) $Bi + H_2SO_4 \rightarrow$
10. Calculate the hydrogen-oxygen bond energy in water from the following data:

    $$H_2 \rightarrow 2H \quad \Delta H = 104.2 \text{ kcal}$$
    $$O_2 \rightarrow 2O \quad \Delta H = 118.3 \text{ kcal}$$
    $$H_2(g) + \tfrac{1}{2}O_2(g) \rightarrow H_2O(g) \quad \Delta H = -57.8 \text{ kcal}$$

11. What weight of ozone can be prepared from 100 g of oxygen?
12. (a) Calculate the amount of heat energy released when 1.0 g of methane is burned in oxygen according to the equation

    $$CH_4(g) + 2O_2(g) \rightarrow CO_2(g) + 2H_2O(l) \quad \Delta H = -213 \text{ kcal}$$

    (b) What weight of oxygen is used in (a)?
    (c) Calculate the amount of heat energy released when 1.0 g of methane is burned in ozone (recall $3O_2 \rightarrow 2O_3 \quad \Delta H = 68$ kcal).
    (d) What weight of ozone is used in (c)?
13. Distinguish between oxidation and combustion. Are all oxidation reactions exothermic? Explain.

14. Formulate an activity series for hydrogen and the hypothetical metals A, B, C, D, E, and F from the data recorded in the reactions below:

$$A + ECl_2 \rightarrow ACl_2 + E$$
$$A + BCl_2 \rightarrow \text{no reaction}$$
$$F + CCl_2 \rightarrow FCl_2 + C$$
$$H_2 + DCl_2 \rightarrow 2HCl + D$$
$$H_2 + ECl_2 \rightarrow \text{no reaction}$$
$$F + DCl_2 \rightarrow \text{no reaction}$$

15. If all manufacturing facilities for the production of elemental hydrogen were permanently lost here and abroad, how might our way of life be affected, now and in the future?
16. Can the relative effectiveness of oxygen and ozone as bleaching agents be related to their bond energies? Explain.
17. Write balanced equations for each of the following. Identify the oxidizing agent and the reducing agent in each case:
    (a) $Al + F_2 \rightarrow$
    (b) $S + O_2 \rightarrow$
    (c) $PbO + H_2 \rightarrow$
    (d) $PtO \xrightarrow{\text{heat}}$
18. Brass is an alloy of zinc and copper. Treatment of 1.500 g of brass with sulfuric acid in apparatus for collecting hydrogen as it forms gave a total of 175 ml of hydrogen (volume corrected to conditions where the density of hydrogen is 0.0899 g per liter). Calculate the percentage composition of the brass.
19. Based on MO theory, should F_2 molecules be paramagnetic or diamagnetic? Explain.
20. Experimentally, it is found that nitrogen is slightly repelled by a magnetic field. What does this suggest about the structure of N_2 molecules?

SUGGESTED READING

Bostrup, O., K. Demandt, and K. O. Hansen, "The Thermal Decomposition of $KClO_3$," *J. Chem. Educ.*, **39**, 573 (1962).

Farber, E., "Oxygen—the Element with Two Faces," *Chemistry*, **39** (5), 17 (1966).

Gale, G., "Phlogiston Revisited—Explanatory Models and Conceptual Change," *Chemistry*, **41** (4), 16 (1968).

Partington, J. R., "The Discovery of Oxygen," *J. Chem. Educ.*, **39**, 123 (1962).

Sanderson, R. T., "Principles of Hydrogen Chemistry," *J. Chem. Educ.*, **41**, 331 (1964).

Soloveichik, S., "The Last Fight for Phlogiston and the Death of Priestley," *J. Chem. Educ.*, **39**, 644 (1962).

Szabadváry, F. (trans. E. R. Oesper), "Jakob Winterl and His Analytical Method for Determining Phlogiston," *J. Chem. Educ.*, **39**, 266 (1962).

9

KINETIC THEORY OF GASES; GAS LAWS

States of Matter. The matter of our planet exists in three *states*: solid, liquid, and gaseous. These three states are so readily recognized that our only need here is to organize the facts that we already know.

A gas has no shape of its own; rather it takes the shape of its container. It has no fixed volume, but is compressed or expanded as its container changes in size. The volume of the container is the volume of the gas.

A liquid has no specific shape; it takes the shape of its container as it seeks its own level under the influence of gravity. But it does have a fixed volume. Although it is not absolutely incompressible, it is compressed only a negligible amount even by moderately high pressures.

A solid has a fixed shape and a fixed volume. Like a liquid, it is not compressed appreciably by moderately high pressures.

BEHAVIOR OF GASES

Each elemental substance may exist as a gas, liquid, or solid, depending on the temperature. Many compounds, however, can exist only in the solid state, or at the most, the solid and liquid states, because they are unstable at elevated temperatures. Ordinary sugar, for example, decomposes instead of melting when heated; potassium chlorate melts to a clear liquid that decomposes rather than vaporizing when heated further. Water is a familiar example of a compound that can exist in all three states.

When substances such as oxygen, iron, water, and carbon dioxide are in the gaseous state, they possess certain properties that are not significant when these same substances are in the liquid or solid state.

COMPRESSIBILITY AND EXPANSIBILITY

All gases are easily compressed. Most of us have had experience in pumping air into bicycle or automobile tires. Air is a mixture of gases that behaves physically in the same manner as pure oxygen, or pure nitrogen, or any other gaseous substance. We can put into a tire a volume of air that is two or three times the volume of the tire. If the tire is punctured, the air, over and above the amount equal to the volume of the tire, will rush out. Such behavior, too, is characteristic of all gases. This, the opposite of compressibility, is called expansibility.

One idea that never occurred to ancient or medieval man was that he was being pressed upon by an enormously heavy blanket of atmosphere. Today we readily accept the fact that the weight of the atmosphere at sea level is 14.7 psi (pounds per square inch), or about 20 tons total pressure on our bodies. The pressure of the atmosphere is that of a " sea of gas " and is similar to the pressure exerted by water on objects beneath its surface.

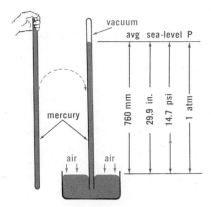

Fig. 9–1. A mercurial barometer.

When a gas is confined in a closed container, it exerts pressure on the walls of the container. The tires of our automobiles are filled with air until the pressure registers about 26 psi on a gauge. This means that the pressure inside the tire is 26 psi *greater* than the pressure outside; since the atmosphere exerts a pressure of about 14.7 psi on the outside of the tire, the total pressure inside is 40.7 psi.

Measuring the Pressure of Gases. There are many ways in which the pressure of a gas can be determined. Some of these involve rather inaccurate instruments such as the dial type of gauge on large cylinders of gas or the tire gauges used in service stations. Another method is shown in Fig. 9-1. The device used for measuring the pressure exerted by the atmosphere is called a **barometer.**

THE BAROMETER. Evangelista Torricelli, a student of Galileo, made the first instrument for measuring the pressure of the atmosphere. His discovery was accidental in part, for he was primarily interested in finding ways of creating a vacuum. One procedure that he employed was to fill a glass tube with mercury and then invert it in a dish of mercury. If the tube was long enough he found that the mercury would fall away from the upper end of the tube. The empty space above the mercury Torricelli considered a vacuum, but today we know that this space is not completely free of particles. Although it is very nearly a perfect vacuum, it does contain a small number of mercury molecules (mercury vapor).

The height of the mercury column is determined by the pressure of the atmosphere. The level in the barometer changes continuously, rising as the atmospheric pressure increases, falling as the atmospheric pressure decreases. The highest and lowest readings over a period of a year may differ by as much as 30 to 40 mm in a given locality.

In addition to the regular changes in pressure due to changing weather conditions, the height of the mercury column in a barometer changes with altitude. At sea level its average height is near 760 mm; on the top of Mount Everest the mercury in a barometer would rise to only about 240 mm. The average pressure at sea level is taken as the **standard atmospheric pressure.** It is expressed in several ways: 14.7 psi, 760 mm of mercury, or simply **1 atmosphere** (atm).

EFFECT OF TEMPERATURE

If a certain quantity of a gas is confined in a vessel in which the pressure can be kept constant, the volume will change with the temperature. When the temperature rises, the volume increases, and vice versa. Fig. 9-2 is a diagram of

3.00 ml at 25°C and atmospheric pressure   4.01 ml at 125°C and same pressure

Fig. 9-2. A given amount of gas is trapped in the tube by the globule of mercury. As the gas is heated, it expands, pushing the globule of mercury outward. The pressure remains constant. The volume can be read at a variety of temperatures to give data like those of Table 9-1 and Fig. 9-3.

an apparatus that may be used to demonstrate this effect. The globule of mercury makes the small glass tube gas-tight, but the mercury moves so easily in this tube that the pressure of the gas from inside must equal the pressure from outside. If the pressure on either side of the mercury changes, the globule moves so as to equalize the pressures. The total volume of gas to be considered is the volume contained in the tube up to the globule of mercury.

If the gas in the bulb is heated, the mercury moves to the right; if the gas is cooled, the globule moves to the left. By noting the position of the globule in the tube, we can tell by how many milliliters the volume changes with each change in temperature. Note that the pressure of the gas in the bulb never changes.

Let us suppose that the tube shown in Fig. 9-2 contains 3.00 ml of dry air at 25°C. Table 9-1 gives the volumes this quantity of air will have at various other temperatures. The data from this table are plotted in the graph in Fig. 9-3. The

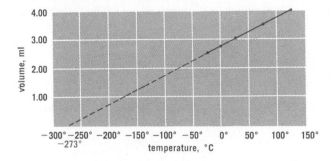

Fig. 9–3. A graph showing that, at constant pressure, the change in volume of a gas is directly proportional to the change in temperature. Extrapolation of the line to a volume of zero suggests that −273°C is the lowest temperature possible.

TABLE 9-1.

Changes in Volume of Air with Temperature (Pressure Remains Constant)

TEMPERATURE, °C	VOLUME, ML
125°	4.01
75°	3.50
25°	3.00
0°	2.75
−25°	2.50

straight-line (linear) relationship of temperature versus volume shows that the changes in volume of a gas are directly proportional to the changes in temperature.

The change in volume of a gas as the temperature changes was first studied experimentally by the French scientist Jacques Charles in 1787.

An Absolute Scale of Temperature. The results of Charles' experiments indicated that there is a lower limit of temperature, a zero temperature in the same sense as there is a zero distance or a zero weight. However, very low temperatures are difficult to obtain; hence the early experimenters were limited to temperatures not too far below 0°C. If we extrapolate the data in Table 9-1, we can predict the behavior of the gas at very low temperatures. At a temperature of $-273°C$ the predicted volume of the gas will be zero.[1] This temperature, $-273°C$, is taken as the zero of the absolute scale, the temperature at which the kinetic energy of molecules is zero. The absolute scale is called the Kelvin scale after Lord Kelvin, who in 1848 convincingly demonstrated the validity of a zero temperature. A comparison of the three temperature scales follows:

	Kelvin	Centigrade	Fahrenheit
Absolute zero	$0°K$	$-273°C$	$-459°F$
Freezing point of water	$273°K$	$0°C$	$32°F$
Boiling point of water	$373°K$	$100°C$	$212°F$

A centigrade temperature is converted to a Kelvin temperature by adding $273°$:

$$°K = °C + 273°$$

It is important to note that temperatures expressed on the centigrade or Fahrenheit scales cannot be used as such in gas law calculations. This is because the numbers on these scales are relative values only; zeros on these scales refer to rather high temperatures rather than no temperature, and negative values actually refer to positive temperatures.

DIFFUSION AND EFFUSION

A drop of perfume slowly evaporates, and the fragrant gas announces the presence of the wearer. One gas can *diffuse* through another quite easily—in this case, perfume through air.

In time, an automobile tire loses air pressure, although there is no detectable leak. The gas *effuses* from the tire—passes through pores or tiny holes in the rubber. The rates or speeds of diffusion and effusion are related to the densities of the gases, light gases diffusing more rapidly than heavy gases. For example, hydrogen will effuse from a child's balloon about four times as fast as air.

THE KINETIC-MOLECULAR THEORY

As information accumulated concerning the behavior of gases, scientists began to theorize on a structure of the gaseous state that would account for the known properties of gases. A successful theory must not only account for the fact that gases diffuse, expand when heated, and are compressible, but also give a logical explanation of why the various laws hold for gases. By the middle of the nineteenth century a **kinetic-molecular theory** of gases had been developed.

---

[1] In actual experiments, we cannot bring the volume of a gas this low because every known gas becomes a liquid at some low temperature. Moreover, we do not think the volume would become zero even if the gas did not liquefy, because as the volume of the gas decreases to a small value, the volume of the molecules themselves becomes appreciable. However, this does not invalidate our graphic extrapolation to a hypothetical volume of zero.

The five essential points of this theory are:

1. Gases are composed of molecules, which are widely separated from one another in otherwise empty space.

2. The gas molecules move about at high speeds, traveling in straight paths but in random directions.

3. The molecules collide with one another, but the collisions are perfectly elastic (result in no loss of energy).

4. The average kinetic energy of gas molecules depends directly on the absolute temperature. At any given temperature the molecules of all gases have the same average kinetic energy.

5. Attraction between molecules of a gas is negligible.

These ideas account for most of the properties associated with gases. Each of these points will now be taken up in detail.

1. Individual Particles in Gases. The first point in the theory accounts for the experimental fact that gases have weight and are compressible. Since the molecules are far apart, they can be forced closer together by an increase in pressure.

The fact that the densities of gases are very low as compared to the densities of liquids and solids is explained by considering the gaseous molecules to be far apart. If the molecules (which represent the weight of a gas) are far apart, a given volume of gas is mostly weightless, empty space. By implication, the particles of liquids and solids must be much closer together.

The individual particles normally present in a gaseous substance are called *molecules*. By far the largest number of gases are covalent compounds (e.g., CO_2, NH_3, SO_2, CH_4). In the special case of the noble gases—helium, neon, argon, krypton, xenon, and radon—the individual particles are single atoms but they are still called molecules. (See definition of molecules in Chap. 5.) The remaining elements that are gases under normal room conditions—hydrogen, nitrogen, oxygen, fluorine, and chlorine—exist as *diatomic molecules*; oxygen also exists as *triatomic molecules*, O_3, called ozone (see Fig. 9-4). Note

Ne Cl_2 O_2 O_3

Fig. 9–4. Molecules of elements may contain one, two, three, or more atoms per molecule.

that a gas may be composed of single atoms, or of molecules composed of several atoms united by covalent bonds. Ionic compounds never exist in the gaseous state under normal conditions of temperature and pressure.

2. The Motion of Gaseous Molecules. The translational motion of gaseous molecules accounts for the property of diffusion (see Fig. 9-5). Because of this motion and the relatively great distances between gaseous particles, molecules move rapidly from one point to another. Furthermore, gases leak (effuse) from a container through pores. Since the molecules themselves are so small, they can move through pores that may be invisible even under a microscope.

Molecular motion is also the basis for explaining the expansibility of gases. A given weight of gas fills any size of container uniformly. Obviously, therefore, the molecules must move from the point at which they are released until they are distributed throughout the container.

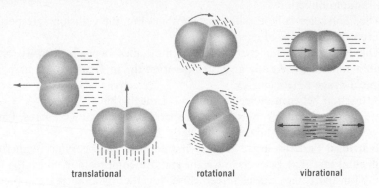

Fig. 9–5. Types of motion of diatomic molecules.

3. The Collision of Molecules. The theory of molecular collision helps account for the rates of diffusion of gases and the pressures they exert. As the molecules move, they encounter obstacles; these obstacles may be other molecules or the walls of a container. Collisions with the walls give rise to gas pressure.

If this explanation of diffusion is correct, a prediction may be made: the fewer the intervening gas molecules present, the more rapidly diffusion will occur, since there will be fewer collisions. This prediction has been verified by numerous experiments with gases at low pressures.

4. Kinetic Energy and Temperature. The walls of the bulb of a thermometer are bombarded by the molecules around it. The more energetic the individual bombardments, the greater the amount of vibration set up in the glass wall, and the greater therefore the amount of movement by the atoms of mercury inside the glass. As the amount of movement increases, each mercury atom bounces into its neighbors more violently, and the volume of the liquid increases. We say that the liquid expands with an increase in temperature. Since the mercury's increase in volume causes it to rise in the fine capillary tube, we can measure the temperature by measuring the height of the mercury in the tube.

This explanation of the action of the common thermometer is based on the idea that temperature is a measure of the degree of movement of molecules (or atoms or ions). The converse statement is thought to be equally true: the degree of movement of the particles of matter determines the temperature.

Since K.E. = $\frac{1}{2}mv^2$, we conclude that an increase in temperature results in an increase in the kinetic energy of the molecules because of the increased velocity. This concept of temperature gives rise to a number of consequences, among them the following:

At higher temperatures the particles of a gas move faster; hence diffusion and effusion should take place more rapidly at high temperatures than at low. This prediction is readily verified by experiment.

Consider two gases, hydrogen and oxygen, at the same temperature. According to point 4 of the kinetic-molecular theory, at the same temperature they will have the same average kinetic energy. Since K.E. $= \frac{1}{2}mv^2$, in order for the lighter hydrogen molecules (mass 2.0 awu) to have the same average kinetic energy as the heavier oxygen molecules (32.0 awu), the hydrogen molecules must move faster (actually they move four times faster).

Also, we can readily account for the fluctuation of gas pressure with temperature in a closed container on the basis of the kinetic energy of molecules. For example, if we check the air pressure in an automobile tire and then recheck it immediately after driving for an hour or so at a high speed, we find that the pressure has risen. If we feel the tire we also note that it is warm or hot to the touch. According to the kinetic theory, as the temperature increases, molecules move about more rapidly. Thus the molecules, with increased velocities, collide with each other and with the tire walls more often; and they hit with a greater force. Both of these effects, more frequent and harder collisions, result in an increased pressure. As the tire cools, the velocities of the molecules decrease and the pressure drops.

5. The Attraction of Molecules. It is apparent that there can be no great attraction between the molecules of a gas. If there were, the molecules would tend to congregate in a pool at the bottom of a container, and there would be no tendency for a gas to expand and fill the container in which it is placed.

THE GAS LAWS

The volume, weight, temperature, and pressure of any gaseous substance bear a simple mathematical relationship to each other. Further, other variables such as rates of diffusion and effusion of different gases are related mathematically. The precise statements of these relationships are known as the **gas laws.** Let us illustrate these by stating the obvious relationship between the weight of a gas and its volume: *The volume of a given gas is directly proportional to the weight of the gas, provided the volumes are measured at the same temperature and pressure.* This statement simply means that if 1 gram of a gas has a volume of 10 liters, 2 grams has a volume of 20 liters, 3 grams a volume of 30 liters, and so on, providing we stay within the stated restrictions of the law.

What restrictions have been set forth in the statement? The first restriction is covered by the term *a given gas*. This means that our calculations must involve a single gaseous substance. We cannot use the law to calculate the volume of two grams of oxygen from the volume of one gram of hydrogen. The second and third restrictions deal with temperature and pressure. These restrictions require that calculations of volume for different weights of a gas be made at the same temperature and pressure. This is necessary, because a change in either temperature or pressure will bring about a change in the volume of a given weight of gas. In the several gas laws to be stated subsequently, we shall carefully set forth the restrictions. It is important to note these restrictions and to understand why they are necessary.

PRESSURE-VOLUME

The volume of a certain weight of gas is inversely proportional to the pressure, at constant temperature. This is **Boyle's law.** Stated mathematically:

$$\frac{V_1}{V_2} = \frac{P_2}{P_1}$$

or

$$P_1 V_1 = P_2 V_2$$

The symbols V_1 and P_1 refer to the original volume and pressure; V_2 and P_2 refer to the volume and pressure under the new or changed conditions.

Although you can memorize mathematical formulas such as the one above for Boyle's law, and then solve problems by substituting numbers for the symbols, you are advised not to follow this approach here. You will make more progress at this stage if you carefully analyze the problems and base your solution on your knowledge of the behavior of gases.

Problem 1: A sample of gas weighing 0.216 g is trapped in a cylinder by a piston, as shown in Fig. 9-6. The volume of the gas is 275 ml when the pressure exerted by the piston is equal to 920 mm. If the pressure is decreased to 780 mm, what will the volume be?

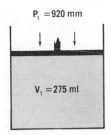
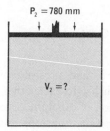

Fig. 9–6. To hold a given quantity of gas in this cylinder at a volume of 275 ml, a pressure of 920 mm must be maintained by the piston. If this pressure is reduced, the gas will force the piston outward and occupy a larger volume.

Analysis: In working gas law problems we are usually interested in four variable quantities: the amount (expressed in weight or in number of molecules), the pressure, the volume, and the temperature. It helps in analyzing a problem to tabulate the data available for the original condition of the gas and for its new or changed condition. If one of the four variables does not change, this is indicated by the symbol k (standing for *constant*). The statement of Problem 1 implies that the cylinder does not leak, so we assume that the weight of gas remains constant at 0.216 g. Since the temperature is not mentioned, we must assume that its initial and final values are the same. We arrange our data as follows:

	weight	P	V	T
original	k	920 mm	275 ml	k
changed to	k	780 mm	V_2	k

We might reason as follows: Only the volume and the pressure change, so Boyle's law applies. The pressure is decreased, therefore the volume must be increased. The changed volume will be greater than 275 ml.

Kinetic Theory of Gases; Gas Laws

For gas law problems in this text, we will find that all calculations involve either a direct or an indirect *proportion*. Such calculations can be made in a straightforward manner by using ratios or factors based on the variables that change in value. In Problem 1,

$$V_2 = 275 \text{ ml} \times \frac{\text{pressure}}{\text{pressure}}$$

└─new   └─original   └─factor to account for
volume   volume   pressure change

We know that the volume depends on the pressure, so our only problem here is to decide which factor to use: $\frac{920 \text{ mm}}{780 \text{ mm}}$ or $\frac{780 \text{ mm}}{920 \text{ mm}}$. The first of these factors has a value *greater* than 1; the second has a value *less* than 1. Since we reasoned that the volume of the gas will be *greater* than 275 ml, we choose the factor $\frac{920 \text{ mm}}{780 \text{ mm}}$. We know that if we use it as a multiplier, the numerical answer will be greater than 275 ml.

Solution:

$$V_2 = 275 \text{ ml} \times \frac{920 \text{ mm}}{780 \text{ mm}}$$

└─factor to correct for pressure change

$$= 324 \text{ ml}$$

Check: We can do much toward improving our efficiency in solving problems by inspecting the answer to see if it is reasonable. Because the answer to a problem usually includes two parts—the units and the number—this inspection should cover both phases. In this problem we note that units have canceled to leave units of volume (milliliters); thus we find that this part is correct. In checking the size of the answer, we note that the new volume is greater than 275 ml, as we predicted it would be. Because the correction factor, $\frac{920 \text{ mm}}{780 \text{ mm}}$, is about 6/5 (divide each member by 15), the new volume should be only about 1/5 larger because of the pressure decrease, that is, about 275 ml + 1/5 of 275 ml, or about 330 ml. Therefore the answer 324 ml seems reasonable.

VOLUME-TEMPERATURE

The volume of a given weight of gas is directly proportional to the absolute temperature, at constant pressure. This is **Charles' law,** discovered by Jacques Charles in 1787. Stated mathematically:

$$\frac{V_1}{V_2} = \frac{T_1}{T_2}$$

Problem 2: A child's balloon has a volume of 3.80 liters (about a gallon) when the temperature is 35°C. What is the volume if the balloon is put into a refrigerator and cooled to 5°C? Assume that the pressure inside the balloon is equal to the atmospheric pressure at all times.

Analysis: Tabulate the data. Because we can use only absolute temperatures in gas-law calculations, other scales of temperature must be converted to the *absolute* scale: °C + 273° = °K.

	weight	V	T	P
original	k	3.80 liters	308°K	k
changed to	k	V_2	278°K	k

We can calculate the new volume by multiplying the old volume by a factor that expresses the influence of temperature on volume:

$$V_2 = 3.80 \text{ liters} \times \frac{\text{temperature}}{\text{temperature}}$$

⌞new   ⌞original   ⌞factor to account for
volume   volume   temperature change

Charles' law is our guide in choosing the proper temperature factor. According to this law, when the temperature decreases, the volume must decrease proportionately. We use the temperature factor $\frac{278°K}{308°K}$, because it is less than 1.

Solution:

$$V_2 = 3.80 \text{ liters} \times \frac{278°K}{308°K}$$

⌞factor to correct for
temperature change

$$= 3.43 \text{ liters}$$

Check: We note that units cancel to give units of volume. Further, the calculated volume is smaller, as was predicted. The factor $\frac{278°K}{308°K}$ is equal to about 9/10, so that we expect the new volume to be about 9/10 of 3.80 liters, or about 3.4 liters.

PRESSURE-TEMPERATURE

The pressure of a given weight of gas is directly proportional to the absolute temperature if the volume does not change.[2] Stated mathematically:

$$\frac{P_1}{P_2} = \frac{T_1}{T_2}$$

Problem 3: A steel cylinder of oxygen gas has a pressure of 135 atm at a temperature of 20°C. Suppose that the cylinder becomes heated to 85°C because it is stored near a steam radiator. Calculate the pressure inside the cylinder at the higher temperature.

Analysis: The gas is stored in a steel cylinder, so we can assume that its mass does not change at all and that the volume does not change appreciably. The two variables are the pressure and temperature.

---

[2]This is sometimes called **Gay-Lussac's law**, sometimes **Amontons' law**, after Guillaume Amontons who related the pressure of a gas to its temperature and constructed a gas thermometer on this basis in 1703.

Kinetic Theory of Gases; Gas Laws

	weight	V	P	T
original	k	k	135 atm	293°K
changed to	k	k	P_2	358°K

It is logical to reason as follows. The temperature is increasing; the molecules move about in the cylinder more rapidly, hitting one another and the walls of the cylinder more often and with greater force. Therefore the pressure increases. We use the temperature factor $\frac{358°K}{293°K}$.

Solution:

$$P_2 = 135 \text{ atm} \times \frac{358°K}{293°K}$$

       └─factor to account for temperature change

$$= 165 \text{ atm}$$

Check: We expect the pressure to be larger by about 1/5, i.e., by about 30 atm, because the factor $\frac{358°}{293°}$ is about 1/5 larger than 1: $\frac{358}{293} = \frac{293 + 65}{293}$, and $\frac{65}{293} \simeq \frac{1}{5}$. The calculated answer agrees with the predicted answer in both magnitude and units.

Standard Conditions. The problems we have just solved make it clear that the amount of gas—i.e., the weight—cannot be specified by giving only the volume. If the weight is not given, the pressure, the temperature, and the volume must all be indicated to reveal an exact quantity of any gas. For this reason and for uniformity in specifying certain properties of gases, such as density and rate of diffusion, scientists the world over have agreed to use a particular temperature and pressure in presenting data on gases. The standard temperature and pressure, commonly referred to as **standard conditions,** are 0°C and 760 mm, respectively. If only the volume of a quantity of gas is given—e.g., 3.5 liters of nitrogen—it is assumed that the temperature is 0°C and the pressure is 760 mm. The volume of a gas at these standard conditions is commonly referred to as the volume at **STP** (standard temperature and pressure).

PRESSURE OF A MIXTURE OF GASES

In a mixture of different gases each gas exerts part of the pressure. This is the same pressure that it would exert if it alone occupied the volume containing the mixed gases. *The total pressure in a mixture of gases is the sum of the individual partial pressures.* This is **Dalton's law,** formulated by John Dalton about 1803. Stated mathematically:

$$P = p_1 + p_2 + p_3 + \ldots$$

P is the total pressure; the small p's refer to the part of the pressure exerted by each different gaseous substance in the mixture.

Problem 4: A sample of nitrogen saturated with water vapor has a volume of 88.3 ml at a temperature of 18.5°C and a pressure of 741 mm. What would the volume of the nitrogen be if it were dry and at the same temperature and pressure? (See Fig. 9-7.)

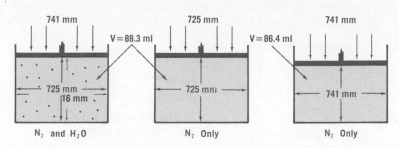

Fig. 9-7. A graphic representation of the solution of Problem 4.

Analysis: The total pressure of 741 mm is the sum of the pressures of the nitrogen plus the water vapor. At 18.5°C the vapor pressure of water is approximately 16 mm (see Table 2 in the Appendix).

$$P_{total} = p_{dry\ nitrogen} + p_{water\ vapor} \quad \text{(Dalton's law)}$$

or

$$p_{dry\ nitrogen} = P_{total} - p_{water\ vapor}$$
$$= 741\ mm - 16\ mm$$
$$= 725\ mm$$

Collecting our data, we have

	weight	V	P	T
original	k	88.3 ml	725 mm	k
changed to	k	V_2	741 mm	k

Note that in the "changed" condition the dry nitrogen exerts the entire pressure of 741 mm. Consequently, the volume of the dry nitrogen must be smaller. The correction factor is $\dfrac{725\ mm}{741\ mm}$.

Solution:

$$V_2 = 88.3\ ml \times \frac{725\ mm}{741\ mm}$$

└─ pressure factor must account for *decrease*, therefore the smaller value is over the larger

$$= 86.4\ ml$$

Check: The fraction $\dfrac{725}{741}$ is only slightly smaller than 1. The new volume is slightly less than 88.3 ml.

Problem 5: A sample of hydrogen is collected in a bottle over water. By carefully raising and lowering the bottle, the height of the water inside is adjusted so that it is just even with the water level outside (see Fig. 9-8). The following

Kinetic Theory of Gases; Gas Laws

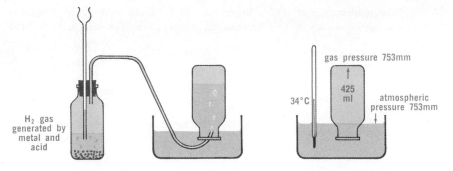

Fig. 9–8. When a sample of gas is collected, we usually measure its pressure, its volume, and its temperature. It is often convenient to measure the latter two when the gas is held at the pressure of the atmosphere. We then read the pressure of the atmosphere on a nearby barometer.

measurements are taken: volume of gas, 425 ml; atmospheric pressure, 753 mm; temperature of water (and of gas also), 34°C. Calculate the volume the hydrogen would have if it were dry and at a pressure of 760 mm and a temperature of 0°C.

Analysis: The hydrogen is mixed with water vapor, therefore not all of the 753 mm pressure is due to hydrogen; a small part is due to the water vapor. The vapor pressure of water varies with the temperature, so that we find from Table 2 in the Appendix that its value at 34°C is 40 mm. The pressure due to hydrogen can now be calculated by means of Dalton's law.

$$P_{total} = p_{H_2} + p_{H_2O}$$

or

$$p_{H_2} = P_{total} - p_{H_2O}$$

$$= 753 \text{ mm} - 40 \text{ mm} = 713 \text{ mm}$$

That is, if the water vapor[3] is removed so that hydrogen alone occupies the entire volume, the pressure will fall from 753 mm to 713 mm. This latter pressure is the one that we must use as the original pressure in solving for the volume of the dry hydrogen at 0°C and 760 mm:

	weight	V	T	P
original	k	425 ml	307°K	713 mm
changed to	k	V_2	273°K	760 mm

$$V_2 = 425 \text{ ml} \times \frac{\text{temperature}}{\text{temperature}} \times \frac{\text{pressure}}{\text{pressure}}$$

└─new volume   └─to correct for temperature change   └─to correct for pressure change

The pressure of 760 mm is greater than the pressure exerted by hydrogen when it was collected over water. The volume of the hydrogen tends to be less because of this change in pressure. The pressure factor is set at $\frac{713 \text{ mm}}{760 \text{ mm}}$ (Boyle's law).

---

[3] The wet gas could be passed through a drying agent such as calcium chloride.

The temperature of 0°C is lower than 34°C; there tends to be a decrease in volume because of this decrease in temperature. The temperature factor is set as $\dfrac{273°K}{307°K}$ (Charles' law).

Solution:

$$V_2 = 425 \text{ ml} \times \underbrace{\frac{273°K}{307°K}}_{\substack{\text{factor to correct} \\ \text{for temperature} \\ \text{change}}} \times \underbrace{\frac{713 \text{ mm}}{760 \text{ mm}}}_{\substack{\text{factor to correct} \\ \text{for pressure} \\ \text{change}}}$$

$$V_2 = 355 \text{ ml}$$

Check: We note that units have canceled to leave units of volume. A rough approximation of the answer can be made as follows: the temperature fraction $\dfrac{273°K}{307°K}$ is about 9/10; the temperature change will cause the volume to be reduced by about 1/10, i.e., 425 ml − 42 ml, or 383 ml. The pressure fraction $\dfrac{713 \text{ mm}}{760 \text{ mm}}$ is about 14/15; the pressure change will cause a further reduction of 1/15 of 383 ml, i.e. 383 ml − 25 ml, or 352 ml. The calculated answer of 355 ml appears to be correct.

Ideal Gases. An **ideal gas** is one whose behavior can be predicted precisely on the basis of the kinetic-molecular theory and the gas laws. Actually, no such gas exists, although gases such as hydrogen and oxygen do not depart greatly from ideal behavior at moderate temperatures and pressures. This subject will be dealt with more fully in the next chapter, after additional gas laws have been discussed.

EXERCISES

1. A liter of liquid mercury at 20°C has a definite weight while a liter of carbon dioxide at 20°C may have any number of weights. Explain.
2. Describe how a barometer gives a measure of the atmospheric pressure.
3. For Boyle's law, what possible variables must be held constant if the gas is to obey the law? Repeat for Charles' law.
4. Convert 20°F to degrees Kelvin.
5. Why does it take longer for a thermometer to reach a constant reading when placed in a low-pressure chamber than it does when placed in a chamber in which the pressure is at atmospheric level? Does a perfect vacuum have a measurable temperature? Explain.
6. Account for Dalton's law of partial pressures in terms of the kinetic-molecular theory.
7. Fifty milliliters of water is placed in a 1-liter metal can and heated to boiling. After boiling a few minutes, the can is sealed with a stopper and set aside to cool. In a short time the can is observed to buckle and collapse. Explain these observations.
8. Explain the following in terms of the kinetic-molecular theory of gases:
   (a) A quantity of gas exerts twice the pressure when compressed to one-half of its original volume.

(b) Hot air rises.
(c) A gas spreads more rapidly through a room when the temperature is 35°C than when the temperature is 0°C.
(d) If a cylinder of gas is placed in the sun on a warm day, the pressure increases.
(e) Although carbon dioxide molecules are 22 times heavier than hydrogen molecules, equal numbers of H_2 and CO_2 molecules exert the same pressure if in the same sized containers and if at the same temperature.

9. A sample of gas has a volume of 250 ml at 20°C and 750 mm. What will be the volume of the gas at 40°C and 750 mm? At 40°C and 375 mm?
10. The temperature of 340 ml of gas is 35°C. If this gas is heated to 100°C in a way so that the pressure remains constant, what is the new volume?
11. One liter of helium, 2 liters of neon, and 5 liters of nitrogen, each gas at STP, are placed in a 3-liter container. What is the resulting pressure if there is no change in temperature?
12. The pressure of a gas in a steel cylinder was determined at several different temperatures. The data collected are: 100°F, 1.00 atm; 25°F, 0.87 atm; 0°F, 0.82 atm; −25°F, 0.78 atm. By plotting these data, determine the value of absolute zero on the Fahrenheit temperature scale.
13. A sample of oxygen has a volume of 584 ml at 28°C and 740 mm. What is the volume at STP.
14. A quantity of hydrogen, measured over water at 22°C and 734 mm, has a volume of 210 ml. Calculate the volume when dry at standard conditions.
15. A sample of carbon monoxide has a volume of 3.0 liters at 27°C. At what temperature does this sample have a volume of (a) 4 liters; (b) 2 liters? What assumptions have you made in performing these calculations?
16. What volume of air at STP must be added to an automobile tire (volume 3.2 gal) to raise the pressure from 15 psi at 70°F to 24 psi at 70°F?
17. If 0.50 g of helium exerts a pressure of 1.40 atm when its volume is 2.0 liters at 0°C, what is the calculated volume of 1 mole of helium at STP?
18. Convert 30 psi to kg/cm².
19. A 2.00-g sample of methane has a volume of 3.58 liters at 77°C and 14.7 psi. What is the density of methane at STP?

SUGGESTED READING

Easley, W. K., and G. F. Powers, "Kinetic Molecular Theory from a Jukebox," *J. Chem. Educ.*, **37**, 302 (1960).

Hecht, C. E., "Negative Absolute Temperatures," *J. Chem. Educ.*, **44**, 124 (1967).

Neville, R. G., "The Discovery of Boyle's Law, 1661–62," *J. Chem. Educ.*, **39**, 356 (1962).

10

GAS LAWS AND WEIGHT RELATIONSHIPS

MOLECULAR WEIGHTS

Not until the early part of this century were methods developed for counting atoms and molecules and for determining the weights of single atoms or molecules. Yet by the middle of the 1800s, methods had been devised for comparing the weights of atoms and molecules. These methods depended upon comparing the weights of equal numbers of atoms or molecules. For example, suppose that we weigh a quantity of A molecules and find their weight to be 0.212 g and a quantity of B molecules which contains the same number of molecules as A and find their weight to be 0.424 g. We conclude from this information that a molecule of B is twice as heavy as a molecule of A. If we represent the weight of a single molecule of A by a number, say 30, then the number 60 would be used to represent the weight of a single B molecule. The fact that we do not know the numbers of molecules in the 0.212 g of A and 0.424 g of B is unimportant to this method; it is only necessary that we know that these quantities contain the same number of molecules. We will describe two methods that were used in the last century for determining the relative numbers of gaseous molecules and the relative molecular weights.

METHOD BASED ON AVOGADRO'S LAW

Gay-Lussac's Law of Combining Volumes. In 1805 and shortly thereafter, the French chemist, J. L. Gay-Lussac performed a series of experiments, which showed that *gases react with one another in simple proportions by volume, when volumes are measured at the same temperature and pressure.* Some examples which illustrate Gay-Lussac's law follow:

$$1 \text{ vol hydrogen} + 1 \text{ vol chlorine} \longrightarrow 2 \text{ vol hydrogen chloride}$$

$$2 \text{ vol hydrogen} + 1 \text{ vol oxygen} \xrightarrow{\text{above } 100°C} 2 \text{ vol water vapor}$$

$$1 \text{ vol nitrogen} + 3 \text{ vol hydrogen} \longrightarrow 2 \text{ vol ammonia}$$

$$1 \text{ vol nitrogen} + 1 \text{ vol oxygen} \longrightarrow 2 \text{ vol nitric oxide}$$

Avogadro's Law. In 1811, Amadeo Avogadro advanced a brilliant hypothesis to account for the findings of Gay-Lussac. He argued that since molecules combine with each other in simple ratios, and since the volumes of gaseous reactants and products are also in simple ratios, equal volumes must contain the same number of molecules. This hypothesis, now called **Avogadro's law**, is stated formally: *Equal volumes of all gases at the same temperature and pressure contain the same number of molecules.*

Molecular Weights. The method for determining molecular weights based on Avogadro's law is outlined as follows:

1. Some substance is chosen as the standard. For many years, the number 32 was assigned to represent the weight of one molecule of oxygen. (One can say that one molecule weighs 32 awu.)

2. For any other substance, the number of molecules to be weighed is equal to the number of oxygen molecules in 32 g of oxygen. This number is called a *mole*. (The number of molecules in a mole is now known to be 6.02×10^{23}. See Chap. 7.)

3. The volume of 32 g of oxygen, 1 mole, is experimentally determined as 22.4 liters at STP.

4. Based on Avogadro's law, 22.4 liters of any gas at STP constitutes 1 mole of that gas; 22.4 liters at STP is called the **molar gas volume,** or the volume of one mole.

5. The weight in grams of 22.4 liters of a gas at STP is the weight of one mole of that gas. This weight in grams is numerically equal to the relative weight of one molecule in atomic weight units. (See Fig. 10-1.)

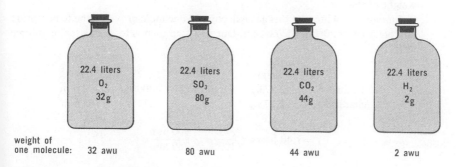

Fig. 10–1. Measured at STP, 22.4 liters of a gas contains a mole (6.0×10^{23} molecules) of any gaseous substance.

6. The experimental determination of a relative molecular weight involves the careful measurement of the *weight* and *volume* of a sample of the gaseous substance at a known *temperature* and *pressure*. Knowing these four quantities we can use the appropriate gas laws to calculate the weight of 22.4 liters of the gaseous substance at 0°C and 760 mm (STP).

7. Based on the new ^{12}C standard, the mole weight of oxygen is 31.9988 g and the molar volume is 22.414 liters. However, the method is not precise enough to warrant the use of more than two or three significant figures—32.0 g of O_2 and 22.4 liters.

Problems 1 and 2 illustrate typical calculations.

Problem 1: A liter container was evacuated and then filled at a temperature of 0.0°C with a gaseous compound until the pressure of the gas was 760 mm. By weighing the gas and container and then subtracting the weight of the container, the weight of the gas was found to be 3.61 g. Calculate in atomic weight units and in grams the weight of one molecule.

Solution: First calculate the weight of one mole of the gas:

$$\frac{3.61 \text{ g}}{1 \text{ liter}} \times \frac{22.4 \text{ liters}}{1 \text{ mole}} = 80.9 \text{ g/mole}$$

Since the weight in grams of one mole is numerically equal to the weight of one molecule in atomic weight units,

Weight of one molecule = 80.9 awu

Since one mole, 80.9 g, contains 6.02×10^{23} molecules, one molecule weighs:

$$\text{Weight of one molecule} = \frac{80.9 \text{ g}}{6.02 \times 10^{23} \text{ molecules}}$$

$$= 1.34 \times 10^{-22} \text{ g/molecule}$$

We may say that one molecule of the compound weighs 80.9 awu or 1.34×10^{-22} g.

Problem 2: A 1.40-liter volume of a gas measured at a temperature of 27°C and a pressure of 900 mm was found to weigh 2.273 g. Calculate the molecular weight of the gas.

Solution: 22.4 liters of this gas will contain one mole provided the temperature and pressure are 0°C and 760 mm. Our first step then is to calculate the volume at STP:

	weight	V	T	P
original	2.273 g	1.40 liters	300°K	900 mm
changed to	2.273 g	V_2	273°K	760 mm

$$V_2 = 1.40 \text{ liters} \times \frac{273°K}{300°K} \times \frac{900 \text{ mm}}{760 \text{ mm}}$$

$V_2 = 1.51$ liters of the gas at 0° and 760 mm.

Gas Laws and Weight Relationships

We see from this calculation that if 2.273 g of the gas is cooled from 27°C to 0°C while the gas is permitted to expand from a pressure of 900 mm to a pressure of 760 mm, the volume changes from 1.40 liters to 1.51 liters. The weight of the gas does not change.

We next calculate the weight of 22.4 liters at STP:

	weight	V	T	P
original	2.273 g	1.51 liters	273°K	760 mm
changed to	M_2	22.4 liters	273°K	760 mm

$$M_2 = 2.273 \text{ g} \times \frac{22.4 \text{ liters}}{1.51 \text{ liters}} = 33.7 \text{ g, weight of 1 mole}$$

The weight of one mole (6.02×10^{23} molecules) in grams is equal numerically to the weight of one molecule in atomic weight units. Therefore, the weight of one molecule is 33.7 awu.

METHOD BASED ON GRAHAM'S LAW OF DIFFUSION

A gas that has a high density diffuses more slowly than a gas with a lower density. *The rates of diffusion of two gases are inversely proportional to the square roots of their densities, providing the temperature and pressure are the same for the two gases.* This is **Graham's law**, discovered by Thomas Graham in 1881. Stated mathematically:

$$\frac{r_1}{r_2} = \frac{\sqrt{d_2}}{\sqrt{d_1}}$$

It follows from Avogadro's law that the densities of gases are proportional to their molecular weights. We may therefore substitute *molecular weight* for *density* in the expression of Graham's law of diffusion:

$$\frac{r_1}{r_2} = \frac{\sqrt{d_2}}{\sqrt{d_1}} = \frac{\sqrt{\text{mol. wt}_2}}{\sqrt{\text{mol. wt}_1}}$$

The relationship also expresses the rate of *effusion* of a gas through a small hole into an evacuated region. That is, if a tiny hole is made in a thin wall of a container filled with a gas, gas molecules pass through the hole into an evacuated space at a rate inversely proportional to the square root of the weight of the molecules.

In order to determine molecular weights by this method, the time required for a given volume of the gas to effuse through a small hole is compared with the time required for the same volume of oxygen to effuse through the same hole. Effusion times can be measured with an apparatus similar to that shown in Fig. 10-2. Very little error is introduced by allowing the gases to effuse into air rather than an evacuated chamber, provided the molecular weights of the two gases to be compared are not greatly different. The following problem serves to show what measurements are made and how the molecular weight is calculated.

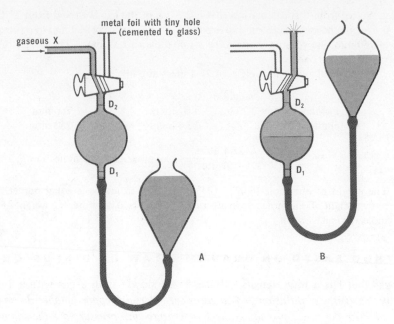

Fig. 10-2. A simple apparatus for comparing the rates of effusion of gases.

Problem 3: The time required for a volume of gas X to effuse through a small hole was 112.2 sec. The time required for the same volume of oxygen was 84.7 sec. Calculate the molecular weight of gas X.

Analysis: Since the time required for effusion is inversely proportional to the rate of effusion (the longer the time, the less the rate), we may further modify the Graham's law equation as follows:

$$\frac{r_1}{r_2} = \frac{t_2}{t_1} = \frac{\sqrt{\text{mol. wt}_2}}{\sqrt{\text{mol. wt}_1}}$$

or

$$\frac{t_{O_2}}{t_x} = \frac{\sqrt{32.0 \text{ awu}}}{\sqrt{\text{mol. wt of } X}}$$

Solution:

$$\frac{84.7 \text{ sec}}{112.2 \text{ sec}} = \frac{\sqrt{32.0 \text{ awu}}}{\sqrt{\text{mol. wt of } X}}$$

$$\text{Mol. wt of } X = \frac{(112.2)^2 \times 32.0 \text{ awu}}{(84.7)^2}$$

$$= 56.2 \text{ awu}$$

It is interesting to note that the molecular weight (or atomic weight) of the noble gas radon was initially determined by the effusion method in 1910, two years after its isolation.

Gas Laws and Weight Relationships

Rates of Diffusion. If the molecular weights of two gases are known, Graham's law may be used to calculate their relative rates in diffusing or spreading into another gas.

Problem 4: Compare the rates of diffusion of helium (He) and sulfur dioxide (SO_2).

Solution: The molecular weights of helium and sulfur dioxide are 4 and 64, respectively. Hence,

$$\frac{r_{He}}{r_{SO_2}} = \frac{\sqrt{64}}{\sqrt{4}} = \frac{8}{2} = \frac{4}{1}$$

Helium diffuses four times as fast as sulfur dioxide.

WEIGHT-VOLUME RELATIONSHIPS

The volume of a mole of any gas measured at STP is 22.4 liters. This relationship is the basis for calculating the volumes of gaseous reactants and products involved in chemical reactions. If a volume is specified at some temperature and pressure other than STP, corrections are made by means of Boyle's and Charles' laws.

Problem 5: Calculate the volume of hydrogen and of oxygen theoretically obtainable by the decomposition of 100 g of water by electrolysis.

Solution: To solve this problem, we first write a balanced chemical equation and calculate the number of moles of hydrogen and oxygen that are formed from the decomposition of 100 g of water. This procedure is exactly like the solutions to the problems discussed in Chap. 7:

$$\underset{\text{2 moles}}{2H_2O} \xrightarrow{\text{elec.}} \underset{\text{2 moles}}{2H_2} + \underset{\text{1 mole}}{O_2}$$

$$\text{Moles of } H_2O = 100 \text{ g of } H_2O \times \frac{1 \text{ mole of } H_2O}{18.0 \text{ g}} = \frac{100}{18.0} \text{ moles of } H_2O$$

$$\text{Moles of } H_2 = \frac{100}{18.0} \text{ moles of } H_2O \times \frac{2 \text{ moles of } H_2}{2 \text{ moles of } H_2O} = \frac{100}{18.0} \text{ moles of } H_2$$

$$\text{Moles of } O_2 = \frac{100}{18.0} \text{ moles of } H_2O \times \frac{1 \text{ mole of } O_2}{2 \text{ moles of } H_2O} = \frac{100}{36.0} \text{ moles of } O_2$$

$$\text{Vol of } H_2 = \frac{100}{18.0} \text{ moles of } H_2 \times \frac{22.4 \text{ liters of } H_2}{1 \text{ mole of } H_2}$$

$$= 124 \text{ liters of } H_2 \text{ at STP}$$

$$\text{Vol of } O_2 = \frac{100}{36.0} \text{ moles of } O_2 \times \frac{22.4 \text{ liters of } O_2}{1 \text{ mole of } O_2}$$

$$= 62 \text{ liters of } O_2 \text{ at STP}$$

Problem 6: If 100 g of water is decomposed (see Problem 5), what volumes of hydrogen and oxygen are obtained if the gases are collected at 700 mm and 26°C?

Solution:

$$\text{Vol of } H_2 = \frac{100}{18.0} \text{ moles of } H_2 \times \frac{22.4 \text{ liters of } H_2}{1 \text{ mole of } H_2} \times \frac{760 \text{ mm}}{700 \text{ mm}} \times \frac{299°K}{273°K}$$

$$= 148 \text{ liters of } H_2$$

Vol of O_2 = one-half the volume of H_2, or 74 liters

Problem 7: What is the theoretical volume of ammonia formed by the action of 10 liters of hydrogen with nitrogen? What volume of nitrogen reacts? (All volumes are measured at the same temperature and pressure.)
Solution:

N_2	+	$3H_2$	→	$2NH_3$
1 mole		3 moles		2 moles
1 molar-volume		3 molar-volumes		2 molar-volumes

We could proceed here as in Problem 6 by first calculating the number of moles of ammonia and of nitrogen and then multiplying the number of moles of each by liters per mole, that is, 22.4 liters/mole. However, a more direct approach can be followed, because the volume of H_2 is given in the problem rather than its weight. Because equal volumes contain the same number of molecules (Avogadro's law), the volume of the nitrogen is one-third the volume of H_2. This follows from the balanced equation that states that one molecule of N_2 reacts with three molecules of H_2. From the same line of reasoning, we see that the volume of NH_3 is two-thirds that of H_2.

$$\text{Vol of } NH_3 \text{ formed} = 10 \text{ liters of } H_2 \times \frac{2 \text{ vol of } NH_3}{3 \text{ vol of } H_2} = 6.7 \text{ liters of } NH_3$$

$$\text{Vol of } N_2 \text{ that reacts} = 10 \text{ liters of } H_2 \times \frac{1 \text{ vol of } N_2}{3 \text{ vol of } H_2} = 3.3 \text{ liters of } N_2$$

Gas Densities. The density of a gas is usually expressed as grams per liter at STP, instead of grams per milliliter. Gas densities can be experimentally determined by carefully weighing a known volume of gas, correcting the volume to STP, and then finding the weight of one liter.

We can also calculate the density of a gas from its molecular formula. This is done by dividing the mole weight by 22.4 to obtain the weight of one liter.

For example, the density of SO_2 is calculated as follows:

$$\frac{64.1 \text{ g}}{22.4 \text{ liter}} = 2.86 \text{ g/liter at STP}$$

Problem 8: Calculate the density of dry air. See Table 30-4 for needed data.
Solution: We first calculate the weight of a mole of dry air. A mole of dry air contains 6.02×10^{23} molecules and has a volume of 22.4 liters at STP. From Table 30-4 we note that 78.08% of the molecules are N_2 molecules, 20.95% are O_2, 0.93% are Ar, and 0.03% are CO_2 (this interpretation follows from Avogadro's law). Hence, in a mole of dry air there are 0.7808 mole of N_2, 0.2095 mole of O_2, 0.0093 mole of Ar, and 0.0003 mole of CO_2. The amounts of Ne, He, and other gases are so small that they have a negligible effect on the total weight. The weight of a mole of dry air is calculated by first calculating the weights of nitro-

Gas Laws and Weight Relationships

gen, oxygen, argon, and carbon dioxide. The sum of these weights is the weight of a mole of dry air:

$$\text{Wt of } N_2 = 0.7808 \text{ mole} \times \frac{28.01 \text{ g}}{\text{mole}} = 21.87 \text{ g}$$

$$\text{Wt of } O_2 = 0.2095 \text{ mole} \times \frac{32.0 \text{ g}}{\text{mole}} = 6.70 \text{ g}$$

$$\text{Wt of Ar} = 0.0093 \text{ mole} \times \frac{39.9 \text{ g}}{\text{mole}} = 0.37 \text{ g}$$

$$\text{Wt of } CO_2 = 0.0003 \text{ mole} \times \frac{44.0 \text{ g}}{\text{mole}} = 0.01 \text{ g}$$

$$\text{Wt of a mole of dry air} = 28.95 \text{ g}$$

The density of dry air at STP is now calculated in the usual way:

$$\text{Density} = \frac{28.95 \text{ g}}{22.4 \text{ liters}} = 1.29 \text{ g/liter}$$

BEHAVIOR OF REAL GASES

In postulating the kinetic-molecular theory, certain assumptions are made about how a gas behaves. A gas that has all the properties described in the kinetic-molecular theory is called an **ideal gas.** (Two significant properties of an ideal gas are that the molecules do not attract each other and they occupy no space.) How close is the behavior of real gases such as hydrogen, nitrogen, and carbon dioxide to that of an ideal gas? Obviously many real gases behave very nearly ideally, for otherwise the kinetic-molecular theory would be of no practical value.

How then can we know whether a real gas is obeying the properties assigned it by the kinetic-molecular theory? Although there are numerous ways, we shall consider only two. The kinetic theory predicts that if the pressure on 2.000 liters of oxygen, originally at 1 atm, is doubled, the volume will become 1.000 liter. Instead, the volume becomes 0.999 liter. Since the oxygen molecules are squeezed closer together by the increase in pressure, their attraction for each other increases; hence their actual volume is 1 ml less than predicted. A Dutch physicist, J. D. van der Waals, first described the effects of these forces, which, in honor of him, are called **van der Waals forces.**

Van der Waals forces appear to be due to the attraction of the nuclei of one molecule for the electrons of its neighbors. As we have seen, the "skin" or outside of all molecules (and indeed of all ions and atoms) consists of electrons. A molecule is made up of atoms that have been joined together by covalent bonds. These atoms may not be able to form any more bonds in the strict sense of sharing more electrons. Even so, there is still a *residual attraction* for other electrons; hence the molecules are attracted to one another. These weak forces are considered to be only physical forces, not chemical bonds.

It is quite clear that the molecules of some substances have stronger mutual attractions than others. In the case of the common cleaning fluid, carbon tetrachloride, CCl_4, there are van der Waals forces which hold the molecules together at room temperature as a liquid.[1] On the other hand, the substances ethane, C_2H_6, and methane, CH_4, are gases at room temperature. The attractive forces between their molecules are probably not as powerful as the forces holding the CCl_4 molecules together.

As a second example of the behavior of real gases, let us consider how closely real gases obey Avogadro's law. The calculated volume of one mole of an ideal gas at STP is 22.414 liters. Tabulated below are the measured volumes per mole of three gases at STP:

Hydrogen	22.43 liters
Oxygen	22.393 liters
Carbon dioxide	22.263 liters

It is apparent from these figures that equal volumes of real gases do not contain precisely the same number of molecules. A volume of carbon dioxide, for example, contains about 0.8 percent more molecules at STP than does an equal volume of hydrogen.

LIQUEFACTION OF GASES

According to the kinetic theory, the molecules in a gas are attracted to one another by van der Waals forces. Because of the high velocity of the gas molecules at ordinary temperatures and the distance between them, these weak attractive forces are not able to pull the molecules together. If the kinetic energy of the gas molecules is lowered by decreasing the temperature, the van der Waals forces can finally pull the particles close to one another and hold them together in the liquid state. The strength of these forces is inversely proportional to a high power of the distance between the molecules. If the molecules are far apart, the attraction is weak; but as they come close to one another, the attraction increases. When an increase in pressure crowds the molecules more closely together, the gas liquefies if the attractive forces are great enough.

However, there is a minimum temperature for each gaseous substance above which the gas cannot be liquefied no matter how much pressure is applied. This temperature is called the **critical temperature.** The pressure that must be applied to bring about liquefaction when a gas is at its critical temperature is called the **critical pressure.** The nonpolar molecules of such gases as hydrogen, oxygen, and nitrogen have relatively small attraction for one another. The kinetic energy of molecules of these gases must be decreased a great deal before the slight attractive forces (made more effective by the application of pressure) can hold them in the liquid form. Their critical temperatures are quite low (Table 10-1). On the other hand, polar molecules of such gases as ammonia and sulfur dioxide have

---

[1] In addition to the attractive forces between molecules, the molecular weight is another important factor in determining whether or not a substance is a gas, a liquid, or a solid at room temperature. In the case of two molecules which are similar chemically, the heavier molecule will have a higher boiling and melting point. (See Fig. 26-7.)

relatively great attraction for one another, great enough to hold them in the liquid state at temperatures well above room temperature if sufficient pressure is applied.

TABLE 10-1.

Critical Temperatures and Pressures

GAS	BOILING POINT, °C	CRITICAL TEMPERATURE, °C	CRITICAL PRESSURE, ATM
Hydrogen	−252.8	−239.9	12.8
Nitrogen	−195.8	−147.1	33.5
Oxygen	−183.0	−118.8	49.7
Methane	−161.5	− 82.5	45.8
Ammonia	− 33.3	132.4	111.5
Sulfur dioxide	− 10.0	157.2	77.7
Water	100.0	374.0	217.7

A gas cannot be liquefied at a temperature above its critical temperature; it can be liquefied at a temperature below its critical temperature; the lower the temperature, the less the pressure required to bring about liquefaction.

EXERCISES

1. Explain how Avogadro was able to account for Gay-Lussac's law of combining volumes.
2. Hydrogen was introduced into a 1000-ml steel cylinder at 0°C to a pressure of 1.0 atm. Oxygen was then added until the total pressure as registered by a gauge was 2.0 atm. This mixture was ignited; next, the cylinder and its contents were cooled to 0°C. The pressure gauge then registered 0.51 atm. Explain why.
3. What volume of air (21% oxygen) is required for the complete combustion of: (a) 250 ml of H_2; (b) 300 ml of CO; (c) a mixture of 600 ml of H_2 and 600 ml of CO? Assume that all volumes are measured at the same pressure and temperature.
4. One liter of ammonia gas was decomposed completely, producing 2.0 liters of a hydrogen–nitrogen mixture. Oxygen was then added until the volume was 3.2 liters. After ignition of the mixture, followed by cooling to condense the water vapor, 0.95 liters of gas remained. Show that these data are in accord with the accepted formula of ammonia.
5. Calculate the densities of the following gases in g/liter at STP: (a) argon; (b) helium; (c) sulfur dioxide; (d) uranium hexafluoride, UF_6.
6. (a) What volume of hydrogen at STP can be obtained by the action of excess sulfuric acid with 2.00 g of zinc?
   (b) What would be the volume of this hydrogen if collected over water at 17°C and 742 mm?
7. A sample of a new gaseous compound was found to have a density of 0.892 g/liter. What is the weight of 1 mole of this gas? Of 1 molecule?

8. The volume of a 1.634-g sample of an organic compound upon vaporization at 100°C and 740 mm was found to be 540 ml. What is the molecular weight of the compound?
9. A 12.5-liter vessel contains 4.0 g of CH_4, 1.8 g of N_2, and 10.0 g of Xe. What is the pressure in the vessel at 0°C?
10. Very careful measurements on the combining volumes of H_2 and O_2 to produce water gave the combining ratio of 200.288 : 100.000. Suggest a reason why this ratio is not exactly 2 : 1.
11. The critical temperatures of ammonia and methane are 132.4°C and −82.5°C, respectively.
    (a) Which gas can be liquified at room temperature by the application of pressure?
    (b) Which will contain more molecules, 10 liters of ammonia or 10 liters of methane at STP? Explain.
    (c) Which has the higher boiling point, ammonia or methane?
12. Compare the rates of diffusion of the following pairs of gases: (a) hydrogen, H_2, and deuterium, D_2; (b) methane, CH_4, and helium, He; (c) nitrogen, N_2, and carbon monoxide, CO; (d) sulfur hexafluoride, SF_6, and sulfur dioxide, SO_2.
13. The time required for a volume of Y to effuse through a small hole was 61 sec. The time for an equal volume of nitrogen was 32 sec. Calculate the molecular weight of Y.
14. Calculate the weight in grams of a SO_2 molecule.
15. Calculate the number of molecules in 1 liter of oxygen at 15°C and 740 mm. What is the number of oxygen molecules in 1 liter of air at these same conditions?
16. A 1.500-g sample of a mixture of $KClO_3$ and NaCl was heated to expel all the oxygen. The oxygen, collected over water, had a volume of 220 ml at 17°C and 732 mm. Calculate the percentage $KClO_3$ in the mixture.
17. The temperature of a certain volume of dry air is 0°C. What is the temperature of this air when heated until the pressure is five times the original pressure? When cooled to a point where the pressure is one-third of the original pressure?
18. Four containers of equal volume are filled as follows:
    No. 1: 2.0 g of H_2 at 0°C
    No. 2: 1.0 g of H_2 at 273°C
    No. 3: 24.0 g of O_2 at 0°C
    No. 4: 16.0 g of CH_4 at 273°C
    Which container (a) is at the greatest pressure; (b) contains the fewest number of molecules; (c) is at the lowest pressure; (d) has the greatest gas density?
19. Dinitrogen tetroxide, N_2O_4, dissociates into nitrogen dioxide as the temperature is increased. Ten grams of N_2O_4 is placed in a 1-liter vessel at 0°C, and the vessel is heated to 50°C. The resulting pressure is observed to be 3.76 atm. What percentage of the original N_2O_4 is dissociated?
20. A quantity of helium has a volume of 200 ml at 700 mm. What volume will this helium occupy at 500 mm? At 5 atm? At 5 psi?
21. Would you predict that a volume of sulfur dioxide contains a small percent more molecules than an equal volume of neon at the same temperature and pressure? Explain.
22. The critical temperatures of hydrogen and methane are −239.9°C and −82.5°C, respectively. Can any conclusions be drawn from these figures about the relative magnitudes of the van der Waals forces of hydrogen and methane?
23. Until the earth cooled to a temperature below 374°C, there was no liquid water on it. Why?

24. Assuming that 1100 ml of carbon dioxide is absorbed by each square meter of leaf surface during 1 hour of sunlight, calculate the weight of CO_2 that will be assimilated by a grove of orange trees consisting of 2000 trees, each possessing 5000 leaves of average area of 15 cm² per leaf, during a day of 12 hours of sunlight.

SUGGESTED READING

Carpenter, D. K., "Kinetic Theory, Temperature, and Equilibrium," *J. Chem. Educ.*, **43**, 332 (1966).

Evans, R. B., L. D. Love, and E. A. Mason, "Graham's Laws: Simple Demonstrations of Gases in Motion," *J. Chem. Educ.*, **46**, 423 (1969).

Feifer, N., "The Relationship between Avogadro's Principle and the Law of Gay-Lussac," *J. Chem. Educ.*, **43**, 411 (1966).

Kirk, A. D., "The Range of Validity of Graham's Law," *J. Chem. Educ.*, **44**, 745 (1967).

11

LIQUIDS AND SOLIDS

The early observations of the properties of gases led to the formulation of the kinetic-molecular theory of gases, which was discussed in Chapters 9 and 10. The concepts of this theory can be applied to the study of liquids and solids. In beginning our study of liquids and solids, we will first describe some of the properties of liquids and then discuss for each property how the kinetic-molecular theory accounts for the observed behavior. Later we will discuss the solid state in a similar manner.

LIQUID STATE OF MATTER

BEHAVIOR OF LIQUIDS

Shape and Volume. FACT: A liquid has no specific shape; it takes the shape of its container as it seeks its own level under the influence of gravity. However, a given quantity of a liquid does have a definite volume.

THEORY: A liquid is composed of molecules that roll over each other so easily that the liquid has little or no rigidity. But the molecules are in contact with one another, like marbles in a bag; the total volume of the collected molecules is definite.

Compressibility. FACT: Liquids are compressed only under extremely high pressures. The fact that liquids are almost incompressible leads to their use in hydraulic brake systems. Pressure exerted on the brake fluid causes the brake to move, since the *volume* of the fluid is not reduced by the pressure.

Liquids and Solids

THEORY: The particles in a liquid are not widely separated; rather, they touch each other. The liquid cannot be compressed much by crowding the particles together, but only by squeezing and deforming the molecules themselves. This requires tremendous pressure.

Change of Volume with Temperature. FACT: A liquid usually expands when heated and contracts when cooled (as does mercury or alcohol in thermometers).

THEORY: The average kinetic energy of molecules depends upon the temperature. The higher the temperature, the greater the vigor of movement and the greater the volume taken up by each molecule as it moves around colliding with its neighbors.

Diffusion. FACT: Colored solutions diffuse through water. Diffusion is much slower in a liquid than in a gas, but it takes place at an easily measured rate. One liquid may diffuse through another, or a solid may dissolve and diffuse through a liquid.

THEORY: The molecules do not move about as freely as gas molecules, but they move *at all times* and slide over one another without much difficulty. They are attracted to each other, but not rigidly held together.

Evaporation and Solidification. FACT: When the temperature of a liquid is raised, it evaporates more quickly. If its temperature is lowered sufficiently, the liquid will freeze to a solid.

THEORY: The energy of the moving particles in a liquid is increased by a rise in temperature. When the molecules have enough kinetic energy, they break away from their neighbors and become gas molecules. A fall in temperature means a decrease in the energy of the moving molecules. If the kinetic energy of the molecules falls enough, even weak van der Waals forces will be able to hold the particles together in the fixed positions of a solid.

The magnitude of the attractive forces between molecules in liquids varies from weak van der Waals forces to stronger dipole attractions. The molecules in a polar liquid may have relatively strong attraction for each other. The dipoles arrange themselves end to end, positive to negative. A considerable amount of energy may be needed to separate them.

EVAPORATION AND VAPOR PRESSURE

Does a liquid vaporize only at its boiling point? Of course not. Water in a glass evaporates slowly at room temperature. Gasoline evaporates quite readily without boiling. A liquid that evaporates readily is said to be **volatile**. Ether is a very volatile liquid, whereas lubricating oil is only slightly volatile.

A liquid will not evaporate if placed in a tightly covered container. The molecules that escape from the surface of the liquid are trapped in the container. As they move about above the liquid, colliding with each other and with the walls of the container, some may hit the surface of the liquid and rejoin it. The right side of Fig. 11-1 represents a system in which the liquid is evaporating and the vapor is condensing at the same rate. When the rate of evaporation equals the rate of condensation, the system is in equilibrium. The **vapor pressure** of a substance is defined as the pressure exerted by the gas of that substance when it is in equilibrium with the liquid or solid phase. The vapor pressure of a liquid (or solid) increases as the temperature increases. In Fig. 11-2 is shown one method of determining the vapor pressure of a liquid.

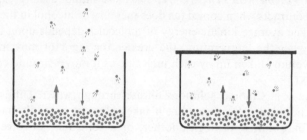

Fig. 11-1. In an open container a liquid will evaporate away, but in a closed container the rate of condensation becomes equal to the rate of evaporation, and the liquid-vapor system is in equilibrium.

The evaporation of liquids and the phenomenon of vapor pressure can be explained in terms of the kinetic-molecular theory. In our discussion of the structure of gases, we said that a molecule has an *average* kinetic energy. This term is used because we think that a given molecule does not have a constant velocity; rather it may move relatively slowly or rapidly as it collides with other molecules. At any instant in a collection of molecules there will be some with low energies, some with high energies, and a large number with nearly average

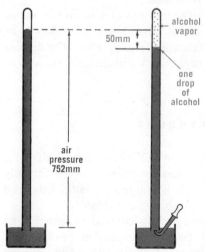

Fig. 11-2. The determination of the vapor pressure of a liquid. As shown in the diagram, when a drop of a liquid is introduced into a barometer it rises to the top of the mercury column and evaporates. The space above the mercury quickly becomes saturated, i.e., an equilibrium is established between the liquid and vapor phases of the sample. The pressure exerted by the vapor forces the mercury column to fall. The fall in the height of the column is a direct measure of the vapor pressure.

energies. Evaporation occurs when high-energy molecules at the surface of the liquid break away from their neighbors and escape into the gas phase. Condensation occurs when molecules in the gas phase strike a surface and are attracted strongly enough to be held.

Liquids and Solids

A volatile liquid is one in which there is little attraction between molecules. Since the attraction of the molecules for each other is slight, a large fraction of them have sufficient energy to escape from the liquid, and therefore evaporation can occur rapidly at room temperature. Liquids that are composed of polar molecules (which strongly attract each other) are less volatile; a smaller fraction of the total molecules have enough energy to overcome their mutual strong attraction.

Another observation can also be explained by this concept. Consider the cooling effect of rubbing alcohol on the skin. The more energetic molecules of the alcohol escape quickly and become gaseous. Those remaining as a liquid are the less energetic particles. As evaporation occurs, the average kinetic energy of the liquid molecules that remain decreases. That is, the temperature of the liquid decreases and the skin is thereby cooled.

When a liquid is heated, its vapor pressure increases and evaporation takes place more rapidly. This is accounted for on the basis that an increase in the temperature increases the average energy of the particles, and thus more particles have the energy necessary for evaporation. The resulting greater number of molecules in the gaseous state exert a greater pressure (the vapor pressure). Figure 11-3 shows the vapor pressures of four common liquids over a range of temperatures.

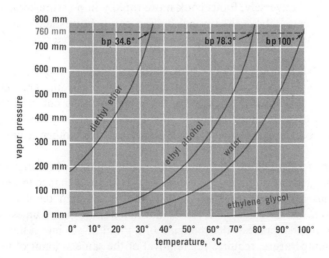

Fig. 11-3. A plot showing the effect of temperature on the vapor pressures of four common liquids. The temperature at which the vapor pressure is 760 mm is the normal boiling point of the liquid.

Boiling. The **boiling point** of a liquid is the temperature at which the pressure of the vapor escaping from the liquid equals the outside pressure. When the vapor pressure equals the outside (or applied) pressure, bubbles are able to form throughout the liquid, and it boils. When the temperature of a liquid is below the boiling point, bubbles do not form within it because the pressure exerted by the gaseous substance is not great enough to maintain the bubbles. The **normal boiling point** of a liquid is defined as the temperature at which its vapor pressure is 760 mm, i.e., standard atmospheric pressure.

The boiling point of water (and other liquids) varies with the atmospheric pressure. In mountainous regions the boiling point of water may be considerably below 100°C, because the atmospheric pressure is below 760 mm (Table 11-1).

TABLE 11-1.

Change in Boiling Point of Water with Altitude

	ALTITUDE	AVERAGE ATMOSPHERIC PRESSURE	BOILING POINT OF WATER
Sea level	0	760 mm	100°C
Mt. Mitchell	6,684 ft	589 mm	93°C
Mt. Whitney	14,495 ft	451 mm	86°C
Mt. Everest	29,141 ft	244 mm	71°C

At high altitudes housewives must boil foods longer because of the low boiling temperatures. Conversely, foods cook more rapidly in pressure cookers because they can be heated above the normal boiling point. For example, a pressure of 10 lb over the standard atmospheric pressure causes water to boil at about 115°C.

Heat of Vaporization. When 1 g of water at 100°C is vaporized, 540 cal of heat must be added to change this water to 1 g of steam. The temperature of the steam formed is 100°; that is, the temperature is not increased at all. This 540 cal of heat energy is used in counteracting the attractive forces so that the molecules of water can break away from each other as gas molecules (steam).

The **heat of vaporization** of a substance is the number of calories required to convert one gram of liquid to one gram of gas without a change in temperature. A liquid has a different heat of vaporization at different temperatures. It is usually given, as in Table 11-2, at its normal boiling point unless otherwise stated. The reverse process, the changing of a gram of gas into a liquid without change in temperature, requires the removal of the same amount of heat energy (*heat of condensation*).

The heat needed to vaporize one mole of a substance is called the molar heat of vaporization or the **molar enthalpy of vaporization,** ΔH_{vap}. The enthalpy of vaporization of water, H_2O, at its boiling point, as calculated from Table 11-2, is $(18.015)(539.6)/1000 = 9.721$ kcal/mole.

Problem 1: Calculate the amount of heat necessary to heat one pound (454 g) of water from 27.0°C to 100.0°C and vaporize the water to steam at 100°C.

Solution: The amount of heat necessary to carry out the above is the sum of the amount necessary to raise the temperature of the water and the amount to vaporize the water. The temperature of the water is changed from 27.0°C to

100.0°C, a total of 73.0°C. Since the specific heat of water is 1.00 cal/(g × deg C), the heat needed to raise the temperature of 454 g is given by

[1.00 cal/(g × deg C)](73.0 deg C)(454 g) = 33,100 cal

The heat of vaporization of water is 540 cal/g (Table 11-2). The heat necessary to vaporize 454 g is then

(540 cal/g)(454 g) = 245,000 cal

The total heat energy for the process described is

33,100 + 245,000 cal = 278,100 cal or 278,000 cal

TABLE 11-2.

Heats of Vaporization and Fusion

SUBSTANCE	HEAT OF FUSION, CAL/G	HEAT OF VAPORIZATION, CAL/G
Ammonia	108.1	327.1
Benzene	30.3	94.3
Carbon dioxide	45.3	87.2[a]
Hydrogen chloride	14.0	98.7
Mercury	2.8	67.8
Oxygen	3.3	50.9
Water	79.7	539.6

[a] From solid carbon dioxide at −60°C.

SOLID STATE OF MATTER

BEHAVIOR OF SOLIDS

Compressibility. FACT: It is even more difficult to compress solids than liquids.
THEORY: The particles that make up solids are in contact with one another. Any significant compression must result in the deforming of the individual particles.

Diffusion. FACT: It is not widely known that diffusion can take place even in a solid body, but this process can be demonstrated in a relatively simple way. If a polished flat piece of pure zinc and a polished flat piece of pure copper are clamped together for a considerable time, say a year, some zinc penetrates the copper, and vice versa (Fig. 11-4).

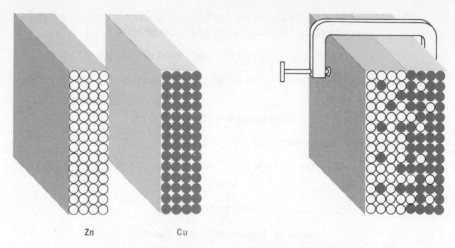

Fig. 11–4. Diffusion of atoms and ions can take place even in solids. The diffusion of zinc and copper atoms is shown schematically here.

THEORY: Movement of the particles in a solid is very limited, but some movement is possible.

Melting. FACT: When a solid is heated, it usually melts to a liquid.

THEORY: A rise in temperature brings about an increase in the kinetic energy of the particles. If the molecular motion increases sufficiently, the attractive forces are largely overcome, so that the particles are no longer held in fixed positions. They slip and slide over one another and display the usual characteristics of a liquid.

Sublimation. Some solids, such as naphthalene (moth balls) and snow or ice below 0°C, may pass directly from the solid state to the vapor state. The reverse process, deposition directly from the vapor state to the solid state, is also quite common. *Sublimation* is the process of passing from a solid state to the gaseous state without going through the liquid state, or the passing from the gaseous state directly to the solid state.[1] An increase in temperature increases the rate of sublimation.

INDIVIDUAL PARTICLES IN SOLIDS

Pure substances that exist as solids at room temperature may consist of ions, molecules, or atoms, as shown in Fig. 11-5. Most elements that are solids at room temperature are considered to consist of atoms, although these atoms may be different from isolated single atoms because of their chemical union with one another (see metallic bond, page 168). For example, the symbols for iron, tin, zinc, copper, and gold are written simply as Fe, Sn, Zn, Cu, and Au, respectively. The symbol without any subscript refers to a single atom.

---

[1] Although the same word has been used for years for both processes, it has been suggested that two terms should be used: *sublimation* for solid → gas and *deposition* for gas → solid.

All strictly ionic compounds are solids at room temperature. This is due to the fact that the ions arrange themselves in certain three-dimensional patterns or crystal forms, in which positive and negative ions alternate. No definite molecules are formed, but the forces of attraction between the unlike charges are so great that the ions are held rather rigidly in place. They probably vibrate, but movement from place to place is difficult. The ionic crystal extends out indefinitely till it is broken or stops growing (it can be considered to be a single huge molecule of indefinite size).

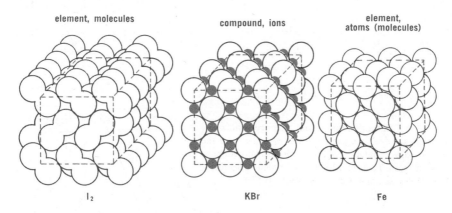

Fig. 11-5. A solid may be considered to be a collection of molecules, or ions, or atoms depending upon the type of attraction that holds the particles in the solid together.

TYPES OF SOLIDS

It has long been recognized that there are two classes of solids. Most solids melt at definite temperatures. A few solids such as tar, glass, gums, and paraffin do not have definite melting points. Instead, as the temperature of such a solid is raised, there is a gradual transition from solid to liquid. Within a certain range of temperature, it may be difficult to classify the substance as a solid or a liquid.

In 1912 Max von Laue discovered that the arrangement of particles in a solid could be determined by means of X rays. In most solids there is a very orderly arrangement of the individual particles that compose them. The atoms, ions, or molecules in the solid are arranged in *layers* or *planes* that are capable of reflecting X rays. Special cameras enable the angles of reflection to be determined, and the distances between the planes to be calculated. A solid whose individual particles are thus arrayed is classified as a **crystalline solid;** such a solid also has a well-defined melting point.

X-ray analysis of an **amorphous solid** shows a quite different picture. There is no orderly array of the individual particles in these solids. Instead, their arrangement is similar to that in liquids—a random array of particles. As a matter of fact, these solids appear to differ from normal liquids only in that the particles of the solids have much less freedom of motion. An amorphous solid has no sharp melting point.

Crystal Arrangements. The individual particles in most solids are arranged in beautifully symmetrical patterns called crystal structures. There are six fundamentally different crystal arrangements. These are the *isometric* (for example, cubic), *tetragonal, monoclinic, triclinic, orthorhombic,* and *hexagonal.* They are sketched in Fig. 11-6.

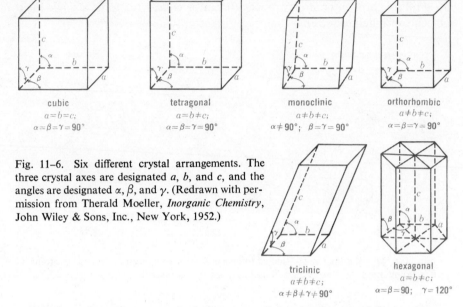

cubic
$a=b=c;$
$\alpha=\beta=\gamma=90°$

tetragonal
$a=b\neq c;$
$\alpha=\beta=\gamma=90°$

monoclinic
$a\neq b\neq c;$
$\alpha\neq 90°;\ \beta=\gamma=90°$

orthorhombic
$a\neq b\neq c;$
$\alpha=\beta=\gamma=90°$

triclinic
$a\neq b\neq c;$
$\alpha\neq\beta\neq\gamma\neq 90°$

hexagonal
$a=b\neq c;$
$\alpha=\beta=90;\ \gamma=120°$

Fig. 11-6. Six different crystal arrangements. The three crystal axes are designated *a, b,* and *c,* and the angles are designated α, β, and γ. (Redrawn with permission from Therald Moeller, *Inorganic Chemistry*, John Wiley & Sons, Inc., New York, 1952.)

The *isometric* arrangement includes some of the simple crystal forms characteristic of many common and important substances. Among these forms are the *cube*, the *tetrahedron*, and the *octahedron*. Figure 11-7 shows the relation of these three figures to one another. A cube could be called a *hexa*hedron, that is, *six* sides; a *tetra*hedron has *four* sides (it is the simplest solid figure); an *octa*hedron has *eight* sides.

cube   tetrahedron   octahedron

Fig. 11-7. Three of the most common crystal forms of solid substances.

Isomorphism. From the Greek word *morphē*, meaning form, is derived the word ending *-morphous*. The word amorphous means without form. Two substances that have the same crystal form and crystallize together in all proportions are said to be **isomorphous.** The formulas of a pair of such substances usually reveal that their atoms are in the same ratios. Examples are:

K_2SO_4 and K_2SeO_4 2:1:4
$NaNO_3$ and $CaCO_3$ 1:1:3
NaF and MgO   1:1
Cr_2O_3 and Fe_2O_3 2:3

Isomorphous substances crystallize together in homogeneous mixtures. However, similarity of both formula and chemical properties is not enough to insure homogeneous crystallization. A famous pair of similar substances that does not crystallize homogeneously is NaCl and KCl.

A single substance that crystallizes in two or more different forms, under different conditions, is said to be **polymorphous** (many forms). Calcium carbonate, $CaCO_3$, silicon dioxide, SiO_2, sulfur, S, and carbon, C, are examples of polymorphous substances. In the case of elements, different crystalline forms have long been referred to as *allotropic* forms; for example, graphite and diamond are allotropic forms of carbon.

Appearance of Crystals. We must not jump to conclusions about the arrangement of the particles inside a large crystal from a consideration of its outer appearance. Because of limited space or for some other reason, a crystal may

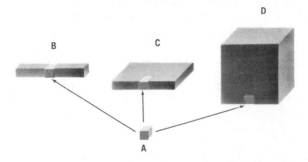

Fig. 11–8. Three habits that may be assumed by a substance that crystallizes in a cubic system. (Based on A. F. Wells, *Structural Inorganic Chemistry*, The Clarendon Press, Oxford, 1962.)

form by growing more in one direction than another. Figure 11-8 shows how a small cube may develop into one of three other possible shapes—a large cube, a flat plate, or a long needle-like structure. All three of these solids have the same crystal structure but their *habits*, or characteristic forms, differ.

BONDING IN SOLIDS

As has been indicated earlier, the relatively strong attractive forces in solids restrict the movement of particles within the crystal and limit their motion almost entirely to vibrations. The nature of the forces operating in each of the three classes—covalent, ionic, and atomic solids—will now be discussed.

Covalent Solids. The individual particles in covalent solids such as camphor, iodine, ice, and solid toluene are molecules. Consequently the binding forces, which hold the molecules in the crystal, must arise from attractions between molecules. The nature and strength of these forces vary widely. For example, camphor and toluene are composed of molecules of approximately the same size, but the latter is a liquid at room temperature. In toluene the van der Waals forces are weak. The much stronger dipolar attraction of the molecules in camphor cause it to be a solid at room temperature.

Ionic Solids. In solids composed of ions, the attractive forces are electrostatic; that is, they are due to the mutual attraction of negative and positive charges. The ions are symmetrically arranged in the solid so that a positive ion has negative ions as its nearest neighbors, and vice versa. Thus each ion is held in place by attraction from all the unlike charged particles that surround it. Since electrostatic forces are strong, an ion must have considerable energy before it can overcome the attractive forces of its neighbors and leave its position in the crystal lattice. For this reason an ionic substance is normally a high-melting solid with a very low vapor pressure at room temperature.

Atomic Solids. A solid metal such as iron, gold, silver, or potassium consists entirely of atoms of the same type. But in order to account for the lack of mobility of the atoms within the crystal, a strong attractive force must operate between the atoms. This force is called the **metallic bond.**

In discussing the metallic bond, we shall use sodium as an example. The crystal lattice of sodium is a type known as *body-centered cubic* (Fig. 11-9). Each

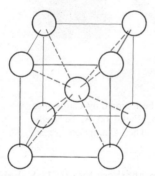

Fig. 11–9. The body-centered cubic lattice.

sodium atom (except those on the surface) is surrounded by eight atoms—its nearest neighbors. The attraction of an atom for its nearest neighbors cannot be described as an ionic or as a simple covalent type of bond. Since no ions or molecules are present, the attractive forces are neither ion-ion nor dipole-dipole. In sodium, as in most metals, the attraction between atoms is much greater than can be accounted for by van der Waals forces. Sodium has a single electron in its outermost shell. This electron is incapable of forming a typical covalent bond for two reasons. (1) The sharing of a pair of electrons between two sodium atoms would not result in a stable octet for either of the atoms. (2) Although such a bond might conceivably bring about a strong attraction between the two sodium atoms involved, it could not account for the experimental observation that each atom is equally bound to *all eight* of its nearest neighbors. It is believed that instead of a single electron from each atom being localized between two atoms to form a bond, the bond *resonates* between an atom and all of its nearest neighbors. In trying to visualize this, imagine that as the outermost electron travels about the sodium nucleus, it spends approximately one-eighth of its time between this one atom and each of the eight neighboring atoms. This idea of the metallic bond emphasizes two important concepts. (1) The electrons in metals are much more mobile than they are in a typical covalent bond, where they are fixed between two atoms. (2) The bond holding two given atoms together is considerably weaker than it would be in a truly covalent bond. Since an electron of a given atom is with any other atom for only one-eighth of the time, we might conclude as a first approximation that the bond uniting these atoms is only one-eighth as strong as it would be if it were typically covalent.

Nonmetallic elements such as carbon (see Chap. 23) and boron also form crystals in which the atom is the unit particle. In these elements, however, the bonds are typical 2-electron covalent bonds.

Metallic and many nonmetallic elemental crystal structures are similar to ionic structures in one important respect: the crystal is a single, very large unit. Any single crystal of these solids, regardless of size, can be considered to be a molecule.

FUSION OF SOLIDS

Melting (Fusion). As the temperature of a crystalline solid rises, a temperature is reached at which it suddenly melts. If the liquid thus formed is cooled, we find that it freezes or solidifies at precisely the same temperature as that at which the solid melted (unless supercooling occurs). If solid and liquid phases of the same substance are in contact and the temperature is precisely at the melting point, the solid melts continually and the liquid freezes continually. When the rate of melting equals the rate of freezing, the system is said to be in equilibrium. The **melting point** of a substance is the temperature at which its solid and liquid phases are in equilibrium; if no heat is added or taken away, the weight of neither the solid nor the liquid changes.

Heat of Fusion. If 1 g of ice at $-10°C$ is carefully heated, we find that 0.478 cal is required for each $1°$ rise. Thus it takes $(10)(0.478) = 4.78$ cal to bring the temperature to $0°C$. As the gram of ice begins to melt, it is found that heat can be added without changing the temperature. In fact, 79.7 cal is required to change the gram of ice at $0°C$ to liquid water at $0°C$. This heat energy is used in counteracting the attractive forces between molecules.

The number of calories needed to change one gram of a solid substance to 1 g of liquid (at the melting point) is called the **heat of fusion.** The heats of fusion for a number of substances are given in Table 11-2.

The same amount of heat is given up when a gram of a substance solidifies (*heat of solidification*) as is taken up when a gram of solid melts. For example, 79.7 cal is liberated to the surroundings when a gram of water freezes.

The heat required to melt one mole of a substance is called the molar heat of fusion or the **molar enthalpy of fusion,** ΔH_{fus}. The molar enthalpy of fusion of ammonia, NH_3, at its melting point, as calculated from Table 11-2, is $(17.0)(108.1)/1000 = 1.84$ kcal/mole.

Problem 2: Compare the amount of heat necessary to change one molar weight of solid mercury at its melting point, $-39.0°C$, to vapor at its boiling point, $357°C$, and the amount of heat necessary to change one molar weight of ice at $0°C$ to steam at $100°C$. The specific heat of liquid mercury is 0.033 cal/(g × deg C). See Table 11-2 for additional data.

Solution:
For mercury:

$$(2.8 \text{ cal/g})(201 \text{ g}) = 563 \text{ cal to melt mercury}$$

$$[0.033 \text{ cal/(g} \times \text{deg C)}](201 \text{ g})(396 \text{ deg C})$$
$$= 2630 \text{ cal to heat from } -39 \text{ to } 357°C$$

$$(67.8 \text{ cal/g})(201 \text{ g}) = 13{,}600 \text{ cal to vaporize mercury}$$

Total heat required for mercury $= 563 + 2630 + 13{,}600$ cal
$= 16{,}793$ or $16{,}800$ cal

For water:

$$(79.7 \text{ cal/g})(18.0 \text{ g}) = 1435 \text{ cal to melt water}$$

$$[1.00 \text{ cal/(g} \times \text{deg C)}](18.0 \text{ g})(100 \text{ deg C})$$
$$= 1800 \text{ cal to heat water from 0 to } 100°C$$

$$(540 \text{ cal/g})(18.0 \text{ g}) = 9720 \text{ cal to vaporize water}$$

Total heat required for water
$= 1435 + 1800 + 9720$ cal
$= 12{,}955$ cal or $13{,}000$ cal

EXERCISES

1. In terms of the atoms, ions, and molecules that comprise substances, why are some materials solids, some liquids, and still others gases at $25°C$?
2. The process of diffusion occurs most rapidly in gases, less rapidly in liquids, and very slowly in solids. Why?
3. Why does water spilled on a floor evaporate more rapidly than the same amount of water in a glass?
4. Why does one feel cooler on emerging from a swimming pool into windy air than into still air at the same temperature?
5. Ethanol, C_2H_5OH, has a density of 0.8063 g/cm² at $0°C$ and a density of 0.7810 g/cm² at $30°C$. Account for this variation in terms of the kinetic-molecular theory of liquids.
6. The average atmospheric pressure on top of Pike's Peak is 407 mm. What is the boiling point of water on Pike's Peak? Would a pressure cooker be of any value on this mountain for cooking food? Explain.

7. Under what conditions will (a) water boil at 30°C; (b) water be a liquid at 110°C; (c) water be a liquid at −5°C?
8. Suggest a method for determining the vapor pressure of a solid such as naphthalene (moth balls).
9. Suggest a method for determining the heat of sublimation of a substance. What data would you collect? What calculations would be necessary?
10. List 4 properties that are generally characteristic of metals but not of nonmetals.
11. How does the metallic bond differ from the covalent bond? How are they similar?
12. Distinguish between crystal structure and crystal habit. Can a crystal built from small tetrahedra have the appearance of a cube or octahedron? Explain.
13. Calculate the amount of heat necessary to change 10.0 g of solid ammonia at −78°C to gaseous ammonia at −33°C. The melting point, boiling point, and specific heat of liquid ammonia are −78°C, −33°C, and 0.590 cal/(g × deg C), respectively.
14. We may say that sodium nitrate and calcium carbonate are isomorphous but that only calcium carbonate is polymorphous. Explain.
15. How is the formation of a dewdrop similar to the formation of a drop of rain? How is the formation of frost similar to the formation of a snowflake? How are sleet and hail formed?
16. What will be the final temperature of the water that results when 5.0 g of steam at 100°C is bubbled into 500 g of water initially at 21°C?
17. The compound difluorodichloromethane, CCl_2F_2, is commonly used as refrigerant in household refrigerators. How much liquid CCl_2F_2 at its boiling point of −30°C must be vaporized in order to cool and freeze 25 lb of water initially at 60°F? The heat of vaporization of CCl_2F_2 is 39.5 cal/g.
18. Heat is supplied uniformly at the rate of 1000 cal/min to a 100-g sample of solid ammonia, initially at −78°C, until the ammonia is vaporized at −33°C. How much time is required for the process? Plot the change in temperature versus time and compare with Fig. 12-1. See Exercise 13 for necessary data.
19. A 6.50-g sample of solid mercury at −50°C is dropped into 20.0 g of liquid mercury at 15°C. The temperature of the final mixture was found to be −21°C. The specific heats of solid and liquid mercury are 0.0336 and 0.0333 cal/(g × deg C), respectively. The melting point of mercury is −39°C. From these data calculate the heat of fusion of mercury.
20. Calculate the molar enthalpy of vaporization, ΔH_{vap}, of (a) oxygen; (b) benzene; (c) hydrogen chloride.
21. No liquid ionic compounds are known, but many of the known covalent compounds are liquids and some are gases. Account for these differences.

SUGGESTED READING

Bernal, J. D., "The Structure of Liquids," *Sci. Amer.*, **203** (2), 124 (1960).

Boer, F. P., and T. H. Jordan, "X-ray Crystallography Experiment," *J. Chem. Educ.*, **42**, 76 (1965).

Gehman, W. G., "Standard Ionic Crystal Structures," *J. Chem. Educ.*, **40**, 54 (1963).

Mason, B. J., "The Growth of Snow Crystals," *Sci. Amer.*, **204** (1), 120 (1961).

McConnell, D., and F. H. Verhoek, "Crystals, Minerals, and Chemistry," *J. Chem. Educ.*, **40**, 512 (1963).

Sime, R. J., "Some Models of Close Packing," *J. Chem. Educ.*, **40**, 61 (1963).

12

WATER; SOLUTIONS

Perhaps no chemical substance is so well known to us as water. It is a major component of all living organisms, constituting about 65 percent of the human body and up to 95 percent of some plants. Variable amounts of water vapor are present in the air. Rivers, lakes, and oceans cover about three-fourths of the earth's surface. Even most of the substances that are ordinarily considered "dry" contain appreciable moisture.

Because of its commonness we tend to overlook the important role water plays in our world. This importance is due both to its abundance and to its unique properties. Many, many compounds are known that are liquids, but not one of them, even though it were as plentiful as water, would serve as a substitute for water in our world.

PROPERTIES OF WATER

Physical Properties. The color, odor, taste, and physical state of water are obvious and familiar properties. Quantitative values for some of its other physical properties are listed in Table 12-1.

Water is unusual in that the amount of heat required to change its state or raise its temperature is quite large when compared with the amount of heat required to bring about the corresponding changes in other substances (see Table 11-2). Because of differences in their specific heats, the amount of heat necessary to raise the temperature of 1 lb of water only 10° is sufficient to raise the temperature of 1 lb of iron 93° or 1 lb of sand 53°. The amount of heat required to boil away 1 lb of water is enough to boil away nearly 6 lb of benzene or 8 lb of mercury. Figure 12-1 shows how the temperature of a body of water changes with time as heat is added at a constant rate.

Another interesting and unusual property of water is the way in which the volume of a given quantity of it changes as the temperature changes. At room temperature, water acts like most other liquids—it expands as the temperature is

TABLE 12-1.

Physical Properties of Water

Freezing point	0.0°C
Boiling point at 760 mm	100.0°C
Density at 4°C	1.0 g/ml
Heat of fusion	79.7 cal/g
Heat of vaporization	539.6 cal/g
Specific heat	1.0 cal/(g × deg C)

raised and contracts as the temperature is lowered. However, at temperatures from 4°C downward, water acts in an abnormal fashion—it expands rather than contracts as the temperature is lowered. Hence, a given weight of water has a minimum volume at 4°C, as is shown in Fig. 12-2. The density of water is at a maximum at 4°C. Practically all other liquids have maximum densities at their freezing temperatures.

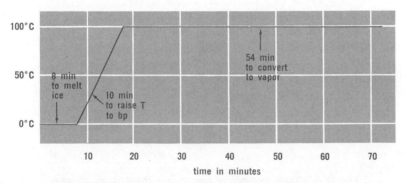

Fig. 12–1. An idealistic representation of temperature changes versus time as 1000 cal per min is uniformly added to 100 g of water. In actual practice it would be difficult to carry out the heating as a separate 3-step process.

When water freezes to ice, a further expansion occurs, to the extent of about 9 percent of the original volume. Ice, therefore, has a considerably lower density than liquid water at the same temperature. Because water does expand when cooled below 4°C and expands further on freezing, ice floats rather than sinks; hence lakes and rivers freeze from the surface downward. The layer of ice insulates the water underneath, so that the transfer of additional heat energy to

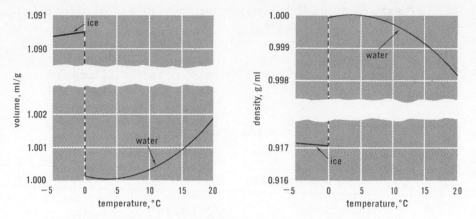

Fig. 12-2. For water, the specific volume (left) is at a minimum and the density (right) is at a maximum at 4°C.

the air (79.7 cal for each gram of ice formed) is not easily made. For this reason a lake may not freeze solid, even though the temperature of the air remains below freezing for several weeks.

The explanation of the so-called unusual properties of water centers around the structure of its molecules. As we see in Fig. 12-3, a water molecule is polar. In fact, water molecules have a greater polar character than most other known molecules. Because of this polar nature, water molecules tend to *associate* or

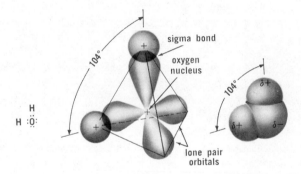

Fig. 12-3. Three methods of representing the structure of water molecules. The oxygen nucleus is thought to be at the center of a tetrahedron with the hydrogen nuclei at two of the corners. This structure gives a high electron density on one side of the molecule and a low density on the other, resulting in a polar molecule.

unite with one another to form a loosely bound lattice of molecules that has a pseudocrystalline form, even in the liquid state. This association or arrangement has been determined experimentally by X-ray studies and molecular weight determinations.

The bond that is needed to unite two simple molecules of water results from the attraction of a positive hydrogen atom of one water molecule for some of the

Water; Solutions

electrons of an oxygen atom of another water molecule. This bond, shown below by the dash, is referred to as the **hydrogen bond.**

$$H:\ddot{O}: + H:\ddot{O}: \rightarrow H:\ddot{O}:-H:\ddot{O}:$$
$$\phantom{H:\ddot{O}:}\ \ \ \ \ \ \ \ \ \ \ H\ H$$
(with H above the second and fourth O)

Although the hydrogen bond is stronger than the van der Waals attraction between two molecules, it is not so strong as an electrovalent or a covalent bond. Since each simple water molecule contains two hydrogen atoms and two unshared pairs of electrons, this hydrogen bonding can continue until large aggregations are formed. For example, H_4O_2 can react with another molecule of H_2O to form H_6O_3, etc. The hydrogen bond is formed as follows:

$$H:\ddot{O}:-H:\ddot{O}: + H:\ddot{O}: \rightarrow H:\ddot{O}:-H:\ddot{O}:-H:\ddot{O}:$$

In ice, the water molecules are united through hydrogen bonds to form a structure in which each oxygen atom, except those in the surface of the crystal, is connected to four other oxygen atoms by hydrogen bridges. A perfect ice crystal or a snowflake may therefore be considered to be a giant molecule of the composition $(H_2O)_x$, where x is a very large number.

The water molecules in ice are not packed closely together but are held in a relatively open crystal structure by hydrogen bonds. This is responsible for the low density of ice as compared to that of water. Many of the hydrogen bonds are broken when ice is melted. As the ice structure collapses, the molecules become more closely packed and the density increases. There is a further breaking up of the associated water molecules as the water is heated, and a further diminishing of the volume, because of the closer packing of the molecules. Above 4°C, however, the expansion due to the increased motion of the molecules becomes greater than the contraction due to the breaking up of the hydrogen bonds, and the liquid expands; i.e., the molecules occupy more space. In the vapor form, the molecules have too much kinetic energy for effective hydrogen bonding to occur. The molecules in water vapor actually conform to the simple formula, H_2O.

The existence of hydrogen bonds between water molecules is responsible for the abnormally large heat of fusion, specific heat, and heat of vaporization. That is, when ice is melted, when water is heated, or when water is vaporized, hydrogen bonds are broken. The breaking of these bonds requires more energy than is required to overcome the attraction between molecules (van der Waals forces) of such substances as benzene and ether where no formal bonding between the molecules exists.

Chemical Properties. Chemically, water is an active substance. We saw in Chap. 8 that water reacts violently with the very active metals. Also, water reacts, usually not violently, with both metallic and nonmetallic oxides. We can account for the tendency for these reactions to occur in terms of the polar structure of water molecules and the relative tendencies of atoms to attract an electron pair. For the reaction with active metals, the strong attraction of oxygen for the electron pair between a hydrogen and oxygen leaves the hydrogen nucleus in a favorable condition for attracting an electron from an atom of the metal:

possible first step, $K\cdot + H:\ddot{\underset{H}{O}}: \rightarrow H\cdot + K^+ + :\ddot{\underset{H}{O}}:^-$

followed by, $H\cdot + H\cdot \rightarrow H:H$

the over-all reaction being,

$$2K + 2H_2O \rightarrow H_2 + 2KOH$$

When one hydrogen is removed from the HOH molecule, the remaining hydrogen is strongly held in the OH⁻ ion. For this reason, the reaction

$$2K + H_2O \rightarrow K_2O + H_2$$

does *not* occur at room temperature and with excess water present.

In the reaction of water with ionic oxides—CaO, Na$_2$O, K$_2$O, BaO, for example—the oxide ion has a greater attraction for the hydrogen nucleus than does the oxygen in the water molecule:

$$:\ddot{O}:^{2-} + H:\ddot{\underset{H}{O}}: \rightarrow :\ddot{O}:H^- + :\ddot{O}:H^-$$

The over-all reaction in the case of sodium oxide is:

$$Na_2O + HOH \rightarrow 2NaOH$$

In the reaction of water with polar covalent oxides—SO$_2$, SO$_3$, CO$_2$, N$_2$O$_5$, for example—there is a reorganization of electrons, so as to best accommodate the attraction of all the atoms for electrons. Although this reaction may also take place in a stepwise fashion, we will simply summarize the over-all process for the reaction with sulfur dioxide as follows:

$$H:\ddot{\underset{H}{O}}: + \ddot{S}::\ddot{\underset{:\ddot{O}:}{O}}: \rightarrow \underset{H:\ddot{O}:}{\overset{H:\ddot{O}:}{:S:\ddot{O}:}}$$

or,

$$\underset{\text{water}}{H_2O} + \underset{\text{sulfur dioxide}}{SO_2} \rightarrow \underset{\text{sulfurous acid}}{H_2SO_3}$$

HYDRATION. The polar nature of water molecules is important when water is used as a solvent. We will see in a later section in this chapter that water readily dissolves many ionic compounds because of the *hydration* of ions. A **hydrated ion** is a cluster of the ion and one or more water molecules. This clustering together is due to the attraction of a positive ion for the negative end of the water molecule, or of a negative ion for the positive end. In solution, the number of

Water; Solutions

water molecules that cluster about many ions appears to be indefinite. Very often, however, when a water solution of a salt is evaporated, the ionic species crystallizes with a precise number of water molecules. Such crystallized salts contain a definite amount of **water of crystallization**. For example, when copper chloride is recrystallized from water, the salt has the composition $CuCl_2 \cdot 4H_2O$. That is, each copper ion in the crystal has a cluster of four water molecules around it. It is to be noted that a pure hydrate appears to be a dry salt; there is no apparent moistness at all.

The force of attraction of an ion for a water molecule depends upon the size of the ion and its charge; the smaller the ion and the greater the charge, the greater is its **charge density** (see Figs. 12-10 and 18-1). We would expect small positive ions with multiple charges—for example, Be^{2+}, Mg^{2+}, Cu^{2+}, Cr^{3+}, Al^{3+}—to have a great attraction for water molecules, and large negative ions with a single charge, such as Br^- or I^-, to have a small attraction. It is found in most cases that the water of crystallization in salts is associated with the positive ions (cations).

When copper chloride and magnesium chloride are crystallized from water solutions, the salts have the composition $CuCl_2 \cdot 4H_2O$ and $MgCl_2 \cdot 6H_2O$, respectively. In the former, the water molecules are thought to be at the corners of an imaginary square, with the Cu^{2+} ion at the center; in the latter the water molecules are held in an octahedral structure with the Mg^{2+} ion at the center (Fig. 12-4). The hydrated Cu^{2+} or Mg^{2+} ions act as units with Cl^- ions to build up crystals of $CuCl_2 \cdot 4H_2O$ or $MgCl_2 \cdot 6H_2O$, respectively.

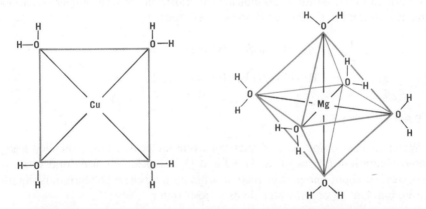

Fig. 12-4. A schematic representation of the shape of a $[Cu(H_2O)_4]^{2+}$ ion (square planar) and a $[Mg(H_2O)_6]^{2+}$ ion (octahedral). The distances between the metal and the oxygen have been exaggerated to show better the spatial relationships.

Often, in naming salts or in writing formulas for them, the name or formula of the nonhydrated salt (*anhydrous* salt) is used to represent the hydrated salt. For example, a water solution of copper sulfate might be represented in an equation by the formula $CuSO_4$, when in reality both the Cu^{2+} ions and SO_4^{2-} are hydrated in the solution. When it is important to take note of the absence or presence of water of hydration, the terms *anhydrous* and *hydrate* are used in the name to distinguish between the two. Examples: Anhydrous copper sulfate, $CuSO_4$; copper sulfate pentahydrate, $CuSO_4 \cdot 5H_2O$; anhydrous zinc chloride, $ZnCl_2$; zinc chloride hexahydrate, $ZnCl_4 \cdot 6H_2O$. Note that *penta-* and *hexa-* denote the number of water molecules. In $CuSO_4 \cdot 5H_2O$, four water molecules are held close to each Cu^{2+} ion and one is held in the crystal between SO_4^{2-} ions.

Many hydrates decompose upon heating to give the anhydrous salt:

$$CaSO_4 \cdot 2H_2O \xrightarrow{heat} CaSO_4 + 2H_2O\uparrow$$
calcium sulfate dihydrate → anhydrous calcium sulfate

The ionization of polar covalent compounds in water affords a special kind of hydration. In Chap. 6, we represented the ionization of hydrogen chloride and hydrogen acetate as:

$$HCl + H_2O \rightleftarrows H_3O^+ + Cl^-$$

$$HC_2H_3O_2 + H_2O \rightleftarrows H_3O^+ + C_2H_3O_2^-$$

In these equations, the proton is shown as having formed the monohydrate, $H^+ \cdot H_2O$, and the anions are not shown as being hydrated. Actually, not one but several water molecules cluster about the proton. The large negative ions do not attract water molecules as strongly as do H^+ ions, but they too are thought to be hydrated in the solution. To indicate this clustering of water molecules about ions in solution the symbol *aq* is sometimes used:

$$HCl(aq) \rightleftarrows H^+(aq) + Cl^-(aq)$$

$$HC_2H_3O_2(aq) \rightleftarrows H^+(aq) + C_2H_3O_2^-(aq)$$

HEAVY WATER

Water that is composed of deuterium oxide molecules, D_2O, is called **heavy water**. Deuterium is present (as HDO and D_2O molecules) in natural water to the extent of about one part in 7000 of ordinary hydrogen. Deuterium is slightly more abundant in ocean water than in fresh water.

Because of the great importance of heavy water as a moderator for nuclear reactors (see Chap. 17), as a source of heavy hydrogen for the hydrogen bomb, and in research, a huge plant for its separation from natural water has been built on the Savannah River in South Carolina. The separation is based on the distribution of deuterium between liquid water and gaseous hydrogen sulfide. The deuterium tends to concentrate in liquid water molecules at low temperatures and in hydrogen sulfide molecules at elevated temperatures.

$$HDO + H_2S \underset{\text{at low T}}{\overset{\text{at high T}}{\rightleftarrows}} HDS + H_2O$$

Thus, if H_2S gas is passed through water at elevated temperatures, HDS and D_2S molecules accumulate in the gas. When this gas is passed through small amounts of cool water the reverse reaction occurs, and the water becomes enriched with HDO and D_2O molecules, yielding a concentrate that is about 15 percent deuterium. This dilute heavy water is further concentrated by fractional distillation and by electrolysis (H_2O molecules are decomposed by electrolysis more rapidly, leaving the remaining water richer in D_2O molecules) until the concentration of D_2O is 99.8 percent. To produce one ton of D_2O the Savannah River plant processes 45,000 tons of water and circulates 150,000 tons of hydrogen sulfide. The cost is estimated to be $13.50 per pound of heavy water produced.

The isotopes of an element exhibit slight differences in ordinary physical and chemical properties. Of all the isotopes known, those of hydrogen show the greatest differences in these properties. Table 12-2 compares certain physical properties of protium, deuterium, protium oxide, and deuterium oxide.

TABLE 12-2.

Differences in Physical Properties Due to Isotopes

	H_2	D_2	H_2O	D_2O
Boiling point, °C	−252.7	−249.6	100.0	101.4
Freezing point, °C	−259.1	−254.4	0.0	3.8
Density, g/ml at 0°C	0.0000899	0.0001798	1.00	1.11

As might be expected, the differences in the chemical properties of isotopes of an element concern mainly the speed of reaction rather than the kind of reaction. These differences in properties are referred to as *isotope effects*. Two examples of isotope effects have been cited: the exchange of protium and deuterium between water and hydrogen sulfide molecules, and the relative production of H_2 versus D_2 in the electrolysis of water.

The hydrogen isotope effect has attracted the attention of many biochemists and biologists, because water is the fluid in which the chemical reactions necessary to life take place; furthermore, it is one of the raw materials from which proteins, fats, carbohydrates, and other body chemicals are synthesized. Enough work has already been done to show that the deuterium isotope effect may have a far-reaching influence on life processes (see Fig. 12-5). For example, when algae are placed in heavy water (99.6 percent D_2O) there is a lag of from several weeks to several months before growth starts.

Recent experiments on the oxidation of glucose (blood sugar) reveal that oxidation with atmospheric oxygen in the presence of enzymes proceeds considerably more rapidly than the oxidation of deuterated glucose, $C_6D_{12}O_6$.[1]

---

[1] Although it is not easy to synthesize complex deuterium compounds, complex substances that are fully deuterated—e.g., carbohydrates, plant pigments, etc.—are now obtained from algae and other organisms grown in heavy water.

Fig. 12-5. Plants of *Atropa belladonna* grown in nutrient solutions containing increasing concentrations of D_2O. (Courtesy of Argonne National Laboratory, Argonne, Ill.)

HYDROGEN PEROXIDE

There are two oxides of hydrogen: hydrogen oxide or water, H_2O, and hydrogen peroxide, H_2O_2. Hydrogen peroxide molecules have the structure

$$H : \ddot{O} : \ddot{O} : H \quad \text{or} \quad H\text{—}O\text{—}O\text{—}H$$

The structure is shown in more detail in Fig. 12-6.

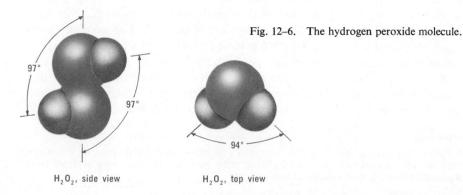

Fig. 12-6. The hydrogen peroxide molecule.

H_2O_2, side view

H_2O_2, top view

Pure hydrogen peroxide is a colorless liquid that melts at 0.9°C and boils at 151°C. Chemically, in the pure form or in water solution, hydrogen peroxide is characterized by its tendency to decompose into water with the liberation of oxygen.

$$2H_2O_2 \rightarrow 2H_2O + O_2 \quad \Delta H \text{ is negative}$$

Finely divided metals, manganese dioxide, and blood catalyze this decomposition. Because of its tendency to liberate oxygen, hydrogen peroxide is an excellent reagent for certain low-temperature oxidation reactions. The dark pigment of hair is readily oxidized at room temperature to a blonde or straw-yellow color by the action of hydrogen peroxide. The protein which comprises the bulk of the hair is unaffected. An equation for the reaction can be formulated as follows:

$$H_2O_2 + \underset{\text{dark pigment}}{X} \rightarrow H_2O + \underset{\text{oxidized pigment}}{XO}$$

The application of arbitrary rules gives the oxygen in hydrogen peroxide an oxidation number of -1. We might expect, therefore, that hydrogen peroxide would act as either an oxidizing agent or a reducing agent, since the oxygen in it is in an intermediate oxidation state as compared with elemental oxygen (with an oxidation number of zero) and with oxygen in water and most oxygen compounds (with an oxidation number of -2). This is true; hydrogen peroxide does act in certain reactions as a reducing agent and in others as an oxidizing agent. For example, silver oxide, Ag_2O, is reduced to elemental silver by hydrogen peroxide.

$$Ag_2O + H_2O_2 \rightarrow 2Ag + H_2O + O_2$$

The preparation of hydrogen peroxide from metallic peroxides is discussed on page 304.

SOLUTIONS

Solutions play an important part in many processes that go on about us. Nutrients are carried in water solution to all parts of a plant; the body fluids of animals are water solutions of numerous substances. The ocean is a vast water solution containing different compounds extracted from the minerals of the earth's crust. Medicines and drugs are frequently aqueous or alcoholic solutions of physiologically active compounds.

Many chemical reactions do not take place at an appreciable rate unless one or more of the reactants is in solution. However, this fact may not be indicated in the written equation. As a general rule, when an equation represents the chemical change of electrolytes, it is assumed that the reaction is carried out in a water solution unless otherwise indicated. The following equations indicate reactions that are carried out in water, although this fact is not stated:

$$NaOH + HCl \rightarrow NaCl + H_2O$$

$$BaCl_2 + Na_2SO_4 \rightarrow BaSO_4 + 2NaCl$$

Definitions. A **solution** is a homogeneous mixture of the molecules, atoms, or ions of two or more different substances.

This definition may be better understood by picturing what is thought to happen when a lump of sugar is dissolved in water. In this process the sugar molecules break away from the lump of sugar and make places for themselves in between the water molecules. The sugar particles (single molecules) are too small to scatter light or reflect it, so they cannot be seen even with the most powerful microscope. If this solution is poured through a filter paper, the sugar molecules pass through the pores of the paper along with the water molecules. Further, the sugar molecules have sufficient kinetic energy to diffuse into all parts of the solution. Therefore, the composition of any part of the solution is identical with that of any other part.

When common salt, NaCl, is dissolved in water, we again have small particles (Na^+ ions and Cl^- ions) breaking away from the attractive forces that hold them in the solid form. They become uniformly mixed with molecules of water. The salt solution differs from the sugar solution in that the latter is a mixture of molecules of two different substances, whereas the former is a mixture of the ions of one substance with the molecules of a second substance.

If copper and gold are melted together, the atoms intermingle so that the molten liquid appears to be a single substance. On cooling and solidifying, the atoms do not aggregate into small lumps of copper and small lumps of gold. Instead, single atoms of copper take positions between atoms of gold (Fig. 12-7)

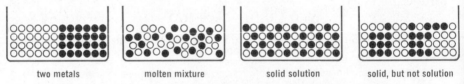

two metals　　　molten mixture　　　solid solution　　　solid, but not solution

Fig. 12-7. When a molten mixture of two metals solidifies, a solid solution may or may not result, depending on the nature of the metals.

to form a mixture called a **solid solution.** In this case, the solution is a mixture of the atoms of two substances. A 12-carat yellow-gold ring is a solid solution containing 50 percent gold and 50 percent copper by weight. (Pure gold is said to be 24 carat.)

Let us now consider a mixture that is not a solution. Clay stirred into water forms a heterogeneous mixture (muddy water) that is fairly uniform but is not a true solution. The clay particles are composed of many molecules and ions clustered together, so they are large enough to scatter and reflect light, making the mixture opaque. In time, the clay particles settle to the bottom, or they may be removed by filtration.

SOLUTE AND SOLVENT. The dissolved substance in a solution is called the **solute** and the dissolving medium is called the **solvent.** In a solution of salt in water, salt is the solute and water is the solvent. These terms have little meaning when applied to certain solutions. For example, solute and solvent apply equally well to either component of a solution containing 50 percent alcohol and 50 percent water.

Water; Solutions

TYPES OF SOLUTIONS

Since there are three states of matter, there are three types of solutions: **gaseous solutions, liquid solutions,** and **solid solutions.** When a solid such as sugar is dissolved in water, the sugar is no longer in the solid state; the sugar molecules are free to move about as the water molecules do. This solution is classified as a *liquid* solution. When water evaporates into a gas—for example, oxygen—the solution formed is a *gaseous* solution. The water molecules are widely separated from other molecules and behave generally as other gaseous molecules do. Actually, we do not usually refer to a gaseous mixture as a solution; a mixture of two or more gases is usually homogeneous, but the attractive forces between molecules are so very weak that the term solution has no special significance.

Solutions may be subclassified from the standpoint of the final state of the solution and the normal state of the pure solute. Six combinations are possible for liquid and solid solutions:

1. The solution is a liquid.
    a. The *solute* was a *solid.* Sugar dissolved in water.
    b. The *solute* was a *liquid.* Alcohol dissolved in water; lubricating oil dissolved in gasoline.
    c. The *solute* was a *gas.* Oxygen dissolved in water; carbon dioxide dissolved in water.
2. The solution is a solid.
    a. The *solute* was a *solid.* Brass (copper and zinc); yellow gold (copper and gold).
    b. The *solute* was a *liquid.* Mercury dissolved in silver. (This solution is an amalgam and is used in filling teeth.)
    c. The *solute* was a *gas.* This type of solution is rare. Hydrogen dissolved in metallic palladium is perhaps the best-known example. There appear to be "holes" in the crystal structure of palladium large enough for single atoms of hydrogen to enter. The number of these "holes" is equal to about one-half the number of palladium atoms present.

WHY SUBSTANCES DISSOLVE

Several factors must be considered in accounting for the intermingling of different species of atoms, molecules, and ions to form solutions, and not all of these factors are thoroughly understood. Here we can discuss only a few of them.

The role of the solvent in the formation of a liquid solution is of primary importance. Sugar molecules do not become homogeneously scattered in just any liquid. Sugar is soluble in some liquids—e.g., water—and insoluble in others—e.g., ether. In general, when a solute A is mixed with a solvent B, three types of attraction exist. These are the attraction of solute particles for each other, A-A; the attraction of solvent particles for each other, B-B; and finally the attraction of solvent particles for solute particles, B-A. If either of the first two is very strong compared to the third, A will not dissolve appreciably in B.

Oil does not dissolve in water because of the strong attraction of water molecules for each other caused by the hydrogen bond, and the weak attraction of water molecules for oil molecules. When oil and water are violently agitated so that the oil molecules are uniformly distributed throughout the water, the oil molecules separate the water molecules. Because of the strong attraction of water molecules for each other, the water tends to coalesce and thus "squeeze out" the oil, and the mixture rapidly separates into two layers.

Salt does not dissolve in gasoline because of the strong attraction of positive

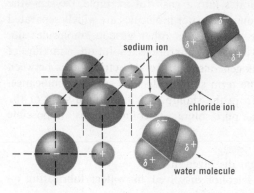

Fig. 12–8. The attractive forces between water molecules and the ions of sodium chloride are responsible for the formation of a salt solution.

ions for negative ions. The attraction of the gasoline molecules for ions is much too weak to separate the sodium and chloride ions of the salt and form a solution.

It is somewhat surprising to find, in view of the strong attraction of oppositely charged ions for each other, that ionic compounds dissolve in any solvent. However, in a polar solvent such as water, there is also a strong attraction between solvent molecules and solute ions. Each negative ion in the face of the crystal attracts the positive end of nearby water molecules, and each positive ion attracts the negative end of water molecules, as shown in Fig. 12-8. This attraction tends to cause ions to break away from the crystal and enter the solution as hydrated ions. Furthermore, once in solution, the sheath of water molecules around the ion tends to prevent it from recombining with other ions to form a crystal (Fig. 12-9). The positive ions with multiple charges hydrate more strongly, and, as was pointed out earlier, may retain a layer of water

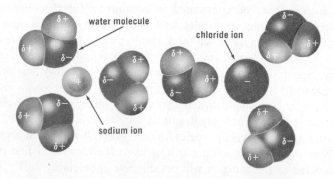

Fig. 12–9. In a water solution of sodium chloride the particles are oriented with respect to their charges.

molecules when crystallized. For example, when anhydrous copper chloride is dissolved in water, and the water is evaporated to the point where crystallization occurs, $[Cu(H_2O)_4]^{2+}$ ions and Cl^- ions aggregate to form crystals of hydrated copper chloride, $CuCl_2 \cdot 4H_2O$. Apparently, although the chloride ion is hydrated in solution, it does not attract water molecules strongly enough to retain them in the crystallized salt. The distribution of the single negative charge throughout the large chloride ion results in a weaker attraction for the positive end of a water molecule, as compared to the attraction of small, bipositive and terpositive ions for water molecules. Large negative ions are said to have a small charge density, while small positive ions have large charge densities (Fig. 12-10).

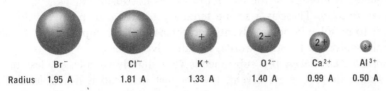

Fig. 12-10. An arrangement of six ions in the order of their increasing charge density. The large bromide ion with a single charge has the least, the small triply charged aluminum ion the greatest charge density.

Entropy. Consider two very similar liquids, for example, hexane (C_6H_{14}) and heptane (C_7H_{16}). These two liquids have densities of 0.659 and 0.684 g/ml, respectively. Let us imagine that we have carefully floated the lighter liquid on top of the heavier one as shown in Fig. 12-11. We find that the molecules diffuse randomly so that, in time, a uniform, homogeneous mixture results. The reverse process has never been observed. That is, the molecules of a mixture of hexane and heptane do not diffuse in an ordered manner so that one species accumulates

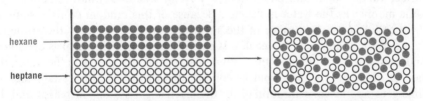

Fig. 12-11. A schematic representation of the tendency of a system to achieve maximum disorder (maximum entropy).

in one part of the container (top, bottom, right, or left) and the other species in the opposite part. This experiment illustrates a general principle: *Physical and chemical systems tend to seek a state of maximum disorder.* In the case of hexane and heptane, the molecules are in a more random or disordered arrangement in the container at the right of Fig. 12-11. A measure of the disorder or randomness of a system is spoken of as the **entropy** of the system. The formation of the solution of hexane and heptane may be attributed, at least in part, to the tendency of a system to achieve maximum entropy (maximum disorder).[2]

---

[2] One statement of the second law of thermodynamics is: *The total amount of entropy in nature is increasing.*

It is important to note that the effect on the solution process of the entropy change is often quite small in comparison with hydration or other effects. Indeed, the strong forces that exist between solute and solvent particles may result in the formation of a solution even though the entropy of the system decreases. Conversely, the strong attractive forces among ions or polar molecules may prevent ionic or polar compounds from scattering to any extent in a nonpolar solvent, or the strong attractive forces among polar solvent molecules may prevent nonpolar compounds from dissolving appreciably, even though the formation of a solution would result in an increase in entropy. For example, if the experiment just described is repeated with water in place of heptane, a homogeneous mixture of the two liquids does not result.

Generalization. There is a strong tendency for ionic or polar covalent compounds to dissolve in polar solvents and for nonpolar compounds to dissolve in nonpolar solvents. In other words, like dissolves like.

Insoluble Substances. If a substance is very slightly soluble, say less than 0.1 g of solute in 1000 g of solvent, the chemist may call it insoluble. Probably nothing is absolutely insoluble in a given solvent, but many things are practically insoluble, for example, glass in water.

When two liquids are mutually insoluble, they are said to be **immiscible**. Such liquids—oil and water, for example—separate into layers. Study of immiscible liquid systems has shed light on problems of likeness and unlikeness of molecular structure.

SATURATED, UNSATURATED, AND SUPERSATURATED SOLUTIONS

When a lump of sugar is placed in water, molecules break away from the surface of the sugar and pass into the solvent, where they move about in the same manner as the water molecules. Because of this random motion, some of them collide with the surface of the sugar and are held there by the attractive forces of the other sugar molecules. It can be shown that the sugar is dissolving and crystallizing at the same time. When the sugar is first placed in the water, the rate of dissolving is very rapid as compared with the rate of crystallizing. As time goes on, the concentration of the dissolved sugar steadily increases, and the rate of crystallizing increases. When the rates of crystallizing and dissolving become the same, the process is said to be *at equilibrium* and the solution is said to be **saturated** (Fig. 12-12). If the excess sugar is left in contact with the solution, this process of dissolving and crystallizing continues at equilibrium, but the

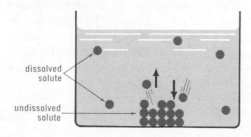

Fig. 12–12. In a saturated solution in contact with excess solute, dissolving and crystallizing occur at the same rate.

concentration of the solution does not change, since the two processes are proceeding at the same rate. A **saturated solution** is defined as a solution that contains the amount of dissolved solute necessary for the existence of an equilibrium between dissolved and undissolved solute. The formation of a saturated solution is hastened by vigorous stirring and an excess of solute.

An **unsaturated solution** is one that is less concentrated (more dilute) than a saturated solution, and a **supersaturated solution**[3] is one that is more concentrated than a saturated solution. Note that a saturated solution is not necessarily a concentrated solution. For example, when limestone rock (calcium carbonate, $CaCO_3$) remains in contact with a quantity of water until an equilibrium is reached between dissolved and undissolved calcium carbonate, the saturated solution is extremely dilute because calcium carbonate is not very soluble.

FACTORS THAT INFLUENCE SOLUBILITY

The amount of solute that dissolves in a given amount of solvent to produce a saturated solution is called the **solubility** of that solute. If large amounts dissolve, the substance is said to be very soluble; if only small amounts dissolve, the substance is described as being slightly or moderately soluble; if no appreciable quantity dissolves, the substance is said to be insoluble. Three factors determine solubility: the nature of the solute and solvent, temperature, and pressure.

Nature of Solute and Solvent. Water acts as an excellent solvent for the electrovalent type of compound because there is a strong attraction between the ions and the polar water molecules. Nonpolar solvents, such as benzene and ether, have little solvent action on ionic compounds. On the other hand, water has little solvent action on fats and waxes, whereas benzene and ether are good solvents for these materials. The nature of the molecules of the solute *and* solvent plays an important part in determining solubility.

Effect of Temperature. SOLIDS IN LIQUIDS. Most solids become more soluble in a liquid as the temperature rises. There are a few solids that actually become less soluble as the temperature increases. Thus, a certain solid may dissolve in water because its molecules react with water to form particles that are strongly attracted by other water molecules. If these particles decompose as they are heated, the solubility of the substance may decrease when the temperature rises. In Fig. 12-13 the solubilities of several compounds in grams per 100 g of water are plotted against temperature. Note that, with the exception of sodium sulfate and cerium sulfate, the solubilities increase as the temperature rises: e.g. the solubility of potassium nitrate, KNO_3, at 0°C is 13 g per 100 g of water, and

---

[3] Many solids become much more soluble in a solvent when the temperature is increased. If a concentrated solution of such a solid is prepared at an elevated temperature and then carefully cooled, the excess solute may remain in solution at the lower temperature. But the solution is unstable, and simply jarring the container may start crystallization. Or if a small crystal of the solute is dropped into the solution, crystallization begins, and the supersaturated solution rapidly becomes a saturated solution.

at 50°C it is 86 g. At about 35°C, the solubility of hydrated sodium sulfate begins to decrease as the temperature rises, because of the decomposition of $Na_2SO_4 \cdot 10H_2O$ to form the less soluble Na_2SO_4.

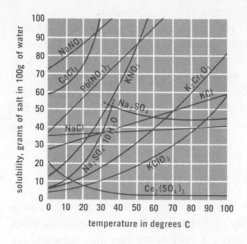

Fig. 12-13. Solubilities of certain solids at different temperatures.

GASES IN LIQUIDS. The solubility of a gas in a liquid usually decreases as the temperature increases. Because of this, carbon dioxide bubbles vigorously out of a carbonated drink if the liquid is warmed. For the same reason, when tap water is heated, bubbles of air begin to appear when the water is lukewarm.

Effect of Pressure. Pressure changes have little effect on solubility if the solute is a liquid or a solid. However, in the formation of a saturated solution of a gas in a liquid, the pressure of the gas plays an important part in determining how much of the gas dissolves (Fig. 12-14). *The weight of a gas dissolved by a given*

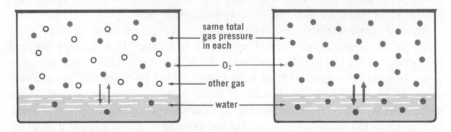

Fig. 12-14. A schematic illustration of Henry's law. Note that the O_2 pressure at the right is double the O_2 pressure on the left. Hence there is twice as much dissolved O_2 in the water on the right. The pressure due to other gaseous molecules has little or no effect on the solubility of O_2.

amount of a liquid is directly proportional to the pressure exerted by the gas when in equilibrium with the solution. This is a statement of **Henry's law.** Henry's law does not hold well for gases that react with the solvent, e.g., hydrogen chloride or ammonia dissolved in water.

EXERCISES

1. Which of the following contains the smallest number of water molecules:
   (a) 1.0 g of liquid water at 0°C;
   (b) 1.0 g of solid water at 0°C;
   (c) 1.0 ml of solid water at 0°C;
   (d) 1.0 ml of liquid water at 0°C;
   (e) 1.0 ml of liquid water at 4°C?
   Which three of the above contain the same number of water molecules?
2. Account for the expansion that occurs when water freezes.
3. If ice were to become more dense than water, how might our environment change?
4. Hydrogen fluoride, HF, has a boiling point of 20°C and a heat of vaporization of 361 cal/g. Hydrogen chloride, HCl, has a boiling point of -85°C and a heat of vaporization of 98.7 cal/g. Account for the differences in properties of these two similar compounds.
5. Write balanced equations for the reactions indicated below:
   (a) lithium + water $\rightarrow$
   (b) barium + water $\rightarrow$
   (c) sodium + ammonia $\rightarrow$
   (Note: Ammonia, NH_3, like water is a polar molecule.)
6. What weight of potassium is required to react with water to form 0.25 g of hydrogen? To form 5.0 liters at STP?
7. The symbol H^+ may be used to represent a certain particle in a cathode-ray tube, or it may be used to represent a particle in water solution. Does it refer to the same particle in both cases? Explain.
8. Write balanced equations for each of the following:
   (a) $CO_2 + H_2O \rightarrow$
   (b) $CaO + H_2O \rightarrow$
   (c) $K_2O + H_2O \rightarrow$
   (d) $N_2O_3 + H_2O \rightarrow$
   (e) $As_4O_{10} + H_2O \rightarrow$
9. By referring to Fig. 18-1, list the following ions in order of their increasing tendencies to form hydrates: Na^+, Al^{3+}, Ca^{2+}, Zn^{2+}, Rb^+.
10. A 3.500-g sample of magnesium sulfate hydrate, $MgSO_4 \cdot xH_2O$ was heated until all of the water of hydration was expelled. The resulting anhydrous magnesium sulfate weighed 1.707 g. Calculate the formula of the hydrate.
11. Starting with heavy water, D_2O, describe how you would proceed to prepare (a) NaOD; (b) DCl; (c) CaD_2.
12. What are isotope effects? Using the separation of heavy water from natural water, illustrate isotope effects.
13. Give specific examples for the following types of solutions:
    (a) liquid solution with gaseous solute;
    (b) gaseous solution with solid solute;
    (c) solid solution with liquid solute;
    (d) liquid solution with liquid solute;
    (e) solid solution with solid solute.
14. Why is solder, which contains tin and lead, called a solution rather than a compound, while copper sulfate pentahydrate is called a compound rather than a solid solution of copper sulfate and water?

15. Define and illustrate entropy.
16. Why is potassium chloride more soluble in water than in gasoline? Why is lubricating oil more soluble in gasoline than in water?
17. Describe the chemical properties of hydrogen peroxide. How are these properties related to the uses of hydrogen peroxide?
18. State Henry's law. Which would more closely follow Henry's law, the water solubility of oxygen or the water solubility of carbon dioxide? Why?
19. Distinguish between a saturated and an unsaturated solution. How could one experimentally test a given solution to determine if it is saturated, unsaturated, or supersaturated?
20. Suppose you were given a solution which contains equal numbers of the ions Na^+, K^+, Cl^-, and NO_3^-. By referring to Fig. 12-13, predict the compound which will first crystallize if the solution is allowed to evaporate at 90°C. What compound will first crystallize if the solution is allowed to evaporate at 15°C?
21. Based upon Fig. 12-13, explain how you might prepare KNO_3 from KCl and $NaNO_3$.
22. Water and alcohol are miscible, but water and carbon tetrachloride are immiscible. Correlate this behavior with the molecular structures of the three compounds.
23. When a solid substance dissolves, two processes are of prime importance: the breaking away of particles from the solid and the joining of solute and solvent particles. In terms of these two processes, tell how the dissolving in water of sodium hydroxide can be exothermic, whereas for ammonium chloride the process is endothermic.

SUGGESTED READING

Katz, J. J., "The Biology of Heavy Water," *Sci. Amer.*, **203** (1), 106 (1960).

Lochmüller, C., and M. Cefola, "Solubility in Mixed Solvents," *J. Chem. Educ.*, **41**, 604 (1964).

Othmer, D. F., "Water and Life," *Chemistry*, **43** (10), 12 (1970).

Snyder, A. E., "Desalting Water by Freezing," *Sci. Amer.*, **207**, (6) 41 (1962).

Taylor, J. E., "An Apparatus for the Continuous Production of Triple Distilled Water," *J. Chem. Educ.*, **37**, 204 (1960).

13

CONCENTRATIONS OF SOLUTIONS; COLLIGATIVE PROPERTIES

EXPRESSING CONCENTRATION

The **concentration** of a solution refers to the weight or volume of the solute present in a specified amount of the solvent or solution. There are several common methods of expressing these amounts.

Percent by Weight. When expressing percent by weight, the percentage given refers to the solute; for example, a 5 percent aqueous NaCl solution contains 5 percent by weight of sodium chloride, the remaining 95 percent being water.

Percent by Volume. The concentration of a solution of two liquids is frequently expressed as a volume percentage, because the volumes of liquids are easily measured. The concentration of alcoholic beverages is usually expressed this way. A wine that is 12 percent alcohol has 12 ml of alcohol per 100 ml of wine.

In chemistry laboratory work, percent by volume is rarely used; unless otherwise stated, a given percent concentration means percent of solute by weight.

MOLAR SOLUTIONS

Knowledge of the number of solute particles in a given quantity of solution is required in many laboratory operations. The usual way of expressing the number of particles present is in terms of the numbers of moles.

Name	Formula	Weight of 1 mole, g	Particles per mole
Alcohol	C_2H_6O	46	6.02×10^{23} molecules
Glycerin	$C_3H_8O_3$	92	6.02×10^{23} molecules
Sugar	$C_{12}H_{22}O_{11}$	342	6.02×10^{23} molecules
Salt	NaCl	58.5	12.04×10^{23} ions

A **one molar** (1 M) **solution** contains 1 mole of solute per liter of solution, a two molar (2 M) solution contains 2 moles per liter of solution, and so on. To make a 0.1 M sugar solution, we weigh a tenth of a mole (34.2 g) of sugar and place it in a volumetric flask (Fig. 13-1); then we add water until the flask is about three-fourths full. The mixture is then stirred until all the sugar is dissolved. Finally,

Fig. 13-1. Volumetric flask calibrated at 20°C.

we add water until the total volume is precisely 1 liter and mix thoroughly. To make a 0.1 M glycerin solution, the procedure is exactly the same, except that 9.2 g of glycerin is used. One liter of 0.1 M glycerin solution contains the same number of dissolved glycerin molecules as there are sugar molecules in one liter of 0.1 M sugar solution. The amounts of water in the two solutions differ to a slight extent.

Problem 1: What weight of calcium bromide, $CaBr_2$, is needed to prepare 150 ml of a 3.5 M solution?

Solution: The weight of one mole of $CaBr_2$ is 200 g. By definition, a 3.5 M solution contains 3.5 moles of $CaBr_2$ dissolved in 1 liter of solution.

$$\text{Wt of } CaBr_2 \text{ per liter} = \frac{200 \text{ g}}{\text{mole}} \times 3.5 \text{ moles} = 700 \text{ g}$$

$$\text{Wt of } CaBr_2 \text{ per 150 ml} = \frac{700 \text{ g}}{1 \text{ liter}} \times \frac{1 \text{ liter}}{1000 \text{ ml}} \times 150 \text{ ml} = 105 \text{ g } CaBr_2$$

MOLAL SOLUTIONS

As just indicated, solutions of the same molarity have slightly different ratios of molecules of solute to molecules of solvent, depending on what solutes are

Concentrations of Solutions; Colligative Properties

involved. In order that he know precisely the solute-solvent ratio, the chemist may express the concentration in terms of the **molality** (m), the number of moles of solute dissolved per kilogram (kg) of solvent.

Problem 2: Explain how you would go about making up approximately 100 ml of a 0.10 m glucose solution. Glucose, $C_6H_{12}O_6$, has a molecular weight of 180 awu.

Solution: By definition, a 0.10 m glucose solution contains 0.10 mole or 18 g glucose in 1 kg of water. Since 1 kg (1000 g) of water has a volume of about 1000 ml, this amount of water is sufficient to make about 10 times as much solution as is required. Take one-tenth of each quantity, i.e., weigh out 1.8 g of glucose and 100 g of water. Dissolve the glucose in the water. The resulting solution will have a molality of 0.10, precise to two significant figures, and will have a volume of approximately 100 ml.

NORMAL SOLUTIONS

The **normality** (N) of a solution is the number of equivalent weights of solute per liter of solution. A one normal (1 N) solution contains one equivalent weight per liter of solution, a 0.5 N solution contains one-half an equivalent weight per liter, etc.

It is especially convenient to express concentrations in terms of normalities when we are dealing with solutions that react with one another (see the discussion of titration in the next section). Study of the following equations will reveal that 1 mole of hydrochloric acid will exactly neutralize 1 mole of sodium hydroxide, whereas only 0.5 mole of sulfuric acid, or only 0.33 mole of phosphoric acid, H_3PO_4, is required to neutralize 1 mole of sodium hydroxide.

$$HCl + NaOH \rightarrow NaCl + HOH$$
$$\text{36.5 g} \quad \text{40 g}$$

$$H_2SO_4 + 2NaOH \rightarrow Na_2SO_4 + 2HOH$$
$$\text{98 g} \quad \text{80 g}$$

$$H_3PO_4 + 3NaOH \rightarrow Na_3PO_4 + 3HOH$$
$$\text{98 g} \quad \text{120 g}$$

It follows that 1 mole of HCl (36.5 g), 0.5 mole of H_2SO_4 (49 g), and 0.33 mole of H_3PO_4 (32.7 g) represent the same quantity of acid. Each will neutralize 40 g of NaOH. These are the **equivalent weights** of these acids. The term is defined as follows:

$$\text{Eq wt of an acid} = \frac{\text{Mole weight}}{\text{No. of available } H^+ \text{ per formula unit}}$$

Following the same line of reasoning, we see that 1 mole (40 g) of sodium hydroxide is equivalent to 0.5 mole (37 g) of calcium hydroxide, $Ca(OH)_2$, in chemical reactions involving bases. Examine the following two equations:

$$NaOH + HCl \rightarrow NaCl + HOH$$

$$Ca(OH)_2 + 2HCl \rightarrow CaCl_2 + 2HOH$$

The equivalent weight of a base is defined as follows:

$$\text{Eq wt of a base} = \frac{\text{Mole weight}}{\text{No. of available OH}^- \text{ per formula unit}}$$

One equivalent weight of any acid can supply a mole of protons; one equivalent weight of any base can accept that number of protons.

Problem 3: Calculate the equivalent weight of barium hydroxide, $Ba(OH)_2$.
Solution: The molecular weight is obtained by adding atomic weights:

$$137.3 + (2 \times 16.0) + (2 \times 1.01) = 171.3 \text{ awu}$$

From the formula, $Ba(OH)_2$, we see that there are two OH^- ions per formula unit:

$$\text{Eq wt} = \frac{\text{mole wt}}{2} = \frac{171.3 \text{ g}}{2} = 85.6 \text{ g}$$

Problem 4: What weight of barium hydroxide is present in 35.0 ml of 0.200 N barium hydroxide?
Solution: Based on the definition of normality, there is 0.200 eq wt of $Ba(OH)_2$ in 1000 ml (1 liter) of 0.200 N solution. The weight is calculated from the eq wt (see Problem 3):

$$\text{Wt of Ba(OH)}_2 \text{ per 1000 ml} = 0.200 \text{ eq wt} \times \frac{85.6 \text{ g}}{1 \text{ eq wt}}$$

$$= 17.1 \text{ g}$$

$$\text{Wt of Ba(OH)}_2 \text{ per 35.0 ml} = \frac{17.1 \text{ g}}{1000 \text{ ml}} \times 35.0 \text{ ml}$$

$$= 0.598 \text{ g}$$

TITRATION OF ACIDS AND BASES

Titration is the process of determining the amount of a solution of known concentration that is required to react completely with a certain amount of a sample that is being analyzed. The solution of known concentration is called a **standard solution,** and the sample being analyzed is referred to as the *unknown*.

In the analysis of acidic and basic solutions, titration involves the careful measurement of the volumes of an acid and a base that just neutralize each other. Suppose that we have a hydrochloric acid solution whose concentration we wish to determine, and that we have on hand in the laboratory a standard base solution with a concentration of 1.20 N. The titration is carried out as follows: Portions of the two solutions are placed in separate burettes (Fig. 13-2), and a convenient quantity of the acid, say 15.0 ml is measured from its burette into the flask. An indicator (litmus or phenolphthalein) is added to the acid, and the flask is placed under the burette containing the base. The base is permitted to run into the flask rather rapidly at first, then slowly, and finally drop by drop until a single final drop causes the indicator to change color. This color change is the signal that reveals the *end point* of the titration. At the end point, an amount of base has been added that *is equivalent in chemical reactivity* to the amount of

Concentrations of Solutions; Colligative Properties

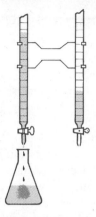

Fig. 13-2. Apparatus for a titration.

acid in the 15.0 ml of the unknown solution. The total volume of base used is read from the burette. Suppose that this volume is 21.2 ml. This means that 21.2 ml of a 1.20 N base is found to just neutralize 15.0 ml of hydrochloric acid of unknown concentration. From consideration of the experimental data:

	Base	Acid
Volume	21.2 ml	15.0 ml
Normality	1.20 N	?

we can see that the acid is more concentrated than the base, because a smaller volume of it was required. That is, even before we calculate it, we know that the normality of the acid is going to be greater than 1.20 N. The concentration of the acid is calculated by using the general relationship that states that the product of the volume times the normality is equal for all solutions that react completely with one another:

$$V_{acid} \times N_{acid} = V_{base} \times N_{base}$$

The volumes may be expressed in liters or milliliters as long as consistent units are used. With this relationship, the calculations are

$$15.0 \text{ ml} \times N_{acid} = 21.2 \text{ ml} \times 1.20 \text{ N}$$

$$N_{acid} = \frac{21.2 \text{ ml} \times 1.20 \text{ N}}{15.0 \text{ ml}} = 1.70 \text{ N}$$

The concentration of a basic solution of unknown concentration would be determined in a similar fashion. A standard acid solution would be required for this titration.

Litmus, a dye extracted from a moss ($litr$ = color + $mosi$ = moss), was one of the first natural organic dyes to be used as an acid-base indicator. Certain vegetable colors—for example, grape juice, elderberry juice, red cabbage—contain molecules that change color, owing to structural changes accompanying the gain or loss of a proton, and can be used as indicators for crude titrations. Today many indicators are synthetic organic compounds. Phenolphthalein is an example.

COLLIGATIVE PROPERTIES OF SOLUTIONS

We are familiar with the use of water solutions of alcohol and ethylene glycol in automobile radiators during the winter. Their use depends on the fact that these solutions freeze at a lower temperature than water alone does. These are not the only substances that lower the freezing point of water. Any soluble compound such as salt, sugar, hydrochloric acid, or baking soda, when dissolved in water, forms a solution that has a lower freezing point than pure water does. Moreover, the boiling points and vapor pressures of these solutions are also different from those of the pure solvent. To what extent do they differ? For solutions of nonvolatile nonelectrolytes, the answer to this question is provided by the following law. **The freezing point, boiling point, and vapor pressure of a solution differ from those of the pure solvent by amounts that are directly proportional to the molal concentration of the solute.**[1]

The kind of solute matters very little so long as (1) it does not have an appreciable vapor pressure and (2) it does not form ions.

The solution properties referred to above—freezing-point depression, boiling-point elevation, and change in vapor pressure—are referred to as **colligative properties.** Osmotic pressure is also a colligative property. Colligative properties are uniformly affected by changes in the molal concentration (changes in the number of solute molecules in a constant quantity of solvent molecules) but not by the specific nature of the solute. See Fig. 13-3. (In a 1 molal solution there is 1 mole of solute per 1 kg of solvent.)

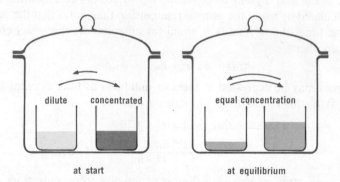

Fig. 13-3. The more dilute a solution, the higher is its vapor pressure. Hence, as long as the concentrations are unequal, evaporation and condensation occur at unequal rates, and a net transfer of solvent occurs from the dilute to the more concentrated solution. In time, the solutions attain the same concentration and no further net change occurs, although evaporation and condensation continue in dynamic equilibrium.

---

[1] For convenience we shall refer to this law as the colligative property law. As it applies to the vapor pressure, this statement is known as Raoult's law (1887).

Concentrations of Solutions; Colligative Properties

FREEZING POINT OF SOLUTIONS

If some sugar is dissolved in water and the solution is then cooled, ice crystals do not form at a temperature of 0°C, but at some lower temperature. At what temperature do ice crystals begin to form in a water solution? According to the colligative property law, the amount by which the freezing point is lowered is determined by the number, not the kind, of dissolved molecules in a given amount of solvent.[2] It has been determined experimentally that 1 mole (6.02 × 10^{23} molecules) of any nonelectrolyte dissolved in 1 kg (1000 g) of water lowers the freezing point by 1.86°C; that is, ice crystals begin to form when the temperature is lowered to −1.86°C. On the other hand, if 0.5 mole (3.01 × 10^{23} molecules) is dissolved in 1000 g of water, the amount by which the freezing point is lowered is 0.93° (that is, 1.86° × 0.5); if 2 moles is dissolved, the amount is 3.72° (that is, 1.86° × 2); etc. The amount by which the freezing point is lowered by dissolving 1 mole of a nonelectrolyte in 1 kg of a solvent is called the **molal freezing-point constant** of the solvent. The values of this constant for three solvents are:

Solvent	Freezing point of pure solvent, °C	Molal freezing-point constant, °C
Water	0.00	1.86
Hydrogen acetate	16.60	3.90
Benzene	5.48	4.90

Since we know the amount by which the freezing point is depressed by one mole of a solute, we can use this information to experimentally determine the molecular weight of a new compound.

Problem 5: A solution made by dissolving 0.912 g of compound X in 10.0 g of water began to form ice crystals when cooled to −0.93°C. Calculate the weight of a mole and of a molecule of X.

Solution: Let us begin by calculating the molality of the solution. The change in freezing point, Δfp, is 0.93°C, that is, 0°C − (−0.93°C). Because 1 mole of solute in 1 kg of water lowers the freezing point 1.86°, it is apparent that the solution of X is less than one molal:

$$\text{Molality of solution} = \frac{\Delta fp}{\text{molal freezing point constant}}$$

$$= \frac{0.93}{1.86} = 0.50 \, m$$

Wt solute = (molality)(no. kg solvent)(wt of 1 mole solute)

$$0.912 \text{ g of } X = (0.50 \text{ mole/kg}) \left(10 \text{ g} \times \frac{1 \text{ kg}}{1000 \text{ g}}\right) (\text{wt of 1 mole solute } X)$$

---

[2] In the case of the freezing-point change, the colligative property law applies not only to nonvolatile but also to volatile solutes, for example, methyl alcohol (wood alcohol) and ethyl alcohol (grain alcohol).

Wt of 1 mole solute $X = \dfrac{0.912 \text{ g}}{0.50 \text{ mole} \times 0.01} = 182$ g/mole

therefore,

Mol wt = 180 awu (rounded to 2 significant figures)

VAPOR PRESSURE AND BOILING POINT OF SOLUTIONS OF NONELECTROLYTES

The boiling point of a solution may be higher or lower than that of the solvent, depending on the volatility of the solute as compared with that of the solvent. If the solute is nonvolatile—sugar is an example—its water solution boils at a temperature higher than water boils; if the solute is quite volatile—for example, alcohol—its water solution boils at a temperature below the boiling point of water (Fig. 13-4).

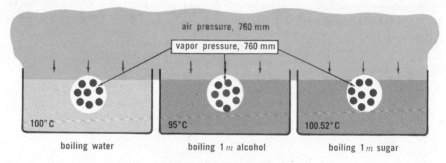

Fig. 13–4. Vapor bubbles in boiling water, alcohol solution, and sugar solution contain enough gaseous molecules to give a pressure of 760 mm.

Consideration of the kinetic theory and the definition of the boiling point make clear why these differences exist. It will be remembered that the boiling point is the temperature at which the vapor pressure of a liquid equals the atmospheric pressure. For pure water, the boiling point is 100°C when the atmospheric pressure is 760 mm. But when the temperature of an aqueous sugar solution is 100°C, the vapor pressure of the water is less than 760 mm. This is due to the fact that the sugar molecules occupy a considerable portion of a unit volume and hence decrease the concentration of water molecules. The vapor pressure of a sugar solution is therefore less than that of pure water at the same temperature; consequently the solution must be heated to a temperature higher than 100°C before it boils (before the vapor pressure reaches 760 mm, see Fig. 13-4).

In the case of the alcohol solution, the presence of the alcohol molecules also lessens the concentration of the water molecules, so that the pressure due to water vapor is lower than that for pure water at the same temperature. However, liquid alcohol has a greater tendency to become a vapor than does water. For this reason, the vapor pressure of the solution (the sum of the alcohol pressure and the water vapor pressure) equals the atmospheric pressure at a temperature below 100°C. The boiling point of the solution is below that of pure water. The colligative property law is applicable in predicting the boiling point of solutions

Concentrations of Solutions; Colligative Properties

involving nonvolatile, nonelectrolytic solutes. The amount by which the vapor pressure is decreased and the boiling point is increased is directly proportional to the number of molecules of solute in a given quantity of solvent (Fig. 13-5). If the number of solute molecules is 6.02×10^{23} (1 mole) and the amount of solvent

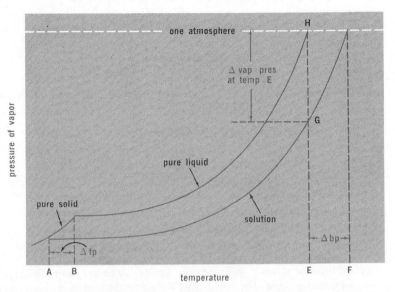

Fig. 13-5. A graphic representation of the change in vapor pressure of pure solvent and solution with change in temperature. Shown also are the changes in freezing point and boiling point. (Δ signifies "change in.")

is 1 kg, the elevation of the boiling point of water is 0.52°C. This is the **molal boiling-point constant** for water.

Problem 6: A solution made by dissolving 6.21 g of Y in 100.0 g of water had a boiling point of 99.85°C. Pure water at the same atmospheric pressure had a boiling point of 99.63°C. Y has no appreciable vapor pressure at 100°C and is known to be a covalent compound. Calculate its molecular weight.

Solution: In this problem, we are concerned with the elevation of the boiling point rather than with the depression of the freezing point, as was the case in Problem 5. Otherwise the solution is similar to that of Problem 5.

$$\Delta bp = 99.85°C - 99.63°C = 0.22°$$

$$\text{molality of solution} = \frac{\Delta bp}{0.52°} = \frac{0.22°}{0.52°} = 0.42\ m$$

Wt of solute = (molality)(kg solvent)(wt of 1 mole solute)

$$6.21\ g = (0.42\ \text{mole/kg})\left(100.0\ g \times \frac{1\ kg}{1000\ g}\right)(\text{wt 1 mole solute})$$

$$\text{Wt 1 mole solute} = \frac{6.21\ g}{0.42\ \text{mole} \times 0.1000} = 148\ g/\text{mole}$$

therefore,

Molecular weight = 150 awu (148 rounded to 2 significant figures)

Problem 7: Compare the boiling points of two solutions, one made by dissolving 100 g of glycerol, $C_3H_8O_3$, in 1000 g of water and the other by dissolving 100 g of glucose, $C_6H_{12}O_6$, in 1000 g of water.

Solution:

Mole wt of glycerol = 92 g (calculated from formula)

$$\text{Moles of glycerol} = \frac{92 \text{ g}}{1 \text{ mole}} \times 100 \text{ g} = 1.1 \text{ moles}$$

$$\text{Elevation of bp} = \frac{0.52°}{\text{mole}} \times 1.1 \text{ moles} = 0.57°$$

Bp of glycerol solution = $100° + 0.57° = 100.57°C$.

Mole wt of glucose = 180 g (calculated from formula)

$$\text{Moles of glucose} = \frac{1 \text{ mole}}{180 \text{ g}} \times 100 \text{ g} = 0.56 \text{ moles}$$

$$\text{Elevation of bp} = \frac{0.52°}{\text{mole}} \times 0.56 \text{ moles} = 0.29°$$

Bp of glucose solution = $100° + 0.29° = 100.29°C$.

Note that the two solutions have the same percentage composition by weight. However, because glycerol molecules are lighter, there are more molecules of glycerol than of glucose. The molality of the glycerol solution is almost double that of glucose, 1.1 : 0.56.

PROPERTIES OF SOLUTIONS OF ELECTROLYTES

In the preceding sections, the discussion of the vapor pressure, boiling point, and freezing point of solutions was restricted to solutions of nonelectrolytes. We shall now see why this restriction was necessary.

It has been emphasized that the extent to which the freezing point, boiling point, and vapor pressure of a solution differ from those of the pure solvent depends upon the number of solute particles in a given weight of the solvent. With nonelectrolytes, 1 mole refers to the same number of particles, namely, 6.02×10^{23} molecules. But in the case of an electrolyte a mole refers to a larger number of particles. The unit NaCl is not a molecule but a pair of ions, Na^+, Cl^-. This means that 58.5 g of NaCl contains not 6.02×10^{23} molecules, but 6.02×10^{23} Na^+ ions and 6.02×10^{23} Cl^- ions. The data in Table 13-1 show that for the electrovalent type of electrolyte, the number of particles in a mole is twice, three times, four times, etc., the number in a mole of covalent compound.

When a mole of an electrovalent type of electrolyte is dissolved in 1000 g of water, the amounts by which the freezing and boiling points are changed are considerably more than $1.86°$ and $0.52°$, respectively. Indeed, it appears at first glance that these constants would be exactly doubled for such compounds as sodium chloride, Na^+, Cl^-, and potassium nitrate, K^+, NO_3^-, and exactly tripled for such compounds as calcium chloride, Ca^{2+}, Cl^-, Cl^-, and sodium sulfate, Na^+, Na^+, SO_4^{2-}. However, laboratory experiments show that the changes are not quite as great as might be expected. The explanation is believed to lie in the fact that each ion strongly attracts the oppositely charged ions in its vicinity, so that no ion can act completely as an independent particle. These interionic attractions are reduced to a minimum when the solution is very dilute,

TABLE 13-1.

Particles Per Mole for Electrolytes

FORMULA	PARTICLES REPRESENTED BY FORMULA	WEIGHT OF 1 MOLE	NO. OF PARTICLES IN 1 MOLE
NaCl	Na^+, Cl^-	58.5 g	$2 \times 6.02 \times 10^{23}$
KNO_3	K^+, NO_3^-	101.1 g	$2 \times 6.02 \times 10^{23}$
$CaCl_2$	Ca^{2+}, Cl^-, Cl^-	111.0 g	$3 \times 6.02 \times 10^{23}$
Na_2SO_4	Na^+, Na^+, SO_4^{2-}	142.0 g	$3 \times 6.02 \times 10^{23}$
AlF_3	Al^{3+}, F^-, F^-, F^-	84.0 g	$4 \times 6.02 \times 10^{23}$

because the ions are separated by many water molecules. Consequently, the observed changes in the freezing and boiling points of very dilute solutions of KCl, NaCl, and $MgSO_4$ approach twice those in a solution of a nonelectrolyte of the same molal concentration, and the change for K_2SO_4 approaches three times the change for a nonelectrolyte.

IDEAL SOLUTIONS

Thus far the discussion has indicated that in solutions of nonelectrolytes, the magnitude of the changes in the freezing point and other properties depends solely on the number of particles, whereas in solutions of electrolytes, the magnitude of the change depends on two factors: namely, the number of particles and the interionic attractions. Actually, 6.02×10^{23} solute molecules of different substances do not produce exactly the same change in the properties of solutions. We think the reason for this is that intermolecular attraction is present in solutions of nonelectrolytes. However, this van der Waals attraction of solute molecules for each other and for solvent molecules has so slight an effect that no great deviations in the freezing-point and boiling-point constants are produced unless the solutions are concentrated. Dilute solutions act so nearly in an ideal way that we need to consider only the number of molecules or ions present, and not their nature, in calculating changes in properties. The data in Table 13-2 show the magnitude of deviation for four moderately concentrated solutions.

TABLE 13-2.

Experimental Molal Freezing-point Depressions

SOLUTE	FORMULA	MOLAR WEIGHT, G	OBSERVED Δfp, °C	DEVIATION
Methyl alcohol	CH_3OH	32	1.86	0.00
Ethyl alcohol	C_2H_5OH	46	1.83	−0.03
Glycerol	$C_3H_5(OH)_3$	92	1.92	+0.06
Urea	$CO(NH_2)_2$	60	1.86	0.00

FRACTIONAL DISTILLATION

Thus far, our discussion has been mostly limited to the type of solution in which the solute has a high boiling point as compared to that of the solvent. Such solutions can be easily separated by distilling the mixture in a distillation apparatus like that shown in Fig. 2-2. But suppose that we have a mixture of two miscible liquids whose boiling points are fairly close together, and we wish to separate them by distillation. What difficulties will we encounter? Let us examine the problem by considering the distillation in the apparatus described in Fig. 2-2 of a solution whose molecular concentration is 50 percent heptane, boiling point 98°C, and 50 percent octane, boiling point 126°C. (These are two of the hydrocarbons present in gasoline.) A solution of these two liquids is very nearly ideal, and because we have an equal concentration of molecules, the solution will begin to boil at a temperature somewhat below the midway point between the two boiling points, say about 110°C.[3] The vapor from the boiling liquid will be richer in heptane, but it will also contain a considerable amount of octane. Let us imagine that we continue the distillation until one-tenth of the mixture has been distilled and collected. We set this aside as fraction number one. This fraction, while containing both heptane and octane, is richer in heptane. We now find that the remaining liquid boils, at a somewhat higher temperature. This is because the liquid is now richer in the higher boiling component. Let us continue the distillation until we have collected a second fraction equal to one-tenth of the initial volume. We find that this fraction is still rich in heptane, but that it is not as rich as was the first fraction. Let us continue the distillation as described so as to completely distill the mixture into ten equal fractions. As the distillation progresses, we find that the boiling point steadily increases and that the distillate contains a diminishing amount of heptane. However, we find that all ten fractions contain both heptane and octane, albeit the first fraction is

---

[3] It might appear that the boiling point would be precisely midway between the boiling points of the two pure liquids. This would be true if the vapor pressure were a linear (straight-line) function of the temperature. However, the vapor pressure is an exponential function of temperature, and the boiling point of the mixture lies on the bottom curve of Fig. 13-6 rather than on a straight line connecting the boiling points of pure heptane and pure octane.

richer in heptane and the last is richer in octane. (See Fig. 13-6.) To bring about a better separation of the two, it is necessary to redistill each fraction into smaller fractions, and so on. Eventually, a first fraction is obtained that approaches 100 percent heptane, and a last fraction that approaches 100 percent octane. This process is one type of fractional distillation. It is called a *batch process*, because each part of the collected liquid is handled as a separate batch.

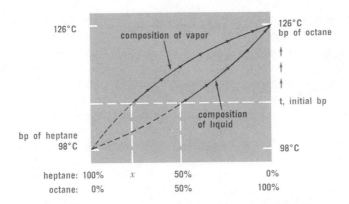

Fig. 13–6. A graph showing the change in composition and boiling point during the distillation of an equi-molecular mixture of heptane and octane. The first drop to distill at bp t has the composition x. As the distillation continues, the compositions of the escaping vapor and remaining liquid change as shown by the arrows. The last drop distills at about 126°C and is virtually pure octane.

Batch processes, be they distillations, chemical syntheses, or other operations, are usually inefficient and tedious. It is often possible to adapt such a process to a *continuous operation*. In the case of fractional distillation, the series of distillations, condensations, redistillations, and recondensations, and so on, can be carried out in an expeditious manner by the use of fractionation columns. In one type of fractionating column, the neck of the flask containing the boiling mixture may extend upward from a foot or so on up to 20 feet or more, and is packed with glass beads or glass helices so that condensation and redistillation occur through the entire length of the column. The vapor that emerges from the top is then passed through a condenser such as that shown in Fig. 2-2. In another type, the vapor passes upward through a series of plates, where condensation and redistillations occur many times before the vapor passes off at the top and is condensed (see Fig. 26-8). In an efficient column, the vapor that first reaches the top of the column is the lower boiling component, and is relatively pure. It must be remembered that we have been discussing distillation of liquid mixtures that do not depart too much from ideal behavior. In some cases, the departure is great enough so that *constant boiling* mixtures result during some phase of the distillation. That is, the distillation continues until the liquid mixture reaches a certain composition, and then the mixture distills without a further change in composition or boiling point. With such a mixture it is not possible to obtain a pure sample of either component by simple distillation. Alcohol and water form such a mixture. In such cases, the boiling point of the constant-boiling mixture might be below the boiling point of either component, or it might be above the boiling point of either.

OSMOSIS

If pure water is separated from a water solution by a membrane through which only water molecules can pass, the water diffuses through the membrane from the pure solvent into the solution more rapidly than it does from the solution into the pure solvent. A membrane that permits the passage of only certain types of molecules is called a **semipermeable membrane;** the passing of the solvent through the membrane is called **osmosis.** Osmosis may be readily demonstrated by fastening a piece of animal bladder or nonwaterproof cellophane over a thistle tube, as shown in Fig. 13-7. An aqueous sugar solution is placed inside

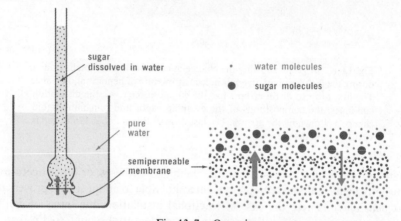

Fig. 13-7. Osmosis.

the thistle tube, which is then immersed in water. At the beginning, water molecules pass more rapidly from the pure water into the sugar solution than in the opposite direction. Because of these unequal rates, water accumulates in the thistle tube, and the water level in it rises. As time goes on, the sugar solution becomes more and more dilute; consequently, the rate at which the water molecules pass back into the solvent steadily increases. There is also an increase in this rate due to the increasing pressure of the water in the column. In time, the rate of flow in both directions becomes the same, and the water level ceases to rise. Actually this equilibrium is very difficult to achieve, because as the water level rises, the pressure of the solution on the membrane may increase till the membrane splits and spills the solution into the solvent.

Biological Importance of Osmosis. Osmosis is of great importance to plants and animals, because it is the process by which water is distributed to all the cells of living organisms. Cell walls are semipermeable membranes through which water passes in both directions in accordance with the principles discussed in the preceding paragraph. The membranes of living cells are also permeable to certain solutes so that nutrients and waste products are exchanged through them. The permeability of cell walls to solutes is frequently selective and to some extent independent of the size of the solute particles and to concentrations. For example, small magnesium ions do not pass through the walls of the gastrointestinal tract to any great extent, whereas much larger glucose molecules pass through at a rate too rapid to be accounted for by simple diffusion.

When solutions to be injected into the blood stream are made up, the osmotic pressure of the solution has to be considered. The average osmotic pressure of the blood is about 7.7 atm (it rises just after meals, then falls). If red blood cells are placed in a solution that has a greater osmotic pressure than that of normal blood, water passes out of the cells till they shrink and settle out of suspension. If the cells are put into a solution that has a lower osmotic pressure than blood, the cells may swell with water till the cell walls burst. Consequently, the osmotic pressure of solutions for injection is adjusted (chiefly with sodium chloride) till it is compatible[4] with blood.

EXERCISES

1. A solution of sulfuric acid was made by dissolving 20.0 g of 98 percent sulfuric acid in 250 ml of water. Express the concentration of this solution in percent by weight.
2. If the solution described in Exercise 1 has a total volume of 262 ml, what is its (a) molarity; (b) molality; (c) normality?
3. What weight of $CaCl_2$ is needed to prepare 100 ml of a 0.20 M solution?
4. What is the normality of a 0.25 M sodium hydroxide solution? Of a 0.25 M barium hydroxide solution? Of a 0.25 M phosphoric acid solution?
5. What volume of 0.15 N HCl is needed to neutralize 20.0 ml of a 0.35 N NaOH solution?
6. What volume of 0.25 M H_2SO_4 is needed to react with 40.0 ml of 0.25 M barium hydroxide?
7. Ten milliliters of vinegar is diluted to 100 ml with water. A 10-ml sample of this diluted solution required 12 ml of 0.05 N sodium hydroxide for neutralization. Calculate the concentration in g/liter of acetic acid in the vinegar. If the vinegar has a density of 1.03 g/ml, calculate the percent by weight of the acetic acid in the vinegar.
8. What weight of alcohol, C_2H_5OH, is present in 1 liter of wine that is 12 percent alcohol by volume? The density of the solution is 0.984 g/ml, and the density of pure alcohol is 0.789 g/ml. What is the molarity of the solution? How many alcohol molecules are present in 1 liter of the wine?
9. What is the boiling point of a solution that contains 60 g of glycerin, $C_3H_8O_3$, in 400 g of water?

---

[4] In biological and medical terminology, solutions that have the same osmotic pressure as blood are called **isotonic** solutions; those which have smaller osmotic pressures are called **hypotonic** solutions; and those which have larger osmotic pressures are called **hypertonic** solutions.

10. An automobile radiator has a capacity of 16 quarts. What volume of antifreeze (ethylene glycol, $C_2H_6O_2$, density, 1.427 g/ml) must be added to this quantity of water so that the freezing point of the solution is $-10°F$?
11. What would be the freezing point of a solution made by dissolving 5.00 g of a compound, molecular weight 75 awu, (a) in 100 g of water; (b) in 100 g of benzene; (c) in 100 g of hydrogen acetate?
12. Calculate the molecular weight of a compound if 2.750 g in 50.0 g of water gives a solution having a boiling point of $100.79°C$. What would be the freezing point of this solution?
13. Four solutions are prepared as follows:
    No. 1: 0.10 mole of ethyl alcohol in 100 g of water
    No. 2: 0.10 mole of sodium chloride in 100 g of water
    No. 3: 0.05 mole of sodium sulfate in 100 g of water
    No. 4: 0.10 mole of magnesium sulfate in 100 g of water
    Which solution has (a) the highest freezing point; (b) the lowest freezing point; (c) the lowest boiling point; (d) the highest boiling point?
14. What is the purpose of adding salt to the ice when making home-made ice cream?
15. A traffic engineer is considering purchasing a deicing salt for spreading on the city streets during winter. He is able to buy NaCl for 70 cents per 100 lb and $CaCl_2$ for 60 cents per 100 lb. Which is the more economical?
16. Hydrogen acetate is a weak electrolyte in water:

$$HC_2H_3O_2 + H_2O \rightleftarrows H_3O^+ + C_2H_3O_2^-$$

A solution of 3.0 g of $HC_2H_3O_2$ in 100 g of water gives a solution that begins to freeze at $-0.99°C$. Calculate the percentage ionization of $HC_2H_3O_2$ in this solution.
17. To determine the molecular weight of a compound, a 1.50-g sample was dissolved in 30.0 g of water. Ice crystals were observed to start to form in this solution at $-0.93°C$. Calculate the molecular weight.
18. Account for the fact that dried beans when placed in water swell and frequently burst.
19. Suggest why salted meat can be stored for longer periods than fresh meat.
20. Consider Fig. 13-7. If the beaker contains an aqueous solution of glycerin instead of pure water, what determines the relative heights of the liquid levels? Discuss fully. (Assume that the membrane is permeable to water only.)

SUGGESTED READING

Babbitt, J. D., "Osmotic Pressure," *Science*, 122, 285 (1955).

14

THE COLLOIDAL STATE

COLLOIDAL DISPERSIONS

When a small lump of sugar is placed in water, single molecules break away from the crystal and are scattered among the water molecules. In time the lump disappears, and the resulting mixture is called a solution (Chap. 12). The size of a sugar particle in the solution is essentially that of a single sugar molecule. Because of this particle size, the sugar cannot be removed by filtration, and the solution appears clear; as a matter of fact, it is as transparent as pure water.

When a small lump of clay is shaken with water that contains a small amount of sodium hydroxide, the clay breaks up into tiny particles that are scattered among the water molecules. The clay cannot be removed by filtration with ordinary filter paper, nor does it settle on standing. In these respects, the clay-water mixture is like the sugar solution. Unlike the sugar solution, the clay-water mixture is not transparent. This is because the clay does not break up into single molecules; rather, the particles are clusters of clay molecules (ranging in size from approximately 10 to 2000 A) that scatter light in random directions. Therefore, the mixture is not properly referred to as a solution. Instead, it is called a **colloidal dispersion** or a **colloidal system.**

It should be noted that, when clay and water are shaken together in the absence of a source of OH^- ions, most of the clay becomes suspended as particles with diameters larger than 2000 A. These particles can be removed by ordinary filtration, and they do settle upon standing. Ordinary muddy water is mostly a suspension of clay in water rather than a colloidal dispersion of clay in water.

We can distinguish between solutions, colloidal dispersions, and suspensions as follows.

Solution	Colloidal dispersion	Suspension
One phase present	Two phases present	Two phases present
Homogeneous	Borderline	Heterogeneous
Does not separate on standing	Does not separate on standing	Separates on standing
Transparent	Borderline	Not transparent

Particle Sizes and Shapes. Matter that is in a state of subdivision with the particle size ranging from approximately 10 A to 2000 A is said to be in the **colloidal state.** The colloidal state is not characteristic of any particular substance; practically all substances, whether normally gaseous, liquid, or solid, can be put into the colloidal state. For matter whose particles have the shape of droplets and granules, the diameter gives a measure of the particle size. For films and filaments, the length, width, and thickness are all needed to indicate the particle size. However, only one of these dimensions has to be in the range of 10 to 2000 A for the material to be classed as colloidal. For example, soap in a soap bubble is classed as colloidal, because the soap film is only a few molecules thick.

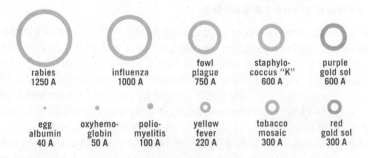

Fig. 14-1. Approximate sizes of some biochemical entities in the colloidal range, compared to particles in two gold sols. Gold sols have been obtained in shades of red, blue, and violet, depending on the size of the dispersed gold particles. (From W. J. Elford, *Trans. Faraday Soc.*, **33**, 1103 (1937) and J. W. McBain, *Colloid Science*, D. C. Heath and Company, New York, 1950.)

Particles in the colloidal size range may be simply clusters of a few hundred or a few thousand simple molecules, or they may be single huge molecules, such as starch, hemoglobin, or viruses. Figure 14-1 shows the relative sizes of some examples.

It should be pointed out that a definition of the colloidal state based entirely on particle size is not completely satisfactory. For one thing, the upper limit of 2000 A is an arbitrary choice. Older literature usually specifies 1000 A as the upper limit, while modern colloid chemists tend to raise the upper limit to 10,000 A, the maximum size of a particle that will remain suspended in a liquid for a reasonable period of time. Thus, colloidal properties, as well as particle size, are helpful in defining the colloidal state. Colloidal properties are discussed in a later section.

Parts of a Colloidal Dispersion. Colloidal dispersions are composed of two parts: the **dispersed phase,** which consists of the colloidal particles, and the **dispersing phase,** which consists of the continuous matter in which the colloidal particles are scattered. For example, in the clay-water dispersion, the clay is the dispersed phase and the water the dispersing phase.

Types of Colloidal Systems. Because both the dispersed and the dispersing phase may be a gas, a liquid, or a solid (except that both may not be gases[1]), there are eight types of colloidal systems. A list of the eight types with examples is given in Table 14-1. The meanings of the terms *foam, solid foam, liquid aerosol, solid aerosol, emulsion, solid emulsion, sol,* and *solid sol* are evident from this table.

TABLE 14-1.

Colloidal Dispersions

DISPERSED PHASE	DISPERSING MEDIUM	TYPE NAME	EXAMPLES
Gas	Liquid	Foam	Whipped cream, beer froth, soap suds
Gas	Solid	Solid foam	Pumice, marshmallow, polyurethane foam
Liquid	Gas	Liquid aerosol	Fog, clouds
Liquid	Liquid	Emulsion	Mayonnaise, milk, highway asphalt
Liquid	Solid	Solid emulsion	Cheese (butter fat dispersed in casein), butter
Solid	Gas	Solid aerosol	Smokes, dust
Solid	Liquid	Sol	Most paints, starch dispersed in water, jellies
Solid	Solid	Solid sol	Many alloys, black diamonds, ruby glass (gold in glass, a supercooled liquid)

Importance of Colloid Chemistry. Inasmuch as most substances can exist in the colloidal state, all fields of chemistry are concerned with colloid chemistry in some way or other. All living tissue is colloidal; hence the complex chemical reactions that are necessary to life must be interpreted in terms of colloid chemistry. The portion of the earth's crust that is referred to as tillable soil is composed in part of colloidal material; therefore soil science must include the application of colloid chemistry to soils. In industry, colloid science is important in the manufacture of paints, ceramics, plastics, textiles, photographic paper and films, glues, inks, cements, rubber, leather, salad dressings, butter, cheese and

---

[1] Only one phase is present in a mixture of gases.

other food products, lubricants, soaps, agricultural sprays and insecticides, detergents, gels and jellies, adhesives, and a host of other products. Such processes as bleaching, deodorizing, tanning, dyeing, and purification and flotation of minerals involve adsorption on the surface of colloidal matter and hence are concerned with colloid chemistry.

Recently a number of products have been marketed that might be described as "instant colloids." A liquid and a gas are confined under pressure in a container; as the mixture is released, the gas expands and forces the liquid out as a foam or an aerosol. Whipped cream and shaving cream are familiar examples of foams; some insect sprays are examples of liquid aerosols. Gases used in forming foams of foods are usually nitrous oxide, carbon dioxide, nitrogen or one of the Freons (the trade name for certain compounds of C, H, Cl, and F). However, it should be noted that in popular usage the term *aerosol* is not necessarily associated with colloids. For instance, a household "aerosol spray" may be merely a gross suspension of liquid droplets in air.

FORMATION OF COLLOIDAL PARTICLES

Matter can be put into the colloidal state by means of **dispersion methods,** in which large pieces of the substance are broken up into particles of colloidal size, and **condensation methods,** in which molecules or ions or atoms are made to cluster together to form particles of the desired size.

Dispersion Method of Forming Colloids. Larger pieces of matter can often be reduced in size by grinding, stirring, beating, or whipping. Specially designed mills called *colloid mills* are used for grinding paint pigments, face powders, and other materials. Beating, stirring, and whipping are employed to form emulsions and foams, such as mayonnaise and whipped cream.

In some cases, larger particles can be made to disintegrate by chemical reagents. As mentioned previously, sodium hydroxide causes clay to break up and become colloidally dispersed in water. Such a process is called **peptization,** and the reagent that brings it about is called a **peptizing agent.** Certain substances —for example, starch, glue, and gelatin—break up spontaneously into colloidal particles when placed in water. Warming and stirring hasten the process.

Condensation Method of Forming Colloids. The formation of fog and clouds by the clustering together of water molecules is a familiar illustration of the formation of a colloid by the condensation process.

The formation of insoluble substances from solutions is another illustration of this process. If a water solution of silver nitrate is added to aqueous sodium chloride, colloidal silver chloride forms briefly. On standing, the colloidal particles coagulate and form a precipitate:

$$AgNO_3 + NaCl \rightarrow AgCl \downarrow + NaNO_3$$

The process is as follows: In the silver nitrate solution, the silver is present as single ions, Ag^+, scattered among the water molecules. The same is true of the chloride ions, Cl^-, in the sodium chloride solution. When the two solutions are mixed, silver ions pair off with chloride ions all through the solution to make particles of insoluble silver chloride. Other silver ions and chloride ions bump into the small particles of silver chloride, and the crystals begin to grow. The mixture then takes on a milky appearance because of the formation of colloidal silver chloride. Unless further particle growth is prevented, the silver chloride particles cluster rapidly together, forming still larger particles that settle to the bottom as a precipitate. Coagulation can be avoided in this case if some gelatin is present to prevent the small silver chloride particles from touching one another.

Another example of the condensation method of preparing colloids is the process for preparing carbon black, the colloidal form of carbon used as a filler for rubber in automobile tires and in dispersions such as printer's ink and India ink. Carbon black is the soot formed when natural gas is burned in a limited amount of air. The carbon particles first formed when the hydrogen burns away aggregate into small particles that collect on cool surfaces from which they are scraped off at suitable intervals.

LIQUID COLLOIDAL SYSTEMS

Perhaps the most important type of colloidal system is that in which colloidal matter is dispersed in a liquid. In order for such a system to be stable, the colloidal particles must interact with the liquid phase sufficiently to prevent their clumping together and forming large particles that settle out. If the interaction of the colloidal particles with the liquid medium produces a stable dispersion, the system is said to be **lyophilic** (solvent loving). If the interaction is not strong enough to prevent clumping, the system is **lyophobic** (solvent hating). When water is used as the dispersing medium, the more specific terms **hydrophilic** (water loving) and **hydrophobic** (water hating) are often used.

Lyophilic (Hydrophilic) Systems. MACROMOLECULES. Often, natural substances derived from living organisms are composed of very large molecules, *macromolecules*, that have at least one dimension in the 10-A to 2000-A range. Noteworthy examples are hemoglobin, starches, gums, viruses, nucleic acids, rubber, and simple proteins such as gelatin. Synthetic plastics are also composed of macromolecules. Macromolecules may be more or less spherical in shape, as are molecules of hemoglobin and influenza virus, or they may be flexibly linear, as are molecules of nylon and cellulose, or they may have various shapes between these two extremes.

Most macromolecules dissolve in appropriate solvents, at least to a small extent, to form true solutions. Because the particle size of the individual solute molecule is in the colloidal range, such solutions are also *lyophilic colloidal dispersions*. The term *colloidal solutions* may be used in these cases.

To explain why the large molecules present in starch, gelatin, and similar organic materials tend to form stable dispersions in water, we need to examine their molecular structures, a subject dealt with in considerable detail in Chaps. 27 and 28. Let us simply point out that large, water-soluble molecules of biologic origin are composed of hundreds of carbon atoms linked together in chain or ring structures. Attached at frequent intervals to the carbon atoms are polar groups such as —OH, —NH_2, —COOH, —NH_3^+, and —COO^-. The strong attractive forces between these groups and water molecules account for the formation of a stable hydrophilic colloid.

Macromolecules of synthetic substances also form stable colloidal solutions. Examples include colloidal dispersions of polyethylene, polystyrene, and Plexiglas in solvents such as acetone, toluene, and ethyl acetate.

The swelling of seeds in water, dry wood in moist air, and rubber in benzene illustrates the natural tendency of macromolecules to become dispersed in a liquid medium. The rayon industry is based on the property of cellulose to form a colloidal solution in a water medium. The cellulose is then precipitated in the form of a fiber or thread (see Chap. 29).

ASSOCIATION COLLOIDS. A second type of lyophilic colloidal system includes dispersions of association colloids. Only examples of such systems in water media will be discussed.

Often, molecules have dual natures with respect to their attraction for water molecules. That is, one part of the molecule may be hydrophilic and another part, hydrophobic. This means that in water, one part of the molecule is strongly attracted by water molecules and tends to bring about the dissolving of individual molecules, while the other part tends to obstruct the solution process. When these opposing forces are properly balanced, solution of individual molecules does not occur; on the other hand, the substance does dissolve. Let us illustrate this apparently contradictory statement by describing the solutions that result when two organic salts, sodium acetate and sodium palmitate (a soap), are dissolved in water. Structures for the salts are shown in Fig. 14-2.

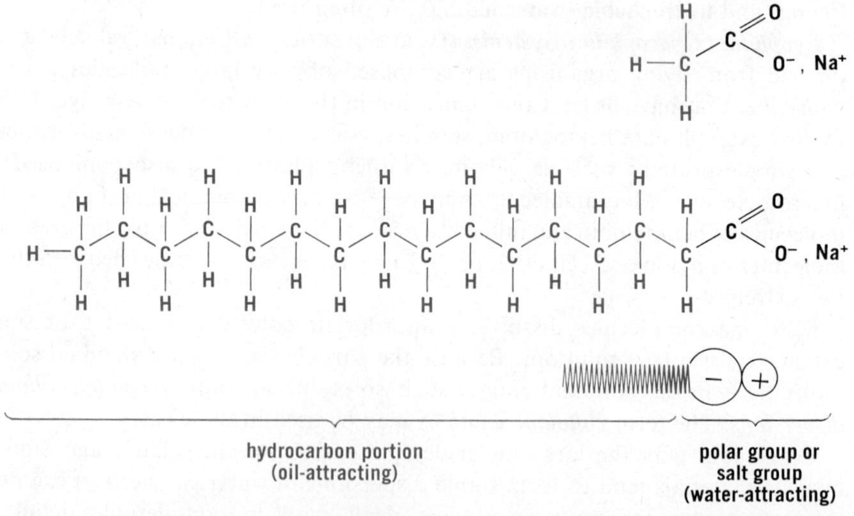

Fig. 14–2. Top, sodium acetate; middle, sodium palmitate; bottom, schematic representation of sodium palmitate and similar hydrophobic-hydrophilic substances.

The Colloidal State

Forces for solution of single sodium and acetate ions sufficiently outweigh the forces against solution to allow sodium acetate to dissolve as solvated sodium and acetate ions. This solution is not colloidal, for the ions are smaller than colloidal particles. In the case of sodium palmitate, no solution of simple ions occurs. Evidently, the energy required to separate water molecules from water molecules to make a place for the large hydrophobic portion of the palmitate ion is too great to allow this type of solution to exist.

However, because sodium palmitate does dissolve in water, some sort of compromise, so to speak, must take place. Several palmitate ions cluster together to form a particle that has the hydrophilic parts at the surface in contact with water and the hydrophobic parts in the interior of the particle where they mutually attract each other (Fig. 14-3a). The sodium ions are in the water phase. The dispersion is stable in the same sense that a sodium acetate solution is stable, but the dispersion is colloidal because of the large particles which have been formed in the solution process. Large particles of this type are called **micelles** or **association colloids**, and their dispersions are properly called colloidal

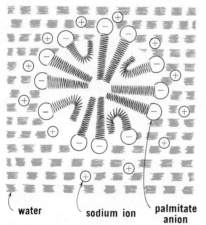
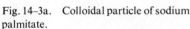
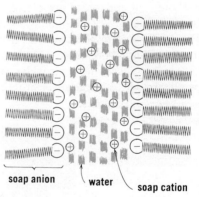

Fig. 14–3a. Colloidal particle of sodium palmitate.

Fig. 14–3b. Soap film of a soap bubble.

solutions. The way in which sodium palmitate and other soaps can form layered micelles in soap films and bubbles is shown in Fig. 14-3b. Here, a thin water layer is sandwiched between two layers of palmitate ions.

It will be recalled from Chap. 13 that the freezing point, boiling point, vapor pressure, and osmotic pressure of a simple solution differ from those of the pure solvent. Further, the amounts by which these values differ from those of the pure solvent are proportional to the molal concentrations (the colligative properties law, see page 196). It is a fact, however, that colloidal solutions do not obey the colligative properties law. To illustrate this point, let us consider two solutions: a 0.5 m solution of potassium chloride and a 0.5 m solution of potassium laurate. (Potassium laurate is a salt similar to sodium palmitate except that the carbon chain of the former contains eleven carbon atoms while that of the latter contains fifteen.) The freezing points of these two solutions are about $-1.8°C$ and $-0.15°C$, respectively. We see then that the potassium laurate lowers the freezing point only about 0.1 as much as is predicted by the colligative properties law. This is understandable, because the lowering of the freezing point is dependent upon the number of dissolved particles in a given amount of solvent, and the number of particles in the potassium laurate solution has been greatly reduced by micelle formation. For the same reason, the boiling point and osmotic pressure are not increased by the amounts predicted by the colligative properties law.

Lyophobic (Hydrophobic) Systems. Despite the fact that lyophobic systems contain dispersed colloidal matter that is normally classed as insoluble in the liquid phase and tends to separate from it, lyophobic systems can be prepared that persist for indefinite periods of time. The hydrophobic system shown in Fig. 14-7, consisting of colloidal carbon dispersed in water, has not noticeably changed in twenty years.

In contrast, some hydrophobic systems have a transitory existence, as was noted for colloidal silver chloride in the section on the condensation method for preparing colloids. As another example, consider the emulsion that is formed when water and an insoluble oil, such as a lubricating oil or an edible oil (Mazola, Wesson, olive, etc.), are vigorously shaken together. In such emulsions with only oil and water present, droplets rapidly coalesce into larger droplets that separate into two liquid layers, an oil layer floating on top of a water layer (Fig. 14-4). Hydrophobic systems that are often encountered in laboratory work or in commercial preparations include oil–water emulsions and water dispersions of metallic oxides and hydroxides, finely divided metals, metallic sulfides, and metallic halides such as silver bromide and silver chloride.

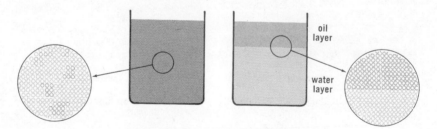

Fig. 14-4. An unstable emulsion of oil in water (left) rapidly changes to form an oil layer floating on a water layer (right).

Since lyophobic colloidal dispersions are intrinsically unstable, *they are not true solutions*. The question naturally arises as to how such dispersions can persist unchanged for long periods of time, as is the case with Faraday's gold colloids. More than one factor may be involved for a given colloid, but among the substances which may stabilize a system are protective colloids, adsorbed ions, and powders.

STABILIZATION BY A PROTECTIVE COLLOID. When oil is shaken with water that contains soap, there results a fairly stable emulsion consisting of oil droplets in the continuous water medium. The soap, a hydrophilic colloid acting as a

The Colloidal State

stabilizing agent for particles of a hydrophobic colloid, is called a **protective colloid**. It is admirably suited for this job because the carbon chain portion (see Fig. 14-2) is attracted to the oil molecules and the salt group at the end of the carbon chain is attracted to the water molecules. This results in the formation of a *skin*, or *film* (Fig. 14-5) around each droplet of oil, with the carbon chains dissolved in the oil droplet and the salt groups protruding through the surface of the droplet where they touch water molecules.

In the stabilization of emulsions (colloidal dispersions of two immiscible liquids), protective colloids are called **emulsifying agents.** Milk is an emulsion of globules of butter fat in water, with casein (a protein) serving as the emulsifying agent. Mayonnaise is an emulsion of a liquid fat (such as olive oil or corn oil) in water, with egg yolk playing the role of the emulsifying agent.

Soap acts as a protective colloid when it is used to wash dirt and grease off objects. The tiny particles of dirt and grease are coated with the soap and are then attracted by the water and carried away.

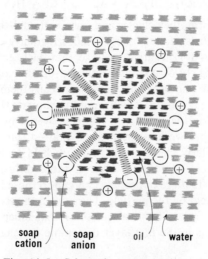

Fig. 14-5. Schematic representation of the stabilization with soap molecules of a particle in an oil–water emulsion.

A protective colloid must have an attraction for both the dispersed and the dispersing substances; it must show a preference for one, but not so great as to make it unattracted by the other. These protective films are examples of adsorbed layers.

Gelatin is used as a protective colloid in manufacturing silver bromide dispersions for photographic films and papers. It is also used in making ice cream; here it prevents the formation of large particles of sugar and ice. Latex, which comes from the rubber tree, is a milky dispersion of rubber particles in water, with protein material acting as the protective colloid. Agar and gums are other substances that can act as protective colloids in water dispersions of hydrophobic colloids.

STABILIZATION BY ADSORBED IONS. **Adsorption** is a phenomenon in which molecules, atoms, and ions become attached to the surfaces of solids and liquids. Adsorption differs from absorption in that the latter involves penetration by the absorbed material, as when water is absorbed by a sponge. Substances that have been truly adsorbed are not obviously present; adsorbed water on glass, for example, does not give the impression of a damp surface, nor can it be removed with a dry towel.

Adsorbed substances are held on surfaces by van der Waals forces or by valence bonds in layers that are usually no more than one or two molecules (or ions) thick. Our discussion is concerned with the adsorption of simple ions by colloidal particles. We shall illustrate this phenomenon with the stabilization of colloidal gold by the adsorption of negative ions. Colloidal gold can be prepared by a condensation process, a water solution of chlorauric acid being subjected to the action of a reducing agent, such as iron(II) chloride:

$$HAuCl_4 + 3FeCl_2 \rightarrow Au + 3FeCl_3 + HCl$$

As the gold atoms form, they cluster and form colloidal-sized crystals of gold. The atoms of gold in the surface of these small, irregularly shaped crystals exert strong attractive forces for negative ions, so that each gold particle becomes negatively charged (Fig. 14-6). Owing to the adsorption of negative ions, the gold particles repel one another. This repulsion keeps them from colliding and thus prevents further coagulation, which would precipitate the colloid. Note that the surface of the colloid particle is, in reality, covered by two layers of ions. The first is the layer that we have been discussing—a thin layer, perhaps only one ion thick, of similarly charged ions that are held to the surface by comparatively strong forces. The second layer is more diffuse and ill-defined. It consists mainly of ions that are oppositely charged to those held in the first layer and are less strongly held by the particle. The ions in the second layer are probably solvated and are free to move about somewhat independently of the colloidal particle. The ions in the first layer determine the charge on the colloidal particle.

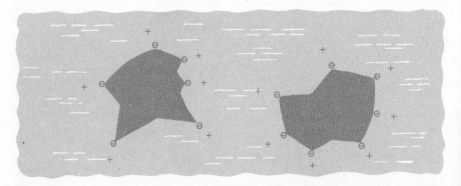

Fig. 14–6. Colloidal gold particles in water stabilized by the adsorption of negative ions.

Colloidal particles of gold, silver, platinum, and many sulfides acquire negative charges when prepared by precipitation from a water solution (condensation method). Colloidal arsenic trisulfide, prepared by passing hydrogen sulfide into aqueous arsenic trioxide, is negatively charged by the adsorption of HS^- ions. Colloidal silver prepared by the arc process adsorbs OH^- ions and is negatively charged.

On the other hand, metallic oxides and hydroxides acquire positive charges. For example, colloidal iron(III) hydroxide[2] particles formed by the following reaction acquire positive charges by the adsorption of either H^+ or Fe^{3+} ions from the solution:

$$FeCl_3 + 3H_2O \rightarrow Fe(OH)_3 + 3HCl$$

When colloidal particles are similarly charged, regardless of whether all are positive or all are negative, they repel one another and thus resist coagulation.

STABILIZATION BY POWDERS. There are many cases in which emulsions are stabilized by fine powders. For example, oil-in-water emulsions are stabilized by clay, lime, calcium carbonate, powdered glass, and pyrites; water-in-oil emulsions are stabilized by carbon.

PROPERTIES OF COLLOIDAL SYSTEMS

Tyndall Effect. All of us have observed the scattering of light by dust particles when a beam of sunlight enters a darkened room through a partly opened door or a slit in a curtain. The dust particles, many of them too small to be seen, look like bright points in the beam of light. If the particles are actually of colloidal size, we do not see the particles themselves; rather we see the light that is scattered by them. This phenomenon, known as the **Tyndall effect,** is due to the fact that small particles scatter light in all directions.

Particles smaller than colloidal size—i.e., molecules—also scatter light. However, unless the layer of molecules through which the light passes is quite thick—the sky or a deep lake, for example—the effect is not apparent to our eye.

The Tyndall effect can be used to differentiate between a colloidal dispersion and a true solution, because the atoms, molecules, or ions that are present in a solution do not scatter light noticeably (Fig. 14-7). This scattering of light accounts for the opacity of colloidal dispersions. For example, although both olive oil and water are transparent, a colloidal dispersion of the two has a milky appearance.

The blue of the sky is due to the fact that the small particles, including molecules of nitrogen and oxygen, scatter the shorter wavelengths (blue) more effectively than the longer wavelengths. If it were not for this scattering, light would reach us only by direct transmission from the sun or by reflection from surfaces; hence the sky would appear as dark in daytime as at night, with the sun acting as a huge spotlight during the day.

---

[2] The composition of colloidal iron(III) hydroxide is probably better represented by $Fe_2O_3 \cdot xH_2O$ than by $Fe(OH)_3$.

218 FUNDAMENTALS OF COLLEGE CHEMISTRY

Fig. 14–7. The Tyndall effect. The bottle on the left contains a fairly concentrated solution of copper sulfate in water. The other bottle contains a colloidal dispersion made by adding a single drop of India ink to water. The path of the beam of light from the spotlight at the left is hardly visible in the true solution but shows brightly in the colloidal dispersion.

Brownian Movement. The *ultramicroscope* is a device for viewing the Tyndall effect through an optical microscope. The optical microscope is focused on the colloidal dispersion at right angles to the light source and with the background in darkness. Under these conditions, the particles are observable not as particles with definite outlines, but as small, sparkling specks that move constantly in random, zigzag paths (Fig. 14-8). This erratic motion is accounted for on the

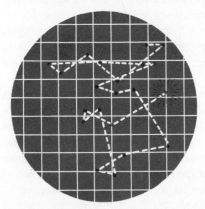

Fig. 14–8. Brownian movement. A plot of the observed position of a colloidal particle at 30-sec intervals. The particle is too small to be seen, except by the sparkle of reflected light. (From Jean Perrin, *Les Atomes*, Presses Universitaires de France, 1948.)

assumption that the molecules of the dispersing medium are constantly colliding with the colloidal particles and imparting motion to them. This random motion of colloidal particles in a dispersing medium is called **Brownian movement,** after the British botanist Robert Brown, who first observed it in 1827. The Brownian movement is one of the most direct pieces of evidence for the kinetic molecular theory. In 1905, in one of his three famous scientific papers of that year, Albert

Einstein showed mathematically how the visible movements of colloidal particles must be due to random collisions with the invisible molecules of the dispersing phase.

Rate of Settling. One of the striking characteristics of colloidal systems is the fact that very dense dispersed particles do not sink but may remain suspended indefinitely in a dispersing medium of much lower density. Gold sols in water can be quite stable; the carbon-water colloid shown at the right in Fig. 14-7 has not noticeably changed for more than twenty years. In this respect, dispersed colloidal particles are like simple molecules in a solution. Both have rotational and translational motions (called Brownian movement for colloidal particles) and both tend to diffuse throughout a medium. However, even molecules tend to settle in a gravitational field, although this settling is not perceptible in a shallow container. This settling is dependent on the gravitational force acting on the particle, the difference in the density of the particle and of the suspending medium, the viscosity[3] of the medium, and the size of the particle. Because the velocity of settling is proportional to the square of the radius, the time required for sedimentation increases rapidly with decreasing particle size. It has been calculated that a silver sphere of radius 1×10^{-3} cm would require 5 sec to settle a distance of 1 cm in water, whereas a silver sphere of radius 1×10^{-6} cm would require 5,000,000 sec (58 days) to settle the same amount. Convection currents would probably be sufficient to keep such colloidal matter suspended indefinitely in spite of this slight tendency to settle.

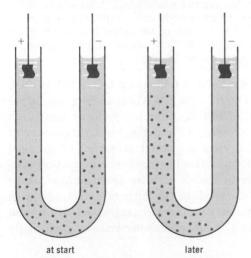

Fig. 14–9. Electrophoresis of a gold sol. The fact that the sol moves toward the positive electrode shows that the sol is negatively charged. On reaching the attracting electrode, some sols deposit and coat the electrode, while others are neutralized and fall coagulated to the bottom of the vessel.

Electrophoresis. The charged, dispersed particles in a colloidal system move toward the electrode that has the opposite charge. This movement is called **electrophoresis.** By putting two electrodes into a colloidal system and connecting them to a source of direct current, we can determine (1) if the particles are charged and (2) whether the charge is positive or negative (Fig. 14-9).

---

[3] The **viscosity** of a liquid is a measure of its resistance to flow. Molasses and tar have high viscosities, whereas water and gasoline have low viscosities.

If the colloid particles reach the electrode, they are neutralized and the colloidal system is precipitated. The dispersed phase usually coagulates and precipitates below the electrode, although in some cases it may be deposited on the electrode in a manner that resembles the electroplating of metal.

Electrophoresis is of importance commercially. For example, natural or synthetic rubber can be electrodeposited on anodes of various shapes. Thus an anode shaped like a hand is used in making rubber gloves. Many types of rubber articles are made in this way. The process is also important in biochemical research. Colloids present in organic materials, for example, proteins and related substances, may be separated and identified by electrophoresis.

Adsorption. Matter in the colloidal state has a tremendous amount of surface. To emphasize the point, let us consider a cube of iron that is 1 cm on an edge (Fig. 14-10). Such a cube would weigh 7.86 g (about 0.25 ounce) and have a

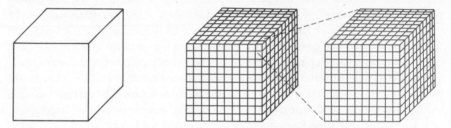

Fig. 14–10. A cube 1 cm on each edge has a surface area of 6 cm². If this cube is subdivided into 1000 cubes and each of these, in turn, is subdivided into 1000 cubes (0.01 cm on edge), the area will be $1000 \times 1000 \times 6 \times (0.01)^2$, or 600 sq cm. If the cubes are further subdivided into cubes of colloidal size (1×10^{-7} cm) (not shown in the diagram), the amount of surface will be 60,000,000 cm² (about 1.5 acres).

surface of 6 cm². If this cube is subdivided into a thousand billion billion small cubes, each of these cubes will be 1×10^{-7} cm (10 A) on an edge and hence will be of colloidal size. The amount of surface of the 7.86 g of iron now is not 6 cm² but 60,000,000 cm²—about 1.5 acres. Considering the large amount of surface, it is not surprising that the most important properties of colloidal matter are those which are dependent on surface interactions such as adsorption.

The forces that cause a substance to adsorb particles on its surface can be either physical (van der Waals) or chemical. In the latter case, the unsatisfied valence forces of surface atoms may lead to rather strong, chemical adsorption (see Fig. 14-11). The huge surface area of colloidal matter makes it particularly

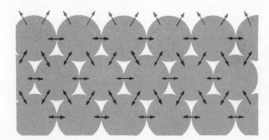

Fig. 14–11. Portion of plane of metal atoms in crystal. In the body of the crystal, neighboring atoms are strongly held by metallic bonds (black arrows), but surface atoms have unsatisfied bonding capacity (colored arrows).

efficient in attracting and holding molecules, atoms, and ions of other substances. Space permits mentioning only a few examples of large-scale applications of this phenomenon.

For use in war and riots, gas masks have been devised that contain charcoal or some other type of adsorbent. These adsorbents tend to adsorb polar molecules much more strongly than the nonpolar major constituents of the atmosphere. In addition to its present use in commercial gas masks, charcoal is widely used in industry and in laboratory work to remove impurities that have objectionable odors, flavors, colors, etc. It is also used in some cigarette filters, presumably with the aim of removing carcinogenic compounds from tobacco smoke.

Raw sugar as it is first obtained from sugar cane juice has a dark-brown color due to the presence of plant pigments. These pigments are removed by dissolving the sugar in warm water and then passing the solution first through adsorbing beds of diatomaceous earth (a porous, finely divided mineral) and next through beds of granular boneblack (charcoal made by the destructive distillation of bones). The colorless solution is then evaporated, yielding pure table sugar.

The separation of trace materials from mixtures—for example, the separation of vitamins and hormones from biological samples—presents an exceedingly difficult problem. Much success has been achieved by taking advantage of the difference in the tendency of the components of a mixture to be adsorbed on certain powdered materials. This method, called **chromatographic separation** or **chromatography,** employs a column of a solid, such as starch, aluminum oxide, sugar, paper (cellulose), etc. (Fig. 14-12). A solution of the mixture is allowed to

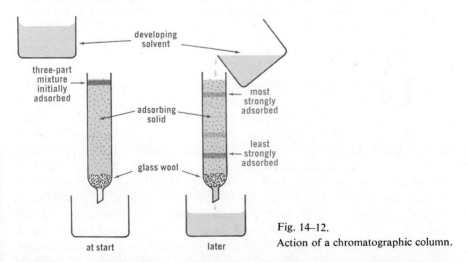

Fig. 14–12. Action of a chromatographic column.

seep downward through the column. The component most strongly adsorbed is held close to the top of the column, the component next most strongly adsorbed is held just below the first component, and so on down the column. If the adsorbed materials are colored, bands of color will denote the location of each component. Frequently, the location of colorless adsorbed material is determined by testing for fluorescent bands with ultraviolet light. The adsorbed material is removed by pouring quantities of fresh solvent down through the column. The component least readily adsorbed is dissolved from the column first.

Not only are different solute compounds adsorbed on solids more or less tenaciously, but solvents also vary greatly in their attraction for cellulose fibers. A solvent that is adsorbed strongly tends to displace adsorbed solutes and move them along efficiently from one site of adsorption to the next. A demonstration of the great difference in attraction for cellulose between water and carbon tetrachloride is shown in Fig. 14-13.

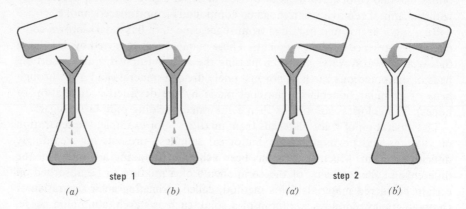

Fig. 14–13. An illustration of the relative attractive forces between cellulose-carbon tetrachloride and cellulose-water. (a) Carbon tetrachloride readily passes through filter paper; (b) Water readily passes through filter paper; (a') Water passes through the filter paper used in the filtration of carbon tetrachloride (and still wet with carbon tetrachloride); (b') Carbon tetrachloride does not pass through the filter paper used in the filtration of water (and still wet with water).

Gas chromatography employs the gas phase as carrier for adsorbates instead of the liquid phase. It is especially valuable in separating and purifying small amounts of gases or liquids, amounts too small to handle by ordinary distillation methods. Mixtures of gases or vaporized liquids are passed through an adsorbing column (at a temperature above the condensation point of any liquid present). Some carrier gas, such as helium, is used to move the vapors through the column. An important feature is that the movement of various compounds through the column can be followed by automatic devices; for example, the thermal conductivity of the gas mixture can be measured. The amount of each item and the time at which it passes out the end of the chromatographic column is plotted as a **chromatogram** on chart paper by an automatic recorder. The left-hand plot in Fig. 14-14 shows the separation of a mixture containing 0.004 ml of ether, 0.007 ml of acetone, 0.009 ml of benzene, and 0.012 ml of 2-butanol. Note that all of the ether came through the column within 2 min after the mixture had been added. Acetone came through about 2 min later, then benzene, and finally 2-butanol, the entire process requiring 8 min. The area under a peak is a measure of the relative amount of the substance causing the peak. A promising application of this technique is in the speedy identification of dangerously infectious bacteria. Each species apparently produces a different mixture of waste products. With the "fingerprint" chromatogram (see Fig. 14-14b) of its waste products, a bacterial strain can be identified in a matter of hours, instead of the days or weeks required for other methods.

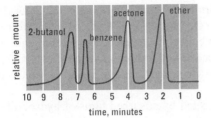
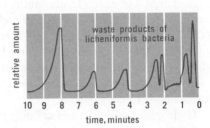

Fig. 14-14. (a) A chromatogram of a laboratory mixture of four vapors. (b) A chromatogram of the mixture of many substances excreted by a colony of bacteria.

Another field in which adsorption is extremely important is catalysis. The speed of many chemical reactions is increased when they are carried out in contact with a solid colloidal material. Finely divided nickel is much used as a catalyst for the hydrogenation of liquid fats to solid fats, and colloidal platinum is an excellent catalyst for the oxidation of sulfur dioxide to sulfur trioxide in the manufacture of sulfuric acid. There is considerable evidence that such reactions often occur while one of the reactants is adsorbed on the surface of the catalyst (see Chap. 15).

COAGULATION OF COLLOIDAL DISPERSIONS

It is often desirable during chemical operations to coagulate colloidal dispersions. Industrial smokes and dusts, colloidal precipitates in laboratory work, and foam formation during distillations are examples of colloidal dispersions that we generally desire to coagulate. Also, we frequently wish to coagulate natural sols and emulsions to obtain valuable products. Such products include butter and cheese from milk, and rubber from the sap of the rubber tree.

There are several methods of bringing about the coagulation of colloidal dispersions. In many cases, heating is all that is necessary. The protein colloid in an egg is coagulated by heating for a few minutes. In laboratory work, a mixture that contains a precipitate is often held at an elevated temperature for a short time before filtering in order to coagulate the fine particles.

"Rain making" has been achieved, but not on a practical scale, by seeding clouds with solid carbon dioxide or with silver iodide crystals. Recently, airport fogs have been successfully dispelled by seeding them with dry ice.

As previously noted, certain amounts of adsorbed ions can stabilize a colloid; however, large amounts of ions can cause coagulation. Perhaps the most effective way of coagulating colloidal dispersions of the sol and emulsion type is by adding an electrolyte. The added ions remove the adsorbed ions, so that the colloidal particles no longer repel one another but coalesce rapidly into larger particles. The choice of electrolyte depends on the type of adsorbed ions to be removed. For colloidal gold, because the adsorbed ions are negative ions, the most effective coagulating electrolytes are those which have positive ions of high charge. For example, aluminum chloride (Al^{3+}, $3Cl^-$) is more effective than an equivalent quantity of sodium chloride (Na^+, Cl^-) or calcium chloride (Ca^{2+}, $2Cl^-$) in coagulating gold sols. On the other hand, colloidal ferric hydroxide (positively charged) is more effectively coagulated by sodium sulfate ($2Na^+$, SO_4^{2-}) than by an equivalent quantity of sodium chloride (Na^+, Cl^-).

The mixing of two colloidal dispersions whose particles are oppositely charged causes both to coagulate.

Hydrophilic colloids, and colloids protected by adsorbed films of hydrophilic colloids, are also coagulated by electrolytes. This is because these colloids—gelatin and other proteins, soaps, starch, and glue—have polar groups that are strongly affected by a high concentration of ions. Raw rubber is obtained by adding acetic acid to the sap from rubber trees. The time-honored method of obtaining butter from milk is to let the milk sour (the dissolved sugar ferments, forming lactic acid). The acid thus formed is effective in removing the protective colloid casein, so that the butter fat globules can cluster together and form larger fat particles. The controlled souring of skim milk is the basis for an industry that produces 800 million pounds of cottage cheese annually. Recently, the high-speed direct coagulation of milk with hydrochloric acid has been used to produce a cheese curd that is superior in many ways to that formed by natural fermentation. The coagulation of skim milk with strong inorganic acid also is used to produce casein, millions of pounds of which are used yearly in making paints, plastics, glue, and paper coatings.

A commonly used procedure for destroying smoke and other types of aerosols is the Cottrell method of electrical coagulation. The smoke is led past a series of sharply pointed electrodes charged to a high potential (20,000 to 75,000 volts). The electrodes discharge and form ions in the air. Smoke particles adsorb these ions and become so highly charged that they are attracted to and held on the oppositely charged electrodes.

Dialysis. Because the presence of excess ions gradually brings about the coagulation of colloids, the removal of some ions is necessary if the dispersion is to be kept for any length of time. Colloidal gold prepared by the reduction of gold chloride soon coagulates, unless most of the Fe^{3+} and Cl^- ions are removed. This can be done by exposing the colloidal dispersion to pure water via a semipermeable membrane. A membrane is chosen through which water molecules and ions can pass, but the colloidal particles cannot. Fresh water flows through the membrane container and gradually extracts the ions from the colloidal dispersion (Fig. 14-15). The separation of ions from colloids by diffusion through a semipermeable membrane is called **dialysis**. Animal membranes, cellophane, and parchment paper are examples of membranes that have pores less than 10 A in diameter, only one-thousandth the diameter of the pores of filter paper.

The Colloidal State

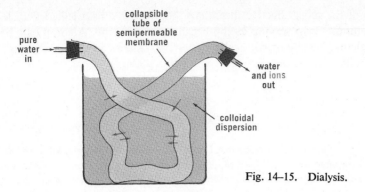

Fig. 14–15. Dialysis.

GELS AND JELLIES

The word **gel** is a general term that includes all such solid and semisolid colloids as jellies, gelatinous precipitates, membranes, synthetic and natural fibers, and similar materials.

The type of gel that is commonly referred to as a **jelly** (fruit jellies, Jell-O, jellied alcohol) belongs to the **sol** type of colloidal dispersion, i.e., a solid dispersed in a liquid. These systems appear to be more or less solid, even though the dispersed solid phase often makes up less than 1 percent of the total. The semisolid structure of jellies is accounted for on the assumption that molecules of the colloidal material aggregate into long, filamentlike structures. These elongated particles become entangled, forming a "brush-heap" structure that is capable of retaining large amounts of liquid, so that the whole appears to be a solid (Fig. 14-16). In fruit jellies, the carbohydrate *pectin* is responsible for forming the brush-heap structure. It is now common practice to use commercial preparations of pectin (Certo, Sure-Gel), for the amount of jelly and jam obtained from a given amount of fruit is considerably increased by their use. The entrapped water contains dissolved sugar, organic acids, and other substances.

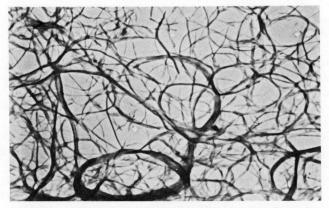

Fig. 14-16. Electron micrograph of sodium soap fibers in a lubricating grease. Oil is held in the tangled brush-heap structure of the soap. (Courtesy of Dr. J. N. Wilson, Shell Development Company.)

One of the most effective incendiary bombs contains a jellied petroleum oil, an aluminum soap serving as the brush-heap structure. Alcohol can be jellied with calcium acetate or with cellulose esters. "Canned heat" is a gel of alcohol and a cellulose ester.

EXERCISES

1. A cube of copper that is 2000 A on an edge contains how many atoms? (Consider that the space occupied by a single atom is equal to the cube of its diameter, 2.6 A.)
2. Could the copper particle described in Exercise 1 be seen clearly with a very powerful optical microscope? Explain.
3. Identify the dispersed and dispersing phases in each of the following: (a) fog; (b) a silver chloride sol; (c) French dressing; (d) tobacco smoke; (e) India ink.
4. For a salt such as $CH_3CH_2CH_2CH_2CH_2CH_2CH_2CH_2CH_2CH_2COOK$, what feature tends to make the salt soluble in water? What tends to make it insoluble? Would this salt probably be soluble in water? Explain.
5. Cite several examples of the use of protective colloids in the preparation of foods and commercial products.
6. Upon reading the labels of the common cleansers—Ajax, Comet, and others—one finds that sodium dodecyl benzenesulfonate is one of the active components. This compound can be thought of as a derivative of $NaHSO_4$ in which an OH of the HSO_4^- has been replaced by a benzene ring (see Chap. 26) to which is attached a 12-carbon chain comparable to the chain in Exercise 4. Explain how this compound aids in removing grease from an object.
7. Colloidal sols cannot be separated by filtration through ordinary filter paper. What does this tell one about the paper?
8. (a) By means of a diagram, show how an aerosol dispenser of whipped cream or shaving cream probably operates.
    (b) Usually, an aerosol dispenser has a label warning against disposing of it by burning, even when it is "empty." Explain.
9. Slightly acidic muddy pond water becomes clear on standing more quickly than slightly basic water. Why?
10. Untreated cotton garments take a much longer time to dry after washing than the "drip dry" fabrics. Suggest an explanation of this difference based on possible molecular structures.
11. Certain fabrics, for example, cotton, can be waterproofed by coating them with appropriate substances. Describe in hypothetical terms, using a diagram, one type of waterproofing substance, hopefully a long-wearing one.
12. The formulas for dodecane (a 12-carbon chain) and potassium dodecanoate are $CH_3(CH_2)_{10}CH_3$ and $CH_3(CH_2)_{10}COOK$, respectively. The former is a liquid and the latter is a solid. (a) Describe the structure and stability of the mixture that results when each is shaken vigorously with water. (b) Could either of the resulting dispersions be referred to as a colloidal solution? Explain.
13. The black ink of this printed page could have been prepared by burning natural gas in a limited amount of air, so that hydrogen burned away leaving colloidal carbon that was then mixed with a liquid to make printer's ink. Is this an example of the condensation or dispersion method for preparing colloids? Explain.

The Colloidal State

14. In some cases a colloidal dispersion may be correctly referred to as a solution, in other cases it may not. Explain why, giving examples.
15. Which of the following electrolytes is most effective in coagulating a sol that has adsorbed negative ions: (a) aluminum sulfate; (b) ammonium phosphate; (c) potassium nitrate; (d) sodium sulfate; (e) tin(IV) chloride? (Equivalent quantities are to be used.) Which is most effective in precipitating a positively charged sol?
16. The mud in a river tends to precipitate and form a delta at the point where it is carried into the ocean. Why?
17. Based on the Tyndall effect, why is yellow light used in preference to white light in fogs?
18. Whole milk tends to curdle when it is poured over sour fruit. Why?
19. A black ink was allowed to flow through a glass tube packed with cellulose particles. Three bands of color formed in the cellulose, a pink, a greenish-yellow, and a blue band. Upon pouring a small volume of colorless acetone through the column, the bands disappeared and the acetone emerged as a black liquid. Account for these observations.
20. A standard puff of cigarette smoke, 35 ml, is said to contain 2×10^{11} particles in the size range of 1500 A to 10,000 A. Compare this number with the number of molecules in pure air at 68°F and 760 mm.
21. The process of dialysis serves what purpose?
22. How might the physical shape of the colloidal particles in a gel differ from that in an oil emulsion?
23. If a spherical particle with a radius of 6×10^{-6} cm requires 40 days to settle 1 cm, how long will it take for a similar particle of 3×10^{-4} cm radius to settle 2 cm?
24. Consider the experiment described in Fig. 14-13. How would the results be changed if alcohol were substituted for carbon tetrachloride? If gasoline were substituted for carbon tetrachloride?

SUGGESTED READING

Anwar, M. H., "Separation of Plant Pigments by Thin Layer Chromatography," *J. Chem. Educ.*, **40**, 29 (1963).

Brewer, J. M., and R. B. Ashworth, "Disc Electrophoresis," *J. Chem. Educ.*, **46**, 41 (1969).

Garvin, J. E., "Student Experiment with Filter Paper Electrophoresis," *J. Chem. Educ.*, **38**, 36 (1961).

Hadorn, E., "Fractionating the Fruit Fly," *Sci. Amer.*, **206** (4), 100 (1962).

Herbener, R. E., "A Demonstration Device for Gas Chromatography," *J. Chem. Educ.*, **41**, 162 (1964).

Kerr, P. F., "Quick Clay," *Sci. Amer.*, **209** (5), 132 (1963).

King, L. C., and E. K. Neilsen, "Estimation of Avogadro's Number," *J. Chem. Educ.*, **35**, 198 (1958).

Ries, H. E., "Monomolecular Films," *Sci. Amer.*, **204** (3), 152 (1961).

Rollins, C., "Thin Layer Chromatographic Separation of Leaf Pigments," *J. Chem. Educ.*, **40**, 32 (1963).

Skomoroski, R. M., "Separation of Surface Active Compounds by Foam Fractionation," *J. Chem. Educ.*, **40**, 470 (1963).

Slabaugh, W. H., "Clay Colloids," *Chemistry*, **43** (4), 8 (1970).

Wilson, J. N., "Colloid and Surface Chemistry in Industrial Research," *J. Chem. Educ.*, **39**, 187 (1962).

15

CHEMICAL KINETICS; EQUILIBRIA

CHEMICAL KINETICS

The study of the mechanisms by which reactions occur and the rates of chemical reactions is known as **chemical kinetics.** By the *mechanism of a reaction* is meant the sequence of steps by which reactants are converted to products. The various steps in a chemical reaction are often quite complicated and cannot be predicted from the over-all equation commonly written for the reaction. For example, the equation we write for the reaction of hydrogen and oxygen to form water,

$$2H_2 + O_2 \rightarrow 2H_2O$$

seems to indicate that two molecules of hydrogen and one of oxygen come together simultaneously, and that two molecules of water are produced immediately from this meeting. It has been shown by numerous experiments that this is not at all the case. This reaction has been the subject of more than a hundred research papers, and, from this work, it is believed that the reaction proceeds through several steps, some of which are as follows:

$$H_2 + O_2 \rightarrow HO_2 + H$$
$$H_2 + HO_2 \rightarrow H_2O + OH$$
$$H_2 + OH \rightarrow H_2O + H$$
$$O_2 + H \rightarrow O + OH$$
$$H_2 + O \rightarrow OH + H$$
$$H + OH \rightarrow H_2O$$

Chemical Kinetics; Equilibria

Very reactive particles such as H, HO_2, OH, and O are believed to exist only momentarily before they collide and react with other particles.

By the **rate** or **speed of a reaction** is meant the amount of reactants converted to products in a unit of time.[1] The amount of reactants is usually expressed in moles per liter; the unit of time may be a second, a minute, an hour, or a day, depending upon whether the reaction is fast or slow.

The instant the button is pushed on a flashlight, a chemical reaction begins inside the cells and electrical energy is released. This is only one example of a chemical reaction that takes place fairly rapidly; there are many such reactions. The combustion of gasoline in an engine cylinder and the explosion of dynamite are chemical reactions that are completed in a fraction of a second. The reactions of ions in a water solution, such as the reaction of an acid and a base, are additional examples of reactions that take place at a very rapid rate. There are also chemical reactions that take minutes, hours, or even months for a detectable amount of change to occur. The rusting of iron under normal conditions usually cannot be noticed until several hours have passed.

Methods for determining the amount of products formed in a unit of time vary according to the nature of the reaction. If a gas is one of the products, its volume can be determined from time to time. If the reaction is being carried out in solution, samples can be taken at suitable intervals for analysis. Often the change in concentration of a reactant or product can be followed conveniently by a modern automatic spectroscope which records the absorption of radiation by a sample.

FACTORS THAT INFLUENCE REACTION RATES

Since all chemical reactions do not proceed at the same rate, and since the same reaction may proceed slowly or rapidly depending on conditions, we turn our attention to the factors that govern these differences. They are: (1) the *nature* of the reactants, (2) the *temperature* at which the reaction proceeds, (3) the presence of a *catalyst*, and (4) the *concentrations* of the reactants.

Nature of Reactants. Substances differ markedly in the rate at which they undergo chemical change. Hydrogen and fluorine molecules react explosively, even at room temperature, producing hydrogen fluoride molecules.

$$H_2 + F_2 \rightarrow 2HF \text{ (very fast at room temperature)}$$

Under similar conditions, hydrogen and iodine molecules react so slowly that no chemical change is apparent.

$$H_2 + I_2 \rightarrow 2HI \text{ (very slow at room temperature)}$$

Nickel and iron in the atmosphere corrode at different rates even when the temperature and concentrations are the same for both. In a relatively short

---

[1] This is a broad definition of the rate and is more appropriately the definition of the *net rate* or *over-all rate*. The *absolute rate* is defined in terms of the disappearance of a specific reactant, or the appearance of a specific product, under conditions such that the products are removed from the reaction medium as rapidly as they are formed.

time, iron oxide (rust) can be seen on the iron, but the surface of the nickel appears to be unchanged.

These differences may be attributed to the structure of the atoms, molecules, and ions that comprise the reacting materials. Thus, if a reaction involves two species of molecules whose atoms are already joined by strong covalent bonds, collisions between these molecules may not provide enough energy to bring about the formation of new bonds, even though the new bonds would represent a more stable condition.

Temperature. During chemical changes, it is necessary for the reacting molecules to approach each other very closely for the reaction to occur. Most frequently this approach is achieved by collisions of the molecules as they move about in a random way. However, in many reactions, even spontaneous exothermic reactions, most of the molecules simply rebound from collisions without reacting, unless the temperature is high. A familiar example is the reaction between hydrogen and oxygen that forms water. If the temperature is high, the reaction will take place explosively with the liberation of a great deal of energy.

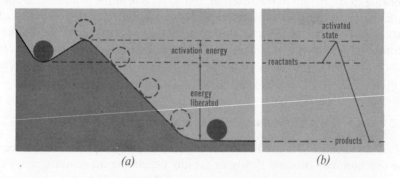

Fig. 15-1. (a) The energy of the ball in the depression must be raised before the ball will give up energy spontaneously and roll down the hill. (b) Many reactants must be given additional energy, e.g., by heating, before their molecules will react to release energy as heat and light and form products with less chemical energy.

But in a mixture of hydrogen and oxygen at room temperature, the molecules collide repeatedly with each other and rebound without change. The situation is similar to that depicted in Fig. 15-1. A ball that is resting in a depression on a hillside will give up energy if it can roll on down the hill. But it will not do this unless its energy is first raised enough to get it out of the depression.

In the case of the hydrogen and oxygen molecules at room temperature, the molecules on the average do not have enough energy to react. But if a lighted match is placed in the container, the molecules in the vicinity of the flame gain enough energy to react when they collide with each other. The energy thus released is transferred to nearby molecules, and the reaction spreads very rapidly to all parts of the container. The added energy needed for the reaction probably serves to weaken the strong covalent bonds that join the atoms in H_2 and O_2 molecules.

The energy that reacting substances must have to take part in a chemical reaction is called the **activation energy** of the reaction. The activation energy for a specific reaction depends primarily on the nature of the reactants.

The speed of chemical reactions is increased by an increase in temperature. A rise of 10°C in temperature usually doubles or trebles the speed of a reaction between molecules. We can account for this increase in reaction rate partly on the grounds that molecules move about more rapidly at higher temperatures and consequently collide with each other more frequently. However, this does not tell the whole story, unless the activation energy for the reaction is essentially zero. As the temperature is increased, not only do the molecules collide more often, but they collide with greater impact because they are moving more rapidly. At the elevated temperature, a larger percentage of the collisions result in a chemical reaction because a larger percentage of the molecules have the necessary activation energy to react (see Fig. 15-2).

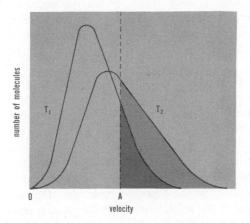

Fig. 15-2. The average velocity of the molecules of a gaseous substance increases as the temperature increases. The number of molecules that have velocity equal to or greater than some value A is greater for the higher temperature T_2, than for the lower temperature T_1.

Presence of a Catalyst. A **catalyst** is a substance that alters the speed of a chemical reaction without itself undergoing a permanent chemical change. The process is called **catalysis.** Catalysts are usually employed to increase the speed of the chemical reaction desired. Not infrequently, however, a substance is used to retard the rate of a chemical reaction. The use of certain substances in antifreezes to retard rusting, in rubber to retard aging, and in hydrogen peroxide to retard decomposition exemplifies the use of substances to slow down undesirable reactions. The substances employed for this purpose are called **inhibitors.**

A catalyst is thought to influence the speed of a chemical reaction in one of two ways: (1) by the formation of intermediate compounds, or (2) by adsorption.

FORMATION OF INTERMEDIATE COMPOUNDS. Let us again refer to Fig. 15-1. The ball will roll down the hill if it is first pushed over the hump. If the ball is very large, it may be desirable to lower the hump first by digging the top off. In chemical reactions that have high activation energies, we can "lift" the reactants over the energy barrier by raising the temperature. Frequently, however, it is undesirable to carry out a given reaction at a high temperature, because the products of the reaction may be unstable. The next approach is to look for a way of lowering the energy barrier, i.e., to provide a path around the barrier so that the molecules with lower energies can react.

How can the path of the reaction be changed? It may be that a substance, i.e., a catalyst, can be found that will react alike with energy-poor and energy-rich molecules to form an **intermediate compound,** which in turn reacts to form the desired substance. If the energy requirements for both these reactions are low, the majority of the molecules can participate at any given time. These reactions may be generalized as follows, C representing the catalyst:

$$A + B \rightarrow AB \text{ (very slow reaction)}$$
high activation energy; AB formed slowly

$$A + C \rightarrow AC \text{ (rapid)}$$

$$AC + B \rightarrow AB + C \text{ (rapid)}$$
low activation energies; AB formed rapidly

Note that C does not undergo a permanent change; it can be used over and over. The intermediate compound AC usually has only a temporary existence, being used up as rapidly as it is formed. Many chemical reactions are known to follow such a path when a catalyst is used. Sometimes the intermediate compound AC can be obtained in small quantities.

ADSORPTION. Many solid substances which act as catalysts can hold appreciable quantities of gases and liquids on their surfaces by adsorption (see Chap. 14). For a solid to be an effective adsorbing agent, it must be in a finely divided or expanded state so that a large amount of surface will be present. Finely divided nickel and platinum are well known for their ability to adsorb large amounts of various gases. The adsorbed molecules are frequently more reactive than the unadsorbed molecules. This increased reactivity can be attributed in some cases to the increased concentration of the adsorbed molecules; they are crowded close together on the surface of the solid, whereas in the gaseous state they are far apart. In other cases, the attractive forces between the molecules of

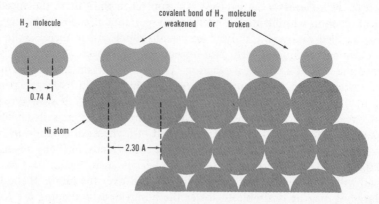

Fig. 15–3. Adsorption of H_2 molecules on surface of nickel catalyst.

Chemical Kinetics; Equilibria

the solid and those of the adsorbed liquid or gas make the adsorbed molecules more active chemically. This causes the reaction between molecules A and B to take place on the surface of the solid at a faster rate than if the catalyst were not present. (See Fig. 15-3.) The catalyst must not strongly adsorb the *product* of the reaction. As the reaction proceeds, the product leaves the surface and more reactants are adsorbed. Thus the surface is used over and over.

The hydrogenation of liquid fats to form solid fats is carried out on the surface of finely divided nickel. The oxidation of sulfur dioxide to sulfur trioxide on a platinum surface (in the production of sulfuric acid) is another case in which a catalyst influences the speed of a reaction by adsorption.

Concentration. We should expect the rate of a reaction to increase as the molecules become more concentrated—crowded more closely together—as the result of the increase in the frequency of collision, and this is usually the case. There are many ways of expressing the concentration of molecules. For reactions carried out in liquid solutions, the concentrations are usually expressed in moles per liter; for reactions involving gaseous mixtures, the concentrations are often expressed in terms of the pressures of the individual gases. Either method, of course, gives a measure of the number of molecules present in a unit of volume.

The amount of change produced in a reaction rate by changes in concentration cannot be predicted from the over-all chemical equation. For example, consider the decomposition of nitrogen pentoxide and nitrogen dioxide:

$$2N_2O_5 \rightarrow 4NO_2 + O_2$$

$$2NO_2 \rightarrow 2NO + O_2$$

It has been determined experimentally that the rate of the first reaction is directly proportional to the concentration of N_2O_5, while the rate of the second reaction is directly proportional to the square of the concentration of NO_2. That is, doubling the concentration of N_2O_5 doubles the rate of the first decomposition, but doubling the concentration of NO_2 quadruples the rate of the second decomposition. These experimental facts may be represented mathematically thus:

$$R_{N_2O_5} \propto [N_2O_5]$$

$$R_{NO_2} \propto [NO_2]^2$$

Where R is the reaction rate, $\propto$ means "is proportional to," and the brackets mean "concentrations in moles per liter." Because the rate is also affected by the temperature, the nature of the reactants, and possibly a catalyst, we may put in a term k to express the effect of these influences, and write

$$R_{N_2O_5} = k \times [N_2O_5]$$

$$R_{NO_2} = k \times [NO_2]^2$$

The quantity k is referred to as the *rate constant* for a reaction. For a given reaction under a specified set of conditions, except that concentrations may be varied at will, the numerical value of k does not change.

As another example of the influence of concentration, let us consider the union of hydrogen and iodine at a temperature at which all the substances are gases:

$$H_2 + I_2 \rightarrow 2HI$$

This reaction has been thoroughly investigated by experiment and the rate has been found to be directly proportional to the concentration of each reactant:

$$R = k \times [H_2] \times [I_2]$$

Accordingly, if the concentration of H_2 were doubled and the concentration of I_2 tripled, the reaction would proceed at six times the original speed. On the other hand, it has been determined that the rate of decomposition of hydrogen iodide,

$$2HI \rightarrow H_2 + I_2$$

is directly proportional to the square of the HI concentration:

$$R = k \times [HI]^2$$

In summary, the rate of a reaction is proportional to the concentration of each reactant raised to some power. The power to which the concentration is to be raised must be determined experimentally. In many cases the power is one (as in the case for the reaction involving H_2 and I_2); in other cases it is two (as for the decomposition of HI); in still other cases it may be fractional, or even zero.

REACTION MECHANISM AND REACTION RATE

The influence of concentration changes on the reaction rate may provide an important clue to the mechanism of the reaction. To illustrate this point, let us consider the hypothetical reaction

$$A_2 + 2B \rightarrow 2AB$$

Suppose we find by experiment that the reaction rate is directly proportional to the concentration of A_2 but is not changed appreciably by changes in the concentration of B, providing of course that there is always some B present. The rate expression is:

$$R = k \times [A_2] \times [B]^0 = k \times [A_2]$$

This might lead us to the assumption that the reaction proceeds through a slow step involving only A_2, followed by a relatively fast step involving B. Accordingly, we might postulate that the reaction proceeds through the following steps:

$$A_2 \rightarrow 2A \quad \text{(slow)} \tag{1}$$

$$A + B \rightarrow AB \quad \text{(fast)} \tag{2}$$

Of course, we would need to devise other experiments, for example, spectroscopic identification of the intermediate particles, so that the postulated steps could be tested by experiment.

For reactions that proceed through a sequence of steps, there is usually a step, called the **rate determining step,** which sets the pace for the over-all rate of the reaction. The rate-determining step for the above hypothetical reaction is (1).

CHEMICAL EQUILIBRIA

When the products of a specific chemical change can react to form the original substances, the change is said to be **reversible**. The pair of reactions, the forward and the reverse reaction, are referred to as two opposing chemical changes. In chemical reactions that are reversible, a condition of chemical equilibrium will result if conditions permit the opposing reactions to occur at the same rate. The following discussion is concerned with opposing chemical changes, reversible reactions, and the establishment of a chemical equilibrium.

Consider the union of hydrogen and nitrogen to form ammonia, Eq. (1), and the reverse or opposing reaction, the decomposition of ammonia to form hydrogen and nitrogen, Eq. (2).

$$3H_2 + N_2 \rightarrow 2NH_3 \tag{1}$$

$$2NH_3 \rightarrow 3H_2 + N_2 \tag{2}$$

When hydrogen and nitrogen are mixed in a 3 : 1 ratio by volume at room temperature, a reaction does not occur at a detectable speed; however, at elevated temperatures, the reaction is rapid. At a temperature of 200°C and a pressure of 30 atm, in the presence of a catalyst, this mixture reacts rapidly until about 67.6 percent of the original two gases is converted into ammonia. No further apparent change occurs in the amount of the three components present so long as the mixture is held at 200°C and 30 atm.

Similarly, ammonia does not decompose at room temperature (reaction 2, above) at a detectable rate. But if the temperature of the ammonia is raised to 200°C, the pressure is raised to 30 atm, and a catalyst is present, its decomposition into hydrogen and nitrogen occurs at a measurable speed. The amount of ammonia diminishes until 32.4 percent of the original amount has decomposed, after which there is no further apparent change.

Regardless of whether we start with pure ammonia or with pure hydrogen and nitrogen, neither reaction goes to completion; each appears to end by forming a mixture that contains by weight:

67.6 percent ammonia
32.4 percent hydrogen and nitrogen

Here we are dealing with two opposing reactions, each of which takes place in such a way that the other can occur at the same time, once the reaction has been started.

If we start with only hydrogen and nitrogen in the container, reaction 2 cannot occur because there is no ammonia. However, as reaction 1 proceeds, ammonia forms and reaction 2 starts, but at a slow rate because not much ammonia is present. Reaction 1 may be proceeding at a very rapid rate. As time goes on, the speed of reaction 1 steadily decreases, because hydrogen and nitrogen are being used up, and the speed of reaction 2 steadily increases, because the amount of ammonia is increasing. Eventually the speeds of the two opposing reactions become equal (Fig. 15-4). Once the reaction rates equalize, the amounts of hydrogen, nitrogen, and ammonia do not change so long as the temperature and pressure do not change and nothing is added to or removed from the container. Both reactions continue, and the system is in a state of equilibrium.

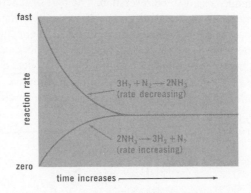

Fig. 15–4. With a mixture of only hydrogen and nitrogen, the reaction to form ammonia occurs at first at a relatively fast rate. However, the reaction rate steadily decreases because the concentrations of hydrogen and nitrogen are being steadily reduced (the gases are being converted into ammonia). At the beginning, the reaction rate for the decomposition of ammonia is zero because no ammonia is present. But this reaction rate increases as more and more ammonia is formed by the forward reaction. In time the concentrations of ammonia, hydrogen, and nitrogen become adjusted so that the two reaction rates are equal. A chemical equilibrium results.

A **chemical equilibrium** is a state in which two opposing reactions are proceeding at the same rate. The equilibrium state is indicated by double arrows, as shown in the following equation:

$$3H_2 + N_2 \rightleftarrows 2NH_3$$

It should be noted at this point that the relative amounts of reactants and products present at equilibrium may vary greatly.

Since most chemical reactions can come to a state of equilibrium under the conditions necessary to start the reaction, an understanding of how to avoid or how to establish an equilibrium is very important. The chemist or chemical engineer who is concerned with the large-scale manufacture of a useful compound is most interested in minimizing the influence of the reverse reaction. In a living organism there are many chemical equilibrium processes that are responsible for the well-being of the organism. For example, the acidity (or alkalinity) of the blood is maintained within very narrow limits by several opposing chemical reactions. Hence the research doctor, the pharmacologist, the nutritionist, the biochemist, and the soil chemist constantly study equilibrium processes in their efforts to solve the problems involved in making plants and animals healthier. The research chemist must delve deeply indeed into this important subject.

We might expect that the amounts of reactants and products at equilibrium would be determined by the same factors that determine reaction rates: nature, concentration, temperature, and catalyst. Surprisingly, perhaps, the presence or absence of a catalyst has no influence on the relative amounts present at equilibrium; a catalyst affects equally the rates of opposing reactions. The amounts of substances present at equilibrium are determined solely by the nature of the reactants, the temperature at which the equilibrium is established, and the initial concentrations. We shall discuss each of these in detail in the following sections.

INFLUENCE OF THE NATURE OF REACTANTS

In the two opposing reactions,

$$3H_2 + N_2 \rightleftarrows 2NH_3$$

the nature of ammonia molecules is such that they have a moderate tendency to decompose at 200°C and 30 atm. On the other hand, hydrogen and nitrogen molecules have a greater tendency to unite under these conditions, forming ammonia. This is the main reason that the equilibrium mixture has a larger proportion of ammonia (67.6 percent NH_3). That is, the rate of decomposition can equal the rate of formation only when the concentration of the ammonia molecules is sufficiently greater than that of the hydrogen and nitrogen molecules to compensate for the natural tendency of ammonia molecules to decompose slowly.

In many reversible reactions, the tendency for one of the opposing reactions to occur is much, much greater than it is for the other. In these cases we frequently ignore the slow one, because it has an insignificant effect on the amount of product obtainable from the reaction. Thus, in such chemical changes as:

$$2H_2 + O_2 \rightleftarrows 2H_2O$$

$$C + O_2 \rightleftarrows CO_2$$

$$2Na + Cl_2 \rightleftarrows 2NaCl$$

the reverse reaction occurs at an extremely slow rate unless the temperature is abnormally high. For all practical purposes we may consider that the forward reaction *goes to completion*, and we indicate this by using a single arrow.

$$2H_2 + O_2 \rightarrow 2H_2O$$

$$C + O_2 \rightarrow CO_2$$

$$2Na + Cl_2 \rightarrow 2NaCl$$

INFLUENCE OF CONCENTRATION

A balanced equation shows the ratio of the moles that react. For example, the equation

$$H_2 + Cl_2 \rightarrow 2HCl$$

states than when one mole of H_2 reacts, one mole of Cl_2 also reacts, and two moles of HCl are formed. However, it is not necessary to bring the reactants together in the molar ratio indicated by the equation for a reaction to occur. The reaction between H_2 and Cl_2 will also take place in mixtures of the two, where the molar concentration of H_2 is small as compared to the concentration of Cl_2, or in mixtures where it is large, as well as in mixtures where it is equal to the Cl_2 concentration.

In this section we will consider how the concentrations at equilibrium are affected by the relative amounts of reactants present at the beginning of the reaction. We will also consider how the amounts at equilibrium are changed when changes are made in the concentrations by adding or removing some of the substances in the equilibrium mixture. To evaluate these effects, we will begin by considering an equilibrium system that has been thoroughly studied by several investigators:

$$H_2(g) + I_2(g) \rightleftarrows 2HI(g)$$

In one set of experiments, A. H. Taylor and R. H. Crist sealed mixtures of H_2 and I_2 in glass tubes and placed these tubes in a liquid bath maintained at 425.4°C. After a suitable period of time to allow the system to come to equilibrium, the amounts of H_2, I_2, and HI were determined by analysis. In a second set of experiments, the procedure was repeated, except that the tubes were filled initially with HI instead of H_2 and I_2. Table 15-1 shows the amounts of H_2, I_2, and HI in each tube at equilibrium.

TABLE 15-1.

Study of the System $H_2 + I_2 \rightleftarrows 2HI$ at $425.4°C^a$

CONCENTRATIONS AT BEGINNING, MOLES/LITER			CONCENTRATIONS AT EQUILIBRIUM, MOLES/LITER			
$[H_2]$	$[I_2]$	$[HI]$	$[H_2]$	$[I_2]$	$[HI]$	K
0.011337	0.007510	None	0.004565	0.0007378	0.01354	54.8
0.011354	0.009044	None	0.003560	0.0012500	0.01559	54.7
0.010667	0.011965	None	0.001831	0.003129	0.01767	54.8
None	None	0.004489	0.0004789	0.0004789	0.003531	54.4
None	None	0.010692	0.001141	0.001141	0.008410	54.4

[a] Data from A. H. Taylor and R. H. Crist, *J. Amer. Chem. Soc.*, 63, 1377 (1941).

A careful study of these results reveals that the quantities present at equilibrium are related to each other by a simple mathematical relationship: *When the concentration of HI is squared and this number is divided by the product of the H_2 and I_2 concentrations, the same number is obtained in each case.* This is illustrated using the concentrations in the first and last sets of data in Table 15-1:

$$\frac{(0.01354)^2}{0.004565 \times 0.000738} = \frac{(0.008410)^2}{0.001141 \times 0.001141} = 54.6 \pm 0.2$$

Chemical Kinetics; Equilibria

We may express the mathematical relationship thus:

$$\frac{[HI]^2}{[H_2] \times [I_2]} = K$$

K, the *equilibrium constant*, has a numerical value of 54.6 *only* if this equilibrium is established at 425.4°C. If the equilibrium is established at another temperature, the numerical value of K will be different, depending on the temperature.

From the experimental study of a great number of chemical equilibria, it has been found that one can write similar mathematical statements relating the concentrations at equilibrium for these systems. The general form of the statement depends on the balanced equation for the equilibrium. If the equilibrium is represented by

$$mA + nB \rightleftarrows yC + zD$$

the general statement is:

$$K = \frac{[C]^y \times [D]^z}{[A]^m \times [B]^n}$$

where m, n, y, and z are the coefficients in the balanced equation and [ ] represents moles/liters of A, B, C, and D. Note that the products on the right of the equation are shown in the numerator and that the concentration of each substance is raised to the power equal to its coefficient in the equation. The following specific examples will serve to further illustrate the relationship.

$$2H_2 + O_2 \rightleftarrows 2H_2O$$

$$K = \frac{[H_2O]^2}{[H_2]^2 \times [O_2]}$$

$$CH_3Cl + HOH \rightleftarrows CH_3OH + HCl$$

$$K = \frac{[CH_3OH] \times [HCl]}{[CH_3Cl] \times [HOH]}$$

$$N_2 + 3H_2 \rightleftarrows 2NH_3$$

$$K = \frac{[NH_3]^2}{[N_2] \times [H_2]^3}$$

The numerical value of K is different for different equilibrium systems. For a given equilibrium system at a specific temperature, the value of K remains the same no matter how the individual concentrations may be changed.

The method followed here to derive the equilibrium constant expression is an *empirical* one. That is, large amounts of experimental results have been considered to reach the conclusion that K remains constant for a given equilibrium system, provided the temperature is not changed. The same conclusions may be reached from rather involved *theoretical* considerations, which we shall not discuss.

Problem 1: A quantity of HI was sealed in a tube, heated to 425.4°C, and held at that temperature until equilibrium was reached. The concentration of HI in the tube at equilibrium was found by analysis to be 0.0353 moles/liter. Calculate the equilibrium concentrations of H_2 and I_2.

Solution: Let x equal the number of moles of H_2 in a liter of the equilibrium mixture. Then x also equals the number of moles of I_2, since H_2 and I_2 are formed in equal molar amounts by the decomposition of HI:

$$2HI \rightarrow H_2 + I_2$$

Inasmuch as the equilibrium constant K is known for the equilibrium $H_2 + I_2 \rightleftarrows 2HI$ (Table 15-1), we use the equilibrium constant expression to solve for x:

$$K = \frac{[HI]^2}{[H_2] \times [I_2]}$$

$$54.6 = \frac{(0.0353)^2}{x \cdot x}$$

$$x^2 = (0.0353)^2 / 54.6$$

$$x = 0.00478 \text{ moles/liter of } H_2 \text{ or } I_2.$$

Problem 2: Into a 5.00-liter flask was placed 35.7 g of PCl_5. The flask and contents were heated to 250°C and then held at that temperature until the following equilibrium was established:

$$PCl_5(g) \rightleftarrows PCl_3(g) + Cl_2(g)$$

It was then shown by analysis that 8.75 g of Cl_2 was present in the equilibrium mixture. Calculate K.

Solution: Since K is defined as

$$K = \frac{[PCl_3] \times [Cl_2]}{[PCl_5]}$$

we need to know the values for the concentrations in moles per liter so that we can solve for the numerical value of K.

$$\text{Moles/liter of } PCl_5 \text{ at start} = \frac{35.7 \text{ g}}{5.00 \text{ liters}} \times \frac{1 \text{ mole}}{208 \text{ g}} = 0.0343$$

$$\text{Moles/liter of } Cl_2 \text{ at equilibrium} = \frac{8.75 \text{ g}}{5.00 \text{ liter}} \times \frac{1 \text{ mole}}{70.9 \text{ g}} = 0.0247$$

Since for each mole of Cl_2 that formed, one mole of PCl_3 must also form and one mole of PCl_5 must disappear, the numbers of moles at equilibrium are:

PCl_5: $0.0343 - 0.0247 = 0.0096$ mole/liter

Cl_2: 0.0247 mole/liter

PCl_3: 0.0247 mole/liter

$$K = \frac{0.0247 \times 0.0247}{0.0096} = 6.4 \times 10^{-2}$$

Pressure Changes. Pressure changes influence relative proportions at equilibrium only insofar as pressure changes bring about changes in concentration. Increasing the pressure on a liquid or solid does not appreciably crowd the molecules closer together and, therefore, has little effect on the concentrations. However, increasing the pressure on a gaseous equilibrium system makes for a

Chemical Kinetics; Equilibria

more crowded condition among the molecules, and may or may not bring about a change in the relative amounts present in the equilibrium mixture. Let us imagine that we have the gaseous system

$$A + B \rightleftarrows C + D$$

at equilibrium and that we have determined by analysis that the concentrations per liter are: A, 0.01 mole; B, 0.01 mole; C, 0.03 mole; D, 0.03 mole (see Fig. 15-5). If we double the pressure, how will these relative amounts be affected?

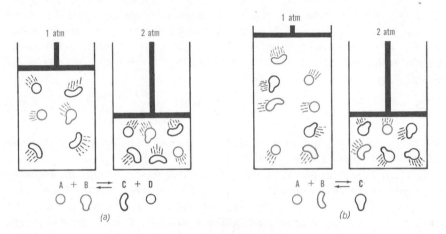

Fig. 15-5. (a) An increase in pressure has no influence on the relative amounts at equilibrium, because the number of molecules is not changed by the reaction. (b) An increase in pressure increases the percentage of C, because the formation of C from A and B reduces the total number of molecules.

By doubling the pressure, the concentration of each is doubled. Upon substituting the new concentrations in the equilibrium expression, we see that these concentrations give the same value for K as the initial concentrations:

$$K = \frac{0.03 \times 0.03}{0.01 \times 0.01} = \frac{(2 \times 0.03) \times (2 \times 0.03)}{(2 \times 0.01) \times (2 \times 0.01)} = 9$$

Hence, no change in relative concentrations is necessary. (Note that the value of K at equilibrium must not change if only the concentrations are being changed.)

Now let us consider a case in which the number of molecules represented on one side of the equation does not equal the number on the other:

$$A + B \rightleftarrows C$$

And let us also assume that at equilibrium at a given pressure and temperature, one liter of the equilibrium mixture contains 0.04 mole of A, 0.04 mole of B, and 0.01 mole of C. How will these proportions be affected if the pressure is suddenly doubled, keeping the temperature constant? As before, all concentrations are doubled. However, the new concentration, when substituted into the equilibrium expression, will not give the initial value for K (the symbol $\neq$ means "does not equal"):

$$K = \frac{0.01}{0.04 \times 0.04} \neq \frac{(2 \times 0.01)}{(2 \times 0.04) \times (2 \times 0.04)}$$

$$K = 6 \neq 3$$

The system cannot be at equilibrium with the new concentrations; there is now too little C and too much A and B present. The forward reaction is proceeding at a faster rate. In time, the amount of A and B present will be reduced sufficiently and the amount of C will be increased sufficiently, so that the system will again be at equilibrium. That is, in the new equilibrium mixture, the moles per liter of C will have increased to about 0.03 mole at the expense of A and B, each of which will have decreased to about 0.07 mole per liter. The value of the equilibrium constant will not have changed:

$$K = \frac{[C]}{[A] \times [B]} = \frac{0.03}{0.07 \times 0.07} = 6$$

We can say for this equilibrium system that an increase in pressure favors the formation of more C from A and B.

Le Chatelier's Principle. We have just seen that we can predict the effects of pressure changes on the amounts in a gaseous equilibrium mixture from a consideration of the equilibrium constant expression. The same prediction can be made from a consideration of **Le Chatelier's principle: When a stress is brought to bear upon a system at equilibrium, the system tends to change so as to relieve the stress.** The effect of increased pressure is to increase the crowded condition of the molecules. According to Le Chatelier's principle, that reaction will be favored which tends to lower the pressure by producing fewer molecules. For the reaction

$$A + B \rightleftarrows C + D$$
2 molecules ⇄ 2 molecules

pressure changes will have no influence on the equilibrium amounts. But for the reaction

$$A + B \rightleftarrows C$$
2 molecules ⇄ 1 molecule

an increase in pressure will favor the forward reaction, producing more C at the expense of A and B.

Let us consider one more example of a gaseous equilibrium system, the reaction of nitrogen and hydrogen to produce ammonia:

$$N_2 + 3H_2 \rightleftarrows 2NH_3$$
4 molecules ⇄ 2 molecules

This is a key reaction in the economy of any industrial country, as we shall see in a later chapter. Here, we will simply make a prediction, based on Le Chatelier's principle, that an increase in pressure will favor the formation of a larger percentage of ammonia, because the formation of ammonia reduces the number of

TABLE 15-2.

Study of the System $N_2 + 3H_2 \rightleftarrows 2NH_3$ *at* $375°C^a$

TOTAL PRESSURE AT EQUILIBRIUM, ATM	CONCENTRATIONS AT EQUILIBRIUM, MOLES/LITER			K	PERCENT NH$_3$
	[NH$_3$]	[H$_2$]	[N$_2$]		
10	0.00988	0.134	0.0446	0.956	5.25
30	0.0755	0.367	0.122	0.972	13.4
50	0.183	0.568	0.190	0.977	19.4
100	0.581	0.975	0.325	1.07	30.9

^{a} Concentrations in moles/liter calculated from data of A. T. Larson and R. L. Dodge, *J. Amer. Chem. Soc.*, **45**, 2918 (1923).

molecules. Table 15-2 shows the experimentally determined amounts of ammonia, nitrogen, and hydrogen in equilibrium mixtures at various pressures. Note that the value of K is essentially constant, while the percent ammonia varies from 5.25 at 10 atm to 30.9 at 100 atm. Figure 15-6 shows this effect of pressure schematically.

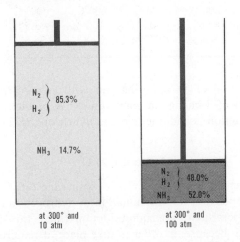

Fig. 15–6. An increase in pressure increases the amount of ammonia and decreases the amount of hydrogen and nitrogen in an equilibrium mixture of the three.

INFLUENCE OF TEMPERATURE

The speed of all chemical reactions is increased by an increase in temperature. However, the speed of different reactions is affected to a different extent by temperature changes. This means that in a reversible reaction, the speed of one of the opposing reactions is increased more than the speed of the other by an increase in temperature. A change in temperature, therefore, changes the relative proportions in an equilibrium mixture.

Let us consider again the equilibrium system involving hydrogen, nitrogen, and ammonia:

$$N_2 + 3H_2 \rightleftarrows 2NH_3$$

Suppose that a mixture of hydrogen and nitrogen in the ratio of three moles of H_2 to one mole of N_2 is placed in a closed container maintained at 30 atm pressure, where it is heated and held at a certain temperature until equilibrium is established. The amounts of ammonia, hydrogen, and nitrogen present at equilibrium (determined by chemical analysis) are shown in Table 15-3 for four different temperatures. Notice that the amount of ammonia in the equilibrium mixture is least at the highest temperature. This means that raising the temperature speeds up the decomposition of ammonia more than it speeds up the synthesis of ammonia. The tendency for ammonia molecules to decompose is so much greater at very high temperatures than the tendency for nitrogen and hydrogen molecules to unite that very little ammonia can be present in the equilibrium mixture.

TABLE 15-3.

Relative Amounts by Volume of Ammonia, Nitrogen, and Hydrogen in Equilibrium Mixtures at Different Temperatures

(Pressure constant at 30 atm)

	200°C	300°C	400°C	500°C
Ammonia, NH_3	67.6%	30.3%	10.2%	3.5%
Nitrogen and hydrogen, $N_2 + 3H_2$	32.4%	69.7%	89.8%	96.5%

How can we predict which reaction of an opposing pair will undergo the greater change in rate as the result of a change in temperature? A prediction can be made if we know which reaction is endothermic and which one is exothermic. In the equilibrium,

$$3H_2 + N_2 \rightleftarrows 2NH_3 \qquad \Delta H = -22.08 \text{ kcal at } 25°C$$

the union of hydrogen and nitrogen to form ammonia is exothermic (heat is liberated); conversely, the decomposition of ammonia into hydrogen and nitrogen is endothermic. If the temperature of the equilibrium mixture is raised, the endothermic reaction is favored, because, in order to take place, it requires heat from the surroundings. The exothermic reaction is favored by a decrease in temperature, because this reaction liberates heat to the surroundings and does so more readily if the surroundings are at a low temperature.

The above predictions are applications of Le Chatelier's principle. The stress being brought to bear in the case under discussion is the addition or removal of heat energy (temperature change).

As a second example, consider the equilibrium system

$$N_2 + 2O_2 \rightleftarrows 2NO_2 \qquad \Delta H = 16.2 \text{ kcal at } 25°C$$

The union of nitrogen and oxygen to form nitrogen dioxide is an endothermic reaction. On the basis of Le Chatelier's principle, we predict that higher temperatures favor the formation of nitrogen dioxide.

A word of caution on predictions: As the temperature goes up, the value of ΔH may change, and the tendency to attain maximum entropy becomes more important. Actually, all compounds tend to decompose if the temperature is high enough. With moderate increases in temperature, the formation of NO_2 is indeed favored, but at very high temperatures the decomposition of nitrogen dioxide is favored.

Before concluding this discussion of the influence of temperature, let us note how the size of the equilibrium constant K changes as the temperature changes. Since the products shown on the right of the equation appear as concentrations in the numerator of the equilibrium constant expression, a temperature change that decreases these concentrations will necessarily bring about a decrease in the magnitude of K. A study of the data in Table 15-3 will help verify this conclusion. The size of K depends upon the concentrations, as defined by:

$$K = \frac{[NH_3]^2}{[N_2] \times [H_2]^3}$$

As the temperature increases $[NH_3]$ decreases, while $[N_2]$ and $[H_2]$ increase. Consequently, K must be smaller at 300°C than at 200°C, still smaller at 400°C, and so on.

Enthalpy and Entropy. In Fig. 15-1, the natural tendency of a ball to roll downhill, thereby achieving a condition of lower potential energy, was pictured. We often observe that a system will spontaneously change to a condition of lower energy: water is constantly flowing downhill to the oceans, a ball in flight quickly comes to rest, a wound watch spring will uncoil, a hot piece of metal gradually comes to the temperature of its surroundings.

A chemical system can also change spontaneously by chemical reaction to a condition of lower energy. In a chemical reaction there are two factors to consider in deciding whether or not the change is spontaneous: (1) does the change result in a decrease in enthalpy, and (2) does the change result in an increase in entropy?

To explore the influence of the two forces, one to minimize enthalpy and the other to maximize disorder, let us consider the simple equilibrium system involving the dissociation of oxygen molecules into oxygen atoms, and the union of oxygen atoms to form oxygen molecules

Fig. 15-7. Orderly arranged balls tend to become randomly arranged when disturbed.

$$O_2 \rightleftarrows 2O \qquad \Delta H = 118.3 \text{ kcal at } 25°C$$

The tendency toward maximum disorder favors the dissociation of oxygen molecules into atoms. That oxygen in the form of randomly distributed atoms represents greater disorder than when in the form of O_2 molecules is apparent when we study Fig. 15-7. The balls on the board at the left are arranged in orderly pairs of two. When the board is vibrated to give the balls some kinetic energy, they tend to become randomly distributed as shown at the right. It is highly improbable that the balls will again pair off in twos, as they were initially, so long as the vibration of the board continues.

In the case of the balls, there is no opposing force to the tendency to achieve maximum disorder, providing the board is perfectly level. In the case of oxygen atoms, the tendency to achieve minimum enthalpy by forming O_2 molecules ($\Delta H = -118.3$ kcal) far outweighs the tendency to achieve maximum disorder (maximum entropy), providing the temperature is not high. A parallel situation can be visualized for the balls shown in Fig. 15-7. If the board is tilted sharply while being vibrated, the balls will collect at the bottom rather than become randomly distributed over the board. Here the tendency to minimize potential energy would outweigh the tendency to maximize disorder.

At equilibrium, then, these two opposing forces become equal.[2] However, since the tendency to minimize energy is so much greater at 25°C, the equilibrium mixture of oxygen will consist almost wholly of O_2 molecules. The equilibrium constant is exceedingly small, as is always the case when the concentrations of the products are small:

$$K = \frac{[O]^2}{[O_2]} = \text{a very small number}$$

As the temperature increases, the tendency to maximize disorder increases, and, ultimately, a temperature is reached, where this force is more important than loss of enthalpy. At a high temperature, say about 4000°C, the concentration of oxygen atoms will be relatively large, the concentration of O_2 molecules will be small, and K will be large.

$$K = \frac{[O]^2}{[O_2]} = \text{a large number}$$

As a second illustration of the influence of the two forces we have been discussing, let us again consider the equilibrium system

$$N_2 + 2O_2 \rightleftarrows 2NO_2 \qquad \Delta H = 16.2 \text{ kcal at } 25°C$$

Since the reaction of $N_2 + O_2$ to form NO_2 is endothermic (ΔH is positive) at 25°C, we predicted on the basis of Le Chatelier's principle that an increase in temperature above 25°C favors the formation of more NO_2. A logical conclusion, based only on Le Chatelier's principle, is that if the temperature is raised sufficiently the amount of NO_2 in the mixture will approach 100 percent. This will not happen, however, because an increase in temperature also increases the tendency toward disorder. In this case, matter in the form of N_2 and O_2 molecules represents greater disorder than in the form of NO_2 molecules. As the

---

[2] The difference in the two driving forces, with due regard to signs, gives a value called the **free energy change, ΔG**, for the reaction. The foregoing statement is usually written thus:

$$\Delta G = \Delta H - T\Delta S$$

At equilibrium, the two driving forces are the same and $\Delta G = 0$.

temperature is raised, the first effect is to increase the amount of NO_2 in the mixture. Eventually, however, a temperature is reached at which the influence of maximizing disorder outweighs the influence to absorb the energy being added to the system. Further increases in temperature will, of course, intensify the approach toward maximum disorder and bring about a corresponding decrease in the NO_2 concentration.

We have discussed two simple equilibrium systems for which it was possible to predict the condition of maximum disorder. For many chemical equilibria, one cannot judge from the equations which reaction will provide greater disorder. In these cases a quantitative experimental approach is necessary. However, we can say that if the temperature is sufficiently high, all polyatomic molecules tend to dissociate into smaller molecules or into atoms.

INFLUENCE OF A CATALYST

In an equilibrium system, a catalyst alters the speed of both forward and reverse reactions to the *same* extent. *A catalyst does not change the relative amounts present at equilibrium; the value of the equilibrium constant is not changed.* The catalyst does *shorten the time* required for establishing the equilibrium.

Reactions that require days or weeks to come to equilibrium may reach it in a matter of minutes in the presence of a catalyst. Furthermore, reactions that proceed at a suitable rate only at very high temperatures may proceed rapidly at much lower temperatures when a catalyst is used. This is especially important if high temperatures decrease the yield of the desired products.

EQUILIBRIUM CONSTANTS IN TERMS OF PARTIAL PRESSURES

Earlier in this chapter it was stated that concentrations of gaseous mixtures could be described in terms of partial pressures as well as in terms of moles per liter. For example, in a mixture of hydrogen, nitrogen, and ammonia that contains 0.100 mole of H_2, 0.100 mole of N_2, and 0.200 mole of NH_3, all in a volume of 1.00 liter, the total pressure at 0°C is 8.96 atm. The partial pressure of each gas is: $p_{H_2} = 2.24$ atm; $p_{N_2} = 2.24$ atm; $p_{NH_3} = 4.48$ atm. Because they are directly proportional to the number of moles, these partial pressures can be used to define the concentration of the mixture. They adequately express the fact that one-fourth of all the molecules in the container are H_2 molecules, one-fourth are N_2 molecules, and one-half are NH_3 molecules.

Measurements of partial pressures in an equilibrium mixture often can be made with simple experimental procedures. Further, the partial pressures give a meaningful statement of the concentration without the necessity of including volume in the statement. For example, if the partial pressure of nitrogen in a gaseous mixture is one-half of the total pressure, nitrogen molecules compose 50 mole percent of the mixture, irrespective of the volume. (Note that we usually define the concentration of water vapor in a gaseous mixture in terms of its vapor pressure.)

The expression for the equilibrium constant K_p in terms of pressure units takes the same form as does the expression for K:

$$2SO_2(g) + O_2(g) \rightleftarrows 2SO_3(g)$$

$$K = \frac{[SO_3]^2}{[SO_2]^2 \times [O_2]} \quad \text{concentrations in moles/liter}$$

$$K_p = \frac{p^2_{SO_2}}{p^2_{SO_3} \times p_{O_2}} \quad \text{concentrations as partial pressures}$$

The principles we have discussed for K are equally valid for K_p. However, K is not numerically equal to K_p for the above equilibrium since different sets of numbers are used to express the same concentrations. In systems where the number of reactant molecules equals the number of product molecules, K equals K_p.

EXERCISES

1. List the factors that influence the rates of chemical reactions. Which of these factors may be varied at will? Which are fixed for a given system?
2. "At equilibrium, all processes come to a halt." What is wrong with this statement when applied to chemical systems?
3. What is meant by "dynamic equilibrium"? Suggest how you might show that chemical equilibria are dynamic.
4. Data on the rate of the reaction $2A + B \to C$ are summarized below.
   (a) What is the mathematical expression for the rate of the reaction?
   (b) What is the numerical value of the rate constant?

Initial concentration of A, moles/liter	Initial concentration of B, moles/liter	Rate of formation of C, moles/(liter-sec)
0.050	0.050	4×10^{-3}
0.100	0.100	16×10^{-3}
0.050	0.025	2×10^{-3}

5. A mixture of hydrogen and oxygen can reside in a vessel for an indefinite period. However, if a nickel wire is introduced into the vessel, an explosion results. Explain.
6. Why do solid fuel stoves and furnaces have a mechanism for adjusting the flow of air into the firebox? Explain in terms of the principles discussed in this chapter.
7. What is meant by energy-rich molecules?
8. Among the factors that influence reaction rates, which is involved in each of the following?
   (a) Milk sours more rapidly in summer than in winter.
   (b) Powdered zinc reacts more rapidly with sulfuric acid than does mossy zinc (chunks of metallic zinc).
   (c) Barium reacts more rapidly with water than does magnesium.
   (d) Combustion of gasoline occurs more rapidly in an internal combustion engine than in an open container.

9. Write an expression for the rate of the following reactions, ignoring the possibility of a rate-determining step:
   (a) $N_2 + O_2 \rightarrow 2NO$
   (b) $N_2 + 3H_2 \rightarrow 2NH_3$
   (c) $2NO + Cl_2 \rightarrow 2NOCl$
   (d) $PCl_5 \rightarrow PCl_3 + Cl_2$
10. The reaction rate, R, for the reaction $2A + B \rightarrow A_2B$ was found experimentally to be given by the expression

$$R = k[A]^2 \times [B]^0$$

   (a) Will k increase, decrease, or remain unchanged if the concentration of A is doubled? If the concentration of B is doubled?
   (b) Will R increase, decrease, or remain unchanged if the concentration of A is doubled? If the concentration of B is doubled?
   (c) Suggest a mechanism that shows a rate-determining step consistent with the observed rate expression.
11. Write expressions for K for the following chemical equilibrium systems:
   (a) $SO_2 + NO_2 \rightleftarrows SO_3 + NO$
   (b) $2Cl_2 + 2H_2O \rightleftarrows 4HCl + O_2$
   (c) $H_2S \rightleftarrows 2H^+ + S^{2-}$
   (d) $2O_3 \rightleftarrows 3O_2$
   (e) $2ClF_3 \rightleftarrows Cl_2 + 3F_2$
   (f) $C_6H_6 + 3H_2 \rightleftarrows C_6H_{12}$
12. The equilibrium constants for three different reactions are 1.3×10^1, 2.6×10^{-21}, and 5.7×10^5. Assuming that each reaction can be represented in the form $A + B \rightleftarrows C + D$, for which of the three will the equilibrium mixture be more completely in the form of just two substances?
13. (a) How, if at all, will the relative amounts be changed when a gaseous equilibrium mixture consisting of hydrogen, oxygen, and water vapor is compressed at a constant temperature of 500°C? Explain.
   (b) Given that the decomposition of water into its elements is endothermic, how will temperature changes affect the relative amounts present in the equilibrium mixture?
   (c) Write the expression for K.
   (d) How will the value of K be affected by temperature changes?
14. If 0.1 mole of $COCl_2$ in a 1-liter vessel is 60 percent dissociated into CO and Cl_2 at a certain temperature, calculate the equilibrium constant.
15. If, in Exercise 14, 4.00 g of $COCl_2$ were initially introduced into a 1-liter container, calculate the concentration of $COCl_2$ at equilibrium at the same temperature as in Exercise 14.
16. The equilibrium constant for the gaseous reaction $H_2 + I_2 \rightleftarrows 2HI$ is 35 at a certain temperature. An equilibrium mixture at that temperature was found by analysis to contain 0.07 mole of HI and 0.015 mole of I_2 in 1 liter. Calculate the moles of H_2 in 1 liter.
17. For the equilibrium $2NO + O_2 \rightleftarrows 2NO_2$, the reaction to the right is endothermic. How does the equilibrium constant for this system vary with temperature? With pressure?
18. Consider the equilibrium system

$$C_2H_4 + H_2 \rightleftarrows C_2H_6 \qquad \Delta H = -32{,}740 \text{ cal}$$

(a) If you were manufacturing ethene, C_2H_4, from ethane, C_2H_6, would you carry out the reaction at a low temperature or a high temperature? Why?

(b) Would you use a low pressure or a high pressure? Explain why.

19. The reaction of ethene with hydrogen to produce ethane as described in Exercise 18 is very slow in the absence of a catalyst such as finely divided platinum. Explain how platinum might catalyze the reaction.

20. Using the data in Table 15-3, calculate K_p for the dissociation of ammonia, $2NH_3 \rightleftarrows 3H_2 + N_2$, (a) at 300°C and 30 atm; (b) at 400°C and 30 atm; (c) at 500°C and 30 atm.

21. Consider the two hypothetical reactions

$$A + B \rightleftarrows AB \quad \Delta H = -100 \text{ kcal}$$
$$C + D \rightleftarrows CD \quad \Delta H = -70 \text{ kcal}$$

If the first reaction involving A and B has an activation energy of 10 kcal/mole while the second reaction has an activation energy of 20 kcal/mole, what can be said about the relative values of the equilibrium constants for these two reactions at a given temperature?

22. The value of K at 77°C for the system $2NO + Br_2 \rightleftarrows 2NOBr$ is 6.4.

(a) Will the value of this constant increase, decrease, or remain unchanged if an equilibrium mixture of NO, Br_2, and NOBr at 77°C is suddenly compressed so that the concentration of each is doubled? Why?

(b) Will the change described in (a) cause the total amount of Br_2 to change? Why?

(c) Will the value of the constant change if Br_2 and NOBr are suddenly added so as to double the amount of each? Explain.

(d) Will the change described in (c) cause the amount of NO to change? Explain.

23. The equilibrium constant for the system

$$SO_2 + NO_2 \rightleftarrows SO_3 + NO$$

is 5 at a certain temperature. An equilibrium mixture at that temperature was found by analysis to contain 0.6 mole of SO_2, 0.3 mole of NO_2, and 1.1 moles of SO_3 in a liter. Calculate the moles of NO per liter.

SUGGESTED READING

Allen, J. A., "Some Applications of Free Energy in the Teaching of Chemistry," *J. Chem. Educ.*, **39**, 561 (1962).

Burton, M., "A Second Lecture in Thermodynamics," *J. Chem. Educ.*, **39**, 500 (1962).

Campbell, J. A., "Kinetics: Early and Often," *J. Chem. Educ.*, **40**, 578 (1963).

Cone, W. H., and R. A. Hermens, "A Simple Kinetics Experiment," *J. Chem. Educ.*, **40**, 421 (1963).

Kokes, R. J., M. K. Dorfman, and T. Mathia, "Chemical Equilibrium: The Hydrogenation of Benzene," *J. Chem. Educ.*, **39**, 91 (1962).

Lindauer, M. W., "The Evolution of the Concept of Chemical Equilibrium from 1775–1923," *J. Chem. Educ.*, **39**, 384 (1962).

MacWood, G. E., and F. H. Verhoek, "How Can You Tell Whether a Reaction Will Occur?," *J. Chem. Educ.*, **38**, 334 (1961).

Mahan, B. H., "Temperature Dependence of Equilibrium," *J. Chem. Educ.*, **40**, 293 (1963).

Matthews, G. W. J., "Demonstrations of Spontaneous Endothermic Reactions," *J. Chem. Educ.*, **43**, 476 (1966).

Miller, A. J., "Le Chatelier's Principle and the Equilibrium Constant," *J. Chem. Educ.*, **31**, 455 (1954).

16

IONIC EQUILIBRIA

In our study of chemical equilibria in the previous chapter we devoted our attention mainly to processes involving molecules. The concept of equilibrium is of equal importance in understanding reactions involving ions, particularly in solution. In this chapter we shall apply the theories of equilibria to solutions of electrolytes in water.

It has already been pointed out that certain covalent compounds react with polar solvents such as water to form ions, and that the process is called *ionization*. Ionization often goes essentially to completion provided the solutions are not too concentrated. In some cases, however, it stops far short of completion because of a strong tendency for the reverse reaction to occur.

When a covalent compound is dissolved in water, an equilibrium between the molecules and ions is rapidly established. The composition of the equilibrium mixture, i.e., the relative proportions of molecules and ions, is of great importance in certain chemical phenomena. We now turn our attention to some of the factors that determine the concentration of ions and molecules in aqueous solutions of electrolytes.

GENERAL ACID-BASE RELATIONSHIPS

According to the Brønsted-Lowry definition (Chap. 6) acids are *proton donors* and bases are *proton acceptors*. So far we have considered only the class of acids which in the pure state are composed of covalent molecules and which have been recognized as acids almost from the beginning of the science of chemistry. These include such acids as hydrochloric, HCl; nitric, HNO_3; sulfuric, H_2SO_4;

phosphoric, H_3PO_4; acetic, $HC_2H_3O_2$; and hydrosulfuric, H_2S. All these compounds react with water to form the hydronium ion, a reaction we have represented in two ways:

$$HA + H_2O \rightleftarrows H_3O^+ + A^-$$
$$HA \rightleftarrows H^+ + A^-$$

Because donating protons is a reversible reaction, every acid must form a base on donating its proton, and every base must form an acid on accepting a proton.[1] The base that results when an acid donates its proton is called the conjugate base of the acid. A^- is the conjugate base of HA, and H_2O is the conjugate base of H_3O^+. Or HA is the conjugate acid of base A^-, and H_3O^+ is the conjugate acid of the base H_2O:

$$\underset{\substack{\text{conjugate} \\ \text{acid I}}}{HA} + \underset{\substack{\text{conjugate} \\ \text{base II}}}{H_2O} \rightleftarrows \underset{\substack{\text{conjugate} \\ \text{acid II}}}{H_3O^+} + \underset{\substack{\text{conjugate} \\ \text{base I}}}{A^-}$$

A number of acids and their conjugate bases are listed in Table 16-1.

Strength of Acids and Bases. An acid that has a great tendency to give up a proton is called a strong acid, and a base that has a great tendency to take up a proton is called a strong base.

The conjugate base of a weak acid is a strong base. For example, water is a very weak acid; its conjugate base, the hydroxide ion, OH^-, is a very strong base, for it has a strong attraction for a proton. The CO_3^{2-} ion is a strong base; therefore, HCO_3^- is a weak acid.

---

[1] Another concept of acids and bases is based on the Lewis definitions. G. N. Lewis defined an acid as any species that acts as an electron-pair acceptor in chemical reactions; a base acts as an electron-pair donor.

According to the Lewis definition, unsolvated protons act as electron-pair acceptors (acids) when reacting with hydroxide ion, water, or ammonia:

$$H:\overset{..}{\underset{..}{O}}:^- + H^+ \rightarrow H:\overset{..}{\underset{..}{O}}:\\ H$$

$$H:\overset{..}{\underset{..}{O}}: + H^+ \rightarrow \left[H:\overset{..}{\underset{..}{O}}:H\right]^+\\ H H$$

$$\underset{H}{\overset{H}{H:\overset{..}{N}:}} + H^+ \rightarrow \left[\underset{H}{\overset{H}{H:\overset{..}{N}:H}}\right]^+$$

This definition, however, extends the concept of the acid-base relationship to a number of reactions not normally classed in this way. For example, in the following reaction, ammonia acts as a base and boron trichloride acts as an acid:

$$\underset{H}{\overset{H}{H:\overset{..}{N}:}} + \underset{Cl}{\overset{Cl}{\overset{..}{B}:Cl}} \rightarrow \underset{HCl}{\overset{HCl}{H:\overset{..}{N}:\overset{..}{B}:Cl}}$$

One advantage of the Lewis concept is that it identifies as acids certain nonhydrogen-containing substances that have the same function in chemical reactions as the common acids have.

Ionic Equilibria

The conjugate base of a strong acid is a weak base. The chloride ion, the conjugate base of hydrochloric acid, displays little tendency to react with protons to form hydrogen chloride molecules and hence is a weak base. Table 16-1 ranks several acids and bases in the order of comparative strength.

TABLE 16-1.
Relative Strengths of Acids and Bases

ACID		CONJUGATE BASE	
$HClO_4$		ClO_4^-	
HCl	Strong	Cl^-	Weak
HNO_3	acids	NO_3^-	bases
H_2SO_4		HSO_4^-	
H_3O^+		H_2O	
H_2SO_3		HSO_3^-	
HSO_4^-		SO_4^{2-}	
H_3PO_4		$H_2PO_4^-$	
$HC_2H_3O_2$	Decreasing	$C_2H_3O_2^-$	Increasing
H_2CO_3	strength	HCO_3^-	strength
H_2S		HS^-	
HSO_3^-		SO_3^{2-}	
HCN		CN^-	
NH_4^+		NH_3	
HCO_3^-		CO_3^{2-}	
HS^-		S^{2-}	
H_2O	Weak	OH^-	Strong
NH_3	acids	NH_2^-	bases
OH^-	↓	O^{2-}	↓

IONIZATION OF WEAK ACIDS

Acids are classified as **monoprotic, diprotic, triprotic,** etc., depending on the number of protons (positive hydrogen ions) that can be donated per molecule. The equations for the ionization in water of some weak, monoprotic acids are as follows:[2]

acetic acid
$$HC_2H_3O_2 \rightleftarrows H^+ + C_2H_3O_2^-$$

formic acid
$$HCHO_2 \rightleftarrows H^+ + CHO_2^-$$

hydrocyanic acid
$$HCN \rightleftarrows H^+ + CN^-$$

---

[2] In Chap. 6 it was pointed out that sometimes it is desirable to accent the interaction of H^+ ions and H_2O molecules by using the hydronium ion, H_3O^+, in equations. In the present chapter we shall use H_3O^+ whenever it contributes to a better understanding of the subject matter, especially when we wish to emphasize the attraction between protons and water molecules.

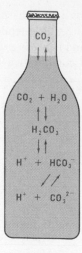

Fig. 16–1. A carbonated beverage affords a good example of equilibria involving molecules and ions.

The ionization of a diprotic acid proceeds in two steps; Fig. 16-1 illustrates a well-known example of such equilibria:

carbonic acid

$$H_2CO_3 \rightleftarrows H^+ + HCO_3^-$$
$$HCO_3^- \rightleftarrows H^+ + CO_3^{2-}$$

The ionization of a triprotic acid takes place in three steps:

phosphoric acid

$$H_3PO_4 \rightleftarrows H^+ + H_2PO_4^-$$
$$H_2PO_4^- \rightleftarrows H^+ + HPO_4^{2-}$$
$$HPO_4^{2-} \rightleftarrows H^+ + PO_4^{3-}$$

IONIZATION CONSTANTS

An equilibrium between ions and molecules can be treated mathematically in the same fashion as an equilibrium in which all species are molecules. The equilibrium constant for an ionization reaction is called the **ionization constant** (K_i).

For the ionization of acetic acid, the expression for the ionization constant follows from the equation

$$HC_2H_3O_2 \rightleftarrows H^+ + C_2H_3O_2^-$$

as[3]

$$K_i = \frac{[H^+] \times [C_2H_3O_2^-]}{[HC_2H_3O_2]} \tag{1}$$

---

[3] It is important to note that this ionization constant is actually the product of two constants, because the ionization proceeds through a reaction with water:

$$HC_2H_3O_2 + H_2O \rightleftarrows H_3O^+ + C_2H_3O_2^-$$

Ionic Equilibria

Because the ionization of a diprotic acid proceeds in two steps, such acids have two ionization constants, one for each step. For the ionization of carbonic acid, these constants are defined as follows:

$$K_{i_1} = \frac{[H^+] \times [HCO_3^-]}{[H_2CO_3]}$$

$$K_{i_2} = \frac{[H^+] \times [CO_3^{2-}]}{[HCO_3^-]}$$

In a solution of a diprotic acid, the concentration of ions formed in the first step is much greater than in the second. Usually the numerical value of K_{i_1} is about 10^5 times that of K_{i_2}. The same ratio holds for the steps in ionization of a triprotic acid.

It is important to note that both steps in the ionization of a diprotic acid are taking place in the same solution. In a solution of H_2CO_3 there is only one $[H^+]$ concentration and only one $[HCO_3^-]$ concentration. The same numerical values for these concentrations are used in calculating either K_{i_1} or K_{i_2}.

DETERMINATION OF IONIZATION CONSTANTS

In order to calculate the ionization constant of a weak electrolyte, we must in some way determine the number of ions present in solution and the number of molecules of the electrolyte that are not ionized. One method of determining the concentration of ions is by measuring the electrical conductivity—the greater the degree of ionization of a dissolved electrolyte, the greater the electrical conductivity of its solutions.

The usual expression for the equilibrium constant in the case above is

$$K = \frac{[H_3O^+] \times [C_2H_3O_2^-]}{[HC_2H_3O_2] \times [H_2O]}$$

However, the concentration of water remains practically unchanged in dilute acetic acid solutions of differing concentrations because of the relatively large amount of water present. So we may regard $[H_2O]$ as constant when we are dealing with solutions of moderate concentration and write

$$K = \frac{[H_3H^+] \times [C_2H_3O_2^-]}{[HC_2H_3O_2] \times k}$$

Then

$$K \times k = \frac{[H_3O^+] \times [C_2H_3O_2^-]}{[HC_2H_3O_2]}$$

Or, because the product of two constants is constant,

$$K_i = \frac{[H_3O^+] \times [C_2H_3O_2^-]}{[HC_2H_3O_2]} \qquad (2)$$

Note that the numerical value of K_i for Eq. (1), which contains $[H^+]$, must be the same as that for Eq. (2), which contains $[H_3O^+]$, because $[H^+]$ and $[H_3O^+]$ have the same value. At present, it is largely a matter of choice whether the expression for K_i for acids is written as in (1), where the proton is not shown attached to a water molecule, or as in (2), where the proton is shown attached to a simple water molecule.

The amount of the electrolyte present as molecules is calculated by subtracting from the total amount of solute the amount that is determined to be present as ions. If our electrical measurements indicate that 5.2 percent of the solute is present as ions, we assume that 94.8 percent of the solute molecules are not ionized.

Measurement of the electrical conductance of 0.1 M acetic acid reveals that it is 1.34 percent ionized. In other words, when 0.1 mole of hydrogen acetate is dissolved in enough water to make 1 liter of acetic acid solution, ionization proceeds rapidly to equilibrium, with the result that 1.34 percent of the hydrogen acetate is in the form of ions, and 98.66 percent is in the form of covalent molecules. Of course, hydrogen acetate molecules ionize continually, but the hydrogen and acetate ions also reunite continually; both reactions take place at the same speed, so that the concentrations of ions and molecules remain constant indefinitely, unless some change is made in the system.

Using these data, we can calculate K_i for the ionization of 0.1 M acetic acid. The concentrations in moles per liter are

$$HC_2H_3O_2 \rightleftarrows H^+ + C_2H_3O_2^-$$

0.09866 mole   0.00134 mole   0.00134 mole
(98.66% of 0.1)   (1.34% of 0.1)   (1.34% of 0.1)

Placing the concentrations in the expression for the equilibrium constant and solving, we obtain

$$K_i = \frac{0.00134 \times 0.00134}{0.09866}$$

$$K_i = 1.82 \times 10^{-5}$$

The ionization constants for weak bases are obtained in similar fashion. In the case of ammonium hydroxide

$$NH_3 + H_2O \rightleftarrows NH_4^+ + OH^-$$

$$K = \frac{[NH_4^+] \times [OH^-]}{[NH_3] \times [H_2O]}$$

With solutions of moderate concentration, the amount of water remains constant for all practical purposes, and

$$K = \frac{[NH_4^+] \times [OH^-]}{[NH_3] \times k}$$

$$K \times k = \frac{[NH_4^+] \times [OH^-]}{[NH_3]}$$

The product $K \times k$ is the ionization constant K_i:

$$K_i = \frac{[NH_4^+] \times [OH^-]}{[NH_3]}$$

The ionization constant for ammonium hydroxide as calculated from conductance measurements is 1.8×10^{-5}. (The fact that it is numerically so nearly the same as that for acetic acid is just a coincidence.) Ionization constants for several weak electrolytes are given in Table 16-2. The magnitude of the constant is indicative of the tendency of the electrolyte to form ions; the smaller the constant, the less this tendency. Arbitrary ranges of K_i values can be used to describe acids or bases, as follows:

very strong K_i greater than 1×10^3
strong K_i in range of 1×10^3 to 1×10^{-2}
weak K_i in range of 1×10^{-2} to 1×10^{-7}
very weak K_i less than 1×10^{-7}

TABLE 16-2.

Ionization Constants of Acids and Bases at $25°C$

NAME	IONIZATION REACTION	K_i
Hydrochloric acid	$HCl \rightleftarrows H^+ + Cl^-$	large
Sulfuric acid	$H_2SO_4 \rightleftarrows H^+ + HSO_4^-$	large
	$HSO_4^- \rightleftarrows H^+ + SO_4^{2-}$	1.2×10^{-2}
Sulfurous acid	$H_2SO_3 \rightleftarrows H^+ + HSO_3^-$	1.5×10^{-2}
	$HSO_3^- \rightleftarrows H^+ + SO_3^{2-}$	1.0×10^{-7}
Phosphoric acid	$H_3PO_4 \rightleftarrows H^+ + H_2PO_4^-$	7.5×10^{-3}
	$H_2PO_4^- \rightleftarrows H^+ + HPO_4^{2-}$	6.2×10^{-8}
	$HPO_4^{2-} \rightleftarrows H^+ + PO_4^{3-}$	2.2×10^{-13}
Hydrofluoric acid	$HF \rightleftarrows H^+ + F^-$	6.7×10^{-4}
Formic acid	$HCHO_2 \rightleftarrows H^+ + CHO_2^-$	1.8×10^{-4}
Acetic acid	$HC_2H_3O_2 \rightleftarrows H^+ + C_2H_3O_2^-$	1.8×10^{-5}
Carbonic acid	$H_2CO_3 \rightleftarrows H^+ + HCO_3^-$	4.3×10^{-7}
	$HCO_3^- \rightleftarrows H^+ + CO_3^{2-}$	5.6×10^{-11}
Hydrocyanic acid	$HCN \rightleftarrows H^+ + CN^-$	4.9×10^{-10}
Zinc hydroxide	$Zn(OH)_2 \rightleftarrows ZnOH^+ + OH^-$	9.6×10^{-4}
Methyl ammonium hydroxide (methyl amine in water)	$CH_3NH_3OH \rightleftarrows CH_3NH_3^+ + OH^-$	4.4×10^{-4}
Ammonium hydroxide	$NH_3 + H_2O \rightleftarrows NH_4^+ + OH^-$	1.8×10^{-5}
Urea	$CON_2H_4 + H_2O \rightleftarrows CON_2H_5^+ + OH^-$	1.5×10^{-14}

Calculation Involving K_i. Once the constant for a given equilibrium is known, the constant can be used in making calculations that involve different concentrations. The following problem is an example, and other examples are given in later sections of this chapter.

Problem 1: Given K_i for acetic acid as 1.8×10^{-5}, calculate the percentage of ionization of $0.50\ M$ acetic acid.

Solution: Let $x =$ number of moles of $HC_2H_3O_2$ ionized per liter.
Then $x =$ moles of H^+ ions and of $C_2H_3O_2^-$ ions per liter. (The small amount of H^+ ions due to ionization of the water itself is insignificant.)
And $0.50 - x =$ moles of $HC_2H_3O_2$ molecules per liter.
At equilibrium

$$\frac{[H^+] \times [C_2H_3O_2^-]}{[HC_2H_3O_2]} = 1.8 \times 10^{-5}$$

$$\frac{x \cdot x}{0.50 - x} = 1.8 \times 10^{-5}$$

To simplify the calculation, we can say that $0.50 - x$ approximately equals 0.50, because acetic acid is a weak acid. Of the original 0.50 mole dissolved to make the solution, only the tiny fraction here represented as x has changed to ions.

$$\frac{x \cdot x}{0.50} \simeq 1.8 \times 10^{-5}$$

$$x^2 \simeq 0.90 \times 10^{-5} = 9.0 \times 10^{-6}$$

$$x \simeq \sqrt{9.0} \times \sqrt{10^{-6}}$$

$$\simeq 3.0 \times 10^{-3} \text{ mole of acetic acid ionized}$$

$$\text{Percentage of ionization} = \frac{\text{no. of moles ionized}}{\text{total no. of moles of solute}} \times 100$$

$$\simeq \frac{3.0 \times 10^{-3}}{0.50} \times 100$$

$$\simeq 0.60 \text{ percent}$$

Comparing this result with the data given previously for the 0.1 M solution of acetic acid illustrates a general rule of great importance: *A dilute solution of a weak electrolyte is more completely ionized than a more concentrated solution.* The 0.1 M solution was 1.34 percent ionized, whereas the much stronger 0.50 M solution was only 0.60 percent ionized.

IONIZATION OF WATER

Electrical conductance measurements and other evidence show that water ionizes to a limited extent in accordance with the following equation:

$$HOH \rightleftarrows H^+ + OH^-$$

or

$$2HOH \rightleftarrows H_3O^+ + OH^-$$

The tendency for water to ionize is very slight as compared with the tendency for the reverse reaction (neutralization) to occur. Consequently, at equilibrium the relative number of ions is very small.

Precise measurement of the electrical conductance of very pure water at 25°C reveals that its ionic composition is as follows:

$$[H^+] = 1 \times 10^{-7} \text{ mole/liter}$$

$$[OH^-] = 1 \times 10^{-7} \text{ mole/liter}$$

At 25°C, there are 55.35 moles of water in 1 liter. Hence only 1 out of every 553,500,000 water molecules is ionized, an amount so insignificant that we probably would not bother to mention the ionization of water if we were dealing only with pure water. However, we shall be much concerned with this equilibrium and with the changes in ionic concentration when substances are dissolved in water. The expression for the equilibrium constant for the ionization of water,

$$K_i = \frac{[H^+] \times [OH^-]}{[HOH]}$$

can be modified to the extent that the concentration of water molecules is considered to be constant.

Ionic Equilibria

The change of a million billion molecules of water into ions, or vice versa, does not produce a detectable change in the enormous number of water molecules in a liter of water, but it does produce a large change in the number of ions. Considering that [HOH] remains constant at 55.35 moles/liter, we may write

$$K_i = \frac{[H^+] \times [OH^-]}{55.35}$$

Then

$$K_i \times 55.35 = [H^+] \times [OH^-]$$

Or, because the product of two constants is also a constant, we may write

$$K_w = [H^+] \times [OH^-]$$

By substituting the hydrogen ion and hydroxide ion concentrations in this expression, we can calculate the value of the constant:

$$K_w = 1 \times 10^{-7} \times 1 \times 10^{-7}$$
$$= 1 \times 10^{-14}$$

This constant, K_w, is called the **ion product for water.** It indicates that in pure water or in any water solution both hydrogen and hydroxyl ions must be present and, moreover, that the product of their concentrations must be constant. It may seem odd, but acid solutions contain a small concentration of hydroxide ions (base), and basic solutions contain a small concentration of hydrogen ions (acid). Further, if the concentration of one of these ions is known, the other can be easily calculated, because the product of the two must equal 1×10^{-14} (K_w at 25°).[4]

For example, if hydrogen chloride, HCl, is added to water till the solution contains 1×10^{-4} mole/liter of the H^+ ion, the OH^- ion concentration is calculated as follows:

$$K_w = [H^+] \times [OH^-]$$
$$1 \times 10^{-14} = 1 \times 10^{-4} \times [OH^-]$$
$$[OH^-] = \frac{1 \times 10^{-14}}{1 \times 10^{-4}}$$
$$= 1 \times 10^{-10} \text{ mole/liter}$$

Conversely, if sodium hydroxide, Na^+, OH^-, is added to water till the solution is 0.001 molar (1×10^{-3}), the hydrogen ion concentration can be calculated in this way:

$$1 \times 10^{-14} = [H^+] \times 1 \times 10^{-3}$$
$$[H^+] = \frac{1 \times 10^{-14}}{1 \times 10^{-3}}$$
$$= 1 \times 10^{-11} \text{ mole/liter}$$

---

[4] The value of K_w increases slightly as the temperature rises. For ordinary purposes, the value is taken as 1×10^{-14} when the temperature is at or near room temperature.

EXPRESSING HYDROGEN ION CONCENTRATION

In water solutions, common acidic properties are due to hydrogen ions, and common basic properties to hydroxide ions. If the concentration of the hydrogen ions is greater than 1×10^{-7} mole/liter, the solution is said to be acidic; if the concentration is less than this, the solution is basic. Because the product of the concentrations of these two ions is constant (1×10^{-14}), it is necessary to state the concentration of only one in order to describe the extent of the acidity or alkalinity of the solution. A widely used method of stating the hydrogen ion concentration of dilute acids, bases, and neutral solutions is in terms of pH. The pH of a solution is defined as follows:

$$p\text{H} = \log \frac{1}{[\text{H}^+]}$$

or

$$p\text{H} = -\log [\text{H}^+]$$

The logarithm of a number is the power to which the number 10 must be raised to give the number. The logarithm of 10 is 1; of 100, 2; of 0.001, -3. A logarithm table is required to find the logarithm if the number is not an exact multiple (or submultiple) of 10. A table of logarithms is given in the Appendix. Four examples will illustrate the use of pH.

1. The hydrogen ion concentration of water is 1×10^{-7} mole/liter at 25°C. The pH of water is

$$p\text{H} = \log \frac{1}{1 \times 10^{-7}} = \log (1 \times 10^7) = \log 1 \times \log 10^7 = 0 + 7 = 7$$

2. A solution of hydrochloric acid is 0.0063 N. If we assume complete ionization, the hydrogen ion concentration is 0.0063 mole/liter. The pH of the solution is

$$p\text{H} = \log \frac{1}{0.0063} = \log (1.59 \times 10^2) = \log 1.59 + \log 10^2 = 2.2$$

Alternate calculation:

$$p\text{H} = -\log 0.0063 = -(\bar{3}.80) = -(-3 + 0.80) = 2.2$$

3. A solution of NaOH is 0.01 N. The OH$^-$ ion concentration follows from the definition of a normal solution as 0.01, or 1×10^{-2} mole/liter. Before calculating the pH, we must obtain the hydrogen ion concentration:

$$K_w = [\text{H}^+] \times [\text{OH}^-]$$

$$1 \times 10^{-14} = [\text{H}^+] \times 1 \times 10^{-2}$$

$$[\text{H}^+] = \frac{1 \times 10^{-14}}{1 \times 10^{-2}} = 1 \times 10^{-12} \text{ mole/liter}$$

We then obtain the pH in the usual manner:

$$p\text{H} = \log \frac{1}{[\text{H}^+]} = \log \frac{1}{1 \times 10^{-12}} = \log (1 \times 10^{12}) = 12$$

For some purposes it is convenient to focus attention on the [OH$^-$] and to speak of the pOH, the negative logarithm of [OH$^-$]. This 0.01 N solution of NaOH has a pOH of 2. The sum of pOH and pH for the same solution is 14.

Ionic Equilibria

4. A solution of ammonium hydroxide is 0.01 N. Although this solution has the same normality as the sodium hydroxide solution above, and potentially as much base is present, the OH^- ion concentration is much less than 0.01, because ammonium hydroxide is a weak base; i.e., much of it is present as covalent molecules. Obtaining the equilibrium constant from Table 16-2, we can proceed with calculating the OH^- ion concentration in the equilibrium mixture:

$$\underset{0.01-x \text{ mole/liter}}{NH_3} + HOH \rightleftarrows \underset{x \text{ mole/liter}}{NH_4^+} + \underset{x \text{ mole/liter}}{OH^-}$$

$$K_i = \frac{[NH_4^+] \times [OH^-]}{[NH_3]}$$

$$1.8 \times 10^{-5} = \frac{x \cdot x}{0.01 - x}$$

$$1.8 \times 10^{-5} \simeq \frac{x \cdot x}{0.01}$$

$$x \simeq 4 \times 10^{-4} \text{ mole/liter concentration of } OH^- \text{ ion}$$

From this point the calculations are similar to those above for 0.01 N NaOH:

$$K_w = [H^+] \times [OH^-]$$
$$1 \times 10^{-14} \simeq [H^+] \times 4 \times 10^{-4}$$
$$[H^+] \simeq \frac{1 \times 10^{-14}}{4 \times 10^{-4}}$$
$$\simeq 0.25 \times 10^{-10} \text{ mole/liter}$$

$$pH \simeq \log \frac{1}{0.25 \times 10^{-10}} = \log \left(\frac{1}{0.25} \times 10^{10}\right) = \log(4 \times 10^{10})$$
$$\simeq \log 4 + \log 10^{10} = 0.6 + 10 = 10.6$$

From the foregoing examples we see that, if the pH is 7, the solution is neutral; if the pH is below 7, the solution is acidic; if the pH is above 7, the solution is basic. Figure 16-2 shows pH values for some common substances.

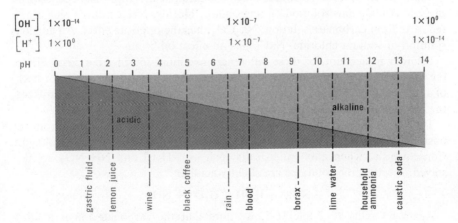

Fig. 16–2. The pH values of some common substances. (Courtesy of Beckman Instruments, Inc.)

INDICATORS

The accurate determination of the hydrogen ion concentration, or the pH, of a solution involves methods that cannot be discussed here. An approximate measurement can be made readily with **indicators**. Indicators are usually complex organic compounds that are one color if the hydrogen ion concentration is above a certain value and a different color if it is lower. The particular pH at which the color change occurs depends on the indicator. By using a variety of indicators and noting their colors in the solution, we can arrive at a fair estimate of the acidity or basicity of soils, water, body fluids, and other types of solutions. Seven of these useful substances are listed in Table 16-3, but many others are available.

TABLE 16-3

Indicators

NAME	pH AT WHICH COLOR CHANGE OCCURS	ACID COLOR	BASE COLOR
Methyl yellow	2-3	Red	Yellow
Methyl orange	3-4.5	Red	Yellow
Sodium alizarin sulfate	4.5-6.5	Yellow	Violet
Litmus	6-8	Red	Blue
Phenolphthalein	8-10	Colorless	Red
Thymolphthalein	10-12	Yellow	Violet
Trinitrobenzene	12-13	Colorless	Orange

HYDROLYSIS

Water solutions of normal salts may be neutral, acidic, or basic, depending on the salt. An aqueous solution of ammonium chloride, NH_4Cl, turns blue litmus red; a sodium carbonate solution, Na_2CO_3, has the opposite effect. An aqueous solution of sodium chloride, NaCl, has no effect on litmus.

In order to account for these differences, we must consider the possibility of the ions of the salt reacting with water. When the cation or the anion (or both) of a salt reacts with water so as to change the pH, the salt is said to be **hydrolyzed**; the reaction is referred to as **hydrolysis**.

Table 16-1 is of help in predicting the acid-base strength of ions. From the position of the NH_4^+ ion in this table, we see that it can act as a weak acid. Consequently, when ammonium salts such as NH_4Cl and NH_4NO_3 are dissolved in water, the solutions are slightly acidic.

$$NH_4^+ + HOH \rightleftarrows H_3O^+ + NH_3$$

Again referring to Table 16-1, we note that the carbonate ion is a fairly strong base. Solutions of carbonates are therefore rather strongly basic.

$$CO_3^{2-} + HOH \rightleftarrows HCO_3^- + OH^-$$

Ionic Equilibria

Both the chloride ion and the nitrate ion are very weak conjugate bases. Water solutions of chlorides and nitrates such as NaCl, KNO_3, $CaCl_2$, etc., are essentially neutral.

COMMON ION EFFECT

The ionization of a weak electrolyte is markedly decreased by adding to the solution an ionic compound containing one of the ions of the weak electrolyte. This effect is called the **common ion effect**.

Let us illustrate this by considering a water solution of ammonia. The following equilibrium exists in such a solution:

$$NH_3 + H_2O \rightleftarrows NH_4^+ + OH^-$$

When solid ammonium chloride (NH_4^+, Cl^-) is added to this solution, the concentration of NH_4^+ ions is greatly increased. The rate of the reverse reaction is therefore increased greatly, whereas the rate of the forward reaction is unaffected at first. But as the ions are consumed by the faster reaction, the concentration of the undissociated molecules (NH_3, H_2O) increases, and the rate of the forward reaction begins to increase. In time, the two rates become equal again, and the equilibrium is reestablished. In the new equilibrium condition, the concentration of the NH_4^+ ions is greater than in the first equilibrium, and the concentration of the OH^- ions is less. The value for K_i will be the same for either equilibrium.

In the same manner, the ionization of weak acids is suppressed by the addition of the salts of these acids. A dilute solution of acetic acid contains a lower concentration of the hydrogen ion when sodium acetate is also present.

The foregoing examples illustrate a general rule: weak acids and bases become even weaker in solutions with their salts.

Problem 2: Calculate the concentration of hydrogen ions in a solution that is 0.1 M with respect to acetic acid and 0.2 M with respect to $NaC_2H_3O_2$. Given: K_i for acetic acid is 1.8×10^{-5}.

Solution: Because sodium acetate is a strong electrolyte (a salt) and because acetic acid is sparingly ionized, the amount of acetate ion present depends almost entirely on the concentration of sodium acetate:

$$[C_2H_3O_2^-] \simeq 0.2 \text{ mole/liter}$$

The concentration of hydrogen acetate molecules may be taken as approximately equal to the concentration of the acetic acid (very little is in the form of ions):

$$[HC_2H_3O_2] \simeq 0.1 \text{ mole/liter}$$

Now we can proceed with calculating the hydrogen ion concentration:

$$1.8 \times 10^{-5} = \frac{[H^+] \times [C_2H_3O_2^-]}{[HC_2H_3O_2]} \simeq \frac{[H^+] \times 0.2}{0.1}$$

$$[H^{+1}] \simeq \frac{1.8 \times 10^{-5} \times 0.1}{0.2} \simeq \frac{1.8 \times 10^{-5} \times 1 \times 10^{-1}}{2 \times 10^{-1}}$$

$$\simeq 0.9 \times 10^{-5} \text{ or } 9 \times 10^{-6} \text{ mole/liter}$$

Note (page 256) that the concentration of H^+ ions in 0.1 M acetic acid (no salt is present) is 1.3×10^{-3} mole/liter, over a hundred times greater than when sodium acetate is present to the extent of 0.2 M.

BUFFERED SOLUTIONS

A solution that contains a weak acid plus a salt of that acid or a weak base plus a salt of that base has the ability to react with both acids and bases. Such a system is referred to as a **buffered solution,** because small additions of either acids or bases produce little change in the pH. Buffered solutions are of importance in many natural and synthetic systems.

For example, if more acid is added to a solution containing acetic acid and sodium acetate, the acetate ions react with the added hydrogen ions to form neutral hydrogen acetate molecules:

$$H^+ + C_2H_3O_2^- \rightarrow HC_2H_3O_2$$

The pH does not appreciably change.

On the other hand, if hydrogen ions are removed by the addition of a base,

$$H^+ + OH^- \rightarrow H_2O$$

the molecular hydrogen acetate ionizes to form more hydrogen ions:

$$HC_2H_3O_2 \rightarrow C_2H_3O_2^- + H^+$$

Again, the pH of the solution does not change significantly unless large quantities of the base are added.

In the blood, carbonic acid and sodium bicarbonate act as one buffer system to help maintain the pH of blood at a constant value (close to 7.4), even though acidic and basic substances continually pass into the blood stream. The buffer action in this case is vital, for if the pH increases or decreases by a small amount, death may result. A solution containing bicarbonate ions may act as a buffer, as is apparent from the following equations.

When an acid is added,

$$HCO_3^- + H^+ \rightarrow H_2CO_3$$

When a base is added,

$$HCO_3^- + OH^- \rightarrow HOH + CO_3^{2-}$$

A substance that, like the bicarbonate ion, can act either as an acid or a base is said to be **amphoteric.** Any amphoteric species, molecular or ionic, can act as a buffer. Most buffers, however, consist of a mixture of at least one acidic and at least one basic species. Buffer mixtures for establishing various pH ranges are listed in reference books.

SOLUBILITY PRODUCT CONSTANT

When a saturated solution of a slightly soluble salt is in contact with undissolved salt, there is an equilibrium between the dissolved ions and the ions in the solid phase of the undissolved salt. For silver chloride, the equilibrium is formulated as follows:

$$\underset{\text{solid}}{AgCl} \rightleftarrows \underset{\text{ions in solution}}{Ag^+ + Cl^-}$$

Ionic Equilibria

The equilibrium constant is given by the usual expression:

$$K = \frac{[Ag^+] \times [Cl^-]}{[AgCl]}$$

But at saturation, the concentration of the dissolved salt is not changed by adding more solid AgCl; so that regardless of the amount of solid silver chloride present at saturation, the effect of the solid is constant:

$$K = \frac{[Ag^+] \times [Cl^-]}{k}$$

$$K \times k = [Ag^+] \times [Cl^-]$$

$$K_{sp} = [Ag^+] \times [Cl^-]$$

The constant K_{sp} is known as the **solubility product constant.** It is equal to the product of the ionic concentrations (in moles per liter of saturated solution), with each concentration raised to the power indicated by the number of ions in the formula. For example, the solubility product for $PbCl_2$ is formulated as follows:

$$\underset{\text{solid}}{PbCl_2} \rightleftarrows \underset{\text{dissolved}}{Pb^{2+} + 2Cl^-}$$

$$K_{sp} = [Pb^{2+}] \times [Cl^-] \times [Cl^-]$$
$$= [Pb^{2+}] \times [Cl^-]^2$$

For $Ca_3(PO_4)_2$,

$$K_{sp} = [Ca^{2+}]^3 \times [PO_4^{3-}]^2$$

Calculation of Solubility Product Constants. The solubility of silver chloride as experimentally determined is 0.0014 g/liter at 25°C. Because the molecular weight of silver chloride is 143.32, its solubility in moles per liter is $0.0014 \div 143.32 = 0.000010 = 1.0 \times 10^{-5}$ mole/liter. Hence the concentration of the Ag^+ ion is 1.0×10^{-5} mole/liter; the concentration of the Cl^- ion is also 1.0×10^{-5} mole/liter, because there is one Cl^- ion for each Ag^+ ion. Substituting these values in the solubility product expression, we have

$$K_{sp} = [Ag^+] \times [Cl^-]$$
$$= 1.0 \times 10^{-5} \times 1.0 \times 10^{-5}$$
$$= 1.0 \times 10^{-10}$$

The solubility product constants for several salts are given in Table 3 in the Appendix. Solubility products are listed only for slightly soluble substances, because the K_{sp} relationship holds precisely only in dilute solutions.

Calculation of Solubility from K_{sp} Value. If we know the value of the solubility product constant, we can calculate the solubility of a salt.

Problem 3: Given that K_{sp} for $Mg(OH)_2$ is 1.2×10^{-11}, calculate the solubility of this compound in grams per 100 ml of solution.

Solution: To see clearly the relationships of molar solubilities, we write the equation for the solubility equilibrium:

$$Mg(OH)_2(s) \rightleftarrows Mg^{2+} + 2OH^-$$

From the equation we see

(if x moles per liter of Mg(OH)$_2$ dissolves) $\xrightarrow{\text{there is formed in solution}}$ (x moles Mg^{2+} per liter) + ($2x$ moles OH$^-$ per liter)

$$K_{sp} = [\text{Mg}^{2+}][\text{OH}^-]^2$$

Letting x equal the moles of Mg(OH)$_2$ dissolved,

$$K_{sp} = (x)(2x)^2 = 4x^3$$
$$4x^3 = 1.2 \times 10^{-11}$$
$$x^3 = \frac{1.2 \times 10^{-11}}{4} = 3 \times 10^{-12}$$
$$x = 1.4 \times 10^{-4} \text{ mole/liter Mg(OH)}_2 \text{ dissolved}^5$$

$$\text{Solubility Mg(OH)}_2 = \frac{1.4 \times 10^{-4} \text{ mole}}{1 \text{ liter}} \times \frac{58.3 \text{ g}}{1 \text{ mole}} \times \frac{1 \text{ liter}}{1000 \text{ ml}} \times 100 \text{ ml}$$
$$= 8.2 \times 10^{-4} \text{ g (in 100 ml of solution)}$$

Common Ion Effect in Precipitation. In the foregoing examples of solubility product calculations, we have considered the salt to be dissolved in pure water. Generally, however, a solution contains an additional source of one of the ions of the insoluble salt. The influence that changing the concentration of one ion has on the concentration of the other ion is another instance of the common ion effect. For example, we calculated in Problem 3 that 1 liter of saturated Mg(OH)$_2$ contains 1.4×10^{-4} moles of Mg^{2+} ions. However, if the OH$^-$ ion concentration is increased by adding a salt containing a common ion—NaOH, KOH, Ca(OH)$_2$—the concentration of Mg^{2+} must decrease in order to keep K_{sp} constant; that is, Mg(OH)$_2$ is less soluble in a solution containing NaOH than in pure water.

As another example of the common ion effect, consider the precipitation of calcium carbonate. Suppose that, to a solution of sodium carbonate, some calcium chloride is added till the concentration of calcium ion in the solution becomes 2×10^{-3} mole/liter. What is the concentration of carbonate ion that can remain in solution? Calcium carbonate is generally thought of as being water-insoluble. Actually, it is soluble to the extent defined by its solubility product constant. At 20°C,

$$K_{sp} = [\text{Ca}^{2+}] \times [\text{CO}_3^{2-}] = 1 \times 10^{-8}$$

If [Ca^{2+}] is 2×10^{-3}, the value of CO$_3^{2-}$ is easily calculated:

$$2 \times 10^{-3} \times [\text{CO}_3^{2-}] = 1 \times 10^{-8}$$
$$[\text{CO}_3^{2-}] = \frac{1 \times 10^{-8}}{2 \times 10^{-3}}$$
$$[\text{CO}_3^{2-}] = 5 \times 10^{-6} \text{ mole/liter}$$

Our first conclusion is that only 5×10^{-6} mole/liter of carbonate ion can remain in solution at equilibrium if the concentration of calcium ion is 2×10^{-3} molar.

---

[5] A convenient way to take the cube root of a number is with logarithms. The log of 3 is 0.4771. The log of the cube root of 3 is 0.4771/3 = 0.1590. The antilog of 0.1590 is 1.443, which we round to 1.4.

Another conclusion that we can draw is that, if the original sodium carbonate is extremely dilute (less than 5×10^{-6} M), no calcium carbonate precipitate forms when calcium ion is added to make $[Ca^{2+}]$ equal to 2×10^{-3}. Suppose $[CO_3^{2-}]$ had been only 3×10^{-6}:

$$[Ca^{2+}][CO_3^{2-}] = 2 \times 10^{-3} \times 3 \times 10^{-6} = 6 \times 10^{-9}$$

which is less than 1×10^{-8}, the value of K_{sp}. If substitution of the ionic concentrations into the expression for the solubility product leads to a calculated value less than the K_{sp} for the salt, no precipitate forms.

It is clear from a consideration of the solubility product principle that an ion cannot be completely removed from solution by forming a so-called insoluble precipitate with another ion. However, the addition of a large excess of the latter ion decreases the concentration of the former ion to the vanishing point (see Fig. 16-3). In the precipitation of CO_3^{2-} ion by the addition of Ca^{2+} ion, if

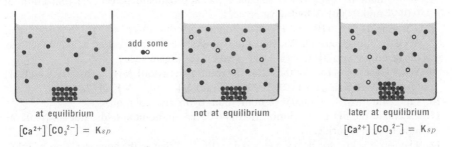

Fig. 16–3. When a salt containing calcium ions (represented as black spheres) is added to a saturated solution of calcium carbonate (black and red spheres), the concentration of the Ca^{2+} ion is increased and the system is no longer at equilibrium (beaker 2). In time, the concentrations of Ca^{2+} and CO_3^{2-} are reduced by precipitation and the system is again at equilibrium (beaker 3). Note that in the new equilibrium system (beaker 3), the concentration of CO_3^{2-} is less and the concentration of Ca^{2+} is greater than in the original solution (beaker 1).

calcium ion is added to the solution till its concentration is 0.5 M, the concentration of the CO_3^{2-} ion is reduced to only 2×10^{-8} mole/liter (that is, only 0.0000012 g of CO_3^{2-} per liter):

$$[Ca^{2+}] \times [CO_3^{2-}] = 1 \times 10^{-8}$$

$$0.5 \times [CO_3^{2-}] = 1 \times 10^{-8}$$

$$[CO_3^{2-}] = \frac{1 \times 10^{-8}}{0.5} = 2 \times 10^{-8} \text{ mole/liter}$$

The decrease in solubility of a salt due to the presence of a common ion is made use of in analysis. To precipitate nearly all the silver ion in a solution as AgCl, we can add HCl to give a rather high $[Cl^-]$ value. If $[Cl^-] = 0.10$, $[Ag^+] = 1.0 \times 10^{-9}$, a very small amount left in solution. However, there can be a limit to the effect of a common ion. In concentrated HCl solutions, silver forms the soluble ion $AgCl_2^-$ and AgCl dissolves. As mentioned previously, the solubility product concept applies only in dilute solution.

EXERCISES

1. Write equations for the ionization of each of the following acids and identify the conjugate base of each: HI, $HClO_3$, HN_3, NH_4^+, HSO_4^-, H_2O, $HC_2H_3O_2$, HNO_3.
2. Write equations to show how each of the following may behave as bases: NH_3, CO_3^{2-}, $H_2PO_4^-$, CrO_4^{2-}, F^-, OH^-.
3. Write expressions for the stepwise ionization constants of the following diprotic and triprotic acids: H_2SO_4, $H_2Cr_2O_7$, H_3PO_4, H_3CrF_6.
4. In a solution of 0.1 M phosphoric acid, H_3PO_4, there are at least five different ions present. What ion is present in the greatest concentration? Support your choice with chemical equations. What are the other four ions?
5. The ionization constant of a weak acid is a special type of equilibrium constant. Explain how it differs from the usual equilibrium constant, K.
6. A weak acid is 1 percent ionized. What other information is needed in order to calculate the ionization constant of the acid?
7. Using data in Table 16-2, calculate the approximate percentage ionization of 0.01 N and 0.0001 N hydrocyanic acid.
8. Three acids, HA, HB, and HC, have ionization constants of 4×10^2, 6×10^{-3}, and 3×10^{-12}, respectively. Which is the strongest acid? The weakest acid? Which has the strongest conjugate base? The weakest conjugate base?
9. What is the pH of each of the following solutions: (a) 0.01 N HCl; (b) 0.01 N NaOH; (c) 0.05 N $Ba(OH)_2$; (d) 0.002 N HNO_3; (e) 3.4×10^{-3} N KOH?
10. Calculate the pH of a 0.05 M ammonium hydroxide solution.
11. Calculate the pH of a solution that is 0.05 M ammonium hydroxide and 0.05 M ammonium chloride.
12. The ion product for heavy water is 1×10^{-15}. What is the concentration of OD^- in a heavy water solution that contains 6×10^{-3} mole of DCl in 500 ml of solution?
13. Calculate the pH of each of the following solutions: 0.001 M hydrochloric acid and 0.2 M acetic acid. Do you find the relatively concentrated solution of hydrogen acetate to be more or less acidic than the dilute solution of hydrogen chloride? Why is this?
14. (a) Bicarbonate of soda ($NaHCO_3$) is sometimes taken in an attempt to relieve "stomach upset." Refer to Fig. 16-2, and then write equations to show the reactions which occur.
    (b) After bicarbonate of soda is taken, belching may occur. Why?
    (c) When bicarbonate of soda is taken, what happens to the pH of the juices in the stomach?
15. Predict if 0.1 M solutions of the following are acidic, basic, or neutral; support your choice by writing equations: NaCN, $NaNO_3$, $NaHSO_4$, NH_4NO_3, $NH_4C_2H_3O_2$.
16. Calculate the H^+ concentration in the following solutions: (a) 0.1 N $KHSO_4$; (b) 0.1 N KF; (c) 0.1 N KCN.
17. Describe a buffer system that would have a pH of (a) about 4; (b) about 7; (c) about 10.
18. (a) Write the expression for the solubility product constant, K_{sp}, of $CaCO_3$.
    (b) The solubility of $CaCO_3$ in water is 1×10^{-4} mole/liter. Calculate K_{sp}.
    (c) What is the maximum concentration of CO_3^{2-} ions in a solution that is 0.02 M in calcium chloride?
19. (a) The solubility of $MgCO_3$ is 0.043 g per 100 ml of water. Calculate K_{sp}.
    (b) The solubility of $Mg(OH)_2$ is 0.00083 g per 100 ml of water. Calculate K_{sp}.
20. A saturated solution of $Ca(OH)_2$ in water has a pH of 12.64. Calculate K_{sp}.

21. Zinc hydroxide, $Zn(OH)_2$, is water-insoluble. Explain with equations why zinc hydroxide dissolves in water as HCl is added to the mixture.
22. The solubility product constant for Bi_2S_3 is 6.8×10^{-97}. How many Bi^{3+} ions are in 1 liter of a saturated solution of Bi_2S_3?
23. The solubility of CaF_2 in pure water is 0.0016 g per 100 ml. What weight of CaF_2 will dissolve in 500 ml of 0.10 N NaF?
24. How is the ionization of H_2S in water solution affected by the addition of (a) 0.1 N HCl; (b) 0.1 N NaOH? Explain.
25. Would methyl orange (Table 16-3) be a suitable indicator for the titration of an acetic acid solution with 0.1 N NaOH? Explain.

SUGGESTED READING

Armstrong, A. R., "Precipitation of the Hydrogen Sulfide Group of Ions Using Thioacetamide," *J. Chem. Educ.*, **37**, 413 (1960).

Goodman, R. C., and R. H. Petrucci, "Is the Solubility Product Constant?," *J. Chem. Educ.*, **42**, 104 (1965).

Grotz, L. C., "Heterogeneous Equilibria in General Chemistry," *J. Chem. Educ.*, **40**, 479 (1963).

Herron, F. Y., "Models Illustrating the Lewis Theory of Acids and Bases," *J. Chem. Educ.*, **30**, 199 (1953).

Jolly, W. L., "The Intrinsic Basicity of the Hydroxide Ion," *J. Chem. Educ.*, **44**, 304 (1967).

Kokes, R. J., M. K. Dorfman, and T. Mathia, "Equilibria in Ionic Solutions," *J. Chem. Educ.*, **39**, 93 (1962).

Mogul, P. H., and J. S. Schmuckler, "Dilute Solutions of Strong Acids: The Effect of Water on pH," *Chemistry*, **42** (9), 14 (1969).

O'Donnell, T. A., "A Simple Separation of the Group IIA Sulfides," *J. Chem. Educ.*, **40**, 415 (1963).

Rainey, R. G., "Making Baking Powder Biscuits," *J. Chem. Educ.*, **39**, 363 (1962).

17

NUCLEAR CHEMISTRY

Research in all branches of chemistry has been pursued with great success since the beginning of the nineteenth century; but the advances that have most effectively caught the imagination of people, chemists included, have been the discoveries that deal with the atom—in particular, its nucleus.

Over a period of several centuries during the Middle Ages, the alchemists tried vainly to convert base metals into gold, a process that would have required the changing of one element into another (*transmutation*). About the beginning of the nineteenth century, scientists generally became convinced that elements were made of atoms that could not be converted one into another. Their conclusions were drawn from observation of ordinary chemical reactions and from interpretation of the laws that govern the weight relationships of these reactions.

For the next hundred years, the atom was thought of as a compact, impenetrable, indivisible object, perhaps spherical, which could be neither synthesized nor decomposed. Then in 1896 came the discovery by Henri Becquerel in Paris that a compound of uranium, potassium uranyl sulfate, emitted a radiation that could pass through paper, or even thin sheets of metals, and affect a photographic plate. This discovery revived interest among scientists concerning the internal structure of atoms and the possibility of the transmutation of elements, an interest that in fifty years led to the transmutation of most of the elements, the production of nuclear bombs, the peacetime use of nuclear energy, and a fuller (though still incomplete) understanding of the internal structure of atoms.

The period just prior to the turn of the present century has been referred to as the golden years of modern chemistry and physics. The comfortable world of matter and energy based on the indestructible atom was literally

Nuclear Chemistry

blasted asunder by a rapid sequence of discoveries—X rays in 1895, radioactivity in 1896, the electron in 1897, and radium in 1898.

Becquerel's description of the phenomenon of radioactivity inspired two of his French contemporaries to launch their famous researches on the ore pitchblende. Pierre Curie and his wife Marie made an indelible record of devotion to and discovery in science that will inspire others for generations to come. Together they isolated and identified two new elements, polonium and radium, from uranium minerals. Their tedious separation of a few hundredths of a gram of radium from about a ton of pitchblende is a classic example of careful, patient effort crowned by well-deserved success.

RADIOACTIVE NUCLEI

Elements that spontaneously give off particles or rays are called **radioactive elements.** Common examples of naturally radioactive elements include radium and uranium. The production of isotopes with unstable nuclei—e.g., sodium with mass number 24—by nuclear bombardment gives rise to radioactive isotopes that do not occur in nature. Bombardment reactions are discussed later in this chapter.

Radioactive changes differ from ordinary chemical reactions in that the former involve transformations that originate in the nucleus of the atom, whereas ordinary chemical reactions involve changes in the outer-shell electrons. Thus radioactivity is solely a nuclear property.

TYPES OF EMISSIONS

Shortly after Becquerel's discovery, it was established that radioactive elements may emit three different types of particles or rays (see Fig. 17-1). Since

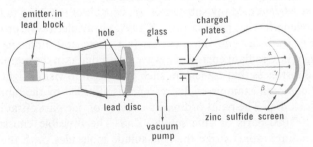

Fig. 17-1. Three kinds of natural radioactivity can be identified by determining the effect of an electrostatic field on their paths of travel.

the nature of these rays was not completely understood, their discoverers named them simply alpha (α), beta (β), and gamma (γ). Some of the characteristics of these three types of radiation are summarized in Table 17-1. All three types of emissions are called **ionizing radiations,** because they are able to cause the formation of ions in matter by knocking electrons off the atoms and molecules in their paths. The chief effects of radioactivity on living plants and animals are traceable to these ionization reactions.

TABLE 17-1.

Types of Natural Emissions

NAME	SYMBOL	UNIT CHARGE	MASS IN AWU	VELOCITY RELATIVE TO VELOCITY OF LIGHT	APPROXIMATE RELATIVE PENETRATING POWER	APPROXIMATE RELATIVE IONIZING POWER	THEORETICAL DESCRIPTION
Alpha particle	α	2+	4	5%	1	10,000	He^{2+} ion
Beta particle	β	1−	0.00055	95%	100	100	An electron
Gamma ray	γ	0	0	100%	10,000	1	Radiant energy

THE DETECTION OF RADIATIONS

Since atomic and subatomic particles cannot be seen, and since radiant energy of short wave length—i.e., gamma rays—is not detectable by the eye, various indirect methods have been developed for detecting the emissions of radioactive substances. Four methods will be described. All are based on the fact that, as the emissions disturb atoms which they bombard, electrons are displaced to higher energy levels or completely removed from the atoms.

Photographic Methods. Photographic film and paper have long been used for the detection of radioactivity. The emissions affect the photographic emulsion in somewhat the same manner as ordinary light does. After exposure, the paper or film is developed in the usual way. The use of photographic film, which is clipped onto the clothing, is the most common way of monitoring the exposure of personnel working with radioactive materials.

Fluorescent Methods. Many substances are capable of receiving radiant energy of short wavelength (e.g., gamma rays, X rays, ultraviolet rays), or of receiving kinetic energy from fast-moving particles (e.g., beta and alpha particles) and transforming it into radiant energy of a wavelength that is in the region detectable by the eye. Substances that transform such energies into visible light are said to **fluoresce.** A common example of fluorescence is afforded by the luminous paint used on a clock dial. The paint ordinarily consists of one part of radium sulfate, $RaSO_4$, to 100,000 parts of zinc sulfide, ZnS. The invisible emissions of the radioactive radium atoms strike the zinc sulfide molecules, with the result that some of the energy is transformed into visible radiant energy.

The Cloud Chamber. An apparatus perfected in 1911 by C. T. R. Wilson, an English physicist, made it possible to *see the path* followed by a single ionizing radiation in its flight through a gas. Wilson made artificial clouds in the laboratory by first saturating a volume of air with water vapor and then causing the moist air to cool itself by expanding it rapidly. If the cooling is sufficient, a cloud will form; if there is somewhat less cooling, the air will only become *supersaturated* with moisture. (Substances other than water—e.g., an alcohol—can be used to supersaturate the air with vapor.)

Wilson found that when a radioactive substance was placed in the supersaturated air of the cloud chamber, thin lines of fog or cloud emanated from the radioactive material (Fig. 17-2).

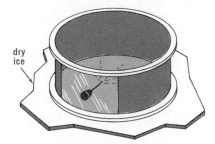

Fig. 17-2. A cloud chamber. The emitter is glued onto a pin stuck into a stopper that is mounted on the chamber wall. The chamber has some methyl alcohol in the bottom and rests on dry ice. The cool air near the bottom becomes supersaturated with methyl alcohol vapor. When a ray speeds through this supersaturated vapor, ions are produced which serve as "seeds" about which the vapor condenses, forming tiny droplets, or fog.

The Geiger Counter. In Fig. 17-3 is shown a simplified schematic drawing of a Geiger tube. The tube operates as follows:

1. The metal guard is made so that a portion of the Geiger tube can be exposed to radiation (at the end in the drawing).
2. A high potential is applied between an inside wire W and the wall of the Geiger tube.
3. The emitted particle or ray comes through the window and causes ionization of some of the gas molecules in the tube. The ions thus formed conduct a pulse of current between the wire and the tube, and this current causes a detector light to flash or a clicker to sound. More elaborate models are provided with automatic recording devices, so that each surge of current is counted.

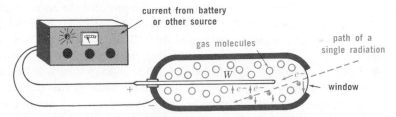

Fig. 17-3. The principle of operation of a Geiger tube. A ray (dashed arrow) emitted by the radioactive substance produces ions along its path.

NATURALLY OCCURRING RADIOACTIVE SERIES

A **radioactive series** is a collection of elements that are formed from a single radioactive element by successive emissions of alpha or beta particles. Since each emission brings about the formation of an atom of a different element, the series begins with the radioactive decay of a *parent element* and continues from atom to atom until some nonradioactive atom is formed. Uranium-238 (uranium whose atoms have a weight of about 238 awu) is the parent element for one naturally occurring series that contains eighteen members (Fig. 17-4). Parent atoms of two other series are uranium-235 and thorium-232.

Alpha Emission. In order to understand how different elements are formed by alpha and beta emissions, let us again note that these particles are ejected from

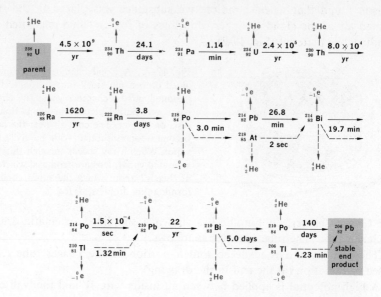

Fig. 17-4. The ^{238}U series of radioactive elements. Note that certain members may decay by the emission of either an α or β particle. In cases of alternative paths or branches, the solid lines show the pathway taken by most atoms.

the nucleus of an atom, so the nucleus no longer has an atomic number characteristic of the original element. For example, a uranium-238 atom that emits an alpha particle suffers a loss of two neutrons and two protons; the atomic number is thereby reduced from 92 to 90, the atomic number characteristic of atoms of thorium, and the atomic weight is decreased from 238 to 234. In other words, the uranium-238 atom becomes a thorium-234 atom by the emission of an alpha particle. The alpha particle is identical with a helium nucleus and is indicated by the symbol $^{4}_{2}$He. The important generalization to be noted at this point is that *the emission of an alpha particle decreases the atomic number by 2 and the atomic weight by 4.* The following equations describe radioactive decay by alpha emission:[1]

$$^{238}_{92}U \rightarrow {}^{234}_{90}Th + {}^{4}_{2}He$$

$$^{226}_{88}Ra \rightarrow {}^{222}_{86}Rn + {}^{4}_{2}He$$

---

[1] In order to focus attention upon the particles of greatest interest, the nuclear chemical equations in this chapter are purposely abbreviated. They include only the atomic nuclei involved and the particles emitted or absorbed during the nuclear changes.

An example of a more complete equation is:

$$^{238}_{92}U \rightarrow {}^{234}_{90}Th^{2-} + {}^{4}_{2}He^{2+}$$

This equation takes into account the electrons around the atoms. The uranium atom had 92 protons and 92 electrons. The loss of an alpha particle leaves a particle with 90 protons and 92 electrons, i.e., Th^{2-}. An isolated negative ion such as this would be unstable and the two electrons would be lost immediately, leaving the thorium atom, $^{234}_{90}$Th. As the alpha particle speeds away, it bumps into molecules and atoms, which slow it down; eventually it gains 2 electrons and becomes a helium atom, $^{4}_{2}$He.

The first equation is read: "Uranium-238 decays by alpha emission to yield thorium-234."

These two equations illustrate a number of important points. Note that the mass number of each particle (number of protons plus neutrons) is placed at the upper left of the chemical symbol, and the atomic number (number of protons or nuclear charges) is placed at the lower left. The sum of the mass numbers on the left side of the arrow equals the sum of the mass numbers on the right side. Also the sum of the nuclear charges on the left side equals the sum of nuclear charges on the right side.

Beta Emission. Another type of nuclear change involves the emission of an electron (beta particle) from an atom. Since the electron comes from the nucleus, its presence is accounted for by the assumption that a neutron changes into a proton and an electron with the emission of the electron. This process increases the number of protons by 1, so *the atomic number increases by 1 on the emission of a beta particle*. However, the weight of the electron is negligible, so *the atomic weight is practically unaltered*, as the following equations[2] show:

$$^{234}_{90}Th \rightarrow {}^{234}_{91}Pa + {}^{0}_{-1}e$$

$$^{218}_{84}Po \rightarrow {}^{218}_{85}At + {}^{0}_{-1}e$$

The many steps required before the radioactive elements of the uranium-238 series finally become a stable element are shown in Fig. 17-4. It is interesting to note that *some isotope of lead is the end product* of all the naturally occurring radioactive series.

NUCLEAR STABILITY

Both stable and unstable nuclei are known for all the common elements, such as sodium, oxygen, nitrogen, chlorine, carbon, potassium, silver, and so on, although many of the unstable nuclei do not occur appreciably, if at all, in nature. In the case of elements with high atomic numbers, beginning with $_{84}Po$, no stable nuclei are known, although some nuclei are much more unstable than others; for example, $^{226}_{88}Ra$ is more unstable than $^{238}_{92}U$.

Apparently one of the factors related to nuclear stability is a favorable **neutron-proton ratio.** Study Fig. 17-5. In this figure the number of protons is plotted against the number of neutrons for the stable nuclei. Note that, as the atomic number increases, the ratio of neutrons to protons increases; that is, neutrons become relatively more numerous in a given nucleus. For the atom $^{12}_{6}C$, the neutron-proton ratio is 6/6 or 1; for $^{207}_{82}Pb$, it is 125/82 or 1.53. That is, although the nuclei of carbon atoms contain an equal number of neutrons and protons, those of lead atoms have approximately 50 percent more neutrons than protons. Each is stable. If the neutron-proton ratio of a nucleus lies outside the so-called favorable belt for nuclear stability, the atom is radioactive. To

---

[2] In nuclear equations the negative charge on the electron is indicated by the -1 at the lower left of the e, instead of by e^-.

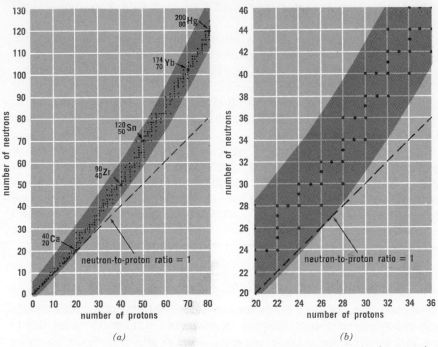

Fig. 17-5. The increase of neutrons over protons as the atomic number increases is shown (left) for stable nuclei through atomic number 80. Note that the atom $^{40}_{20}$Ca is the largest atom with a neutron-proton ratio of 1. On the right, a portion of the data is replotted to a larger scale to show more clearly that stable nuclei with even numbers of either protons or neutrons are more numerous than those with odd numbers.

provide for a more favorable neutron-proton ratio, radioactive nuclei emit particles and, in so doing, achieve neutron-proton ratios that provide for greater stability.

If a nucleus has a neutron-proton ratio greater than that expected for a stable nucleus, the ratio is lowered by the emission of a beta particle. For example, $^{14}_{6}$C has a neutron-proton ratio of 8/6 or 1.33. Upon emission of a beta particle $^{14}_{7}$N is formed in which the neutron-proton ratio is 7/7 or 1. Emission of an alpha particle increases the neutron-proton ratio for atoms with ratios greater than 1. Thus if a nucleus has a neutron-proton ratio less than that necessary for stability, the emission of an alpha particle will bring the ratio closer to that of a stable nucleus. Consider the decay of $^{212}_{84}$Po to $^{208}_{82}$Pb. The neutron-proton ratio of $^{212}_{84}$Po is 128/84 or 1.52. The stable $^{208}_{82}$Pb formed by the alpha decay of $^{212}_{84}$Po has a ratio of 126/82 or 1.54.

HALF-LIFE

How long does it take for a given sample of uranium to change into lead, or, for that matter, how long does it take for any radioactive substance to change into other substances? *The time required is independent of the amount of radioactive material.* For example, if it takes 1620 years for half of a 1-g sample of radium to change into radon, the same period of time (1620 years) will be required for half of a 2-g sample to change into radon, because the 2-g sample, with twice as many radium atoms, will in all probability emit twice as many alpha particles per unit of time as the 1-g sample does. For this reason, the rate of radioactive change is expressed in terms of the **half-life** ($t_{1/2}$) of the radioactive substance, i.e., the length of time required for one-half of a given starting weight of a substance to change into other substances.

The half-lives are determined experimentally by counting the number of emissions in a suitable period of time by a given weight of the radioactive sample (a Geiger counter might be employed for this). Note that certain radioactive elements have very short half-lives, whereas others have very long ones. For example, protactinium-234 is a beta emitter that emits *half* of the total possible beta particles in 1.14 minutes. Of a sample of pure ^{234}Pa weighing 0.008 g, only 0.004 g would be left after 1.14 minutes, because half of the original material would have turned into ^{234}U.

$$^{234}_{91}Pa \xrightarrow[1.14 \text{ min}]{t_{1/2}} {}^{234}_{92}U + {}^{0}_{-1}e$$

In 1.14 minutes more there would be only 0.002 g of ^{234}Pa; and after 3 × 1.14, or 3.42 minutes elapsed, only 0.001 g of the original 0.008 g would be left.

The activity of a radioactive substance theoretically never falls to zero. However, after a period of time equal to about ten half-lives, the radioactivity is so weak that it can hardly be measured. The half-lives of the members of the uranium-238 series are shown in Fig. 17-4.

Age of the Earth. One of the very fascinating calculations based on the concept of half-lives is the *minimum* age of the earth. As shown in Fig. 17-4, the end product of the uranium-238 decay series is stable lead-206. In nature where lead-206 and uranium-238 are found together in certain minerals, it is assumed that the lead has been formed as the result of radioactive decay during the passage of many years. The amounts of uranium and lead present can be determined by chemical analysis. On the basis of a knowledge of the half-lives of all the intermediate elements involved,[3] one can calculate the time required to establish the uranium-to-lead ratio found in the minerals.

Such calculations indicate that many rocks have existed in much their present state for billions of years. These and other studies support the theory that the earth is about four to six billion years old.

Age of Organic Materials. An interesting method of dating ancient objects of an organic nature is based on the fact that the preserved object, if not too old, contains a measurable amount of radioactive carbon-14. Carbon-14 is present in the atmosphere (as $^{14}CO_2$) in a constant small amount, because it is continually being produced by cosmic ray activity that results in neutron capture by a nitrogen atom and the subsequent expulsion of a proton:

$$^{14}_{7}N + {}^{1}_{0}n \rightarrow {}^{14}_{6}C + {}^{1}_{1}H$$

---

[3] For ordinary calculations, we need consider only the half-life of ^{238}U, because the next longest-lived member (^{234}U) has a half-life of 2.4×10^5 yr, only about 1/10,000 of the ^{238}U half-life. The amount of material that exists as ^{234}U, ^{226}Ra, and other intermediate members of the series at any one time is relatively negligible.

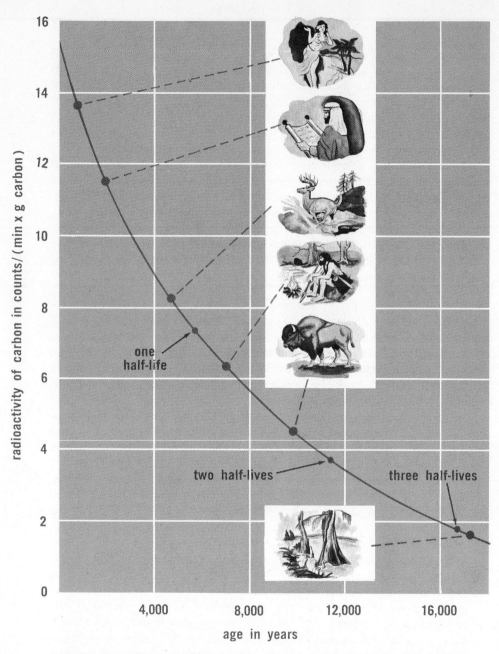

Fig. 17–6. Radiocarbon dating has been of great value to archaeologists and geologists in establishing the ages of prehistoric Indian villages, glacial deposits, etc. For this work, carbon-containing materials must have been preserved, usually by being covered with earth, to the present time. Reading from the top of the graph downward along the carbon-14 decay curve are shown the ages of an early Polynesian culture established from charcoal found in Hawaii, age 950 years; the Dead Sea scrolls, age about 1900 years; deer antler from Indian mound near Indian Knoll, Kentucky, age 4900 years; charcoal from a hearth at Long Site, South Dakota, age 7100 years; bison bone from near Lubbock, Texas, age 9900 years; cypress wood from a buried stump near Santee, South Carolina, older than 17,000 years. (Data from W. F. Libby, *Radiocarbon Dating*, courtesy of the University of Chicago Press.)

Therefore, at all time, a constant, small quantity of $^{14}CO_2$ is available to growing plants. Once the growing process stops and carbon-14 is no longer taken up by the plant, the amount of it in the plant tissues begins to diminish through radioactive decay. By measuring the radioactivity due to carbon-14 in living wood and comparing this to the radioactivity in preserved wood or charcoal, we can calculate from the known half-life of carbon-14 (5770 years) the time that must have elapsed to reduce the radioactivity to that of the preserved object. The interesting dates in Fig. 17-6 have been determined in this manner.

BOMBARDMENT REACTIONS

All the nuclear changes in the uranium-238 radioactive series are examples of natural radioactivity; they are changes that involve only a single reactant atom and the products formed when this single atom emits a particle. Man can observe these radioactive changes, he can even concentrate and purify large quantities of radioactive materials, but he cannot control the nature of the particles emitted or the rate at which they are emitted.

However, there is a second class of nuclear reactions, called **bombardment reactions,** which can be controlled. These reactions result when particles of atomic or subatomic size strike atoms of matter and permanently change these atoms. The idea that an atom of one element could be purposely changed into an atom of another violated a theory that had been universally accepted since Dalton's time, but in 1919 Sir Ernest Rutherford reported that this indeed happened when ordinary nitrogen was exposed to the alpha particles emitted by a small sample of radium. The equation for the reaction is:

$$^{14}_{7}N + ^{4}_{2}He \rightarrow ^{17}_{8}O + ^{1}_{1}H$$

The high-speed alpha particle evidently meets a nitrogen nucleus head-on with such force that it momentarily fuses with it, making an unstable intermediate nuclear particle, which then ejects a proton, $^{1}_{1}H$, and becomes an atom of oxygen, $^{17}_{8}O$. The proton was identified by means of cloud chamber photographs (similar to those in Fig. 17-2), which provided the data necessary to calculate its mass and charge.

Nuclear reactions of this kind generally cannot be carried out so as to give appreciable yields, because only occasionally do the speeding alpha particles collide head-on with an atomic nucleus. This is due to the fact that the nucleus of an atom is extremely small as compared to the total size of the atom. In the case of Rutherford's work, most of the alpha particles emitted by the radium simply passed through the nitrogen gas without causing a reaction.

A most important bombardment reaction was the reaction between beryllium atoms and alpha particles reported in 1930 by W. Bothe and H. Becker in Germany. This bombardment resulted in the appearance of a very penetrating radiation, which had the characteristics of a stream of particles but was not deflected on passing through a magnetic field. In 1932 James Chadwick, of England, showed that the particles had masses nearly equal to that of protons but were uncharged. Thus the **neutron** was discovered.

$$^{9}_{4}Be + ^{4}_{2}He \rightarrow ^{12}_{6}C + ^{1}_{0}n$$

ACCELERATION OF CHARGED PARTICLES

As more bombardment reactions were studied, it was found that atoms of low atomic number, e.g., $^{14}_{7}N$ and $^{9}_{4}Be$, could be altered when struck by natural alpha particles, but these particles had no effect on atoms of high atomic number, e.g., $^{197}_{79}Au$ and $^{206}_{82}Pb$. These results were explained by assuming that the atoms with highly charged nuclei (many protons) repelled the alpha particles and prevented collisions. Scientists the world over became concerned with the problem of increasing the speed of alpha and other subatomic particles. Soon there were invented a number of devices called **accelerators** that could increase the velocity of a charged particle, thereby increasing its effect on an atom struck by it. Only one of these devices will be described.

The Cyclotron. The **cyclotron** is perhaps the most useful of all the accelerators of positive particles presently employed. Its operation is based on two well-known physical laws: *A charged particle is repelled by a like charge and attracted by an unlike charge*, and *a charged particle which is moving in a magnetic field follows a curved path.*

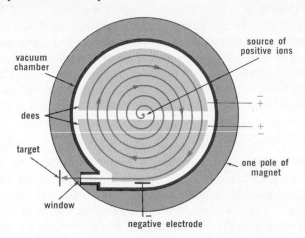

Fig. 17–7. A schematic diagram of a cyclotron. One of the poles of the magnet is shown underneath the dees; the other pole (not shown) is above the dees.

The schematic drawing of a cyclotron in Fig. 17-7 shows six of its principal parts: (1) one of two poles of a huge electromagnet (only the bottom one is shown), (2) a hollow disk split into halves (called *dees* because of their shape), (3) a source of charged particles in the center of the apparatus, (4) a vacuum chamber, which fits between the poles of the magnet and completely encloses the dees, (5) a negatively charged plate on the rim of the vacuum chamber and (6) a "window" of thin material in the side of the vacuum chamber through which a beam of speeding particles can emerge.

Not indicated in the figure is the electrical system that controls the electrostatic charge on the dees. When one dee is positively charged, the other is negative; but these charges can be automatically and rapidly alternated.

The cyclotron operates as follows:

1. A positively charged particle produced at the source in the center is attracted toward the negatively charged dee and enters it at high speed.

2. Because of the magnetic field maintained by the huge magnets above and below the dees, the speeding positive particle is forced to move in a curved path (as shown in Fig. 17-7). While inside a dee, the particle travels in a semicircle and is soon heading back toward the gap between the dees.

3. While the positive particle is inside a dee, the charges on the dees are reversed. When the particle reaches the gap, the dee ahead of it is negative, the dee behind is positive; pulled by one dee, repelled by the other, the speeding particle is accelerated while it is in the gap.

4. Particles that are traveling in phase with the rapid reversals of the charges on the dees are accelerated each time they go from one dee to the other. As their velocity steadily increases, the radius of their spiral path increases until it is almost as large as the radius of the dees.

5. As the path of the speeding positive particles nears the rim of the vacuum chamber, the attraction of a negative electrode deflects the particles from their spiral path and causes them to fly out of the cyclotron through a window. This window is actually a thin sheet of material through which the particles can pass easily. At the end of the exit tube, in front of the window, are placed samples of different elements to be used as *targets*. The speeding protons, deuterons, or alpha particles strike the atoms of the target element and cause nuclear chemical reactions such as:

$$^{209}_{83}Bi + ^{2}_{1}H \rightarrow ^{210}_{84}Po + ^{1}_{0}n$$

$$^{7}_{3}Li + ^{4}_{2}He \rightarrow ^{10}_{5}B + ^{1}_{0}n$$

E. O. Lawrence, the inventor of the cyclotron, reported his discovery in 1930. One of his first machines was built to fit between magnet pole pieces only 4 in. in diameter. The cyclotron has proved so useful that a number of larger ones have been built. The larger and more powerful the electromagnet, the greater the energy of the bombarding particles can be. One of these machines, built about two decades after Lawrence's invention at the University of California at a cost of $1,500,000, has magnet pole pieces 15 ft across. Even larger more expensive cyclotrons have been built at other institutions. However, there is a practical limit to the size of a cyclotron. This is because an accelerating particle gains in mass as it gains in energy, thus causing it to get out of phase with the oscillating current when its velocity has increased greatly. To cope with this problem, other types of circular accelerators have been built, such as the **synchrocyclotron**[4] and the **synchrotron**.

---

[4] In some of the modern cyclotrons, only one dee is used, the oscillating current being applied between it and a ground connection. As a charged particle enters or leaves the dee, it receives additional energy, so that it accelerates along a spiral path just as if two dees were used. The purpose of this arrangement is to make it possible to vary the frequency of oscillations in order to compensate for the increase in mass of the positive ion as it accelerates to very high speeds.

NEUTRON BOMBARDMENTS

One of the most effective "bullets" for bringing about nuclear changes is the neutron. Unlike the positively charged protons, deuterons, and alpha particles, which are repelled by the positive nuclei of atoms, the uncharged neutrons do not require high velocities to drive them into a nucleus. Indeed, a slowly moving neutron passing close to a nucleus has a greater chance of being attracted into the nucleus than a high-speed neutron, which might race on past.

Neutrons are easily generated by bombarding light elements such as lithium and beryllium with alpha particles, or by the reactions occurring in a nuclear reactor (discussed later in this chapter). The neutrons emitted as a result of the interaction of ^{7}Li with alpha particles may cause additional reactions in materials placed in the target area. Two examples are:

$$^{14}_{7}N + ^{1}_{0}n \rightarrow ^{14}_{6}C + ^{1}_{1}H$$

$$^{35}_{17}Cl + ^{1}_{0}n \rightarrow ^{35}_{16}S + ^{1}_{1}H$$

THE SYNTHESIS OF ELEMENTS

When it was found that atoms of one element could be changed into atoms of another by nuclear bombardment, there were still a few blank spaces in the periodic table between elements 1 and 92. Numbers 43, 61, 85, and 87 either had not been discovered[5] or the evidence for their discovery was open to doubt. All of these have been made since 1937 by cyclotron bombardment or by nuclear fission, and some of them have been found in minute amounts in nature. The possibility that they could exist at all was suggested by the blank spaces in the periodic table. The synthesis of these unknown elements was guided by the principle that an atom that is bombarded with a small particle may absorb all or part of the particle and become an atom with greater atomic weight and atomic number. In attempting to make the unknown element 43, it was logical to use a sample of molybdenum (number 42) as a target. The successful reaction was:

$$^{98}_{42}Mo + ^{2}_{1}H \rightarrow ^{99}_{43}Tc + ^{1}_{0}n$$

The new element was named *technetium*, from the Greek word for "artificial," because this was the first artificially made element.

THE TRANSURANIUM ELEMENTS

The heaviest atom known to occur naturally in appreciable quantities is radioactive uranium-238.

With the discovery of the transmutation of atoms by bombardment with subatomic particles, the bombardment of the heaviest known atom was an especially exciting experiment. These studies, initiated by the natural curiosity

---

[5] One of the interesting facts about the natural occurrence of the elements is that elements with odd atomic numbers tend to be less abundant than those with even atomic numbers (when both atomic numbers have about the same value). See Fig. 17-5 and Fig. 30-3.

of men with merely an "academic interest" in nature, have already resulted in scientific and political upheavals that are shaping man's destiny on earth.

Not only were new isotopes of known elements produced, but also atoms of *unknown elements*. For example, when uranium-238 is bombarded by neutrons, uranium-239 is formed. This latter isotope is unstable and undergoes beta decay (half-life of 23.5 minutes), producing atoms of neptunium, Np, an element that has since been found only in the slightest traces in nature. The reactions are:

$$^{238}_{92}U + ^{1}_{0}n \rightarrow ^{239}_{92}U$$

$$^{239}_{92}U \rightarrow ^{239}_{93}Np + ^{0}_{-1}e$$

Neptunium-239 is itself a radioactive element that decays, producing plutonium, Pu, another synthetic element with an even higher atomic number.

$$^{239}_{93}Np \rightarrow ^{239}_{94}Pu + ^{0}_{-1}e$$

When the atoms of plutonium-239 are bombarded by alpha particles, neutrons are produced. By balancing the nuclear equation it can be shown that an atom of another new element has been formed.

$$^{239}_{94}Pu + ^{4}_{2}He \rightarrow ? + ^{1}_{0}n$$

This previously unknown element has been shown to be an atom with a mass number of 242 and with 96 protons in its nucleus. This new element has been named curium, Cm, in honor of the Curies.

These synthetic elements with atomic numbers above 92 (uranium) are called the **transuranium elements.** Much of the work on them was done by Dr. Glenn T. Seaborg, A. Ghiorso, and their co-workers at the University of California. In addition to the three elements mentioned above, the California group is credited with the first synthesis of americium, berkelium, californium, mendelevium, and lawrencium. The first synthesis of elements 104 and 105 is claimed by both the California group and G. N. Flerov of U.S.S.R. Ghiorso suggested the names rutherfordium and hahnium for 104 and 105, respectively, while Flerov suggested the name kurchatovium for 104. Although the initial discoveries of the transuranium elements were usually based on amounts too small to be seen, several have now been produced in gram and kilogram amounts.

INTERCONVERSION OF MATTER AND ENERGY

In 1905, Albert Einstein formulated an equation that can be used to relate the energy change of a reaction to the difference in the masses of the reactants and products:

$$E = mc^2$$

where E is the change in energy, m is the change in mass, and c^2 is the square of the velocity of light.[6]

---

[6] It might be preferable to write this equation as $\Delta E = \Delta m \times c^2$ to emphasize that we are dealing with the change in mass of a system that is exchanging (gaining or losing) energy with its surroundings.

When energy is expressed in ergs, mass in grams, and the speed of light in centimeters per second,

$$E \text{ (ergs)} = m \times (3.00 \times 10^{10})^2$$
$$= m \times (9.00 \times 10^{20})$$

If we think about just 1 g of matter being converted entirely into energy,

$$E = 1 \text{ g} \times (3.00 \times 10^{10} \text{ cm/sec})^2$$
$$= 9.00 \times 10^{20} \text{ g} \times \text{cm}^2/\text{sec}^2 = 9.00 \times 10^{20} \text{ ergs}$$

An **erg** is a very small unit of energy. It requires 4.184×10^7 ergs to equal 1 calorie. Hence, when 1 g of matter is converted into energy,

$$E = 9.00 \times 10^{20} \text{ ergs} \times \frac{1 \text{ cal}}{4.184 \times 10^7 \text{ ergs}} = 2.15 \times 10^{13} \text{ cal}$$

This tremendous amount of energy, over 21 trillion calories, would be enough to heat and vaporize 9,500,000 gal of water. It is equivalent to the explosion of 20,000 tons of TNT or to one nuclear bomb of the type used in World War II.

According to Einstein's theoretical equation, when 2380 tons of gasoline and 8370 tons of oxygen burn and give off 2.15×10^{13} cal, the loss in mass is about 1 g. Such a weight change in a total weight of 10,750 tons is much too small to be detected by any known instrument. It is believed that *minute changes in mass do occur in every chemical reaction* (whether exothermic or endothermic), but that these changes are too small to measure by weighing. In chemical reactions the weight of the reactants equals the weight of the products for all practical calculations.

However, in nuclear reactions, where the energy change is much greater, the difference in mass between reactants and products is sufficient to provide an *experimental test* of Einstein's theoretical equation.

MASS LOSS AND BINDING ENERGY

The method of determining the masses of atomic particles by means of a mass spectrograph was described in Chap. 3. It is a fact, although a difficult one for the beginner in science to accept, that atoms always have slightly smaller masses than the sum of the masses of the particles composing them. As an example, consider the atom $^{20}_{10}$Ne, which contains 10 protons, 10 neutrons, and 10 orbital electrons. Let us compare its weight with the weight of these 30 particles.

10 protons:	10.073	
10 electrons:	0.0055	one $^{20}_{10}$Ne atom: 19.993 awu
10 neutrons:	10.087	
	20.166 awu	

$$20.166 - 19.993 = 0.173 \text{ awu}$$

That is, a neon-20 atom is 0.173 awu lighter than one would expect from the masses of the electrons, protons, and neutrons that compose it. It is now believed

Nuclear Chemistry

that this **mass loss** is associated with the organization of neutrons, protons, and electrons into atoms; that is, a certain amount of energy, called binding energy, is necessary to break an atom into separated parts. The greater the mass loss, the greater is the binding energy.

$$\text{protons} + \text{neutrons} + \text{electrons} \rightarrow \underbrace{\text{atom}}_{\substack{\text{weight is } less \\ \text{than that of} \\ \text{reactants}}} + \text{binding energy}$$

The **binding energy** is defined as the energy that would be needed to break up an atom into single protons, neutrons, and electrons. In this process there would be a *gain* in mass.

$$\text{atom} + \text{binding energy} \rightarrow \underbrace{\text{protons} + \text{neutrons} + \text{electrons}}_{\substack{\text{weight of parts is } greater \text{ than that of} \\ \text{original atom}}}$$

It is common to calculate both mass loss and binding energy for atoms on a per nuclear particle basis, in order that atoms of differing masses be compared. For neon-20, which has 20 nuclear particles, the mass loss per nuclear particle is 0.173 awu/20, or 0.0086 awu. The binding energy per nuclear particle for neon-20 is:

$$\text{Binding energy} = \left(0.0086 \text{ awu} \times \frac{1 \text{ g}}{6.02 \times 10^{23} \text{ awu}}\right) \times (3.00 \times 10^{10} \text{ cm/sec})^2$$

$$= 1.3 \times 10^{-5} \text{ erg}$$

Since one mev (1×10^6 ev) is equivalent to 1.602×10^{-6} erg, the binding energy per nuclear particle is:

$$\text{Binding energy} = 1.3 \times 10^{-5} \text{ erg} \times \frac{1 \text{ mev}}{1.6 \times 10^{-6} \text{ erg}}$$

$$= 8.1 \text{ mev}$$

The graph in Fig. 17-8 reveals that the binding energy is greatest for elements with intermediate mass numbers from about 20 to 150. The atoms of these elements are held together more strongly than either the very light or the very heavy atoms.

Study of Fig. 17-8 helps us understand several concepts.

1. The mass losses and binding energies are greatest for atoms of intermediate weights (masses 20 to 150).

2. If the nucleus of a very heavy atom were to split into two or more nuclei of intermediate weight (of mass 20 to 150), these two nuclei would weigh less than the original nucleus, even though all protons, neutrons, and electrons were accounted for. This process can occur; it is called **nuclear fission.**

A well-known fission reaction is the splitting of a uranium-235 atom into two smaller atoms when struck by a neutron. Fission reactions will be discussed in detail in connection with nuclear bombs and nuclear-power reactors.

3. If two or more nuclei of light atoms (mass less than 20) were to join to make a nucleus of a heavier atom, there would be a loss in mass, even though the resulting atom contained all the parts of the smaller atoms. This process can occur; it is called **nuclear fusion**. An example is:

$$^{12}_{6}C + ^{4}_{2}He \rightarrow ^{16}_{8}O + energy$$

Fusion reactions will also be discussed in greater detail in connection with hydrogen bombs and stellar energy.

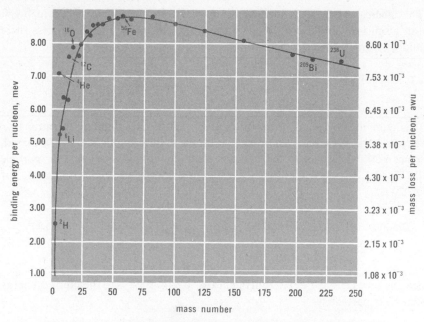

Fig. 17–8. A graph showing binding energy per nuclear particle (left ordinate) and mass loss per nuclear particle (right ordinate) versus mass number.

4. Since a loss of mass must result in the appearance of an equivalent amount of energy, both nuclear fission and nuclear fusion can be violently exothermic reactions. Of course, either fission or fusion could be endothermic, with products having lower binding energies than reactants, but such reactions would absorb rather than emit energy, and could not be used for production of power.

NUCLEAR FISSION

Fission Bombs. The nuclear bombs developed toward the close of World War II were of the fission type. Uranium-235 constituted the fissionable material in the first bomb used in warfare, and plutonium-239 was the fissionable material in the second one.

Of interest is the fact that although all the elements heavier than bismuth are radioactive, only the nuclei of one natural isotope, uranium-235, split almost

Nuclear Chemistry

instantaneously into two fragments (undergo fission) when struck by thermal neutrons.[7] Among the synthetic isotopes, plutonium-239 and uranium-233 have the same property. Their production is discussed later in this chapter.

The fissioning process for uranium-235 nuclei is represented schematically in Fig. 17-9. The equations below represent two of the thirty or more ways that uranium atoms may split. Of great importance is the fact that each fission produces more neutrons.

$$^1_0n + {}^{235}_{92}U \rightarrow {}^{103}_{42}Mo + {}^{131}_{50}Sn + 2{}^1_0n$$

$$^1_0n + {}^{235}_{92}U \rightarrow {}^{139}_{56}Ba + {}^{94}_{36}Kr + 3{}^1_0n$$

The fission products, $^{103}_{42}Mo$, $^{131}_{50}Sn$, $^{139}_{56}Ba$, and $^{94}_{36}Kr$, emit beta and gamma rays until stable isotopes are formed.

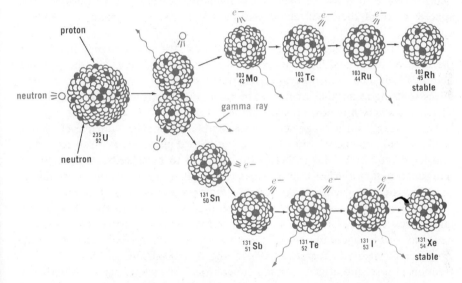

Fig. 17-9. A uranium-235 nucleus after capturing a neutron (left) splits into two smaller nuclei with the emission of gamma rays and two or three neutrons. The uranium-235 nuclei can split in over 30 ways, producing a total of about 200 radioactive species, generally with atomic numbers 30 to 64 and masses 72 to 161.

CRITICAL MASS. With the discovery that two or three neutrons are released when a uranium-235 atom fissions, it was realized that, under the proper conditions, neutrons could multiply rapidly so that billions would become available, and the fission process would then proceed on a huge scale with the release of a tremendous amount of energy. To see what these conditions are, we begin by considering natural uranium.

In nature, only a low concentration of uranium-235 is dispersed in uranium-238, so there is small probability of a neutron hitting a ^{235}U atom. Some fissions

---

[7] Neutrons that have kinetic energies comparable to the kinetic energies of gas molecules at ordinary temperatures are referred to variously as slow-moving neutrons, low-energy neutrons, or thermal neutrons.

occur because of neutrons that are set in motion by cosmic radiation, but natural fissions are relatively rare and are so widely separated that the energy is easily dissipated and the extra neutrons are absorbed by the nonfissionable atoms in the mineral.

Next, we consider a sphere of *pure* uranium-235. When the sphere is small, about the size of a marble, most of the neutrons that are produced by occasional fissioning of uranium-235 atoms induced by cosmic rays escape from the sphere. This is so because the nucleus of the atom is extremely small compared to the total volume of the atom. It is quite improbable, therefore, that a neutron will accidentally hit a tiny nucleus as it passes through the atoms in a thin piece of uranium. Under these conditions, the chain reaction is not self-sustaining.

Now we will consider a larger sphere of pure uranium-235. Because the neutrons have to travel through a larger number of atoms before leaving this sphere, more of them will collide with uranium-235 nuclei than in a smaller sphere.

As the sphere increases in size (perhaps to a little larger than a baseball in the case of uranium-235), a mass will be reached in which, on the average, one and only one neutron from each nucleus that is undergoing fission produces splitting in another nucleus. The remaining neutrons escape or are lost.[8] The chain reaction is now self-sustaining.

The amount and arrangement of fissionable material in which each fission produces only one new fission is called the **critical mass;** the neutron reproduction factor equals 1. Masses below this are said to be **subcritical;** the neutron reproduction factor is less than 1. Masses larger than the critical mass are said to be **supercritical;** the neutron reproduction factor is greater than 1 (see Fig. 17-10).

In supercritical masses of fissionable substances, the reproduction of neutrons takes place at a fantastic rate, so the supercritical mass is virtually a beehive of neutrons within one-millionth of a second. Under these conditions, fissioning is so rapid that the temperature rises to about 10,000,000°C or more. Within a fraction of a second after becoming supercritical, the whole mass explodes into many subcritical masses.

To cause a nuclear bomb to explode, sufficient subcritical masses are suddenly brought together to make a supercritical mass (see Fig. 17-11). The neutrons and fissioning atoms must be held together long enough for the chain reaction to build up to a considerable rate. This is done by enclosing the uranium (or plutonium) in a strong dense case that reflects neutrons back into the fissionable material, and also retards the bomb burst.

FALL-OUT. After the explosion, the highly compressed gases (even uranium and the metal casing are vaporized at 10,000,000°C) expand very rapidly over an area of several hundred square miles or so. Although a proportion of the bomb "ashes" may fall to the earth fairly close to the explosion site, much is

---

[8] The loss of neutrons may be due to a variety of factors in addition to the obvious one of missing all surrounding nuclei and speeding off into space. Bumping into the nuclei of impurities uses up some neutrons, and ineffective collisions with ^{235}U nuclei use up a considerable number. Approximately 5/6 of the collisions with ^{235}U produce fission, 1/6 produce ^{236}U nuclei.

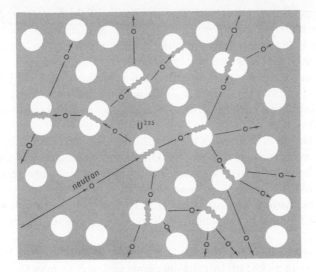

Fig. 17-10. A schematic representation of how neutrons become very numerous in a supercritical mass of ^{235}U. (Only nuclei are represented; because a nucleus comprises only about 1/10,000th of the diameter of the atom, its size is necessarily exaggerated here relative to the space between it and other nuclei.)

carried by air currents to all parts of the globe. Gradually, rain and snow carry this to the earth, where it constitutes a hazard because of the radioactive material present.

To understand why so much radioactive material is present in the bomb ashes, we again consider the two equations given earlier for typical fission reactions. The atoms produced were $^{103}_{42}$Mo, $^{131}_{50}$Sn, $^{139}_{56}$Ba, and $^{94}_{36}$Kr. When we compare the weights of these atoms with the weights of the common isotopes

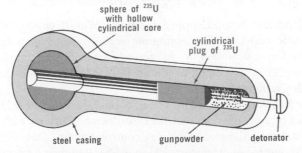

Fig. 17-11. A possible method for exploding a fission bomb. Two subcritical masses—a hollow sphere of uranium-235 and a cylinder of uranium-235 machined so as to fit the hollow sphere—are located at opposite ends of a steel casing. When the gunpowder is exploded, the uranium-235 cylinder is forced through the open bore of the steel casing into the hollow uranium-235 sphere to form a supercritical mass of uranium-235.

of these elements, we see at once that the atoms of the fission products weigh much more than the common nonradioactive isotopes do. Another way of putting this is that the neutron-proton ratio is much higher in atoms of the fission products than it is in common nonradioactive atoms.

These unstable nuclei are radioactive and achieve stability by emitting beta particles (and gamma rays) until a stable neutron-proton ratio is attained (see Fig. 17-9). Recall that each beta emission is accompanied by the loss of a neutron and the gain of a proton.

Since the half-lives of many of the unstable isotopes that result from fissioning range from months to several years, some of the radioactivity of the approximately 300 bombs already exploded will be with us for many years to come. All but two of these bombs have been exploded in bomb-testing programs of the United States, Russia, and other nations. Testing in the atmosphere has been discontinued by most of the major world powers because of the danger of fall-out.

Certain unstable isotopes from the bomb ashes—strontium-90 and cesium-137, for example—are picked up by growing plants and later by man from such foods as cereals, milk, tea, etc. Strontium-90, like calcium, accumulates in bones, especially in those of children, where it is reported to cause bone cancer and leukemia.

Source of Fissionable Material. *Plutonium-239 and Uranium-233.* Because of the scarcity of uranium-235 in nature, great efforts have been made to obtain other fissionable material. This has been accomplished by converting nonfissionable uranium-238 into fissionable plutonium-239 and converting thorium-232 into uranium-233 by neutron capture in nuclear reactors. The equations are as follows for plutonium production:

$$^{238}_{92}U + ^{1}_{0}n \longrightarrow ^{239}_{92}U$$

$$^{239}_{92}U \xrightarrow[23 \text{ min}]{t_{1/2}} ^{239}_{93}Np + ^{0}_{-1}e$$

$$^{239}_{93}Np \xrightarrow[2.3 \text{ days}]{t_{1/2}} ^{239}_{94}Pu + ^{0}_{-1}e$$

The second nuclear bomb dropped in World War II contained plutonium as the fissionable material. The plutonium was produced in a huge nuclear reactor at Hanford, Washington, in accordance with the reactions in the equations above. It is also being produced in plants located on the Savannah River in South Carolina.

The Nuclear Reactor. The fission of uranium-235, uranium-233, and plutonium-239 can be controlled so that the chain reaction occurs without a disastrous explosion. Under these conditions of control, the heat that is liberated in a fraction of a second by an atomic explosion is liberated over a period of several days or weeks. One apparatus that makes this possible is the atomic pile, or **nuclear reactor.** A number of these reactors have been built in this country, and more are being constructed. Although probably no two are exactly alike, the original one built at Oak Ridge in 1943 will serve as an example (Fig. 17-12). This reactor, now obsolete, has been designated a National Historical Landmark. The following points serve to illustrate its operation.

1. The reactor is essentially a huge cube made of bricks of graphite, surrounded by a shield of concrete 7 ft thick. The graphite is called the **moderator,** because it slows down the neutrons produced by the fission of uranium. Slow-moving (thermal) neutrons are more efficient than fast neutrons in producing fissions.

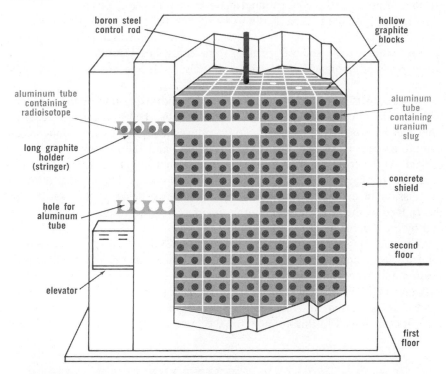

Fig. 17–12. Schematic drawing of the first nuclear reactor at Oak Ridge, Tennessee.

2. Cylindrical ingots of elemental uranium are rammed into holes in the graphite.

3. Two kinds of rods, cadmium encased in steel and special boron steel, are mounted so that they can be dropped into vertical holes or pushed into horizontal holes in the reactor cube. Both cadmium and boron steel are efficient neutron absorbers. If the rate of fission becomes too high, these control rods are inserted in the holes so that they will absorb some neutrons and thus slow down the chain reaction.

4. There are hollow channels into which long graphite sample holders are pushed. Anything placed in these holders can be subjected to intense neutron bombardment near the center of the reactor.

5. During operation of the reactor, the system is cooled by air drawn through the reactor at the rate of 100,000 cu ft per min. This flow of air holds the temperature of the uranium slugs at around 280°C.

Nuclear Reactors for the Production of Power. The principles involved when a reactor is operated to produce electrical power are quite simple. An arrangement of uranium, or uranium and thorium, with moderator and control rods supplies heat energy as described in the preceding section on the Oak Ridge reactor. This heat energy is carried away by circulating a coolant such as water, molten sodium, or carbon dioxide through the reactor. The coolant gives up the heat in a heat exchanger to water, which is converted to steam. The steam in turn operates turbogenerators in the conventional manner to produce electricity.

Although graphite is used as the moderator in the gas-cooled (CO_2) reactors that have been built in Great Britain, and in the molten sodium-cooled reactors in this country, ordinary water (or in some cases, heavy water) serves as both moderator and coolant for most reactors now producing power in the United States.

The reactor is controlled by inserting and withdrawing one or more control rods, usually of cadmium. Enough neutrons are absorbed so that on the average only one neutron from a given fission produces a second fission. Under these conditions, the chain reaction is self-sustaining, and the reactor is said to be **critical.** At the critical stage, all through the fissionable fuel, nuclei are fissioning at a fairly regular rate with the liberation of much energy. The number of nuclei that fission in a unit of time, and hence the power level of the reactor, are determined by the total amount of uranium present, by the ratio of ^{235}U to ^{238}U, and by the reactor design.

After operating at the critical stage for several months, the reactor fuel becomes diluted with fission products to such an extent that it must be removed for purification and processing, so it can be reused.

Figure 17-13 shows schematically how the energy from the pressurized water reactor is converted into electricity. The pressurized water reactor was the type first employed in this country for the propulsion of submarines; recently,

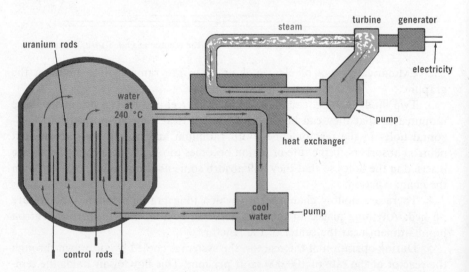

Fig. 17–13. A pressurized power reactor. In this type of reactor, natural uranium is usually enriched with ^{235}U so that the reactor can be more readily brought to a critical condition. Water circulating through the core removes the heat of fission.

the boiling-water type has also become important. Other types include the breeder reactor, which utilizes a large proportion of nonfissionable fuel such as thorium-232 and uranium-238, sodium-cooled reactors, gas-cooled reactors, and homogeneous reactors in which the fuel (uranium compounds) is in solution.

In 1970, the nuclear power capacity in the United States was about 6 billion watts. According to United States Atomic Energy Commission estimates, about one-half of the electrical power consumed in the United States will be produced by nuclear power in the year 2000. Due to the scarcity of fossil fuels, other advanced nations expect to be producing at least half of their electrical power with nuclear energy by then. Aside from the necessity of producing more energy for an expanding population, the matter of air pollution is an added incentive for the production of nuclear energy. The burning of coal, petroleum oils, and natural gas is now adding such large amounts of contaminants to the atmosphere that the air is actually unsuitable for breathing in certain densely populated areas.

FUSION REACTIONS

In discussing mass loss we noted that matter is converted into energy when certain small atoms fuse to make new atoms. For example, in the fusion of four hydrogen atoms to form one helium atom, 0.7 percent of the matter is converted into energy. In the fission process, 0.1 percent of the matter of a uranium-235 atom is converted into energy. The following are several examples of fusion reactions that release energy.

$$4{}_1^1H \rightarrow {}_2^4He + 2{}_{+1}^{0}e + \text{energy}^9 \tag{1}$$

$$_1^2H + {}_1^2H \rightarrow {}_2^3He + {}_0^1n + \text{energy} \tag{2}$$

$$_1^2H + {}_1^3H \rightarrow {}_2^4He + {}_0^1n + \text{energy} \tag{3}$$

$$_2^3He + {}_1^2H \rightarrow {}_2^4He + {}_1^1H + \text{energy} \tag{4}$$

$$3{}_2^4He \rightarrow {}_6^{12}C + \text{energy} \tag{5}$$

$$_6^{12}C + {}_2^4He \rightarrow {}_8^{16}O + \text{energy} \tag{6}$$

Reactions such as these may take place spontaneously when the temperature is of the order of 100 million degrees or more. At these high temperatures, atoms do not exist as such; instead, there is a plasma of nuclei and of electrons. In this plasma, nuclei merge or interact. Fusion reactions that take place under conditions of high temperatures are often referred to as *thermonuclear* reactions.

Stellar Energy. Spectroscopic examination of the light from our sun indicates that extremely large amounts of hydrogen and helium are present in its atmosphere. The fusion of hydrogen-1 to helium (reaction (1) above) is thought to be the over-all reaction taking place and is thought to be responsible for the tremendous amount of energy released by the sun. This radiant energy is calculated

---

[9] The positron, $_{+1}^{0}e$, is a subatomic particle with the same mass as an electron but a unit positive charge. On the average, a positron exists for only 1×10^{-9} second before colliding with an electron and turning into energy.

to be equivalent to the loss of 5×10^6 tons of matter per second. Reactions (5) and (6) are believed to occur in stars in which all the hydrogen in the central region has been converted into helium. Thus, the helium nuclei are thought to be converted into heavier nuclei in a stepwise fusion process, the nuclei of iron being the end product. (In iron nuclei the mass defect is at a maximum.) According to one theory, atoms heavier than iron atoms are built up in stars by neutron capture, a process which is *endothermic*.

Hydrogen Bombs (Thermonuclear Bombs). Thermonuclear reactions, such as those in (1), (5), and (6) above, take place slowly at stellar temperatures. Reaction (3), involving tritium, takes place very rapidly at temperatures that can be produced by a fission bomb composed of uranium-235 or plutonium. However, tritium ($t_{1/2} = 12$ years) is difficult to produce and store. One type of hydrogen bomb depends on the production of tritium in the bomb. In this type, lithium deuteride (^{6}Li^{2}H, a solid substance) is placed around an ordinary ^{235}U or ^{239}Pu fission bomb. The fission bomb is set off in the usual way. Lithium-6 absorbs some of the neutrons produced and splits into tritium and helium-4:

$$^6_3\text{Li}^2_1\text{H} + ^1_0n \rightarrow ^3_1\text{H} + ^4_2\text{He} + ^2_1\text{H}$$

The temperature reached by the fissioning of ^{235}U or ^{239}Pu is sufficiently high to bring about the fusion reaction of tritium and deuterium:

$$^3_1\text{H} + ^2_1\text{H} \rightarrow ^4_2\text{He} + ^1_0n + 17.6 \text{ mev}[10]$$

The largest nuclear bomb exploded to date, a Russian one, was a 60-megaton[11] fission-fusion bomb. A 20-megaton bomb usually contains about 300 lb of lithium deuteride as well as a considerable amount of plutonium and uranium.

Fusion Power. The controlled release of energy by fusion reactions, comparable to the controlled fission reactions, would provide a solution to our ever increasing need for energy on a planet that has a limited supply of coal and petroleum and a growing problem with the pollution of the atmosphere from the burning of these fuels. Compared with power from nuclear fission, fusion power would be free of radioactive waste and would have a lasting supply of cheap fuel in the form of heavy hydrogen from the waters of the earth. The problem is how to conduct a process in a controlled way at a temperature of above 100,000,000°C. Today, much research is being carried out over the world toward the solution of this problem, and there seems little doubt that the dream of unlimited energy from fusion reactions will eventually be a reality.

THE PRODUCTION OF RADIOACTIVE ISOTOPES

Most elements do not have naturally occurring radioactive isotopes. Some isotopes that do occur in nature are so rare and hence so expensive that their use in tracer work, treatment of disease, and other practical applications is neces-

---

[10] When the energy of a reaction is expressed in electron volts, the amounts of reactants and products shown in the equation are to be interpreted as individual particles rather than mole quantities.

[11] A megaton bomb releases the energy equivalent to the explosion of one million tons of TNT.

sarily severely restricted. In the majority of cases, however, the desired radioactive isotope can be produced by bombardment in a cyclotron or nuclear pile. Two examples will be discussed.

The element sodium has no known radicactive form that occurs naturally. But sodium-24, a beta emitter, can be made by neutron bombardment in the nuclear reactor.

$$^{24}_{12}Mg + ^{1}_{0}n \longrightarrow ^{24}_{11}Na + ^{1}_{1}H$$

Then

$$^{24}_{11}Na \xrightarrow[14.8 \text{ hr}]{t_{1/2}} ^{24}_{12}Mg + ^{0}_{-1}e$$

The artificial isotope, cobalt-60, emits gamma radiation, which is used in medical research and in the treatment of disease. Slugs containing cobalt-60 encased in thick lead shields are being used in hospitals in place of X-ray machines. The isotope is produced by the neutron bombardment of natural cobalt.

$$^{59}_{27}Co + ^{1}_{0}n \longrightarrow ^{60}_{27}Co$$

Then

$$^{60}_{27}Co \xrightarrow[5.3 \text{ yr}]{t_{1/2}} ^{60}_{28}Ni + ^{0}_{-1}e + ^{0}_{0}\gamma$$

In addition to these methods, which are designed to produce specific isotopes, nuclear reactors operated for the production of electricity, plutonium, etc., produce a great variety of unstable isotopes.

Uses of Radioactive Isotopes. The story is told that one of the first practical applications of radioactivity was made by the youthful Ernest Rutherford. Suspicious that his landlady was serving again scraps of food left on plates the day before, Rutherford used radioactive lead to label some meat he left on his plate. Sure enough, the stew next day was radioactive.

Since World War II, there has been a tremendous increase in the use of radioisotopes in industry, medicine, and research. Each year we now use radioisotopes equivalent in radioactivity to about 200,000 g of radium. These uses are so varied that we can mention only a few.

IN INDUSTRY. A trace of radioactive substance can be dissolved in a fluid that is being pumped through a long pipe line. The location of the fluid containing it can be determined at any time by using a Geiger counter on the outside of the pipe line. Thus the progress of any sample of fluid can be followed in gas or oil pipe lines hundreds of miles long.

As it flows from the huge rollers in a mill, sheet metal can be led between a gamma-emitting isotope and a Geiger counter. If the metal becomes too thick, the gamma radiation that reaches the counter will be decreased. The counter than sends out a signal, which automatically increases the pressure on the rollers and thins the sheet metal.

Other industrial uses of radioisotopes include preservation of foods, control of uniformity in cigarette manufacture, radiographic inspection of locomotives and other types of machinery for flaws in the metal, evaluation of dirt-removing efficiency of detergents, and removal of static charges from paper to prevent sticking during printing, handling, etc.

IN MEDICINE. One of the most widely publicized applications of radioactive isotopes is in the treatment of disease in plants and animals. Beta and gamma emitters such as cobalt-60 and cesium-137 are used to irradiate diseased portions of living tissue in much the same way that X rays have long been employed. Some seven hundred gamma-ray teletherapy units are now in use in medical institutions in this country. Radioactive radiation is often more effective than X rays, because its action can be limited to the afflicted tissue more easily, thus preventing radiation damage to healthy tissue. For example, radioactive iodine is used in the treatment of diseased thyroid glands, radioactive phosphorus in cancer of the bone, and radioactive sodium (in capsules) in localized malignancies.

A compound called iodofluorescein is known to concentrate in the area of brain tumors. After a patient is injected with iodofluorescein containing radioactive iodine, the surgeon can locate the tumor by moving the tube of a Geiger counter over the outside of the patient's head. This means that painful and dangerous exploratory operations are minimized.

It has been suggested that the heat energy released during radioactive decay might someday be utilized to operate a small steam engine, which could be used as an artificial heart.

Overexposure to radiation can have serious consequences for living organisms, because ionizing particles can destroy cells. Persons working with radioactive substances must take every precaution to shield themselves from harmful amounts of radiation. We should remember, however, that we live in an environment in which we are continuously subjected to weak *natural* radiation. Small amounts of radium and uranium are common in the minerals around us; cosmic rays bombard us ceaselessly; and radioactive atoms are present even in our own bodies. The chief sources of radiation in the human body are carbon-14 and potassium-40. In the average adult, about 20,000 beta particles per second are emitted by atoms of these two elements.

IN RESEARCH. The application of isotopes as tracers for studying the mechanism of chemical reactions and investigating life processes is now regarded as one of the most important advances in the history of chemistry and biology. Either a radioisotope or an uncommon stable isotope is used as a *tracer*.

EXERCISES

1. The rate with which radioactive chlorine-36 emits beta particles is the same, irrespective of whether the chlorine is combined in NaCl, in CCl_4, or in chlorine molecules (Cl_2). Give a reason for this behavior.
2. Helium gas is often found trapped in rocks or minerals that contain radioactive substances. Account for its presence.

Nuclear Chemistry

3. Describe the operation of the cloud chamber. Show with a well-labeled diagram how you might construct a cloud chamber from materials found in your home.
4. Explain the operating principles of the Geiger counter. Why must a battery or some other source of electricity be used to make the Geiger counter operate?
5. Thorium-232 forms a radioactive series similar to the uranium-238 series. If the stable end product of the series is lead-208, how many alpha particles are emitted in the radioactive decay process of changing thorium-232 to lead-208? How many beta particles are emitted?
6. Refer to Fig. 17-5 and predict whether the following nuclei will be more likely to emit a beta particle or an alpha particle: $^{52}_{22}Ti$, $^{48}_{24}Cr$, $^{56}_{25}Mn$, $^{68}_{32}Ge$, $^{72}_{33}As$. Write equations showing your predicted reactions.
7. Refer to Fig. 17-5b. For atomic numbers 20 through 36, compare the number of stable nuclei that have an odd atomic number with the number of stable nuclei that have an even atomic number. Compare the number of stable nuclei that have an odd number of neutrons to the number of stable nuclei that have an even number of neutrons. Formulate a general rule that summarizes the above observations.
8. What product is formed when an atom of ^{220}Rn emits an alpha particle? When an atom of ^{228}Ac emits a beta particle?
9. Starting with a 0.50-g sample of ^{210}Bi, how much remains after 15 days? The half-life of ^{210}Bi is 5 days.
10. A Geiger counter showed that a radioactive substance gave a reading of 5180 counts per minute in January 1955; by January 1970, the reading had fallen to 81 counts per minute. What is the half-life of the substance?
11. A certain radioactive isotope has a half-life of 8 hours. If a chemist wants to prepare a compound containing 0.50 g of this isotope, with how much of the isotope must he start if the preparation requires 16 hours of working time? The chemist works from 9.00 A.M. to 5.00 P.M. each day while preparing the compound and will use it immediately on finishing. Assume a 100 percent efficiency in the preparation, other than loss from radioactive decay.
12. Only protons and helium nuclei of high energy are effective in bringing about nuclear reactions. On the other hand, low-energy neutrons may be more effective than fast neutrons. Account for this difference.
13. Why is exposure to gamma rays likely to be more dangerous than exposure to alpha particles?
14. State in your own words the physical laws on which the operation of the cyclotron is based.
15. Balance the following nuclear equations by filling in data for the question marks; some are equations for natural radioactivity, but most are not:
    (a) $^{14}_{7}N + ^{1}_{0}n \rightarrow ? + ^{1}_{1}H$
    (b) $^{80}_{35}Br + ^{0}_{0}\gamma \rightarrow ^{79}_{35}Br + ?$
    (c) $^{16}_{8}O + ? \rightarrow ^{17}_{9}F$
    (d) $^{3}_{1}H \rightarrow ? + _{-1}^{0}e$
    (e) $^{23}_{11}Na + ^{2}_{1}H \rightarrow ^{24}_{11}Na + ?$
    (f) $^{63}_{29}Ca + ^{1}_{1}H \rightarrow ^{63}_{30}Zn + ?$
    (g) $^{214}_{82}Pb \rightarrow ? + _{-1}^{0}e$
    (h) $^{54}_{26}Fe + ^{2}_{1}H \rightarrow ^{4}_{2}He + ?$
    (i) $^{59}_{27}Co + ? \rightarrow ^{58}_{27}Co + 2^{1}_{0}n$
    (j) $? + ^{1}_{0}n \rightarrow ^{32}_{15}P + ^{1}_{1}H$
    (k) $^{60}_{28}Ni + ^{4}_{2}He \rightarrow ? + ^{1}_{0}n$
    (l) $^{238}_{92}U \rightarrow ^{4}_{2}He + ?$
    (m) $^{252}_{99}Es \rightarrow ^{248}_{97}Bk + ?$
    (n) $? + ^{1}_{1}H \rightarrow ^{28}_{14}Si + ^{0}_{0}\gamma$
16. (a) Write a hypothetical equation for the fission of ^{233}U as the result of neutron capture. Assume that isotopes of lanthanum and bromine and three neutrons are produced.
    (b) Could a reaction such as that described in (a) be the basis for a chain reaction? Explain.

17. Based on data read from Fig. 17-8, calculate the average energy evolved in ergs per nucleon for the reaction $^2H + {}^6Li \rightarrow 2\,{}^4He$. Also, calculate the energy in calories per gram of helium produced and in kilocalories per mole of helium.
18. (a) The storage and disposal of fission products from nuclear power plants is a pressing problem. Some have been put deep underground, some have been sunk in concrete containers in the sea. Suggest possible objections to these methods.
    (b) Harmful wastes from ordinary chemical plants are sometimes disposed of by diluting them or by changing them chemically and pumping them into a river or into the sea. Discuss the application of these techniques to fission-product wastes.
19. How does the purpose of the moderator in a nuclear reactor differ from that of the control rods?
20. The age of an old carbon-containing object is determined simply by the measurement of the radioactivity present in the carbon. How is this procedure the same as, or how does it differ from, the determination of the age of minerals?
21. Would the splitting of a ^{12}C atom into two ^6Li atoms be an endothermic or exothermic process? Give a reason for your answer.
22. Discuss the origin of solar energy.
23. In what way is the production of electrical energy by utilizing fissionable materials similar to the production of electrical energy by utilizing coal or petroleum products? In what ways are the two methods different?
24. What are the hurdles to be overcome if electrical energy is to be produced from energy made available by thermonuclear reactions? What advantages, if any, would this method for producing electrical energy have over existing methods?

SUGGESTED READING

Anders, O. U., "The Place of Isotopes in the Periodic Table," *J. Chem. Educ.*, **41**, 522 (1964).

Bauman, R. P., "Can Matter Be Converted to Energy?," *J. Chem. Educ.*, **43**, 366 (1966).

Badash, L., "How the Newer Alchemy Was Received," *Sci. Amer.*, **215** (2), 88 (1966).

Fowler, T. K., and R. F. Post, "Progress Toward Fusion Power," *Sci. Amer.*, **215** (6), 21 (1966).

Ginzton, E. L., and W. Kirk, "The Two-mile Accelerator," *Sci. Amer.*, **205** (5), 49 (1961).

Keller, O. L., Jr., "Predicted Properties of Elements 113 and 114," *Chemistry*, **43** (10), 8 (1970).

Seaborg, G. T., "From Mendeleev to Mendelevium—and Beyond," *Chemistry*, **43** (1), 6 (1970).

Trower, W. P., "Matter and Antimatter," *Chemistry*, **49** (2), 8 (1969).

Turner, G., "Argon-40/Argon-39 Dating of Lunar Rock Samples," *Science*, **167**, 466 (1970).

Woodwell, G. M., "The Ecological Effects of Radiation," *Sci. Amer.*, **208** (6), 40 (1963).

Young, G., "Dry Lands and Desalted Water," *Science*, **167**, 339 (1970).

Zimmerman, J., "Answering the Question When," *Chemistry*, **43** (7), 22 (1970).

18

INORGANIC CHEMISTRY: THE ALKALI AND ALKALINE EARTH ELEMENTS

SYSTEMATIC STUDY OF INORGANIC CHEMISTRY

This is the first of several chapters devoted to the study of inorganic chemistry, that branch of chemistry which treats of all substances that do not contain carbon, plus a very few earthy or rock-like carbon-containing substances. It has long been customary to undertake this study by examining in turn different families of elements in the periodic table. This method has the great advantages of enabling one to study similar elements and their compounds together and of relating families of dissimilar elements to one another by way of the well-known fundamental periodic relationships.

The chemistry of the elements in the groups at the left and right of the periodic table is less complicated than that in the groups toward the middle of the table. Therefore, we will introduce the study of the table by taking up in this chapter groups IA and IIA and then turn in Chap. 19 to groups VIIA and VIIIA. In Chaps. 21 through 23, groups VIA, VA, and part of IVA are discussed, and, finally, in Chap. 24 we will take up the reactions of many of the metals from families in the middle region of the periodic table.

To begin our systematic study, let us first review the division of the periodic table into general areas and examine some of the trends in behavior exhibited in the periods and groups of the table.

General Divisions of the Periodic Table. The elements can be divided into four major classes, each with important physical and chemical characteristics. These classes occupy four reasonably distinct areas on the long-form periodic table.

NOBLE GASES. The six elements that comprise group VIIIA are known as the **noble gases.** They are colorless, monatomic gases whose atoms show little tendency to gain or lose electrons. The electronegativities of the noble gases are not well established, because so few of their compounds are known.

METALS. About eighty elements are classified as metals, including some from every group except VIIA and VIIIA. These elements are at the left and in the center of the periodic table. As a class, they are malleable, ductile, good conductors of heat and electricity, and have a metallic luster.

In chemical reactions, the atoms of metals tend to donate electrons. Their electronegativities are low, ranging from about 2.2 down. An important property of metals with very low electronegativities is that their oxides react with water and yield OH^- ions. Metal oxides that act in this way are called *basic oxides*, because they are good proton acceptors and because in water they form the proton acceptor ion, OH^-. Metals whose hydrated oxides are *acidic* in water solutions are called *metallo-acid elements*.

NONMETALS. The nonmetals, consisting of about a dozen relatively common and important elements, are to the right on the periodic table, with the exception of hydrogen. These elements are characterized by their lack of the metallic properties in the solid state: they are, with few exceptions, nonconductors, nonmalleable, nonductile, and have no metallic luster. The nonmetals that are gases under normal conditions are composed of diatomic molecules.

The atoms of the nonmetals tend to accept electrons in chemical reactions with metals, but they also readily react with one another by forming covalent bonds, for example in SO_3, CO_2, CCl_4, and H_2O.

Electronegativities of nonmetals range from about 2.2 to 4.0. An important property is that nonmetal oxides react with water and yield H_3O^+ ions. Nonmetal oxides are *acidic oxides*.

Hydrogen is classed by some chemists in group IA, by others in group VIIA. One reason for this is that toward some elements it acts as an electron donor (like a metal) and toward other elements it acts as an electron acceptor. With an intermediate electronegativity of 2.1, it is between the metals and nonmetals in this respect. Because of its physical characteristics and the types of compounds it forms, hydrogen is known as a nonmetal.

BORDERLINE ELEMENTS. Borderline elements are elements that to some extent exhibit both metallic and nonmetallic properties; they usually act as electron donors with nonmetals and as electron acceptors with metals. Chemists do not agree exactly which elements should be included in this class, but they do agree that such elements lie close to the zigzag line in the periodic table inside the front cover. The names of some of these elements are boron, B, silicon, Si, germanium, Ge, and tellurium, Te. They are all solids at room temperature, somewhat brittle, and rather poor conductors of heat and electricity.

The electronegativities of the borderline elements range between 1.8 and 2.1. Their oxides react with water to yield solutions that are either weakly acidic or weakly basic.

The Periodic Table and Trends in Behavior. Perhaps the most important use of the table in predicting chemical properties is based on the relations between chemical properties and the structures of atoms. It will be helpful to refer to Fig. 18-1 as we take up each of the following trends in behavior.

Trend 1. There is a tendency for atoms to become smaller from left to right in a period. The electrons are pulled in closer to the nucleus, because the increase in the positive charge on the nucleus exerts a continually greater attraction, and because this effect outweighs the repulsion of outer electrons by each other and by electrons in lower levels.

There is a great increase in size with the addition of an *s* electron to start a new main energy level (see Li, Na, K, Rb, and Cs). Note, however, that the minimum size is reached near the middle of the long periods and that there is even a slight increase in size toward the middle-right side. In periods 5 and 6, indium (In, 49) and thallium (Tl, 81), respectively, are seen to be slightly larger than their neighbors. This is apparently related to the fact that with these elements the 5*p* and 6*p* sublevels begin to be filled (see Fig. 4-12). Starting with yttrium (Y, 39) and lanthanum (La, 57), respectively, the elements in these two periods are adding electrons in inner levels; with indium and thallium, electrons again are added to the outside main levels, 5 and 6, respectively. The atoms then become progressively smaller as each period fills to a noble gas atom.

Trend 2. The tendency of atoms in A families to gain electrons increases from left to right in a period. The smaller the atom, the closer a valence electron can come to the positive nucleus, and the more tightly it is held. This trend does not apply well to the B families, as can be seen by checking their relative electronegativities in Fig. 18-1.

Trend 3. The tendency for atoms to lose electrons increases from top to bottom in an A family.[1] A tabulation of the electronegativity of elements in a few representative families shows this trend clearly. These data are listed in Table 5-2 and plotted in Fig. 5-9.

TRANSITION SERIES. In the long periods the regular addition of electrons to *d* sublevels begins with the third element in the period and continues till the eighth element from the end is reached (that is, an element in group IB). Examine the electron arrangements of scandium, Sc, and yttrium, Y, in periods 4 and 5 (see periodic table inside the front cover). In these cases the electron that was added as a result of the increase in atomic number went into the main energy level next to the outer-most one, that is, into a 3*d* or 4*d* orbital, respectively. In the elements after scandium and yttrium a *d* sublevel is built up from atom to atom roughly as the atomic number increases, till it contains its maximum of 10 electrons. These elements are also referred to as **metalloids.**

The elements in a series in which electrons are being added to *d* orbitals are called *transition elements.* Chemists differ as to precisely what elements this definition includes. Some would include the ten B family elements in each period (for example, see Fig. 4-12); perhaps the majority of chemists would not include

---

[1] In some families the trend is just the reverse. For example, group IB consists of Cu, Ag, and Au, of which copper loses its electrons most easily; and group IIB consists of Zn, Cd, and Hg, of which zinc has the greatest tendency to lose electrons. Both gold and mercury are relatively inactive electron donors.

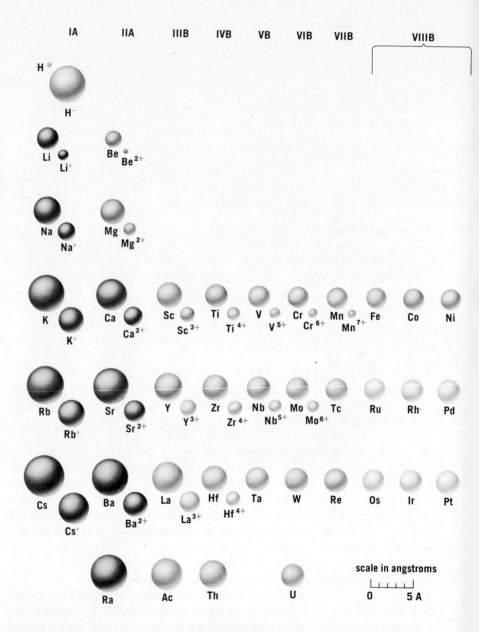

Fig. 18–1. The relative sizes of the atoms and ions of most of the elements. Electronegativities are indicated by the color code in the lower right corner. (Radii of noble gas atoms are from Bing-Man Fung, *J. Phys. Chem.*, **69**, 596 [1965].)

| IB | IIB | IIIA | IVA | VA | VIA | VIIA | VIIIA |

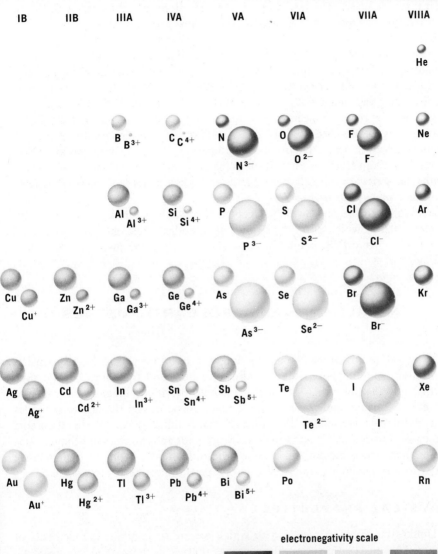

electronegativity scale

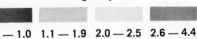

0.7 — 1.0   1.1 — 1.9   2.0 — 2.5   2.6 — 4.4

the IIB elements; and a third view is to include just the eight elements in families IIIB through VIIIB. The last choice is the one adopted in this text, with the IB and IIB families referred to as neighbors of transition elements.

INNER TRANSITION SERIES. The third elements in periods 6 and 7 (lanthanum, La, and actinium, Ac) have outer electron configurations similar to those of scandium and yttrium. But with the fourth element in each of these last two periods, a series of fourteen elements begins in which the electron configurations of the outer two main levels remain very nearly constant. With cerium, Ce, begins a series in which electrons are added to the seven $4f$ orbitals of the fourth main energy level. (See Fig. 4-12.) A similar series is apparently developed beginning with thorium, Th, in the seventh period, in which f electrons are added in the fifth main level (thorium itself being an exception). These two series, the **lanthanide series**[2] in the sixth period and the **actinide series** in the seventh period, are called *inner transition series*.

According to the most widely accepted theory, the actinide series ends with element 103, at which point the $5f$ orbitals are filled. Element 104 should be a member of group IVB, being placed in the periodic table just under hafnium, number 72.

ALKALI AND ALKALINE EARTH METALS

The alkali family of metals, located in group IA of the periodic table, consists of lithium, sodium, potassium, rubidium, cesium, and francium. The alkaline earth family, the group IIA elements, includes beryllium, magnesium, calcium, strontium, barium, and radium. The families are so named because their oxides and hydroxides are among the strongest bases (alkalis) known. In discussing these two families, particularly in describing trends and extremes in behavior, it is understood that francium and radium are not included; owing to their relative rarity and radioactivity, many of their properties have been little studied.

PHYSICAL PROPERTIES

Unlike the common metals with which we are so familiar, the elements of groups IA and IIA have several properties that we do not commonly associate with metals: they have low densities and most of them are relatively soft and have low melting points. These properties are especially typical of the alkali metals, one being a liquid at room temperature and three having densities less than that of water. All of them, from lithium to cesium, can be deformed by squeezing between thumb and forefinger (with proper protection for one's skin). The alkaline earth elements are somewhat harder, ranging from barium, which is about as hard as lead, to beryllium, which is hard enough to scratch most other metals.

---

[2] The elements of the lanthanide series were for many years referred to as the *rare earths*.

Inorganic Chemistry: The Alkali and Alkaline Earth Elements

In Tables 18-1 and 18-2 are listed some of the important physical properties of the elements in groups IA and IIA, respectively. The elements in both of these families have the silvery luster of typical metals on freshly cut surfaces, but they tarnish rapidly on exposure to air. They also have the high electrical and thermal conductivities characteristic of metals.

TABLE 18-1.

Physical Properties of Group IA Metals (Excluding Francium)

	Li	Na	K	Rb	Cs
Melting point, °C	180	98	63	39	29
Boiling point, °C	1326	889	757	679	690
Density, g/ml	0.53	0.97	0.86	1.53	1.90
Electron distribution	2,1	2,8,1	2,8,8,1	2,8,18,8,1	2,8,18,18,8,1
Ionization energy, ev	5.39	5.14	4.34	4.18	3.89
Atomic radius, A	1.22	1.57	2.02	2.16	2.35
Ionic radius, A	0.60	0.95	1.33	1.48	1.69
Electronegativity	1.0	0.9	0.8	0.8	0.7

An understanding of the data in the latter part of Tables 18-1 and 18-2 is prerequisite to understanding the chemical behavior of the two families. The electronic structure is of prime interest. The relative simplicity of the chemical reactions of groups IA and IIA is associated with their simple electronic structures. Outside a stable core structure similar to that of a noble gas element, they have, respectively, one or two s electrons that are lost relatively easily.

TABLE 18-2.

Physical Properties of Group IIA Metals (Excluding Radium)

	Be	Mg	Ca	Sr	Ba
Melting point, °C	1280	650	850	770	704
Boiling point, °C	2970	1120	1487	1384	1638
Density, g/ml	1.86	1.74	1.55	2.6	3.59
Electron distribution	2,2	2,8,2	2,8,8,2	2,8,18,8,2	2,8,18,18,8,2
Ionization energy, ev	9.32	7.64	6.11	5.69	5.21
Atomic radius, A	0.89	1.36	1.74	1.91	1.98
Ionic radius, A	0.31	0.65	0.99	1.13	1.35
Electronegativity	1.5	1.2	1.0	1.0	0.9

The elements in groups IA and IIA have the lowest average ionization energies and electronegativities of all the families of elements. These properties are related to the sizes of the atoms (Fig. 18-1) and the relatively great distances from the nuclei of the outer s electrons.

In both families, as the atomic number increases the atoms increase in size and the bonding between atoms apparently weakens, as indicated by the decreases in melting and boiling points within each family. The lower melting and boiling points of the IA as compared to the IIA elements indicate that the bonds between IA atoms are weaker. This difference in bond strengths is probably due to the fact that the IA metals have only one electron per atom for bond formation, whereas the IIA metals have two valence electrons per atom.

Flame Spectra. As pointed out in Chap. 4, the elements of groups IA and IIA impart characteristic colors to an ordinary flame. In analytical laboratory work, flame tests are often used to reveal the presence of various alkali and alkaline earth elements. The yellow flame test for sodium is one of the most sensitive, 1 part of sodium in 10,000,000 parts of solvent being detectable.

CHEMICAL PROPERTIES

Activity. The most striking characteristic of these elements is their extreme activity. The reason that most people are not familiar with the appearance of the very common metals sodium, potassium, and calcium is that these metals are so active. Samples of the metals must be stored in sealed containers or under oil to protect them from the attack of air and moisture. None of the elements in group IA or IIA exists in nature as an element. All the alkali metals exist in natural compounds as unipositive ions; all the alkaline earth metals exist as dipositive ions.

Characteristic Reactions. The alkali and alkaline earth metals are powerful reducing agents, because they lose electrons so readily. They combine energetically with most nonmetal elements, forming ionic compounds, such as halides, hydrides, oxides, and sulfides. Because lithium and the alkaline earth metals react directly with nitrogen at high temperatures, they continue to burn in air even after they have combined with all the oxygen available.

The alkali metals react spectacularly with water; calcium, strontium, and barium react less violently. Sample reactions are:

$$2K + 2H_2O \rightarrow 2KOH + H_2 \uparrow$$
$$Ca + 2H_2O \rightarrow Ca(OH)_2 + H_2 \uparrow$$

In the case of potassium, rubidium, and cesium the reactions with water are so rapid and so exothermic that the hydrogen evolved usually bursts into flame. Lithium reacts much more slowly than the other group IA elements but still rapidly enough so that the lithium-water reaction has been studied for use as a propellant reaction for torpedoes. Of the elements in IIA, beryllium and magnesium do not react appreciably with water at room temperature.

All the elements tarnish quickly in air, and all except beryllium and magnesium corrode steadily till they are completely converted into oxides or hydroxides or carbonates. Beryllium and magnesium oxidize readily, but the tough oxide film that is formed tends to protect the underlying metal from further attack at room temperature. When strongly heated, even these two metals burn violently. At high temperatures magnesium reacts with nitrogen or carbon dioxide, and so will continue to burn in air after it has combined with the oxygen. Lithium is the only one of the alkali metals that combines directly with nitrogen.

Lithium is similar to magnesium in other chemical reactions, and beryllium is similar to aluminum. This diagonal relationship of similarities involving a period 2 element and a period 3 element in the next group shows up in other cases also. For instance, boron is similar to silicon.

The principal chemical reactions of the elements in groups IA and IIA are listed in Tables 18-3 and 18-4. It is evident from these data that the elements

TABLE 18-3.

Reactions of Group IA Metals (M Represents an Alkali Metal)

$4M + O_2 \to 2M_2O$	Limited amount of O_2
$2M + O_2 \to M_2O_2$	In air; M = K, Rb, or Cs
$2M + X_2 \to 2MX$	X = F, Cl, Br, or I
$2M + S \to M_2S$	Se and Te also react
$2M + 2H_2O \to 2MOH + H_2$	Violent, except with Li
$2M + 2NH_3 \to 2MNH_2 + H_2$	With catalyst
$2M + H_2 \to 2MH$	Dry H_2 gas
$2M + 2H^+ \to 2M^+ + H_2$	Violent

in these two groups have many common chemical characteristics. Their reactions with oxygen, sulfur, the halogens, hydrogen, water, and the hydrogen ion follow the same pattern. Later in the chapter we shall find that they form similar compounds with most negative ions.

TABLE 18-4.

Reactions of Group IIA Metals (M Represents an Alkaline Earth Metal)

$2M + O_2 \to 2MO$	Be and Mg must be heated
$M + O_2 \to MO_2$	Ba, Sr; high oxygen pressure helps
$M + X_2 \to MX_2$	X = F, Cl, Br, or I
$M + S \to MS$	Se and Te also react
$M + 2H_2O \to M(OH)_2 + H_2$	Mg and Be react only with steam to give oxides
$3M + N_2 \to M_3N_2$	Heated
$M + H_2 \to MH_2$	Heated; Be and Mg do not react
$M + 2H^+ \to M^{2+} + H_2$	Rapid

COMPOUNDS OF IA AND IIA ELEMENTS

Occurrence. The most common materials in the earth's crust are the igneous rocks, that is, rocks that exist in the form they assumed on solidifying from a molten condition. Members of the alkali and alkaline earth families are found in such rocks as components of practically insoluble minerals, e.g., the complex silicates. Table 18-5 lists the amounts of the various elements in the earth's crust.

TABLE 18-5.

Estimated Percentage Composition of Group IA and IIA Elements in Earth's Crust

Lithium	2×10^{-3}	Beryllium	4×10^{-4}
Sodium	2.6	Magnesium	1.9
Potassium	2.4	Calcium	3.4
Rubidium	4×10^{-3}	Strontium	1×10^{-2}
Cesium	1×10^{-4}	Barium	1×10^{-2}
Francium	Trace	Radium	8×10^{-12}

Simple compounds of the alkali metals, formed when igneous rocks are broken down by weathering, are so soluble that they tend to be leached out of the soil by rain water and carried by streams and rivers to the sea. Sea water contains about 3 percent by weight of alkali metal compounds, such as NaCl, KCl, etc. Large quantities of alkali metal salts are found in salt lakes and salt flats, and in underground pools of brine and beds of sedimentary rock left by the evaporation of seas in times long past. There are rich sedimentary deposits of alkali metal salts all over the world that serve as ores.

The simple alkaline earth metal compounds are not quite so soluble as the alkali metal compounds; indeed, some of the common ones—the carbonates and certain of the sulfates—are insoluble enough to exist as minerals that resist strongly the weathering action of rain water. The magnesium ion, however, is one of the very common ions in sea water, and a continuing supply of calcium ion is available in the sea for building the carbonate shells of crustaceous sea animals.

General Considerations. Because of the striking chemical similarity of the alkali metals, their compounds tend to be so much alike that we really need to discuss only the sodium and potassium compounds. Sodium compounds are most widely used, because they are cheapest.

The chemical differences between the various alkali metal ions are so slight that one can be substituted for another in most laboratory and industrial reactions. In some cases, however, substitution of one ion for another is not desirable. For instance, plants must secure potassium from the soil for proper growth, and hence fertilizers must contain potassium compounds; sodium compounds cannot be used for this purpose.

All the alkali metal ions are colorless and quite inactive. Solutions of these compounds are typically strong electrolytes. Their simple salts—such as LiCl, KNO_3, Cs_2SO_4, and Rb_2CO_3—are usually very soluble in water. Lithium compounds, however, do resemble magnesium compounds; for example, the solubilities of their carbonates and phosphates are low.

Among the alkaline earths, calcium, strontium, and barium form compounds that are very similar to one another. Magnesium, and more particularly beryllium, form compounds that differ from the other three in properties. Because of its small ionic size (hence, a large charge density), beryllium forms covalent-ionic and covalent bonds with a number of other atoms. Beryllium compounds are used with caution today because of their poisonous character, in ironic contrast with one of their earliest discovered properties, that of their sweet taste.

The compounds of beryllium tend to hydrolyze in water, partly because of the formation of the insoluble hydroxide, $Be(OH)_2$. The high charge density of the tiny Be^{2+} ion enables it to attack water; in this and in some other ways it resembles the Al^{3+} ion.

The ions of the alkaline earth elements are colorless and fairly inactive. Many of their simple salts—such as $MgSO_4$, $CaCl_2$, $Ba(NO_3)_2$, and $BeSO_4$—are very soluble. However, the sulfates, normal carbonates, and normal phosphates of calcium, strontium, and barium are only slightly soluble.

Oxides. The group IA oxides of the M_2O type (Na_2O, K_2O, etc.) are white solids that are extremely sensitive to moisture and carbon dioxide, reacting to form the hydroxides, MOH, and the carbonates, M_2CO_3, respectively.

The monoxides of the alkali metals can be obtained by heating the metals in a limited supply of dry air at relatively low temperatures (below about 180°C).

The common oxides of group IIA have the expected formula, MO. Both lime, CaO, and magnesia, MgO, are made by the high-temperature decomposition of naturally occurring carbonate rocks in *lime kilns*. The reaction for lime is

$$CaCO_3 \xrightarrow[900°C]{about} CaO + CO_2$$

Magnesia is used for firebrick and as insulation for steam pipes. Lime is used to make mortar and plaster and to neutralize acid soils; it also is the cheapest industrial source of hydroxide ions (slaked lime).

Its extreme affinity for water makes calcium oxide widely used for the dehydration of such liquids as alcohol, and for the drying of gases.

The group IIA oxides are white solids with very high melting points. They tend to react slowly with the water and carbon dioxide in the air to form the hydroxide and the carbonate, respectively:

$$BaO + H_2O \to Ba(OH)_2$$

$$CaO + H_2O \to \underset{\text{slaked lime}}{Ca(OH)_2}$$

$$CaO + CO_2 \to CaCO_3$$

$$SrO + CO_2 \to SrCO_3$$

The reaction with water is an exothermic process called **slaking**. In the case of barium oxide the heat of slaking is so great that, if only a little water is used, the mass may become visibly red hot. When slaked lime is used in a mortar in laying bricks, the setting process involves drying and crystallization, followed by the slow conversion of the slaked lime to calcium carbonate by the action of carbon dioxide from the atmosphere.

The more active alkali and alkaline earth metals form peroxides that are useful oxidizing agents. Sodium peroxide, Na_2O_2, potassium peroxide, K_2O_2, and barium peroxide, BaO_2, are well-known examples. Sodium peroxide is formed when sodium is burned in a stream of dry oxygen; barium peroxide is formed when barium oxide is heated in air. These two peroxides are sometimes used as a source of oxygen and of hydrogen peroxide:

$$Na_2O_2 + H_2SO_4 \rightarrow Na_2SO_4 + H_2O_2$$

$$BaO_2 + H_2SO_4 \rightarrow BaSO_4 \downarrow + H_2O_2$$

Superoxides and ozonides, such as KO_2 and KO_3, have been studied as possible *air revitalizers* for manned satellites and rockets. Not only can they be decomposed to yield oxygen for breathing, but the simple oxides produced can then be used to remove exhaled carbon dioxide from the air. First, $4KO_2(s) \rightarrow 2K_2O(s) + 3O_2(g)$; then $K_2O(s) + CO_2(g) \rightarrow K_2CO_3(s)$.

Hydroxides. Two of the most widely used strong bases are sodium hydroxide and calcium hydroxide. Because of its lower solubility, calcium hydroxide does not form a concentrated solution, but it is a typical strong base. The equations for preparing calcium hydroxide were given in the preceding section. About one-fifth of the sodium hydroxide (lye) used in this country is made by a process in which calcium hydroxide is a raw material:

$$Na_2CO_3 + Ca(OH)_2 \rightarrow CaCO_3 \downarrow + 2NaOH$$

or

$$2Na^+ + CO_3^{2-} + Ca^{2+} + 2OH^- \rightarrow CaCO_3 \downarrow + 2Na^+ + 2OH^-$$

Of the four possible compounds, calcium carbonate is the least soluble, so that it precipitates from the concentrated solution, leaving the sodium hydroxide in solution.

Sodium hydroxide is made in huge quantities by the electrolysis of sodium chloride brine (see Fig. 19-3). This process is the cheapest source of *concentrated* hydroxide solutions for the chemical industry. Sodium hydroxide is used in the production of soap, petroleum, rayon, textiles, and paper pulp.

Large quantities of calcium hydroxide are used in the manufacture of mortar, bleaching powder, and ammonia and in water softening.

Magnesium hydroxide is the familiar *milk of magnesia*, the anti-acid slurry long used as a household remedy.

Halides. A number of the alkali and alkaline earth halides occur so abundantly in nature that they serve as the raw material for making other compounds of the metals and halogens. Sodium chloride and potassium chloride are mined directly. During the purification of their ores or solutions, other halides—such as lithium, rubidium, and cesium chlorides, and some bromides and iodides—that are present to a small extent in the ores are recovered. Most of the potassium chloride thus obtained, about 3 million tons annually in the United States, is used in fertilizers.

Magnesium chloride is produced from salt wells and from sea water, as one step in the production of elemental magnesium. Calcium chloride, although found in nature, is synthesized on a large scale as a by-product of the Solvay process for making sodium carbonate. Used as a drying agent, it is also put on dusty roads because of its tendency to *deliquesce*, that is, to remove moisture from the air and form droplets of concentrated solution.

Carbonates. The alkali carbonates, M_2CO_3, are much more soluble than the alkaline earth carbonates, MCO_3, a fact that explains why the latter are more commonly found as beds of sedimentary rock. Carbonates are among the most abundant natural group IIA compounds. Calcium carbonate is deposited on the ocean floor as the lowly oyster shell, as lacy coral, or in other forms. Geological metamorphosis then produces great beds of limestone, or marble, or even beautifully transparent, colorless crystals of calcite. Though their appearances differ, all these forms are essentially $CaCO_3$. Magnesium carbonate, calcium carbonate, calcium bicarbonate, and sodium bicarbonate are discussed in Chap. 23.

Of all the alkali compounds, sodium carbonate, Na_2CO_3, is second only to sodium chloride in terms of tons used, almost 6 million tons annually in the United States. It is one of the cheapest very soluble strong bases and is a source of carbonate ions for the chemical industry. Large amounts are used in making glass, soap, and paper and in cleaning preparations (washing soda) and water softeners. Although produced commercially from natural salt lakes, the demand for sodium carbonate is so great that about 90 percent of it is made indirectly from NaCl and $CaCO_3$ by the multistep Solvay process.

Sulfates. Sodium sulfate, Na_2SO_4, is used in making glass and in one process for pulping wood. Potassium sulfate, K_2SO_4, is a desirable ingredient of certain types of fertilizer.

The solubility of the alkaline earth sulfates decreases markedly, from the very soluble beryllium sulfate to the practically insoluble barium sulfate. It is common in chemical analysis to add soluble barium chloride to a sulfate-containing solution and then collect and carefully weigh the barium sulfate precipitate. From the weight of the barium sulfate can be calculated the weight of the sulfate in the original solution.

Magnesium sulfate, $MgSO_4$, and calcium sulfate, $CaSO_4$, are mentioned in Chap. 21.

Other Compounds. The elements of groups IA and IIA form stable compounds with practically all the known negative ions. Their exceptional solubility in water and their low cost make the sodium compounds the most widely used sources of specific anions for chemical reactions.

Because of their low electronegativities, these elements (except beryllium and magnesium) react at high temperatures with hydrogen to form saltlike compounds. Some of these hydrides can be melted and electrolyzed, the metal ion going to the negative pole, and the hydride ion, H^-, going to the positive pole. In such compounds hydrogen behaves like a halogen, the family name for the group VIIA elements, to which we turn our attention in the next chapter.

EXERCISES

1. (a) Using the data early in the chapter for electronegativity ranges for classes such as metals and nonmetals, compare these ranges with Fig. 18-1 and point out some instances in which it is difficult to say to which class an element belongs.

(b) Do the borderline elements divide the periodic table into two distinct areas of differing electronegativities?
(c) Are there examples of metals that have higher electronegativities than certain nonmetals?

2. Both chromium and sulfur form acidic oxides of formula RO_3. What is the formula of the acid formed when each of these oxides is dissolved in water? Chromium is a metallic element while sulfur is a nonmetallic element. From the data on electronic structure inside the front cover, explain how the two elements can form such similar compounds. What nonmetal might form a few compounds similar to the metallic element, vanadium, in a manner analogous to the similarity of certain compounds of chromium and sulfur?

3. There are three main factors that determine how tightly a valence electron is held in an atom. Discuss these factors, using the elements magnesium, aluminum, and calcium as examples, and correlate your discussion with ionization energy data.

4. As pointed out in the text, hydrogen is sometimes placed in group IA and sometimes in group VIIA of the periodic table. Write an equation that would support its placement in group IA. Write an equation indicating that it should be in group VIIA.

5. In periods 5 and 6, at the point at which the $5p$ and $6p$ sublevels begin to fill, the atoms increase in size. How might this be accounted for theoretically?

6. How does a ton of magnesium compare in volume with a ton of sodium? With a ton of gold?

7. On the basis of physical properties, explain why strontium should be intermediate in chemical activity between calcium and barium.

8. One should not handle the alkali metals with bare hands. Write equations for possible reactions and specify the danger to one's skin.

9. Write equations for the following reactions and state what conditions are necessary for the reactions to occur:
   (a) $Rb + H_2O \rightarrow$
   (b) $Sr + O_2 \rightarrow$
   (c) $Be + O_2 \rightarrow$
   (d) $Mg + Br_2 \rightarrow$
   (e) $K + S \rightarrow$
   (f) $Ba + N_2 \rightarrow$
   (g) $Ca + H_2 \rightarrow$
   (h) $Na + O_2 \rightarrow$

10. Calcium is about twice as abundant as magnesium in the earth's crust, yet magnesium ions are more than three times as abundant in seawater as are calcium ions. Why is this?

11. Which of the following compounds are water-insoluble: $CaCO_3$, Na_2CO_3, $BaSO_4$, KCl, $Ca(HCO_3)_2$, $CsNO_3$, $Sr(NO_3)_2$, $LiCl$, $MgCO_3$, K_2CO_3?

12. Explain why sodium compounds can be and are used more commonly in industry than the corresponding potassium or rubidium compounds. What is one major use of potassium compounds in contrast to sodium compounds?

13. Write an equation for the formation of a hydrate formed during the deliquescence of calcium chloride.

14. To analyze a batch of commercial potassium sulfate, three samples were taken, dissolved in water, and added to separate solutions of barium chloride. The barium sulfate precipitates were filtered, dried, and weighed. The weights of the samples were 0.546, 0.623, and 0.578 g; and the weights of the corresponding precipitates were 0.675, 0.777, and 0.720 g. Calculate the percentage potassium sulfate in the commercial material.

15. Of all the metals in groups IA and IIA, only magnesium is used as a structural metal in such items as ladders and airplanes. Why?

16. Marble can be cut into slabs and polished to make very attractive table tops. Why are marble tops not widely used on benches in chemical laboratories?
17. Which alkaline earth metal ion has the strongest attraction for water molecules? Why?
18. The chemistry of the alkali metals is primarily the chemistry of ions that have a charge of $+1$. How would you justify this statement?
19. Elemental sodium, potassium, and calcium were unknown as elements until about the beginning of the nineteenth century, although their compounds were abundantly scattered in nature and some forty elements had already been characterized. What delayed the discovery of these elements?
20. Note the densities of the group IIA metals as listed in Table 18-2. Account for the fact that calcium is the least dense of the metals by calculating the density of single atoms from the atomic weights and atomic radii. Why are your calculated values different from those reported in the table?
21. Crushed limestone is widely used by farmers to change the pH of soil. Is it used to increase or decrease the pH? Using an equation, show how the change is produced.
22. A paper bag of lime, after standing for a year or so, is found to be rock-like. Write equations to show possible chemical reactions that account for the hardening.

SUGGESTED READING

Chave, K. E., "Chemical Reactions and the Composition of Sea Water," *J. Chem. Educ.*, **48**, 148 (1971).

Duke, F. R., "Acid-Base Reactions in Fused Alkali Nitrates," *J. Chem. Educ.*, **39**, 57 (1962).

Markowitz, M. M., "Alkali-Metal Water Reactions," *J. Chem. Educ.*, **40**, 633 (1963).

Navratil, J. D., "Magnesium," *Chemistry*, **44** (5), 6 (1971).

Petrocelli, A. W., and D. L. Kraus, "The Inorganic Superoxides," *J. Chem. Educ.*, **40**, 146 (1963).

19

GROUPS VIIA AND VIIIA: THE HALOGENS AND THE NOBLE GASES

We began our study of families of elements at the left of the periodic table with groups IA and IIA. In this chapter we go to our first families at the right of the table, groups VIIA and VIIIA. As mentioned earlier, family resemblances are strongest in the groups at the far left and far right of the periodic table.

THE HALOGENS

The elements fluorine, chlorine, bromine, and iodine were known as the *halogen family* of elements long before the formulation of the modern atomic theory that groups them together in the periodic table. However, the similarities of these elements are best explained and studied against the background of their atomic structure. Each element in family VIIA has seven electrons (s^2p^5) in its outside main energy level. In addition to the four common elements, there is a rare halogen, astatine, which has been known since 1940, when it was made by means of nuclear bombardment experiments. Since then it has been found to occur in nature, but only in extremely minute quantities.

PHYSICAL PROPERTIES

Table 19-1 lists some of the important physical properties of the halogens. The striking generality evident from the tabulated data is that any given property changes in a regular way from one element to the next.

TABLE 19-1.

Physical Properties of the Halogen Family

	FLUORINE F	CHLORINE Cl	BROMINE Br	IODINE I
Appearance at room temperature	Yellowish gas	Greenish gas	Deep red liquid	Purple, almost black, solid
Molecular formula	F_2	Cl_2	Br_2	I_2
Melting point, °C	−218	−101	−7	114
Boiling point, °C	−188	−34	59	184
Ionization energy, ev	17.34	12.95	11.80	10.6
Radius of atom, A	0.72	0.99	1.14	1.33
Radius of ion (X^-), A	1.36	1.81	1.95	2.16
Electronic structure	2,7	2,8,7	2,8,18,7	2,8,18,18,7
Electronegativity	4.0	3.0	2.8	2.5

The increase in melting and boiling points with atomic number is explained by the fact that larger molecules have greater masses and greater intermolecular attraction than do small ones. The I_2 molecules have the greatest number of electrons far away from their respective atomic nuclei; and because these electrons are attracted by the nuclei of other iodine atoms, the van der Waals forces in iodine are stronger than in the smaller molecules of the other halogens.

Except for the noble gases, the halogens have the highest ionization energies of any family of elements. These high values are associated with the nearly complete *p* sublevels, as pointed out in the discussion of orbitals and ionization energies in Chap. 4. The trend in ionization energies in group VIIA reveals that the fluorine atom holds most tightly to its electrons, the iodine least tightly. This trend may be correlated with the sizes of the halogen atoms by referring to Fig. 19-1.

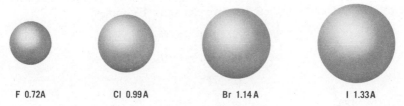

F 0.72A   Cl 0.99A   Br 1.14A   I 1.33A

Fig. 19–1. Relative sizes of halogen atoms.

CHEMICAL PROPERTIES

All four elements are extremely irritating to the nose and throat. Although solid iodine has a low vapor pressure, enough of it evaporates so that its odor is easily detected in a closed space. Liquid bromine is one of the most dangerous of the common laboratory reagents, because of its effect on the eyes and nasal passages and because it causes severe burns on contact with the skin. Chlorine and fluorine, usually handled as gases, should be used only in hoods and in rooms with good ventilation. All the halogens must be kept out of contact with substances that can be oxidized.

There is a regular decrease in chemical activity from fluorine to iodine, as shown by the trend in oxidizing strengths. The trend in standard oxidation potentials (Table 20-2) should also be noted. The diatomic fluorine molecule, F_2, is a stronger oxidizing agent than any other element in its normal state. Both fluorine and chlorine support combustion reactions in the same manner as does oxygen. Hydrogen and the active metals burn in either gas with the liberation of heat and light.

Principal Reactions. Here we list some of the reactions of halogens with other substances that are of considerable importance in industry or research and hence are of interest. In these reactions the symbol X is sometimes used in formulas to indicate any halogen.

HALOGENS WITH HALOGENS. In a reaction between two halogens, the one with the lower atomic number will be the oxidizing agent and will be assigned a negative oxidation state in the compound. Consider the reaction of bromine with fluorine:

$$Br_2 + F_2 \rightarrow 2BrF$$

Although the bond is largely covalent (no discrete ions are formed), the electron pair is drawn over toward the fluorine ($\delta-$) and away from the bromine ($\delta+$). The molecule is polar.

Many interhalogen compounds are known, the fluorides being most common. In the fluorides the maximum number of fluorine atoms per molecule increases from chlorine to iodine, for example, ClF_3, BrF_5, and IF_7. Interhalogen compounds are very active oxidizing agents, behaving chemically much like mixtures of halogens.

WITH METALS. The halogens react readily with most metals. Bromine and iodine do not react with gold, platinum, or some of the other noble metals, but fluorine and chlorine attack even these inactive elements. Examples:

$$3Br_2 + 2Al \rightarrow 2AlBr_3$$

$$F_2 + Cu \rightarrow CuF_2$$

$$3Cl_2 + 2Fe \rightarrow 2FeCl_3$$

WITH WATER.

$$Cl_2 + HOH \rightleftarrows \underset{\text{hypochlorous acid}}{HClO} + HCl$$

$$Br_2 + HOH \rightleftarrows \underset{\text{hypobromous acid}}{HBrO} + HBr$$

Fluorine gives hydrofluoric acid but no hypofluorous acid:

$$2F_2 + 2H_2O \rightarrow 4HF + O_2$$

Solutions of chlorine or bromine in water are strong oxidizing agents. In these solutions HClO or HBrO usually functions as the oxidizing agent. The bleaching action of chlorine is accounted for on the assumption that chlorine first reacts with water to form hypochlorous acid (HClO), which then oxidizes colored compounds to colorless ones. In the compound HXO, X has an oxidation state of $+1$ and has a strong attraction for one or two electrons.

WITH HYDROGEN.
$$X_2 + H_2 \rightarrow 2HX$$

The reaction takes place with explosive violence for fluorine and even chlorine, but bromine and iodine react slowly. (See hydrogen halides, pages 318–319.)

The reactions are said to be **photochemical,** because they occur much more rapidly on exposure to radiant energy. A mixture of hydrogen and chlorine can be kept in the dark for a long time, but when the mixture is exposed to sunlight a violent combustion occurs. The sunlight causes some of the chlorine molecules to dissociate into (very active) atoms of chlorine. The minimum energy for this decomposition is possessed by radiant energy of 4900 A wavelength (see Figs. 1-1 and 4-3):

$$Cl_2 \xrightarrow{\text{sunlight}} Cl + Cl$$

The chlorine atoms can then react with hydrogen molecules to form hydrogen chloride and (active) hydrogen atoms:

$$Cl + H_2 \rightarrow HCl + H$$

Next, hydrogen atoms can react with other chlorine molecules. Note that a chlorine atom is again formed:

$$H + Cl_2 \rightarrow HCl + Cl$$

This is a **chain reaction.** Presumably there are relatively few atoms of hydrogen and chlorine in existence at any one time; but as the reaction proceeds, additional atoms are successively formed in the manner indicated by the equations above.

The chain is broken (1) when atoms of chlorine react with atoms of hydrogen, or (2) when atoms of hydrogen react with one another, or (3) when atoms of chlorine react with one another:

$$H + Cl \rightarrow HCl$$
$$H + H \rightarrow H_2$$
$$Cl + Cl \rightarrow Cl_2$$

The breaking of the chain tends to stop the reaction between the hydrogen and chlorine. However, all three reactions liberate heat; hence, the temperature may rise sufficiently so that hydrogen and chlorine molecules will react directly. The reaction, which eventually goes to completion, is written simply as

$$Cl_2 + H_2 \rightarrow 2HCl$$

WITH CERTAIN NONMETALS (Z).
$$3X_2 + 2Z \rightarrow 2ZX_3$$

Z can be B, P, or As.

WITH COMPOUNDS OF CARBON AND HYDROGEN.
$$X_2 + CH_4 \rightarrow CH_3X + HX$$
methane

This very important type of reaction is discussed in detail in Chap. 26.

WITH COMPOUNDS OF OTHER HALOGENS. This type of reaction can be thought of as a displacement reaction, with the more active halogen displacing a less active one from its compounds. Examples:

$$F_2 + 2NaBr \rightarrow Br_2 + 2NaF$$
$$Br_2 + CaI_2 \rightarrow I_2 + CaBr_2$$
$$Cl_2 + CaF_2 \rightarrow \text{no reaction}$$

OCCURRENCE

Although traces of elemental iodine have been found in nature, the halogens are chemically too active as electron acceptors to exist as free elements in this world of ours, comprised as it is of the other common atoms, ions, and molecules. The state in which an element exists in nature reveals something about its chemical activity. Elements with the highest electronegativity usually exist as negative ions, and those with the lowest electronegativity usually exist as positive ions. Elements with intermediate electronegativity are often found as ions or in molecules, but may also be found in the free (elemental) form.

Like other active elements, the halogens exist in nature primarily in compounds. Their most common state is as the halide ions F^-, Cl^-, Br^-, and I^-. These ions always occur in association with positive ions (compounds). Because the compounds of the halides are usually soluble in water, the halide ions are found concentrated in the sea, in salt lakes, and in underground beds of salt that were formed ages ago by the evaporation of bodies of salt water.

Listing them in order of their abundance in the earth's crust gives Cl, 0.2 percent, F, 0.1 percent, and Br and I, 0.001 percent each. Because the halogens, taken together, comprise about 3 out of every 1000 lb of the earth's crust, they are rather common elements. Some of the abundant halogen-containing minerals are listed in Table 19-2. Sodium chloride is one of the most important of all

TABLE 19-2.
Abundant Halogen-Containing Minerals

FORMULA	CHEMICAL NAME	MINERAL NAME
CaF_2	Calcium fluoride	Fluorspar
Na_3AlF_6	Sodium aluminum fluoride	Cryolite
NaCl	Sodium chloride	Halite
KCl	Potassium chloride	Sylvite
KCl and NaCl	Potassium and sodium chloride	Sylvinite
$MgBr_2 \cdot KBr \cdot 6H_2O$	Magnesium potassium bromide	Bromo-carnallite
$NaIO_3$	Sodium iodate	
$NaIO_4$	Sodium periodate	

inorganic raw materials. Not only is it used as salt, but it is the starting material for the commercial manufacture of most sodium and chlorine compounds. Bromine is produced commercially from sea water; iodine is found concentrated in sea weed, although sodium iodate is the chief commercial raw material.

The halogens have important roles in the chemical reactions of our bodies. Traces of chloride ion are found in ordinary drinking water. Chloride is one of the essential ions in the blood and gastric juices. Iodide ion is usually present in water in minute amounts. In localities where it is not present, the prevalence of goiter, a condition of the thyroid gland, is increased. Iodide ion tends to concentrate in the thyroid gland and in certain other human tissues. Traces of fluoride ion in drinking water are effective in preventing tooth decay, but too much fluoride may cause the teeth to be mottled or chalky.

PREPARATION AND USE OF THE HALOGENS

The halogens can be prepared by suitable chemical or electrochemical procedures from naturally occurring compounds. In most cases the chemical change that is involved results in electrons being taken away from a halide ion. That is, an oxidation reaction must be brought about. In the case of chloride, bromide, and iodide ions there are strong chemical oxidizers that are effective, but the electronegativity of the fluorine atom is so high that chemical oxidizing agents are ineffective. The common way to produce fluorine and chlorine is by *electrolysis*. In electrolysis, chemical reactions are brought about by forcing a current of electricity through a solution between two electrodes (usually made of metal). Electrons enter the solution at one electrode and leave it at the other. Electrolytic reactions are described in detail in Chap. 20.

Fluorine. The commercial source of fluorine is the mineral fluorspar, CaF_2. By treating the calcium fluoride with sulfuric acid, hydrogen fluoride is produced (see page 370). Elemental fluorine is produced by the electrolytic decomposition of hydrogen fluoride:

$$2HF \xrightarrow[\text{molten KF·2HF}]{\text{electrolysis in}} H_2 + F_2$$

Oxidation at one electrode:

$$2F^- \longrightarrow F_2 + 2e^-$$

Reduction at the other:

$$2H^+ + 2e^- \longrightarrow H_2$$

A view of industrial cells for fluorine production is shown in Fig. 19-2.

IMPORTANCE OF FLUORINE PRODUCTION. Elemental fluorine is needed in the manufacture of many useful compounds, among which are the Freon refrigerant gases, such as CCl_2F_2, and the heat-resistant plastic Teflon. Fluorine plays a key role in the nuclear energy field, being used to make uranium hexafluoride, UF_6, the volatile uranium compound used in the gaseous diffusion method of separating the ^{235}U and ^{238}U isotopes.

Liquid fluorine has been used as the oxidizing agent in some rockets. Very high energy yields per pound of propellant are obtained when fluorine acts as the oxidant.

Fig. 19–2. Interior view of a fluorine plant showing electrolysis cells. The cutaway drawing shows the carbon anodes and steel cathodes. (Courtesy Allied Chemical Corporation.)

Chlorine. COMMERCIAL PREPARATION. Chlorine, to the extent of about ten million tons annually in the United States, is produced commercially in several ways. We shall describe in detail only one of the two important electrolytic processes, the electrolysis of a concentrated sodium chloride solution with unreactive electrodes. The following reactions occur:

$$2NaCl + 2H_2O \xrightarrow[\text{conc soln}]{\text{electrolysis}} Cl_2 + H_2 + 2NaOH$$

Oxidation at one electrode:

$$2Cl^- \longrightarrow Cl_2 + 2e^-$$

Reduction at the other:

$$2H_2O + 2e^- \longrightarrow H_2 + 2OH^-$$

These reactions can be carried out as shown schematically in Fig. 19-3. Major problems are to keep the chlorine and hydrogen gases separated, because they react explosively, and to keep the chlorine away from the sodium hydroxide solution, because they will also react. These separations are achieved by use of an asbestos cloth diaphragm through which solutions can pass en route to electrodes but which, when wet, stops gas bubbles. Hydrogen gas and sodium hydroxide solution are formed at the electrode on one side of the diaphragm and are taken out of the cell via conduits; chlorine gas, formed on the opposite side of the diaphragm at the other electrode is taken out from the other side of the cell.

This process is a continuous one, with fresh concentrated NaCl solution flowing into the cells and the three products, Cl_2, H_2, and NaOH solution, flowing out. Hundreds of large cells are commonly used in a single plant. Individual cells are shut down as needed for cleaning and replacing worn electrodes, especially the ones that are attacked by the chlorine.

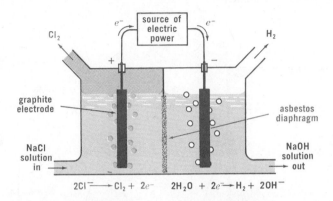

Fig. 19-3. Schematic representation of the electrolysis of concentrated aqueous sodium chloride.

IMPORTANCE OF CHLORINE PRODUCTION. On the basis of amounts used, chlorine is by far the most important of the halogens—about 100 pounds annually being produced per person in the United States. Elemental chlorine is used for purifying drinking water, sanitizing swimming pools, for bleaching paper pulp, in making insecticides and germicides, and in manufacturing organic compounds such as dyes, drugs, plastics, and solvents. Many of these organic compounds are produced to the extent of millions of pounds each year.

LABORATORY PREPARATION. The laboratory preparation of chlorine can be carried out in a variety of ways, practically all of which involve the oxidation of the chloride ion. We shall describe two of these procedures. First,

$$MnO_2 + 4HCl \xrightarrow[\text{in water}]{\text{heat}} Cl_2 + MnCl_2 + 2H_2O$$

Manganese in the +4 oxidation state is a strong oxidizing agent. The reduction reaction

$$MnO_2 + 2e^- + 4H^+ \rightarrow Mn^{2+} + 2H_2O$$

takes place as the Mn^{4+} acquires $2e^-$ from two Cl^- ions (the latter is the oxidation reaction):

$$2Cl^- \rightarrow Cl_2 + 2e^-$$

If needed, the solution is heated slightly to speed up the reaction and to force the Cl_2 out of solution (see Fig. 19-4). The reaction is carried out in an acid solution so that H^+ ions will be available to combine with the O^{2-} ions, which are initially associated with Mn^{4+} ions.

The gas is bubbled through cool water to be stripped of any HCl contaminant; then the wet Cl_2 is bubbled through concentrated sulfuric acid to be stripped of the water vapor.

Historically, this is the reaction investigated by Wilhelm Scheele in 1774, when he discovered chlorine. It is still used for preparing small quantities of Cl_2 gas from easily available chemicals.

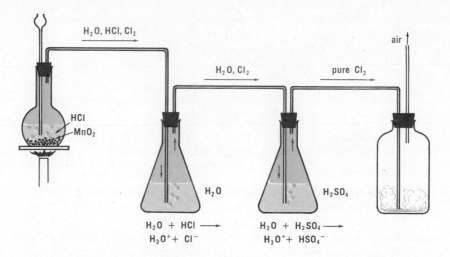

Fig. 19–4. Laboratory preparation of chlorine. The dense chlorine displaces the air in the collecting bottle.

For a second laboratory procedure, we note that the source of Cl^- need not be hydrogen chloride; it can be a salt containing the chloride ion. The reaction is carried out in an acid solution. For example:

$$MnO_2 + 2NaCl + 3H_2SO_4 \rightarrow Cl_2 + MnSO_4 + 2NaHSO_4 + 2H_2O$$

Several different oxidizing agents can be used to attract electrons from the chloride ion and cause the formation of elemental chlorine. Examples of such agents are permanganates and dichromates:

$$2KMnO_4 + 16HCl \rightarrow 5Cl_2 + 2MnCl_2 + 2KCl + 8H_2O$$

$$Na_2Cr_2O_7 + 14HCl \rightarrow 3Cl_2 + 2CrCl_3 + 2NaCl + 7H_2O$$

Bromine. COMMERCIAL PREPARATION. Our most abundant sources of bromine compounds are brines from salt wells and from the ocean. Because water from these sources usually contains large proportions of chloride ions in comparison with the amount of bromide ions, the preparation of bromine must involve an oxidizing agent that does not simultaneously oxidize chloride ions. Chlorine itself is admirably suited for this job:

$$2Br^- + Cl_2 \xrightarrow[\text{solution}]{\text{water}} Br_2 + 2Cl^-$$

Chlorine has a greater electron affinity than the bromine:

$$Cl_2 + 2e^- \rightarrow 2Cl^-$$

$$2Br^- \rightarrow Br_2 + 2e^-$$

IMPORTANCE OF BROMINE PRODUCTION. In the past, a large part of bromine production was used to make ethylene dibromide, a component of ethyl gasoline. Lead compounds such as lead tetraethyl are present in ethyl gasoline as antiknock agents, but the lead formed during their combustion tends to foul the spark plugs. To prevent this, ethylene dibromide is also added to the gasoline; volatile lead bromide forms during the combustion and escapes with the exhaust gases. Because lead bromide is a dangerous air pollutant, leaded gasolines are being gradually discontinued.

A great amount of bromine is used in the manufacture of silver bromide, one of the light-sensitive compounds of photographic film and paper. Certain bromine compounds are also used in medicines.

LABORATORY PREPARATION. Any of the reactions used for the preparation of chlorine can be modified for preparing bromine; for example:

$$MnO_2 + 4HBr \rightarrow Br_2 + MnBr_2 + 2H_2O$$

$$2KMnO_4 + 16HBr \rightarrow 5Br_2 + 2MnBr_2 + 2KBr + 8H_2O$$

Iodine. Methods used in preparing chlorine can be used to prepare iodine if the element is present as the iodide ion. The cheapest source, however, is the compound sodium iodate, $NaIO_3$, which is found associated with large deposits of sodium nitrate in Chile.

HYDROGEN HALIDES

The hydrogen halides are so important to the study and practice of chemistry that they are given special attention. Their properties, summarized in Table 19-3, have two important characteristics: (1) the trends from chloride to bromide to

TABLE 19-3.

Properties of Hydrogen Halides

	HF	HCl	HBr	HI
Boiling point, °C	20	−85	−67	−36
Melting point, °C	−83	−114	−87	−51
% Dissociation at 1000°C	Too slight to measure	0.0014	0.50	33
Solubility at 20°C and 1 atm, g/100 g H_2O	35.3	42	49	57
$\Delta H°$ per mole for $\tfrac{1}{2}H_2 + \tfrac{1}{2}X_2 \rightarrow HX$, kcal	−64	−22	−8.7	+6.2
Color	None	None	None	None
Odor	All very irritating; attack delicate nasal tissues			

iodide are as predicted from their location in the periodic table; and (2) the fluorine compound does not conform to the boiling and melting point trends, probably because of the small size and high electronegativity of the fluorine atom.

For covalent compounds that are of the same general nature it is usual for the boiling and melting points to decrease as the molecular weights decrease. But hydrogen fluoride does not show this behavior, because the HF molecules are so strongly attracted to one another (via hydrogen bonds) that they form zigzag chains that act like larger molecules:

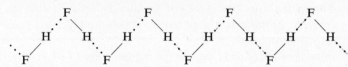

Therefore, the boiling and melting points of HF are abnormally high, as shown in Table 19-3. This behavior is shown graphically in Fig. 19-5, where it is related to similar effects for H_2O and NH_3.

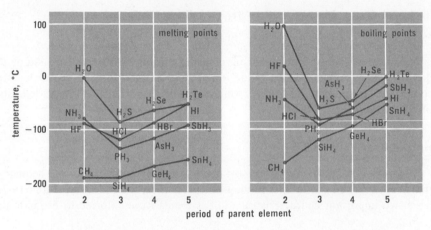

Fig. 19-5. The melting and boiling points of the hydrogen compounds of groups IVA, VA, VIA, and VIIA. (Redrawn by permission from L. Pauling, *The Nature of the Chemical Bond*, Cornell University Press, 1940.)

Strong hydrogen bonds are formed only by small atoms, such as fluorine, oxygen, and nitrogen. It appears that a free pair of electrons in a small atom is more available to the hydrogen of another molecule than is a free pair in a large atom. In the larger orbitals of atoms, such as chlorine, sulfur, or phosphorus, the two electrons are probably not sufficiently localized to attract a hydrogen atom. Noting positions for CH_4 in Fig. 19-5, we realize that the carbon, though a small atom, has no free electron pair and can form no hydrogen bonds. The low melting and boiling points of methane indicate that CH_4 molecules are attracted to one another only by van der Waals forces.

Each of the HX compounds can be prepared by a reaction between hydrogen and the corresponding elemental halogen, $H_2 + X_2 \rightarrow 2HX$. This reaction, explosively exothermic in the case of F_2, decreases in violence from fluorine to iodine, as can be seen by comparing the heat evolved per mole of HX formed (Table 19-3).

The stability of the compounds can be correlated with the energy of formation. The greater the energy of formation, the more stable the compound—HF is more stable than HI. The percentage dissociation at 1000°C is a quantitative measure of the strength of the various H—X bonds; the H—I bond is weakest of the four.

All the hydrogen halides are shown to be covalent compounds by the fact that the pure liquid substances do not conduct an electric current. However, when dissolved in water, HCl, HBr, and HI form strong acid solutions; HF forms a weak acid solution, because the strong HF bond prevents a complete reaction with the H_2O molecules to form H_3O^+ and F^-. The water solutions of all these compounds are called *hydro- -ic* acids; they conduct an electric current because of the presence of hydronium ions and halide ions:

$$HX + H_2O \rightleftarrows H_3O^+ + X^-$$
$$HX \rightleftarrows H^+ + X^-$$

All the hydrogen halides form constant-boiling mixtures with water.

The solubilities of the hydrogen halides in water are quite striking when the volume of gas is considered (Fig. 19-6). At 20°C and 1.0 atm, 264 liters of gaseous

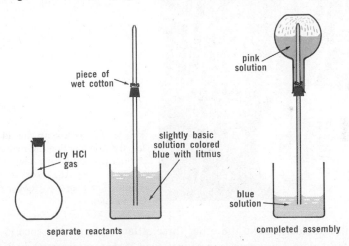

Fig. 19–6. The hydrogen chloride "fountain." To make the fountain operate, the stopper is removed from the flask of dry gas and the flask is then quickly inverted and set tightly on the rubber stopper on which there is some wet cotton. The water in the cotton dissolves the hydrogen chloride gas so completely that the pressure inside the flask decreases sharply. Atmospheric pressure forces the water in the beaker up the tube and into the flask. If the solution in the beaker is slightly basic, a little litmus indicator will color it blue, and the hydrochloric acid solution that is formed in the flask will turn the litmus pink.

hydrogen chloride will dissolve in 1 liter of water. Still there are almost five H_2O molecules for every HCl molecule that has dissolved.

Hydrochloric acid, often called muriatic acid commercially, is by far the most important of the four acids. It is used to remove rust from scrap iron, to clean mortar from masonry, to prepare metal surfaces for electroplating, and to neutralize basic solutions. Hydrobromic and hydriodic acids are not widely used, because, although chemically almost identical to hydrochloric, they are much more expensive. Their solutions act as typical sources of hydrogen ions in neutralizing bases and dissolving active metals and oxides.

Fig. 19-7. Hydrogen chloride burner for the combination of hydrogen and chlorine gas. The burner is water-jacketed to carry away the heat of reaction. (Photograph courtesy Hooker Chemical Corporation.)

Hydrofluoric acid, unlike the other three, attacks glass and hence is used to etch and frost glass. Very precise markings such as thermometer graduations and even pictures can be etched out of the glass surface. The glass is first completely covered with a thin film of melted paraffin; the wax is then carefully cut away so as to expose the parts that are to be etched. Fumes of HF do not attack the paraffin but they do attack the silicon-containing compounds in the glass. The chemical reaction destroys the glass as the result of the formation of the gaseous silicon compound, SiF_4. The reactions with silicon dioxide and sodium silicate, two compounds that react in a way similar to glass, are as follows:

$$SiO_2 + 4HF \rightarrow SiF_4 + 2H_2O$$

$$Na_2SiO_3 + 6HF \rightarrow SiF_4 + 2NaF + 3H_2O$$

Concentrated solutions of HCl, HBr, and HI can be shipped in steel or glass containers, but HF solutions must be kept in wax-lined vessels or special plastic tanks.

The modern process of making hydrogen chloride involves a direct combination of hydrogen and chlorine (Fig. 19-7). Both of these gases are produced during the electrolysis of aqueous sodium chloride; and in regions where electric power is cheap, it is economical to produce these gases for the purpose of making hydrogen chloride.

The chlorine and hydrogen gases are mixed in a burner in the same way that methane and air or acetylene and oxygen are mixed in laboratory burners. The hydrogen can be said to be burning in an atmosphere of chlorine. Only a slight excess of hydrogen is present, so the gas produced runs as high as 99.0 percent HCl.

OXY-ACIDS AND OXY-SALTS

Compounds containing the ions ClO_4^-, ClO_3^-, ClO_2^-, and ClO^- are well known. In these compounds the oxidation state of chlorine is $+7$, $+5$, $+3$, and $+1$, respectively (see Fig. 6-1). Bromine and iodine form some similar oxy-ions, as shown in Table 19-4, but fluorine, which shows only the -1 state in any of its compounds, does not. The hydrogen compounds HXO_4, HXO_3, HXO_2, and HXO are acids and oxidizing agents. The salts of these acids, especially the sodium and potassium salts, also frequently serve as oxidizing agents.

TABLE 19-4.

Group VIIA Oxy-Acids and Oxy-Ions

OXIDATION STATE	NAME OF ACID		EXAMPLES	NAME OF ANION	
$+1$	Hypo-	-ous	HClO, HBrO, HIO	Hypo-	-ite
$+3$		-ous	$HClO_2$		-ite
$+5$		-ic	$HClO_3$, $HBrO_3$, HIO_3		-ate
$+7$	Per-	-ic	$HClO_4$, HIO_4, H_5IO_6	Per-	-ate

Naming of Oxy-Ions and Their Compounds. A number of elements other than the halogens form oxy-acids. Structural models of three representative oxy-ions are shown in Fig. 19-8. Because of the great number and importance of oxy-ions, a system of naming them has been worked out. According to this system, the names of the oxy-acids and oxy-salts in a given family of elements include appropriate suffixes (and prefixes, if needed) to indicate the oxidation state of the parent element. The system, which is illustrated in Table 19-4 for the halogen compounds, is as follows:

1. Some common acid in the family is arbitrarily called the *-ic* acid. (In the case of the halogens, $HClO_3$ is designated as chloric acid.) The name of a salt of this acid ends in *-ate*.

2. An acid whose parent atom has an oxidation state next lower than the -ic acid (one less oxygen atom) is called the *-ous* acid. The name of a salt of this acid ends in *-ite*.

3. When more than two oxygen acids of an element exist, the acid in which the oxidation state of the parent atom is higher than it is in the -ic acid (one more oxygen atom) is called the *per- -ic* acid. A salt of this acid is a *per- -ate* salt.

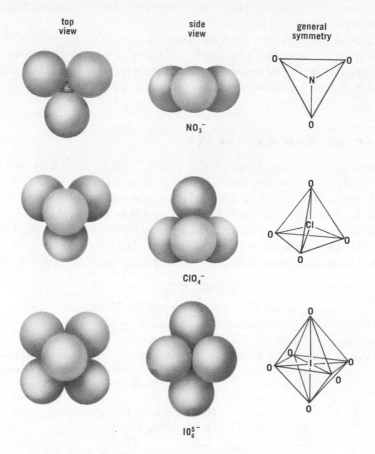

Fig. 19–8. Representatives of three common oxy-ion structures.

4. An acid in which the oxidation state of the parent atom is lower than it is is in the -ous acid is called the *hypo- -ous* acid. A salt of this acid is a *hypo- -ite* salt.

5. The root of the name of the acid or the salt indicates the parent atom.

The following examples will help clarify the nomenclature: hypoiodous acid is HIO; perchloric acid is $HClO_4$; sodium iodate is $NaIO_3$; and calcium hypochlorite is $Ca(ClO)_2$. Although oxy-compounds of nonhalogen elements will be taken up in other chapters, a few examples of naming are: nitric acid is HNO_3; potassium nitrite is KNO_2; sulfurous acid is H_2SO_3; and magnesium sulfate is $MgSO_4$.

Note in Table 19-4 that the central atom changes in oxidation state by two units each time it gains or loses one oxygen atom. When an element forms a number of oxy-acids, the most stable one usually has a structure in which the central atom is surrounded by as many oxygen atoms as can fit around it. Also, the strengths of acids in such a series increase as the number of coordinated oxygen atoms increases. The following three series bear out these statements:

$$\begin{array}{c} \text{less} \\ \text{stable} \end{array} \Bigg\downarrow \quad \begin{array}{c} HClO_4 \\ HClO_3 \\ HClO_2 \\ HClO \end{array} \quad \begin{array}{c} H_2SO_4 \\ H_2SO_3 \end{array} \quad \begin{array}{c} HNO_3 \\ HNO_2 \end{array} \Bigg\uparrow \begin{array}{c} \text{acid} \\ \text{strength} \end{array}$$

Important Salts. The salts of the halogen oxy-ions include a number of common, useful substances. Chemically speaking, these salts can be thought of as sources of the various oxy-ions. Because of the relative costs, the chlorine compounds are much more widely used than the bromine and iodine compounds.

Sodium hypochlorite, NaClO, is the source of the active ion in Clorox, a household disinfectant and bleach. It is formed by passing chlorine into dilute sodium hydroxide.

$$Cl_2 + 2NaOH \rightarrow NaCl + NaClO + H_2O$$

No effort is made to separate the NaCl from the reaction product. Instead, the product is bottled and marketed as a mixture. NaClO and NaCl each constitutes about 5 percent of the mixture, the remainder being water.

Potassium chlorate, $KClO_3$, is a compound that serves as a convenient laboratory source for oxygen gas because it is so easily decomposed:

$$2KClO_3 \xrightarrow[\text{MnO}_2 \text{ or Fe}_2\text{O}_3]{\text{heat to } 200°C} 2KCl + 3O_2$$
$$\text{(catalysts)}$$

An interesting application of this characteristic is the inclusion of potassium chlorate with the combustible material in the head of a match (see Fig. 22-5). The initial combustion of the match does not depend on atmospheric oxygen; instead, the oxygen comes from the potassium chlorate. This compound is extremely dangerous when mixed with such combustible materials as paper, cloth, and sulfur. Under certain conditions combustion may begin without warning and proceed rapidly enough so that an explosion results.

Red flares (which can be explosive) sometimes contain a mixture of potassium chlorate, sugar, and strontium nitrate. The potassium chlorate acts as the oxidizer for the combustion of the sugar. The heat from the reaction causes the strontium compound to emit red light. (Review flame spectra in Chap. 4.)

Potassium chlorate is formed by heating a solution of potassium hypochlorite:

$$3KClO \rightarrow KClO_3 + 2KCl$$

or

$$3ClO^- \rightarrow ClO_3^- + 2Cl^-$$

This reaction involves internal oxidation-reduction in which some of the chlorine atoms gain electrons and others lose. In the reduction equation, the chlorine oxidation state changes from $+1$ to -1; in the oxidation equation, from $+1$ to $+5$.

Reduction equation:

$$ClO^- + 2H^+ + 2e^- \rightarrow Cl^- + H_2O$$

Oxidation equation:

$$ClO^- + 2H_2O \rightarrow ClO_3^- + 4H^+ + 4e^-$$

Potassium perchlorate, $KClO_4$, is a more stable compound than potassium chlorate. Unlike the latter, it does not decompose unless heated to a high temperature. Flares, rocket propellant mixtures, fuse mixtures, and fireworks can be made from a very combustible material and powdered potassium perchlorate. Powdered active metals, such as aluminum and magnesium, burn fiercely when mixed with $KClO_4$ and ignited.

From the formulas $KClO_3$ and $KClO_4$, we might think that the perchlorate would liberate oxygen more readily than the chlorate, but this is not the case. Evidently the bonds in the tetrahedral ClO_4^- make possible a more tightly bonded packing, whereas those in the pyramidal ClO_3^- are strained and easily broken by a little energy of activation. Potassium chlorate is much more dangerous than potassium perchlorate for use in flares and pyrotechnics.

THE NOBLE GASES

HISTORICAL BACKGROUND

The family of noble gases consists of the six elements in family VIIIA of the periodic table. It is an important fact of chemical history, however, that none of these elements had been identified chemically at the time Mendeleev constructed the table. At that time (1869) elements representative of all the other families were known, but no such unreactive elements as the noble gases had been studied.

In the century before Mendeleev, the English chemist Henry Cavendish recorded an observation that might have led to the discovery of the most common of the noble gases. He found that when all the nitrogen and oxygen was removed from air, about one part in 120 of gas remained. It was not until a hundred years later, in 1894, that another Englishman, William Ramsay, identified the unreactive remaining gas as a new element; he named it argon, meaning "lazy one." Four years later, Ramsay prepared another unreactive gas by heating the mineral cleveite. The spectrum of this new gas was similar to certain lines in the spectrum of the sun that had been attributed in 1868 to an element helium, then known only as a substance in the sun. With his discovery of helium on earth, and the finding that it was unreactive like argon, Ramsay pursued the idea that a new group of elements in the periodic table must be located between the halogens and the alkali metals. By the fractional distillation of liquid argon prepared from liquid air, Ramsay separated and identified three more members of the family, neon, krypton, and xenon. Later he also studied the properties of the radioactive member of the family, radon.

For many years the new group of elements was referred to as the *inert gases*, because each seemed to be completely unreactive chemically. It was a shock and a thrill to the chemical world when, in 1962, the Canadian chemist Neil Bartlett prepared a stable compound thought to have the formula $XePtF_6$. The spell of the inert elements was broken. Very shortly others showed that xenon would react directly with fluorine to form simple binary compounds such as XeF_2, XeF_4, and XeF_6. The term "inert" no longer seemed appropriate; most chemists began to refer to the family as the *noble gases*, just as the rather unreactive and chemically aloof elements such as gold and platinum are referred to as the *noble metals*.

Although the direct combination of xenon and fluorine had been attempted in the 1930s, the results were negative, probably because of the reaction of the fluorine with the containing vessel of quartz. Many attempts were made to form special types of compounds, for example, hydrates of the noble gases at extremely low temperatures. Also, transitory compounds appeared to be formed in gaseous mixtures subjected to high-energy electrical discharge. However, the repeated failures to prepare any ordinary compounds led to the general acceptance of, first, the idea that these elements were inert to chemical reaction, and, second, the theory that an atom with eight electrons in the outside main level was incapable of chemical reaction.

The downfall of these views in 1962 has three lessons for the student of science that are of general applicability. First, Dr. Bartlett's test of the accepted theory was not as direct a challenge as the attempt, say, to make a simple binary compound. He had prepared the compound O_2PtF_6, in which a molecule of O_2 loses an electron to form the PtF_6^- ion; but the O_2^+ unit retains its diatomic "molecular" character. By analogy, he decided to see if the Xe molecule might react similarly to the O_2 one, because the ionization energy of the former (12.1 ev) is less than the latter (12.2 ev). The rather exotic compound he formed, $XePtF_6(?)$, has not been fully identified as yet. Second, it must be noted that able chemists had shown that it was a "fact" that xenon and fluorine would not react; this fact was later shown to be in error. Thirdly, after xenon was shown to react in a direct simple way with fluorine, the theory of the "inert eight electrons" was easily dropped; the reactivity of xenon was explained in terms of promoting outer electrons to higher energy orbitals, a well-developed theory, which had already been applied to many other elements. In summary our three lessons are these: A well-entrenched, useful theory is rarely challenged directly; facts are the basis of all science, but, like theories, they are open to question; and, when the facts change, the theory must change accordingly.

As a postscript to this case history in scientific method, it must be mentioned that there was at least one seriously considered theoretical prediction that xenon and other members of the family should react chemically. Dr. Linus Pauling, later to win a Nobel prize in chemistry, held in 1933 that the noble gases possibly could react with fluorine. When laboratory tests of his predictions turned out negatively, attempts at direct combination were essentially abandoned for thirty years. His theory, which did not lead to a fruitful test of the facts, is seen now as a suggestion that should have been hammered at more thoroughly. But who is to say when a theory has been tested thoroughly?

COMPOUNDS OF THE NOBLE GASES

Only three of the noble gases have been shown to combine—xenon, radon, and krypton. Radon's chemistry has been studied very little because of its intense radioactivity and short half-life of four days. Krypton, to date, appears to be much the least reactive of the three, and few of its compounds are known. A number of xenon compounds have been prepared in which xenon is attached either to fluorine or oxygen, the two most electronegative of all elements.

Xenon shows the even oxidation numbers +2, +4, and +6 in the compounds XeF_2, XeF_4, and XeF_6. The compounds in which an oxygen replaces two fluorines in the hexafluoride are known: $XeOF_4$, XeO_2F_2, and XeO_3. The last compound, the trioxide, is formed by the hydrolysis of solutions of the +4 or +6 fluorides. XeO_3 solutions in water are stable, but the dry white solid is a violent explosive and dangerously sensitive to detonation, as a number of accidental laboratory explosions testify.

Important salts are the xenates, e.g., Na_2XeO_4, and the perxenates, e.g., Na_4XeO_6. The perxenates, in which xenon exhibits its maximum oxidation state of +8, are among the most powerful oxidizing agents known and will probably be used commonly by chemists as soon as they are readily available commercially.

It is known that radon will react directly with fluorine, although the formula for the compound is not determined. Krypton forms KrF_2 when mixed with fluorine and subjected to radiation or an electric discharge. Attempts to make compounds of the three lightest group VIIIA elements have so far been unsuccessful.

The chemistry of xenon is similar to that of those halogens that exhibit positive oxidation states. All the heavier halogens react with fluorine to form interhalogen compounds such as chlorine fluoride or bromine fluoride. The fluorides of xenon are similar to chlorine and bromine fluorides, e.g., ClF and BrF_3; these covalent compounds, like the xenon fluorides, are very reactive as fluorinating agents. The xenate ion, XeO_4^{2-}, and the perxenate, XeO_6^{4-}, have certain properties like the chlorate, ClO_3^-, or perchlorate, ClO_4^-, ions. Finally, although not as sensitive to detonation as XeO_3, both Cl_2O and ClO_2 are explosive oxides of chlorine.

In summary, the chemistry of the group VIIIA elements, though very abbreviated, is similar, so far as it goes, to that shown by group VIIA elements in their positive oxidation states.

USES OF THE NOBLE GASES

The foregoing discussion of the chemical reactions of the noble gases should not blind us to the fact that the most important property of these elements is their extreme resistance to any chemical combination. Most of the present uses of these gases are related to their chemical inactivity.

Helium, least dense of all elements except hydrogen, is preferred for filling meteorological balloons or dirigibles because, unlike hydrogen, it is not combustible. Another interesting use of helium depends upon its very low solubility in water. Because helium is quite insoluble even at elevated pressures, a helium-oxygen mixture instead of compressed air is breathed by divers and others who work under high pressures. When a person breathes compressed air, considerable nitrogen dissolves in the blood (Henry's law). When he begins to breathe air again at normal pressure, the dissolved nitrogen comes out of solution and may form tiny bubbles in the blood. As these bubbles pass through the capillaries there is extreme pain; this is the condition, suffered by divers and sandhogs, known as the *bends*.

Helium is found in fairly high concentrations in certain natural gases, largely from wells in Kansas and Texas. The U.S. Bureau of Mines for several years has been recovering helium from natural gas and storing it for future use. Till recently, this source provided an adequate supply of the element for the few commercial uses that had developed for it. After World War II, the use of helium in atomic energy and missile programs, as well as in certain industrial processes, brought about a sharp increase in the consumption of the element. The Bureau of Mines, to support its helium conservation program, recently increased its recovery facilities and drastically raised the price of the gas (to \$35/1000 ft^3).

Argon, making up about 1 percent of the atmosphere, has been steadily increasing in industrial importance and is now being produced to the extent of about 1 billion ft^3 annually. Argon can be used interchangeably with helium in many processes. Welding of titanium and other exotic metals in aircraft and rocket construction requires an inert atmosphere, and argon serves this purpose. It is also used in filling incandescent light bulbs, because it does not react with the white-hot tungsten wire as oxygen or nitrogen does. Neon, helium, and argon are used in "neon signs."

EXERCISES

1. Explain, on the basis of atomic theory, why the elements within a group at the left or right of the periodic table should be more similar to each other chemically than the elements in a group toward the middle of the table.
2. The ionization energies of the halogens decrease from 17.34 ev for fluorine to 10.60 ev for iodine. Account for this trend in terms of atomic structure.
3. Which is the most active oxidizing agent among the following: Rb, I_2, I^-, Rb^+, Cl_2, Cl^-, or F^- ?
4. Which is the most active reducing agent among the following: Cs^+, Li^+, F_2, F^-, Br_2, I_2, Br^-, or Na^+ ?
5. Complete each of the following equations for which a reaction tends to occur:
   (a) $Cl_2 + F_2 \rightarrow$
   (b) $Br_2 + NaCl \rightarrow$
   (c) $Br_2 + NaI \rightarrow$
   (d) $Al + Cl_2 \rightarrow$
   (e) $Pt + I_2 \rightarrow$
   (f) $CH_4 + Cl_2 \rightarrow$
   (g) $H_2O + F_2 \rightarrow$
   (h) $H_2O + Br_2 \rightarrow$
6. Arrange the halogens in order of (a) reactivity as oxidizing agents; (b) electronegativity; (c) ease of preparation from X^- ion; (d) boiling points; (e) strength of H—X bond.

7. Write a balanced equation for each of the following:
    (a) a laboratory preparation of chlorine from calcium chloride and potassium permanganate;
    (b) the etching of quartz, SiO_2, with hydrogen fluoride.
8. Describe and illustrate a chain reaction.
9. Could bromine be produced by the electrolysis of a sodium bromide solution? What are some of the reasons why this is not the principal commercial method of production?
10. For the electrolysis of a concentrated sodium chloride solution, calculate:
    (a) moles of Cl_2 per mole of NaCl used;
    (b) moles of NaOH formed per mole Cl_2 formed;
    (c) cubic foot H_2 formed per cubic foot Cl_2 formed;
    (d) cubic foot H_2 per pound NaCl used (see Table 1 in Appendix);
    (e) tons of 50 percent NaOH solution produced per ton of NaCl electrolyzed, assuming 60 percent efficiency.
11. Write equations to show two ways of preparing HBr.
12. Hydrochloric acid is used to remove rust, Fe_2O_3, from iron before coating the iron with zinc or other protective agents. Write an equation that accounts for the cleaning process.
13. (a) Fluorine or chlorine could be used as oxidizing agent for hydrogen in a rocket engine. Using data in Table 19-3, calculate the heat evolved when 1 ton of hydrogen reacts with chlorine, and then calculate the same for fluorine.
    (b) If the reactants just described are held as liquids in the rocket, will the heat evolved in the overall reaction to yield hot gases be more or less than if they are held as gases? Explain.
14. If you were given a sample solution known to contain either NaCl, NaBr, or NaI and a water solution of chlorine, describe how you would determine which one of the three salts is in the sample solution.
15. The enthalpies of vaporization, ΔH_{vap}, of HF, HCl, and HBr are 7.24, 3.85, and 4.21 kcal/mole, respectively. Account for the fact that the enthalpies of vaporization of HF and HBr are larger than that of HCl.
16. In Fig. 19-5, the anomalous positions of HF, H_2O, and NH_3 are attributed to hydrogen bonding. What seems to be the explanation for the decreased importance in this type of bonding in HCl, H_2S, PH_3, and CH_4?
17. (a) Name the following compounds: LiIO, $HBrO_3$, $Ca(ClO)_2$, RbCl, $KMnO_4$, K_2MnO_4.
    (b) Write formulas for the following: sodium chlorate, barium hypobromite, chlorous acid, potassium perbromate.
18. Which do you think will be the more ionic, HF or HCl? $BeCl_2$ or $BaCl_2$? $AlCl_3$ or CsCl? Explain in each case.
19. Green flares sometimes contain potassium chlorate, sugar, and barium nitrate. What is the purpose of each of these compounds in the flare?
20. On the basis of their ionization energies only, which should be the more electronegative, fluorine or xenon? Fluorine or neon?
21. Write an equation for the explosion of sodium xenate.
22. What is the oxidation state of each element in the following compounds: Na_4XeO_6, KrF_2, $Ca(ClO_4)_2$?
23. The commercial uses of argon, helium, and neon depend on certain properties of these elements. Are these properties the same for all three so that the elements can be used interchangeably? Elaborate, citing examples to illustrate your discussion.

24. "Hydrofluoric acid is a weak acid; therefore, it does not attack containers so readily as a strong acid such as hydrochloric acid." Discuss this statement.
25. If bromine could be purchased at the same price per pound as chlorine, would it be economical to use HBr in a neutralization reaction instead of HCl? Explain.
26. Efforts to prepare stable compounds of helium or of neon have been unfruitful so far. Why should compounds of these elements be less stable than compounds of xenon?

SUGGESTED READING

Bloch, M. R., "The Social Influence of Salt," *Sci. Amer.*, **209** (1), 88 (1963).

Hyde, E. K., "Astatine and Francium," *J. Chem. Educ.*, **36**, 15 (1959).

Hyman, H. H., "The Chemistry of the Noble Gases," *J. Chem. Educ.*, **41**, 174 (1964).

Mellor, D. P., "The Noble Gases and Their Compounds," *Chemistry*, **41** (10), 16 (1969).

Sanderson, R. T., "Principles of Halogen Chemistry," *J. Chem. Educ.*, **41**, 361 (1965).

Selig, H., J. G. Malm, and H. H. Classen, "The Chemistry of the Noble Gases," *Sci. Amer.*, **210** (5), 66 (1964).

Siegfried, R., "Humphry Davy and the Elementary Nature of Chlorine," *J. Chem. Educ.*, **36**, 568 (1959).

20

ELECTROCHEMISTRY; OXIDATION-REDUCTION

Oxidation was defined in Chap. 8 as the increase in oxidation state of atoms or ions, and reduction as the decrease in oxidation state of these particles. For example, when sodium and chlorine are brought together, they combine to form sodium chloride, a compound composed of Na^+ ions and Cl^- ions. This is an oxidation-reduction reaction, because sodium atoms increase in oxidation state from 0 to $+1$, and chlorine decreases from 0 to -1.

$$\begin{aligned} \text{Over-all reaction:} \quad & 2Na + Cl_2 \rightarrow 2NaCl \\ \text{Oxidation:} \quad & 2Na \rightarrow 2Na^+ + 2e^- \\ \text{Reduction:} \quad & Cl_2 + 2e^- \rightarrow 2Cl^- \end{aligned}$$

In the example just cited, the increase in oxidation state is brought about by the loss of an electron by a sodium atom; the decrease, by the gain of an electron by a chlorine atom. Hence, for reactions which involve the actual transfer of electrons from one particle to another, we can define oxidation as the loss of electrons by atoms, molecules, or ions; and reduction as the gain of electrons by these particles.

Electrochemical reactions are oxidation-reduction reactions too, but they differ from ordinary oxidation-reduction reactions in that the electrons being lost and gained by reactants travel through some conducting material, e.g., a metal electrode. For convenience, the field of electrochemistry may be subdivided into two sections; one deals with chemical reactions that *are produced* by a current of electricity (electrolysis), and the other with chemical reactions that *produce* a current of electricity (this process occurs in a battery). The first type

Electrochemistry; Oxidation-Reduction

of reaction is usually endothermic; the second is usually exothermic. Before we discuss these, we need to consider briefly some of the units for measuring different characteristics of electric currents.

ELECTRICAL UNITS

The flow of electricity through a metallic conductor such as a copper wire is thought to be the movement of the outermost electrons from atom to atom. In becoming familiar with the units that are used to describe this flow, a comparison of the motion of electrons in a conductor with the flow of water through a pipe will be helpful. Table 20-1 lists units that can be used to describe the flow of water, and comparable units that describe the flow of electricity.

TABLE 20-1.

Units of Measurement of the Flow of Water and of Electricity

	QUANTITY	RATE	PRESSURE	POWER[a]	ENERGY OR WORK
Water	Gallons	Gallons per sec	Pounds per sq in.	Horsepower[b]	Horsepower-hours
Electricity	Coulombs or faradays[d]	Amperes[c]	Volts	Watts or kilowatts[e]	Watt-hours, kilowatt-hours[f]

[a] **Power** is the rate of doing work or supplying energy.
[b] **Horsepower** is a function of the quantity of water pumped per second × pressure.
[c] 1 **ampere** = 1 coulomb per second.
[d] 1 **faraday** = 6.02×10^{23} electrons = 96,500 coulombs.
[e] **Watts** = coulombs per second × volts.
[f] 1 ampere flowing at a potential (pressure) of 1 volt for 1 hour supplies 1 **watt-hour** of energy.

CHEMICAL REACTIONS PRODUCED BY A CURRENT OF ELECTRICITY

Many chemical reactions that are important commercially do not occur spontaneously. Among these nonspontaneous reactions are the following:

$$2NaCl \rightarrow 2Na + Cl_2$$

$$2Al_2O_3 \rightarrow 4Al + 3O_2$$

$$MgCl_2 \rightarrow Mg + Cl_2$$

These three are of particular commercial importance, for they represent how metallic sodium is obtained from salt, how aluminum is obtained from aluminum oxide, and how magnesium is obtained from a salt that is produced using magnesium from ocean water.

Such reactions may be forced to occur by using a direct current of electricity whose voltage is high enough to remove electrons from the negative ions (oxidation) and add electrons to the positive ions (reduction). This process is called

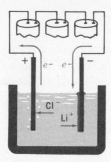

Fig. 20-1. The electrolysis of molten lithium chloride. The salt must be in a liquid condition so that the ions can migrate to the charged electrodes.

electrolysis (see Chap. 6). We shall describe what happens during the electrolysis of a salt such as lithium chloride (Fig. 20-1).

ELECTROLYSIS OF A MOLTEN SALT

Lithium chloride is a solid at ordinary temperatures; the ions are firmly fixed in the crystal lattice. In order for the ions to diffuse or migrate to the oppositely charged electrodes, the salt must be in liquid form. This can be achieved by heating the pure compound until it is molten, or by dissolving it in an appropriate solvent. The procedure shown in Fig. 20-1 is carried out at a temperature above the melting point of the lithium chloride. In this state, negative ions can migrate freely to the positive electrode, the anode, and positive ions to the negative electrode, the cathode. When sufficient voltage is applied to the electrodes, the following changes occur:

At the cathode (−): $\quad 2Li^+ + 2e^- \rightarrow 2Li \quad$ (reduction)

At the anode (+): $\quad 2Cl^- \rightarrow Cl_2 + 2e^- \quad$ (oxidation)

Over-all reaction: $\quad 2LiCl \rightarrow 2Li + Cl_2$

ELECTROLYSIS IN WATER SOLUTIONS

When a metal is *above hydrogen in activity*, electrolysis of an aqueous solution of one of its salts usually liberates hydrogen gas at the negative electrode. For example, if a water solution of magnesium chloride is electrolyzed, hydrogen rather than magnesium is formed.

$$2HOH + 2e^- \rightarrow H_2 + 2OH^- \quad \text{(reduction)}$$

This behavior may be explained on the basis of the activity series (see Chap. 8). The metals that are more active than hydrogen are those that are more easily oxidized than hydrogen. Conversely, the positive ions of these active metals have less tendency to undergo reduction than hydrogen ions do. In other words, when atoms of active metals such as sodium or magnesium lose electrons (undergo oxidation) and form Na^+ or Mg^{2+}, these ions have little tendency to regain electrons (undergo reduction) to form neutral sodium or magnesium atoms. On the other hand, atoms of inactive metals—for example, gold and silver—are oxidized with difficulty, but their positive ions can be reduced with ease. In commercial processes for producing an active metal such as lithium, sodium, potassium, magnesium, or calcium, not a water solution of the salt but the molten salt is subjected to electrolysis.

When a metal is below hydrogen in activity, electrolysis of an aqueous solution of one of its salts results in the metal being deposited at the negative electrode. Thus, when a direct current is passed through an aqueous solution of gold chloride, $AuCl_3$, gold plates out on the negative electrode; i.e., gold ions have a greater tendency to undergo reduction than hydrogen in water.

$$Au^{3+} + 3e^- \to Au \qquad \text{(reduction)}$$

Chlorine is formed at the positive electrode.[1]

$$2Cl^- \to Cl_2 + 2e^- \qquad \text{(oxidation)}$$

The over-all reaction is

$$2AuCl_3 \to 2Au + 3Cl_2$$

Electroplating. In the manufacture of metallic articles, an article that is fabricated from metal or an alloy of metals is frequently covered with a thin plate of some other metal. This is generally done to protect it against corrosion and to make it more attractive.

One method of plating is by electrolysis. The article to be plated is the cathode, and a block of the plating metal is the anode. Both electrodes are immersed in an aqueous solution of a salt of the plating metal and connected to a source of direct current.

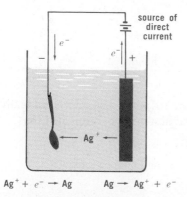

Fig. 20–2. In silver plating base metals, the object to be plated is made the cathode (−); the anode (+) is made of pure silver.

The plating of pure silver on a spoon made of a base metal is shown in Fig. 20-2. Many factors are involved in getting a plate of uniform thickness which adheres strongly to the base metal. Among the important variables that must be

---

[1] Oxygen may also form during the electrolysis of dilute aqueous solutions of chlorides.

controlled are the cleanness of the surface to be plated, the voltage, the temperature and purity of the solution, the concentration of the ion being plated out, and the total concentration of ions in the solution.

PRODUCTION OF ELECTRICITY BY CHEMICAL REACTIONS

SPONTANEOUS CHEMICAL REACTION

All the reactions discussed in this chapter in connection with electrochemical changes are oxidation-reduction reactions. As a simple example of oxidation and reduction let us consider the reaction that occurs when a strip of zinc is immersed in a solution of copper sulfate (Fig. 20-3). There is a spontaneous

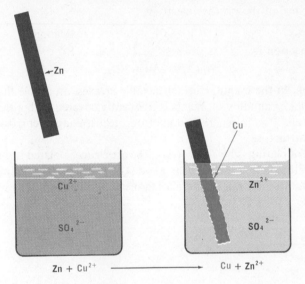

Fig. 20–3. Electrons are spontaneously transferred from zinc atoms to copper ions.

reaction; metallic copper plates out on the zinc strip, the zinc strip is gradually dissolved, and heat energy is liberated. The reaction may be formulated thus:

$$Zn + CuSO_4 \rightarrow ZnSO_4 + Cu$$

Actually, the reaction occurs between zinc atoms and copper ions:

$$Zn + Cu^{2+} \rightarrow Zn^{2+} + Cu$$

When we inspect this equation we see that each atom of zinc must lose 2 electrons to become a zinc ion, and each ion of copper must gain 2 electrons to become a copper atom:

$$Zn \rightarrow Zn^{2+} + 2e^- \quad \text{(oxidation)}$$

$$Cu^{2+} + 2e^- \rightarrow Cu \quad \text{(reduction)}$$

Electrochemistry; Oxidation-Reduction

Although this phenomenon is electrical in nature, no flow of electrons can be detected, because the copper ions touch the zinc atoms, and the electrons are transferred directly from the atoms to the ions.

The fact that this is a spontaneous reaction means there is a considerable driving force or potential behind the transfer of valence electrons from the zinc atoms to the copper ions. Actually this force is sufficient to cause the electrons to flow from the zinc atoms to the copper ions through a conductor.

VOLTAIC CELL

In order for the reaction above to occur when the zinc is not in direct contact with the copper sulfate solution, provisions must be made for (1) the flow of electrons from zinc atoms to copper ions and (2) diffusion of the positive and negative ions so that the solution will remain essentially neutral.

There are several methods of providing for the diffusion of the ions. A common laboratory method is to immerse the zinc strip in a solution of a zinc salt, such as zinc sulfate, and to immerse a piece of copper in a solution of copper sulfate. The zinc sulfate solution is connected to the copper sulfate solution by a *salt bridge* (see Fig. 20-4), which provides for the diffusion of the ions. The salt

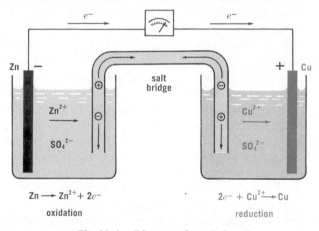

Fig. 20-4. Diagram of a voltaic cell.

bridge is filled with a solution of an electrolyte that does not change chemically in the process. Potassium sulfate (K^+, K^+, SO_4^{2-}) serves nicely for this purpose. Such salts as NaCl, KCl, and KNO_3 are also satisfactory.

When the reaction

$$Zn + CuSO_4 \rightarrow ZnSO_4 + Cu$$

proceeds as indicated in Fig. 20-4, the action will continue till one of the reactants, i.e., either the zinc atoms or the copper ions, is used up. Valence electrons will flow from the zinc atoms into the conducting wire, and the Zn^{2+} ions, as they form, will enter the solution and diffuse away from the zinc strip.

$$Zn \rightarrow Zn^{2+} + 2e^-$$

Negative ions will also diffuse through the salt bridge toward the zinc electrode. In time, it can be seen that the zinc strip is disappearing.

The electrons given up by the zinc atoms enter the connecting wire and cause electrons at the other end of the wire to collect on the surface of the copper electrode. These electrons react with copper ions to form copper atoms that adhere to the electrode as a copper plate:

$$Cu^{2+} + 2e^- \rightarrow Cu$$

In time, the metallic copper electrode increases in size and the blue color of the copper sulfate solution fades because of the decreasing concentration of the copper ions. The SO_4^{2-} ions that are left behind by the copper ions diffuse away from the copper electrode. The K^+ ions also diffuse out of the salt bridge toward the copper electrode. Thus, while the reaction is in progress, there is an over-all motion of negative ions toward the zinc strip and an over-all motion of positive ions toward the copper electrode. The pathway for this directional flow of ions through the solution may be thought of as the **internal circuit,** and the pathway for the flow of electrons through the conducting wire as the **external circuit.** Unfortunately, the internal circuit also provides a pathway by which the copper ions can reach the zinc atoms by diffusion. In time, therefore, the reaction will go to completion (and the battery will run down), even though the external circuit has not been closed.

Because the flow of electrons through the external circuit constitutes an electric current, we have harnessed the reaction between the zinc and copper sulfate so that a current of electricity will be produced. The energy that is given off as heat energy when these two react by direct contact is now evolved mostly as electrical energy. The apparatus and chemicals (Fig. 20-4) are referred to collectively as a **battery** or **voltaic cell.** The zinc strip is the negative electrode, because electrons originate at it; the copper is the positive electrode, because electrons are attracted to it. A battery in which zinc and zinc sulfate and copper and copper sulfate are used is known as a Daniell cell, after its inventor. The voltage of the Daniell cell is about 1.1 volts, depending on the concentrations.

Electricity is not stored, as such, in this cell or any other voltaic cell, such as a flashlight battery or an automobile battery. These batteries provide for exothermic oxidation-reduction reactions in which the chemical change takes place by the transfer of electrons through an external circuit. Oxidation—i.e., the loss of electrons—occurs at one electrode, the anode; reduction occurs at the other electrode, the cathode. The **cathode** is defined as the electrode at which reduction occurs and therefore in voltaic cells is the positive electrode, whereas in electrolytic cells the cathode is the negative electrode. The **anode** is defined as the electrode at which oxidation occurs.

Gravity Cell. Another way of utilizing the reaction between zinc and copper sulfate to supply a current of electricity is shown in Fig. 20-5. Metallic copper, immersed in a saturated solution of copper sulfate, serves as one of the electrodes. Because copper ions are removed as the current is drawn from the battery, solid copper sulfate is placed in the bottom of the jar so that it will dissolve as needed. Floating on this dense solution of copper sulfate is a dilute solution of zinc

Electrochemistry; Oxidation-Reduction   343

sulfate in which is immersed a large piece of zinc metal. The zinc is branched in the form of a *crowfoot* to provide a large surface, and is quite heavy, because it will be used up as the battery provides current. The zinc is not in direct contact with the copper ions. No reaction can occur till the external circuit is closed, thus permitting the passage of electrons from the zinc to the copper ions.

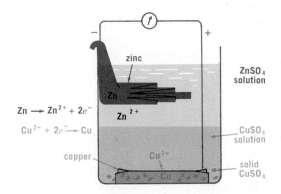

Fig. 20–5. The gravity cell.

No salt bridge is needed in the gravity cell, because the zinc sulfate solution is in direct contact with the copper sulfate solution, and ions can diffuse freely from one to the other, so that electroneutrality is maintained in each. The reactions at the two electrodes are identical with those in the Daniell cell. At one time the gravity cell was widely used as the source of electricity for telegraph instruments.

Dry Cell. The construction of the ordinary dry battery[2] used in flashlights and portable radios is shown in Fig. 20-6. Zinc metal again acts as the negative

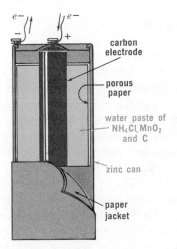

Fig. 20–6. The dry cell.

---

[2] The word *battery* was originally used to designate a *series* of voltaic cells, but it is now popularly used to denote any voltaic source of current, either a single dry-cell "battery" or the several cells of an automobile "battery."

electrode and is also used as a container for the other components of the battery. The can is lined with moist porous paper that prevents the zinc from coming in contact with the other reactants but permits the diffusion of ions. The positive electrode is an inert carbon rod located in the center of the can. Surrounding this rod is a watery paste of manganese dioxide and ammonium chloride, with some carbon added to improve the conductivity. This cell is called "dry" because the amount of water is relatively small and the cell contents do not splash about. However, some moisture is essential to provide a solution for the diffusion of ions between the electrodes.

When the cell is delivering current, the reaction at the negative electrode is one that involves the oxidation of zinc:

$$Zn \rightarrow Zn^{2+} + 2e^-$$

The reaction at the positive electrode appears to involve two or more reactions, depending in part on the conditions under which the cell is operated. If the cell is used for extremely short periods alternating with extended idle periods, the diffusion of reactants and products is such as to maintain fairly constant concentrations in the vicinity of the electrodes. On the other hand, if the cell is operated for longer periods and the idle periods are shorter, the concentrations change and the reaction changes accordingly. Under conditions of moderate use, hydrous manganous oxide, $MnO(OH)$ or $Mn_2O_3 \cdot H_2O$, appears to be the main reduction product. The reaction for its formation at the positive electrode is

$$MnO_2 + H_2O + e^- \rightarrow MnO(OH) + OH^-$$

A single dry cell has a voltage of 1.5 volts. By connecting several cells in series, higher voltages can be obtained.

HYDROGEN ELECTRODE

No difficulties are encountered in making an electrode of one of the common metals; all that is necessary is to immerse a strip of the metal in a solution of its ions. However, building an electrode involving one of the gaseous elements and its ions presents more of a problem. Obviously, we cannot take a "piece" of gas, insert it in a solution of its ions, and connect a wire to it, thus making it part of a voltaic cell. Yet methods have been worked out that, in principle, accomplish just this. Elemental hydrogen, maintained in contact with a solution of its ions, can be made one electrode of a voltaic cell by the method shown in Fig. 20-7.

The **hydrogen electrode** consists of a skirted glass tube through which hydrogen gas can be continuously passed over a platinum foil (left electrode in Fig. 20-7). Coming down through the tube is a platinum wire that is connected to the platinum foil. The foil itself is covered with finely divided platinum to provide a large surface. Platinum adsorbs hydrogen; hence in effect the electrode is a core of platinum with an adsorbed film of hydrogen exposed to a solution that contains H^+ ions, i.e., an acid. When the voltaic cell operates, if hydrogen is used to produce H^+ ions, the film of H_2 is constantly replenished from the hydrogen that passes through the glass tube; if hydrogen is produced from H^+, H_2 joins the stream and passes from the tube. The platinum, being much less active than hydrogen, does not lose or gain electrons so long as hydrogen and hydrogen ions are present. The chlorine electrode, the other electrode in Fig. 20-7, is constructed in a similar manner.

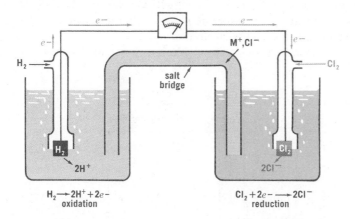

Fig. 20–7. A voltaic cell which depends on the reactions of two gaseous elements.

In the voltaic cell shown in Fig. 20-7, the over-all cell reaction is

$$H_2 + Cl_2 \rightarrow 2HCl$$

When the battery is delivering current, hydrogen molecules give up electrons to form hydrogen ions, and chlorine molecules take up electrons to form chloride ions. The hydrogen electrode is the negative electrode, because electrons are originating at this point. The electrode reactions are:

At − pole (anode):

$$H_2 \rightarrow 2H^+ + 2e^- \quad \text{(oxidation)}$$

At + pole (cathode):

$$Cl_2 + 2e^- \rightarrow 2Cl^- \quad \text{(reduction)}$$

STANDARD ELECTRODE POTENTIALS

The tendency of a substance to undergo oxidation or reduction reactions can best be expressed in terms of oxidation (or reduction) potentials. To do this, an electrode to be used for comparison is arbitrarily selected. This comparison electrode is used as half of a voltaic cell, the cell being completed with an electrode of the substance being compared. The voltage of this battery is read with a voltmeter. The magnitude of the voltage and the direction of the electron flow are indicative of the tendency of the substance to undergo oxidation by giving up electrons to the comparison electrode, or reduction by receiving electrons from the comparison electrode, as the case may be.

In actual practice, the hydrogen electrode is often used as the comparison electrode. This is fortunate for us, for our study of the reactions of metals and acids (Chap. 8) introduced us to the differences in the tendencies of metals to lose electrons to hydrogen ions.

Because temperature and concentration influence voltage, these variables must be held constant for the comparisons. The temperature is held at 25°C, the concentration of the ions in contact with the elemental electrodes is held at one molal,[3] and the gas pressure at 1 atm. Electrodes maintained under these conditions are called **standard electrodes**.

The comparisons are made in a voltaic cell set up like that in Fig. 20-8.

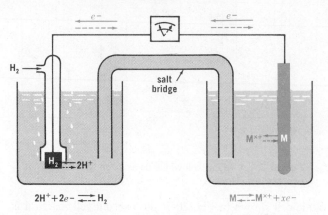

Fig. 20-8. A voltaic cell for determining standard oxidation potentials.

Remember that in these comparisons the hydrogen electrode is always half of the cell and that the M electrode, the standard electrode of the element being compared, is the other half. If the M electrode is a standard copper electrode (a strip of copper immersed in a 1 molal solution of Cu^{2+} ions), the voltmeter will indicate a voltage of 0.337 volt. The reactions at the two electrodes are:

At − pole (anode):

$$H_2 \rightarrow 2H^+ + 2e^- \quad \text{(oxidation)}$$

At + pole (cathode):

$$Cu^{2+} + 2e^- \rightarrow Cu \quad \text{(reduction)}$$

Cell reaction:

$$H_2 + Cu^{2+} \rightarrow 2H^+ + Cu$$

Because electrons originate at the hydrogen electrode, it is negative; the copper electrode is positive.

If the M electrode is a standard silver electrode, the voltage of the cell as

---

[3] The ionic concentration is not precisely 1 mole per 1000 g of solvent but is a concentration that behaves as an ideal 1 molal solution should behave.

Electrochemistry; Oxidation-Reduction

indicated by the voltmeter will be 0.799 volt. The fact that this voltage is higher than the hydrogen-copper cell voltage indicates that silver ions have a greater tendency to undergo reduction (gain electrons) than copper ions do.

With an M electrode of magnesium, the deflection of the voltmeter is in the opposite direction; the reading is 2.37 volts. This opposite deflection means that magnesium atoms rather than hydrogen atoms are giving up electrons, and magnesium is acting as the negative electrode. Hydrogen, therefore, acts as the positive electrode, and the reactions are:

At − pole (anode):

$$Mg \rightarrow Mg^{2+} + 2e^- \quad \text{(oxidation)}$$

At + pole (cathode):

$$2H^+ + 2e^- \rightarrow H_2 \quad \text{(reduction)}$$

Cell reaction:

$$Mg + 2H^+ \rightarrow Mg^{2+} + H_2$$

If the M electrode is nickel, the deflection of the voltmeter will be in the same direction as that obtained with a magnesium electrode; the reading is 0.25 volt. This smaller voltage indicates that nickel has less tendency than magnesium to give up electrons to hydrogen ions. The reactions are:

At − pole (anode):

$$Ni \rightarrow Ni^{2+} + 2e^- \quad \text{(oxidation)}$$

At + pole (cathode):

$$2H^+ + 2e^- \rightarrow H_2 \quad \text{(reduction)}$$

Cell reaction:

$$Ni + 2H^+ \rightarrow Ni^{2+} + H_2$$

By comparing the voltage readings and the direction of current flow in the four cells described above, we can list the five elements—copper, silver, nickel, magnesium, and hydrogen—in a double column, with the most easily oxidized element at the top of one column and the most easily reduced ion at the bottom of the other column.

more easily oxidized ↑	Mg Ni H_2 Cu Ag	Mg^{2+} Ni^{2+} H^+ Cu^{2+} Ag^+ ↓		more easily reduced

We now compare the over-all reactions that occur spontaneously in the comparison cells:

$$Mg + 2H^+ \rightarrow Mg^{2+} + H_2 \quad \text{(oxidation of Mg, +2.37 volts)}$$
$$Ni + 2H^+ \rightarrow Ni^{2+} + H_2 \quad \text{(oxidation of Ni, +0.25 volt)}$$
$$H_2 + Cu^{2+} \rightarrow 2H^+ + Cu \quad \text{(reduction of } Cu^{2+}, +0.34 \text{ volt)}$$
$$H_2 + 2Ag^+ \rightarrow 2H^+ + 2Ag \quad \text{(reduction of } Ag^+, +0.80 \text{ volt)}$$

Study of these equations indicates that:

1. Metals above hydrogen in activity undergo oxidation in the comparison cell. (The M electrode is the anode.)
2. Metal ions of metals below hydrogen undergo reduction in the comparison cell. (The M electrode is the cathode).

We would infer that the reverse reactions below would not occur spontaneously in voltaic cells, but that they could be made to occur by passing a direct current of sufficient voltage through the cell, that is, by electrolysis:

$H_2 + Mg^{2+} \rightarrow 2H^+ + Mg$ (reduction of Mg^{2+}, -2.37 volts)

$H_2 + Ni^{2+} \rightarrow 2H^+ + Ni$ (reduction of Ni^{2+}, -0.25 volt)

$Cu + 2H^+ \rightarrow Cu^{2+} + H_2$ (oxidation of Cu, -0.34 volt)

$2Ag + 2H^+ \rightarrow 2Ag^+ + H_2$ (oxidation of Ag, -0.80 volt)

TABLE 20-2.

Standard Oxidation Potentials, E^0

ANODE[a]	ANODE REACTION (OXIDATION)	OXIDATION POTENTIAL, VOLTS (STANDARD HYDROGEN ELECTRODE = 0)
$K;K^+$	$K \rightarrow K^+ + e^-$	$+2.93$
$Ca;Ca^{2+}$	$Ca \rightarrow Ca^{2+} + 2e^-$	$+2.87$
$Na;Na^+$	$Na \rightarrow Na^+ + e^-$	$+2.71$
$Mg;Mg^{2+}$	$Mg \rightarrow Mg^{2+} + 2e^-$	$+2.37$
$Al;Al^{3+}$	$Al \rightarrow Al^{3+} + 3e^-$	$+1.66$
$Zn;Zn^{2+}$	$Zn \rightarrow Zn^{2+} + 2e^-$	$+0.76$
$Fe;Fe^{2+}$	$Fe \rightarrow Fe^{2+} + 2e^-$	$+0.44$
$Pb,PbSO_4;Pb^{2+}$	$Pb + SO_4^{2-} \rightarrow PbSO_4 + 2e^-$	$+0.36$
$Co;Co^{2+}$	$Co \rightarrow Co^{2+} + 2e^-$	$+0.28$
$Ni;Ni^{2+}$	$Ni \rightarrow Ni^{2+} + 2e^-$	$+0.25$
$Sn;Sn^{2+}$	$Sn \rightarrow Sn^{2+} + 2e^-$	$+0.14$
$Pb;Pb^{2+}$	$Pb \rightarrow Pb^{2+} + 2e^-$	$+0.13$
$D_2;D^+$	$D_2 \rightarrow 2D^+ + 2e^-$	$+0.003$
$H_2;H^+$	$H_2 \rightarrow 2H^+ + 2e^-$	0.00
$Cu;Cu^{2+}$	$Cu \rightarrow Cu^{2+} + 2e^-$	-0.34
$I_2;I^-$	$2I^- \rightarrow I_2 + 2e^-$	-0.54
$Hg;Hg^{2+}$	$Hg \rightarrow Hg^{2+} + 2e^-$	-0.79
$Ag;Ag^+$	$Ag \rightarrow Ag^+ + e^-$	-0.80
$Br_2;Br^-$	$2Br^- \rightarrow Br_2 + 2e^-$	-1.07
$Cl_2;Cl^-$	$2Cl^- \rightarrow Cl_2 + 2e^-$	-1.36
$PbO_2;Pb^{2+}$	$Pb^{2+} + 2H_2O \rightarrow PbO_2 + 4H^+ + 2e^-$	-1.46
$Au;Au^{3+}$	$Au \rightarrow Au^{3+} + 3e^-$	-1.50
$F_2;F^-$	$2F^- \rightarrow F_2 + 2e^-$	-2.65

[a] The inert part, if any, of the electrode is not shown. For example, in the hydrogen and chlorine electrodes platinum may be used as shown in Fig. 20-7. Such electrodes may be represented as $Pt,H_2;H^+$ and $Pt,Cl_2;Cl^-$, for anodes; and $H^+;H_2,Pt$, and $Cl^-;Cl_2,Pt$, for cathodes.

Because the most active elements, such as potassium and fluorine, react with water, water solutions cannot be used in constructing their electrodes; E^0 in such cases is determined indirectly.

Electrochemistry; Oxidation-Reduction

Data such as the foregoing for Mg, Ni, Cu, and Ag have been determined for practically all of the elements. These data are recorded for some of the common elements in Table 20-2. Because they are abbreviated, we shall examine the table carefully.

1. The voltage of the entire cell has been assigned to the M electrode and called the *standard oxidation potential*, E^0. This is for comparison purposes only, because the potential of a single electrode cannot be measured. However, the hydrogen electrode is always half of each cell that is being compared. In each case therefore, the effect of the hydrogen electrode on the voltage is constant; the different voltages observed actually reflect the different tendencies of the M electrodes to lose or gain electrons.

2. Only the anode or oxidation reaction is shown. When the electrode acts as the cathode and undergoes reduction, the reaction is the reverse of the one in the table.

3. Whether or not the anode reaction will occur spontaneously when the electrode is connected to a hydrogen electrode can be inferred from the sign of the oxidation potential in the table. If the sign is positive, the reaction will occur as written, and the electrode will act as the anode, the hydrogen electrode acting as the cathode. If the sign is negative, the reverse reaction will occur spontaneously, and the hydrogen electrode will act as the anode (undergo oxidation).

4. When a hydrogen electrode acts as the cathode, the reaction is

$$2H^+ + 2e^- \rightarrow H_2 \quad \text{(reduction)}$$

When it acts as the anode, the reaction is

$$H_2 \rightarrow 2H^+ + 2e^- \quad \text{(oxidation)}$$

5. The oxidation potential decreases from $+2.93$ volts for potassium to -2.65 volts for fluorine. This means that there is a decreasing tendency from top to bottom to lose electrons (undergo oxidation) and an increasing tendency to gain electrons (undergo reduction).

POTENTIALS OF VOLTAIC CELLS

From the oxidation potentials in Table 20-2, we can predict the voltage of any voltaic cell that consists of two standard electrodes listed in the table. Certain conventions used by electrochemists will be of value here; these are itemized below. We then show how the conventions are used to predict cell reactions, voltages for given combinations, and spontaneity of oxidation-reduction reactions.

Representation of a Voltaic Cell. 1. To represent the anode (the electrode at which oxidation occurs), the solid phase is written first and the electrolyte last. The two are separated by a semicolon. (A vertical line is also commonly used.) The following examples illustrate this:

$Zn;Zn^{2+}$ an anode consisting of metallic zinc immersed in a water solution of zinc ions, for example, zinc sulfate, zinc chloride, or other salt of zinc

$Pt,H_2;H^+$ an anode consisting of solid platinum in contact with hydrogen —i.e., hydrogen adsorbed on platinum—immersed in a water solution of hydronium ion, for example, an acid such as HCl or H_2SO_4

2. To represent the cathode (the electrode at which reduction occurs), the electrolyte is written first and the solid phase last. Examples:

$Cl^-;Cl_2,Pt$ solid platinum cathode, with chlorine adsorbed on the platinum surface, immersed in a water solution of a chloride

$Zn^{2+};Zn$ a cathode consisting of zinc immersed in a solution containing zinc ions

3. To represent the complete cell, the anode is written first and the cathode last. If a salt bridge is used, this is indicated by separating the two with two vertical parallel lines. Thus, the Daniell cell in Fig. 20-4 is represented by

$$Zn;Zn^{2+} \parallel Cu^{2+};Cu$$

A cell in which the over-all reaction is $H_2 + Cl_2 \rightarrow 2HCl$ could be represented by

$$Pt,H_2;H^+ \parallel Cl^-;Cl_2,Pt$$

Use this mnemonic aid to keep track of the order:

solid; ions in solution ∥ ions in solution; solid
anode   cathode
(oxidation)   (reduction)
higher in Table 20-2   lower in Table 20-2

CELL REACTION. The *cell reaction* is the algebraic sum of the reactions that take place at the electrodes. The reaction at an electrode is frequently referred to as a *half-reaction*. For the cell $Zn;Zn^{2+} \parallel Cu^{2+};Cu$, the half-reactions and cell reaction are

$$\begin{array}{ll} Zn \rightarrow Zn^{2+} + 2e^- & \text{(oxidation at anode)} \\ \underline{Cu^{2+} + 2e^- \rightarrow Cu} & \text{(reduction at cathode)} \\ Zn + Cu^{2+} \rightarrow Zn^{2+} + Cu & \text{(cell reaction)} \end{array}$$

VOLTAGE OF CELL. The *voltage of the cell* is the algebraic sum of the oxidation potential and the reduction potential. (If we are dealing with standard electrodes, the voltage is designated by E^0.) The voltage for the cell above is

$$\begin{aligned} E^0 &= E^0_{oxidation} + E^0_{reduction} \\ &= E^0_{Zn;Zn^{2+}} + E^0_{Cu^{2+};Cu} \\ &= 0.76 \text{ volt} + 0.34 \text{ volt} \\ &= 1.10 \text{ volts} \end{aligned}$$

Note that E^0 for the electrode at the left has the same sign as that given in Table 20-2 and that E^0 for the electrode at the right has the sign opposite to that in the table. That is, reduction occurs at the latter electrode. Because Table 20-2

Electrochemistry; Oxidation-Reduction

gives only oxidation potentials, the sign must be changed if a reduction potential is needed.

Spontaneity of Reaction. If the voltage of the cell as calculated above is positive, the cell reaction will take place spontaneously as written, and the cell will provide current. Finding that $E^0_{Zn;Zn^{2+}} + E^0_{Cu^{2+};Cu}$ gives a positive 1.10 volts reveals that $Zn + Cu^{2+} \to Zn^{2+} + Cu$ is a spontaneous exothermic process. Conversely, $E^0_{Cu;Cu^{2+}} + E^0_{Zn^{2+};Zn}$ gives a negative 1.10 volts, showing that the chemical reaction, $Cu + Zn^{2+} \to Zn + Cu^{2+}$, does not take place spontaneously.

Problem 1: For each of the following voltaic cells, write the half-reactions, designating which is oxidation and which is reduction; then write the cell reaction. Next, calculate the voltage of the cell made from standard electrodes.

(a) $Co;Co^{2+} \| Ni^{2+};Ni$

Solution: Half-reactions and cell reaction:

$$\begin{array}{ll} Co \to Co^{2+} + 2e^- & \text{(oxidation)} \\ \underline{Ni^{2+} + 2e^- \to Ni} & \text{(reduction)} \\ Co + Ni^{2+} \to Co^{2+} + Ni & \text{(cell reaction)} \end{array}$$

Voltage of cell:

$$E^0 = E^0_{oxidation} + E^0_{reduction}$$
$$= E^0_{Co;Co^{2+}} + E^0_{Ni^{2+};Ni}$$

(Now refer to Table 20-2 for oxidation potentials; the reduction potential is opposite in sign to the oxidation potential.)

$$= 0.28 \text{ volt} + (-0.25 \text{ volt})$$
$$= 0.03 \text{ volt}$$

(b) $Cu;Cu^{2+} \| Ag^+;Ag$

Solution: Half-reactions and cell reaction:

$$\begin{array}{ll} Cu \to Cu^{2+} + 2e^- & \text{(oxidation)} \\ \underline{2Ag^+ + 2e^- \to 2Ag} & \text{(reduction)} \\ Cu + 2Ag^+ \to 2Ag + Cu^{2+} & \text{(cell reaction)} \end{array}$$

Voltage of cell:

$$E^0 = -0.34 \text{ volt} + 0.80 \text{ volt}$$
$$= 0.46 \text{ volt}$$

Although the equation for the silver reaction is written as $Ag \to Ag^+ + e^-$ in Table 20-2 and as $2Ag^+ + 2e^- \to 2Ag$ here, E^0 has its usual value. That is, the value of E^0 does not depend upon the number of electrons involved in balancing the equations.

(c) $Pt,D_2;D^+ \| I^-;I_2,Pt$

Solution: Half-reactions and cell reaction:

$$\begin{array}{ll} D_2 \to 2D^+ + 2e^- & \text{(oxidation)} \\ \underline{I_2 + 2e^- \to 2I^-} & \text{(reduction)} \\ D_2 + I_2 \to 2D^+ + 2I^- & \text{(cell reaction)} \end{array}$$

Voltage of cell:

$$E^0 = 0.003 + 0.54$$
$$= 0.543 \text{ volt}$$

(Note that, if ordinary hydrogen is used as the anode instead of heavy hydrogen, the voltage is $E^0 = 0 + 0.54 = 0.54$ volt.)

Problem 2: Represent the electrochemical cell based on each of the chemical reactions below. Then predict whether the cell could be used to supply current.

(a) $Pb + Br_2 \rightarrow PbBr_2$

Solution: Because the lead changes from an oxidation state of 0 to $+2$ (oxidation), our representation would show $Pb;Pb^{2+}$ as the anode. At the same time bromine changes from 0 to -1 (reduction). Hence, the cell is represented thus:

$$Pb;Pb^{2+} \parallel Br^-;Br_2,Pt$$

Table 20-2 shows that the sum of $E^0_{oxidation}$ and $E^0_{reduction}$ has a positive value. Hence, the reaction $Pb + Br_2 \rightarrow PbBr_2$ proceeds spontaneously and could be used in a voltaic cell to supply current.

(b) $Cu + 2HCl \rightarrow CuCl_2 + H_2$

Solution: In this reaction copper atoms undergo oxidation to copper ions, and hydrogen ions undergo reduction to hydrogen molecules. The cell is written thus:

$$Cu;Cu^{2+} \parallel H^+;H_2,Pt$$

From Table 20-2 we see that the sum of the oxidation and reduction potentials is minus 0.34 volt. Hence, this reaction requires energy, and it could not be used to supply current.

LEAD STORAGE CELL

The storage battery of an automobile is both a voltaic and an electrolytic cell. When it is being used to start the car, it acts as a voltaic cell to supply electrical energy; while the motor is running, it functions as an electrolytic cell, receiving electricity from the generator, which restores, by electrolysis, the original reactants of the battery.

The lead storage battery is constructed of alternate plates of spongy lead and lead dioxide, separated by wood or glass-fiber spacers, and immersed in an electrolyte, an aqueous solution of sulfuric acid (Fig. 20-9). When the battery supplies current, the lead plate (Pb) is the negative pole and the lead dioxide plate (PbO_2) is the positive pole. The following changes take place:

At the − pole (anode):

$$Pb \rightarrow Pb^{2+} + 2e^- \quad \text{(oxidation)}$$
$$Pb^{2+} + SO_4^{2-} \rightarrow PbSO_4 \downarrow$$

At the + pole (cathode):

$$PbO_2 + 4H^+ + 2e^- \rightarrow Pb^{2+} + 2H_2O \quad \text{(reduction)}$$
$$Pb^{2+} + SO_4^{2-} \rightarrow PbSO_4 \downarrow$$

The complete reaction that occurs when current is drawn from the battery is given by the following equation:

$$Pb + PbO_2 + 2H_2SO_4 \rightarrow 2PbSO_4 \downarrow + 2H_2O$$

Electrochemistry; Oxidation-Reduction

Note that lead sulfate is formed at each electrode. Being insoluble, it becomes part of the electrode at which it forms, rather than dissolving in the solution. Note also that sulfuric acid is used up and water is formed. Because the diluted sulfuric acid is less dense than the original concentrated acid, the density of the electrolyte is commonly measured to determine the extent to which the battery is "run down."

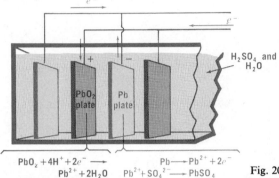

$PbO_2 + 4H^+ + 2e^- \longrightarrow$
$\qquad Pb^{2+} + 2H_2O \qquad Pb \longrightarrow Pb^{2+} + 2e^-$
$Pb^{2+} + SO_4^{2-} \longrightarrow PbSO_4 \quad Pb^{2+} + SO_4^{2-} \longrightarrow PbSO_4$

Fig. 20–9. Schematic drawing of the lead storage battery.

Recharging the battery consists of "pumping" the electrons through the battery in the opposite direction; in other words, all the chemical changes above are reversed. The lead sulfate and water are changed back to lead, lead dioxide, and sulfuric acid:

$$Pb + PbO_2 + 2H_2SO_4 \underset{\text{charge}}{\overset{\text{discharge}}{\rightleftarrows}} 2PbSO_4 + 2H_2O$$

When fully charged, a single cell of a lead storage battery has a potential of about 2.1 volts. The usual automobile battery has three or six of these cells connected in series and thus has a potential of 6 or 12 volts.

FUEL CELLS

Traditionally, voltaic cells are constructed in a manner in which a limited amount of the reactants are available. This arises because usually the cell consists of an arrangement of materials in a conveniently sized container, as in the flashlight battery. After depletion of the reactants, the cell is discarded. Much research has recently been devoted to cells in which the reactants can be stored outside the cell itself and utilized as needed (much like the storage of gasoline in the tank of an automobile). Cells of this type, which use an oxidizing agent and oxidizable fuels such as hydrogen, methane (CH_4), or even coal, are called fuel cells. The fuel cell is not new. In 1839, W. R. Grove built the first fuel cell by supplying oxygen and hydrogen to two platinum electrodes immersed in a sulfuric acid solution. With this arrangement, he observed the flow of a current in an external circuit. Many investigations have been made since. However, it is only recently that the utilization of fuel cells as a practical source of electrical energy has become a reality. A hydrogen-oxygen fuel cell has been successfully used as the power source in manned space capsules.

A hydrogen-oxygen fuel cell is shown schematically in Fig. 20-10. The two gases, hydrogen and oxygen, are led into the cell, where each comes in contact with a porous electrode of either nickel or graphite. The two electrodes are

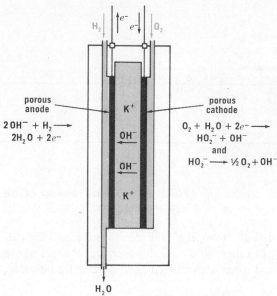

Fig. 20–10. Schematic representation of one type of a fuel cell.

separated by an electrolyte such as potassium hydroxide. A theoretical reaction at the cathode is

$$O_2 + 2H_2O + 4e^- \rightarrow 4OH^-$$

However, the actual reaction results in the formation of hydrogen peroxide ions:

$$O_2 + H_2O + 2e^- \rightarrow HO_2^- + OH^-$$

Catalysts are embedded in the porous electrodes to hasten the decomposition of the peroxide ions:

$$2HO_2^- \xrightarrow{\text{catalysts}} O_2 + 2OH^-$$

Hydroxide ions then diffuse from the cathode to the anode, where they participate in the oxidation of hydrogen:

$$H_2 + 2OH^- \rightarrow 2H_2O + 2e^-$$

The over-all reaction is

$$O_2 + 2H_2 \rightarrow 2H_2O$$

Note that the hydroxide ion is produced at the cathode and is consumed at the anode. Hence, the KOH used as the electrolyte is not depleted; however, it

Electrochemistry; Oxidation-Reduction

is required in order that there be a high concentration of OH^- ions at the anode at any time the cell operates. This concentration is maintained as the cell produces electricity by the diffusion of OH^- ions from the cathode. Fuel cells in which acid electrolytes are used produce hydrogen ions at the anode and consume hydrogen ions at the cathode. In addition to hydrogen, which may be used in cells with either alkaline or acid electrolytes, other fuels may also be used, for example ammonia with alkaline electrolytes and hydrocarbons with acid electrolytes.

FARADAY'S LAW

The passage of a current of electricity along a metallic conductor, such as a copper wire, involves no chemical reaction, but its passage through a solution of an electrolyte is achieved only by the loss and gain of electrons by the ions in the solution. Therefore, the flow of electricity brings about a definite amount of chemical change in the electrolyte, the amount depending on the amount of electricity that passes through the circuit.

The experimentally determined fact that the amount of chemical change produced is directly proportional to the quantity of electricity passed was discovered by Michael Faraday before the electron nature of an electric current was known. Our knowledge today enables us to see the basis for this definite relationship. For example, when a current is flowing through molten sodium chloride, for each electron that enters from the cathode there must be a sodium ion, Na^+, to receive it; the sodium ion changes to an atom of sodium. Similarly, each electron that leaves at the anode comes from a chloride ion, Cl^-, with the resultant change of that chloride ion to an atom of chlorine. Therefore, the passage of one *faraday* (6.02×10^{23} electrons) of electricity through molten sodium chloride is always accompanied by the formation of 6.02×10^{23} atoms (1 mole, 22.990 g) of sodium and the same number of chlorine atoms (1 mole, 35.453 g). However, this same amount of electricity passing through a water solution of copper chloride will plate out only one-half of this number of copper atoms (0.5 mole, $63.54 \div 2$, or 31.77 g), because each ion of copper, Cu^{2+}, gains two electrons when it changes to a copper atom. The weight of chlorine formed at the anode remains the same as in the case of sodium chloride. It follows, then, that *during electrolysis, or while a voltaic cell is discharging, the passage of 1 faraday (96,500 coulombs) through the circuit is accompanied by the oxidation of 1 equivalent weight of matter at one electrode and the reduction of 1 equivalent weight at the other.*[4] This is a statement of **Faraday's law.** The calculations in the following problems are based on this law.

> *Problem 3:* What weight of sodium and of chlorine will be formed when 10,000 coulombs of electricity is passed through molten sodium chloride? The electrode reactions are as follows.

---

[4] In oxidation-reduction reactions, the equivalent weight is that weight which supplies or accepts the Avogadro number of electrons. One mole of Na^+ ions, $\frac{1}{2}$ mole Cu^{2+} ions, or $\frac{1}{3}$ mole of Au^{3+} ions represent equivalent quantities, because each of these quantities can accept one mole of electrons to form one mole of sodium, $\frac{1}{2}$ mole of copper, or $\frac{1}{3}$ mole of gold.

$$Na^+ + e^- \rightarrow Na$$

$$2Cl^- \rightarrow Cl_2 + 2e^-$$

Solution: The equivalent weights (mole weight ÷ change in oxidation state) of sodium and of chlorine are 22.990 and 35.453 g, respectively. Therefore, these amounts will be formed by the passage of 96,500 coulombs. The amounts formed by the passage of 10,000 coulombs will be

$$\frac{10,000 \text{ coulombs}}{96,500 \text{ coulombs}} \times 22.990 \text{ g} = 2.38 \text{ g of Na}$$

$$\frac{10,000 \text{ coulombs}}{96,500 \text{ coulombs}} \times 35.453 \text{ g} = 3.67 \text{ g of Cl}_2$$

Problem 4: How much electricity will be required to decompose 1 lb (454 g) of water by electrolysis? The over-all reaction is

$$2H_2O \rightarrow 2H_2 + O_2$$

Solution: One mole of water (18 g) contains 2 equivalent weights of hydrogen (2 g) and 2 equivalent weights of oxygen (16 g). The equivalent weight of water is therefore 9 g. This weight of water will be decomposed by the passage of 1 faraday of electricity (yielding 1 g of hydrogen and 8 g of oxygen). To decompose 1 lb of water, the following amount of electricity is needed:

$$\frac{454 \text{ g}}{9 \text{ g}} \times 1 \text{ faraday} = 50.4 \text{ faradays}$$

OXIDATION-REDUCTION EQUATIONS

The discussion in this chapter of some oxidation and reduction reactions that take place at electrodes provides a good background for a more detailed study of the general subject of oxidation-reduction reactions. In the two preceding and in the next five chapters there are many inorganic reactions that involve oxidation-reduction. In studying these reactions we need to learn (1) how to recognize the oxidizing agent and the reducing agent and (2) how to balance the equation for the reaction. If we master these two problems, we will be able to understand many chemical reactions that otherwise we might just memorize. We shall begin by reviewing oxidation state numbers.

OXIDATION STATE NUMBERS AND BALANCING EQUATIONS

Oxidation state numbers are arbitrary numbers based upon rules such as the following.

1. The oxidation state of any uncombined element is zero.

2. The common oxidation state of hydrogen in compounds is $+1$. For oxygen it is -2.

3. The common oxidation state for group VIIA elements in binary compounds is -1.

Electrochemistry; Oxidation-Reduction

4. The common oxidation state for group IA elements in compounds is $+1$; for group IIA elements it is $+2$; and for group IIIA elements it is $+3$.

5. The oxidation state of an element in a compound is readily calculated if the oxidation states of all the other elements in the compound are known, since the sum of all oxidation states in any compound is zero.

Problem 5: Calculate the oxidation state of tin in $SnCl_2$ and in $SnCl_4$.

Solution: Since the oxidation state of chlorine in binary compounds is -1, it follows that the oxidation state of tin in $SnCl_2$ is $+2$ and in $SnCl_4$ is $+4$, i.e., $\overset{+2(-1)_2}{SnCl_2}$ and $\overset{+4(-1)_4}{SnCl_4}$.

Problem 6: Calculate the oxidation state of manganese in $KMnO_4$.

Solution: Since the oxidation state of oxygen is -2 and the oxidation state of potassium is $+1$, the oxidation state of manganese must be $+7$: $\overset{+1\ +7(-2)_4}{KMnO_4}$.

Problem 7: Balance the equation for the following reaction:
$$K_2Cr_2O_7 + HCl \rightarrow$$

Solution: Step 1. From a reference text we find that the reactants and products are:
$$K_2Cr_2O_7 + HCl \rightarrow KCl + CrCl_3 + H_2O + Cl_2$$

The ionic equation is:[5]
$$2K^+ + Cr_2O_7^{2-} + H^+ + Cl^- \rightarrow K^+ + Cl^- + Cr^{3+} + 3Cl^- + H_2O + Cl_2$$

Step 2. Calculate the oxidation states.

Elements	Rule	Oxidation state in reactant	Oxidation state in product	Electron change
Oxygen	Oxygen is -2 in compounds.	-2	-2	None
Hydrogen	Hydrogen is $+1$ in compounds.	$+1$	$+1$	None
Potassium	Elements in group IA are $+1$ in compounds.	$+1$	$+1$	None
Chlorine	Chlorine is -1 in most binary compounds; any element is in the zero state.	-1	$-1;0$	Loss
Chromium	Oxidation state of Cr is calculated from those assigned to oxygen and chlorine, respectively.	$\frac{(14-2)}{2} = +6$	$+3$	Gain

The chromium is reduced, since its oxidation state changes from $+6$ to $+3$. Some of the chlorine is oxidized (it changes from -1 to zero), but some of it is not changed.

---

[5] It is sometimes difficult for the student to decide what ions exist in solution on the basis of the formula of the compound. The following system is used for most ionic compounds containing oxygen and two other elements. Consider $K_2Cr_2O_7$ as an example. The first element in the formula forms a simple positive ion, i.e., K^+; the second element is combined with oxygen in an oxy-ion, i.e., $Cr_2O_7^{2-}$.

Step 3. For the reduction equation:

$$Cr_2O_7^{2-} \rightarrow 2Cr^{3+} \qquad \text{(incomplete)}$$

Then, knowing that the oxygen goes to form water, we have

$$Cr_2O_7^{2-} \rightarrow 2Cr^{3+} + 7H_2O \qquad \text{(incomplete)}$$

And knowing that hydrogen ions must join in the oxygen to form water, we have

$$Cr_2O_7^{2-} + 14H^+ \rightarrow 2Cr^{3+} + 7H_2O$$
(balanced as to atoms)

Adding enough electrons to the left side to balance the equation electrically gives:

$$Cr_2O_7^{2-} + 14H^+ + 6e^- \rightarrow 2Cr^{3+} + 7H_2O$$
(balanced as to atoms and charges)

$$(-2) + (+14) + (-6) = (+6) + (0)$$

For the oxidation equation:

$$2Cl^- \rightarrow Cl_2 \qquad \text{(balanced as to atoms)}$$

To balance this equation electrically, $2e^-$ must be added to the right-hand side.

$$2Cl^- \rightarrow Cl_2 + 2e^- \qquad \text{(balanced as to atoms and charges)}$$

$$-2 = (0) + (-2)$$

Step 4. Add the balanced reduction equation to the balanced oxidation equation. Multiply the second equation by 3 so the number of electrons lost equals the number gained. Six electrons are needed, so 6 chloride ions change into 3 chlorine molecules and in so doing set free 6 electrons.

$$Cr_2O_7^{2-} + 14H^+ + 6e^- \rightarrow 2Cr^{3+} + 7H_2O$$
$$3[2Cl^- \rightarrow Cl_2 + 2e^-]$$
$$\overline{Cr_2O_7^{2-} + 14H^+ + 6Cl^- + 6e^- \rightarrow 2Cr^{3+} + 7H_2O + 3Cl_2 + 6e^-}$$

Then cancel electrons.

$$Cr_2O_7^{2-} + 14H^+ + 6Cl^- \rightarrow 2Cr^{3+} + 7H_2O + 3Cl_2 \qquad \text{(balanced ionic equation)}$$

$$(-2) + (+14) + (-6) = 2(+3) + 0 + 0$$

Step 5. In the ionic equation there is no mention of K^+ ions. Nor are enough negative ions shown to make the net charges on both sides of the equation equal zero. Both of these omissions must be taken care of by including two K^+ ions for each $Cr_2O_7^{2-}$ and one Cl^- for each H^+ in writing the over-all balanced equation. The final equation is:

$$K_2Cr_2O_7 + 14HCl \rightarrow 2CrCl_3 + 2KCl + 7H_2O + 3Cl_2$$

Drill Exercises: You must be able to write Step 1 for any equation you propose to balance. The following equations are for practice work. Try to balance them, and then check your answers by referring to the appropriate chapter (Chap. 19 for the first three, 21 for the next, and 22 for the last two).

$$MnO_2 + HCl \rightarrow MnCl_2 + H_2O + Cl_2$$
$$KMnO_4 + HCl \rightarrow KCl + MnCl_2 + H_2O + Cl_2$$

$$HClO \rightarrow HCl + HClO_3$$

$$Cu + H_2SO_4 \rightarrow CuSO_4 + SO_2 + H_2O$$

$$Cu + HNO_3 \rightarrow Cu(NO_3)_2 + NO_2 + H_2O$$

$$Mg + HNO_3 \rightarrow Mg(NO_3)_2 + NH_4NO_3 + H_2O$$

An Alternate Method of Balancing Equations. Consideration of the change in oxidation state numbers alone enables us to obtain a ratio showing the relative quantities of oxidizing agent and reducing agent. Once the numbers that express this ratio are put into the equation, balancing is completed by inspection. We use this method to balance the equation:

$$\overset{+7}{K}MnO_4 + H_2SO_4 + Na_2\overset{+4}{S}O_3 \rightarrow K_2SO_4 + \overset{+2\ +6}{MnSO_4} + Na_2\overset{+6}{S}O_4 + H_2O$$

The oxidation numbers are deduced as previously described. Only those which change are shown; we need not consider those which remain unchanged. We note that manganese changes from $+7$ to $+2$ and that some sulfur changes from $+4$ to $+6$.

$$\overset{+7}{Mn} \rightarrow \overset{+2}{Mn} \quad \text{(decrease of 5 units)}$$

$$\overset{+4}{S} \rightarrow \overset{+6}{S} \quad \text{(increase of 2 units)}$$

It follows that the numbers of "apparent" molecules of $KMnO_4$ and Na_2SO_3 must be in the ratio of 2 : 5 in order for the decrease in oxidation state to be balanced by the increase.

$$\overset{+7}{2Mn} \rightarrow \overset{+2}{2Mn} \quad \text{(decrease of 10 units)}$$
$$\overset{+4}{5S} \rightarrow \overset{+6}{5S} \quad \text{(increase of 10 units)}$$
$$\overline{\overset{+7}{2Mn} + \overset{+4}{5S} \rightarrow \overset{+2}{2Mn} + \overset{+6}{5S}}$$

Placing the numbers 2 and 5 at the appropriate places in the equation, we have:

$$2KMnO_4 + ?H_2SO_4 + 5Na_2SO_3 \rightarrow K_2SO_4 + 2MnSO_4 + 5Na_2SO_4 + ?H_2O$$

The equation is almost balanced. All that we need to do now is to determine by inspection the numbers to be placed in front of the H_2SO_4 and the H_2O. The balanced equation is:

$$2KMnO_4 + 3H_2SO_4 + 5Na_2SO_3 \rightarrow K_2SO_4 + 2MnSO_4 + 5Na_2SO_4 + 3H_2O$$

Note that once the numbers expressing the ratio between oxidizing agent and reducing agent are correctly placed in the equation (2, 5, 2, 5 in the above case), they must not be changed unless all are changed proportionately.

EXERCISES

1. For each of the following, name an appropriate electrical unit: energy, power, rate, amount of electricity.
2. In which of the following reactions do oxidation and reduction occur? Is the "loss or gain of electrons" definition satisfactory for each reaction shown here? Explain.
   (a) $2CO + O_2 \rightarrow 2CO_2$
   (b) $Mg + H_2SO_4 \rightarrow H_2 + MgSO_4$
   (c) $CaCl_2 + Na_2CO_3 \rightarrow CaCO_3 + 2NaCl$
   (d) $H_2O + SO_2 \rightarrow H_2SO_3$
   (e) $H_2 + S \rightarrow H_2S$
3. Predict what is liberated at each electrode when an aqueous solution of each of the following is electrolyzed between inert electrodes: $CuBr_2$, KBr, H_2SO_4, K_2SO_4, $AgNO_3$, $Ba(NO_3)_2$.
4. Why is pure molten potassium chloride a better conductor of an electric current than solid potassium chloride?
5. Explain why a direct current rather than an alternating current is necessary to electroplate a spoon with silver.
6. For each of the following indicated reactions, write three balanced equations to show (1) the overall reaction, (2) the loss of electrons, and (3) the gain of electrons. In each case indicate which partial equation shows the oxidation and which the reduction:
   (a) $Al + CuSO_4 \rightarrow$
   (b) $Ca + Cl_2 \rightarrow$
   (c) $Fe + HCl \rightarrow FeCl_2 +$
   (d) $Li + O_2 \rightarrow$
7. Using mercury, iron, and any other chemical needed, show with a diagram how a cell similar in purpose to the Daniell cell could be constructed.
8. The gravity cell is of interest historically, and it illustrates some principles of cell design; but it is of limited utility. Why? Contrast the possible use of a series of gravity cells with that of lead storage cells for use in automobiles.
9. Diagram an apparatus that might be used in the electrolysis of an aqueous solution of $SnCl_2$. Label the diagram to show the charge on each electrode, the direction of electron flow, the direction of ion migration, the electrode at which oxidation is occurring, and the electrode at which reduction is occurring. Write equations for the reaction at each electrode and for the overall reaction.
10. On the basis of the data in Table 20-2, predict which of the following reactions will occur as written. For each of those that will occur, complete the equation and calculate the potential of a standard voltaic cell utilizing the reaction:
    (a) $Cu + H^+ \rightarrow$
    (b) $Co + Cl_2 \rightarrow$
    (c) $Hg + I_2 \rightarrow$
    (d) $Au + F_2 \rightarrow$
    (e) $Zn + SnCl_2 \rightarrow$
    (f) $D_2 + H^+ \rightarrow$
11. A solution contains dissolved salts of copper, zinc, cobalt, nickel, and silver. Indicate the order in which these metals will be plated out at the cathode if the voltage applied to the electrolysis cell is gradually increased.
12. Two dry cells are used in a flashlight until 0.50 g of zinc is converted to Zn^{2+}. How many electrons pass through the light bulb of the flashlight?
13. Is gold(III) iodide, AuI_3, an active reducing agent? Explain. Is the compound likely to be stable or unstable?

Electrochemistry; Oxidation-Reduction

14. Cells for the electrolytic production of chlorine require 1000 amperes. What weight of chlorine should be produced in 24 hours in one of these cells?
15. Represent in conventional fashion the standard cells in which the following overall reactions are occurring spontaneously. What is the cell voltage in each case?
    (a) $2Al + 3Pb^{2+} \to 2Al^{3+} + 3Pb$
    (b) $Fe + 2H^+ \to Fe^{2+} + H_2$
    (c) $Sn + Cl_2 \to Sn^{2+} + 2Cl^-$
    (d) $Co + 2Ag^+ \to Co^{2+} + 2Ag$
16. In parts (a), (b), and (d) of Exercise 15, anions are not indicated in the equations because they do not participate directly in the electrode reactions. However, the cells would not operate without anions. What role do the anions play in these cells?
17. Choose the proper chemicals for constructing four cells, all alike, that will produce a voltage between 6.0 and 6.6 volts. Diagram the equipment, show the chemicals involved, and show the direction of electron flow.
18. How does the density of the electrolyte indicate whether a lead storage battery is charged or discharged?
19. In all the batteries that we have described except the lead storage battery and the hydrogen-oxygen fuel cell, special provisions were made to prevent certain ions from coming in direct contact with certain electrodes. How was this accomplished in the gravity cell? In the dry cell? In the comparison cell? Why are no provisions necessary in the lead storage cell? In the fuel cell?
20. How does a fuel cell differ from a voltaic cell? How are they alike?
21. (a) A current rated at 3.0 amperes was passed through a Au^{3+} solution for 30 minutes. What weight of gold should be plated out?
    (b) If the measured weight of gold plated out under the conditions in (a) is 3.4 g, what is the efficiency of this process?
    (c) List several reasons why the efficiency is less than 100 percent.
22. Calculate the number of coulombs of electricity necessary to produce 1 mole of each of the following: Cl_2 from Cl^- solution; Zn from Zn^{2+} solution; Bi from Bi^{3+} solution; ClO_3^- from Cl^- solution.
23. For each of the following reactions, write balanced partial equations for both oxidation and reduction. Then balance the overall equation:
    (a) $Cu + H_2SO_4 \to CuSO_4 + SO_2 + H_2O$
    (b) $PbS + O_2 \to PbO + SO_2$
    (c) $Na_3AsO_3 + SnCl_2 + HCl \to SnCl_4 + As + NaCl + H_2O$
    (d) $S + HNO_3 \to H_2SO_4 + NO$
    (e) $I_2 + HNO_3 \to HIO_3 + NO + H_2O$
    (f) $K_2Cr_2O_7 + HBr \to Br_2 + KBr + CrBr_3 + H_2O$
    (g) $Bi + HNO_3 \to Bi(NO_3)_3 + N_2O + H_2O$
    (h) $(NH_4)_2Cr_2O_7 \to Cr_2O_3 + H_2O + N_2$
24. (a) Identify with symbols and charges the ions present in a water solution of each of the following: KNO_3, $(NH_4)_2CO_3$, Na_3PO_4, $CoCl_2$, $NiSO_4$, $Ca(ClO_3)_2$, Rb_2SO_3, $KMnO_4$, Na_2CrO_4, and $SnCl_4$.
    (b) Does each of the above solutions contain equal numbers of positive and negative ions? Explain.
25. (a) A solution of a metal in an unknown oxidation state was electrolyzed at 0.268 ampere for 30.0 minutes, resulting in the plating out of 0.266 g of metal on the cathode. Calculate the equivalent weight of the metal.
    (b) If the specific heat of the metal in (a) is 0.0565 cal/(g × deg C), what is its oxidation state and atomic weight?

SUGGESTED READING

Anson, F. C., "Electrode Sign Conventions," *J. Chem. Educ.*, **36**, 394 (1959).

Bishop, J. A., "Redox Reactions and the Acid-Base Properties of Solvents," *Chemistry*, **43** (1), 18 (1970).

Dillard, C. R., and P. H. Kammeyer, "An Experiment with Galvanic Cells," *J. Chem. Educ.*, **40**, 363 (1963).

Ehl, R. G., and A. J. Ihde, "Faraday's Electrochemical Laws and the Determination of Equivalent Weights," *J. Chem. Educ.*, **31**, 226 (1955).

Garrett, A. B., "Nuclear Batteries," *J. Chem. Educ.*, **33**, 446 (1956).

Mathur, P. B., and N. J. Paul, "Magnesium Cell for Demonstration," *J. Chem. Educ.*, **40**, 43 (1963).

Petrucci, R. H., and P. C. Moews, Jr., "A Simple Quantitative Electrolysis Experiment for First Year Chemistry," *J. Chem. Educ.*, **41**, 552 (1964).

Weissbart, J., "Fuel Cells: Electrochemical Converters of Chemical to Electrical Energy," *J. Chem. Educ.*, **38**, 267 (1961).

Williams, L. P., "Humphry Davy," *Sci. Amer.*, **202** (6), 106 (1960).

21

GROUP VIA: THE SULFUR FAMILY

In Chaps. 18 and 19 we discussed the chemistry of two families at the left and two at the right of the periodic table. Now we move to the group to the left of the halogens, group VIA. In order of increasing atomic number, these elements are oxygen, sulfur, selenium, tellurium, and polonium.

Typical of a period 2 element, oxygen is so different in physical and chemical properties from the other members of its group that it is conveniently studied separately. Also, it is so important early in the study of chemistry that it was treated in some detail in Chap. 8. In contrast, polonium is so rare that its chemistry is relatively unimportant. It is of some interest because of its radioactivity (see Fig. 17-4).

In the present discussion we shall consider sulfur, S, selenium, Se, and tellurium, Te, as a family, oxygen being mentioned from time to time for comparison. Each element in family VIA has six electrons (s^2p^4) in its outside main energy level.

PHYSICAL PROPERTIES

Some of the physical properties of elements in the sulfur family are listed in Table 21-1. It can be seen at a glance that oxygen is quite unlike the other three. As far as trends are concerned, oxygen does not fit well into the series because it has too low a melting point, boiling point, and atomic radius; it has too high an ionization energy and electronegativity. The important trends to note in Table

21-1 are (1) an increase in melting and boiling point and atomic radius as the atomic number increases and (2) a decrease in ionization energy and electronegativity as the atomic number increases. The sizes of these atoms are shown to scale in Fig. 21-1.

Fig. 21–1. Relative sizes of the atoms in the VIA family.

Group VIA provides a good example to support the generalization that, as the atomic number increases in an A family, the elements become more metallic in character. Whereas oxygen and sulfur are typical nonmetals with low electrical and heat conductivities, tellurium approaches some metals in electrical conductivity. Also, tellurium and one form of selenium look like metals. Selenium has a rare property that is noteworthy—photoconductivity; that is, its electrical conductivity, though low, is greatly increased when light shines on it. Hence this element is used in instruments designed for measuring the intensity of light (even that from stars) and in automatic switches that turn lights on when the sun sets and turn them off again at daybreak.

One of the most popular copying devices, that of the Xerox Corporation, depends on the photoconductivity of selenium (Fig. 21-2). A rotating aluminum

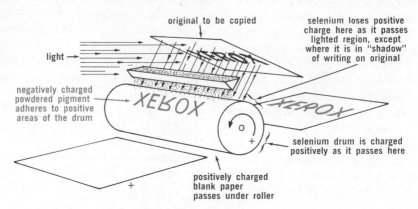

Fig. 21–2. Schematic diagram of copying machine that takes advantage of the photoconductivity of elemental selenium.

drum coated with amorphous selenium is given a positive charge while protected from light. Next the drum is exposed to light reflected from the document to be copied; where irradiated by light, the selenium drum conducts electricity from the supporting aluminum and loses its charge. Negatively charged powdered pigment is sprinkled on the drum and adheres to the positively charged areas; then the drum is rolled over a positively charged sheet of paper, which attracts the (negative) powdered pigment to give an image of the original document. The last step involves heating the paper and pigment to fuse the latter to the paper.

As noted in Table 21-1, selenium may be either red or gray. This suggests that this element exists in more than one crystalline form. Similar polymorphism is exhibited by sulfur, but both common crystalline forms are yellow in color. Actually more than 30 allotropes of sulfur are known, but most are unstable. In addition to existing in one or more crystalline modifications, sulfur, selenium, and tellurium can be prepared in the plastic or amorphous condition by suddenly

TABLE 21-1.

Physical Properties of the Sulfur Family

	OXYGEN O	SULFUR S	SELENIUM Se	TELLURIUM Te
Appearance at room temperature	Colorless gas	Yellow brittle solid	Red or gray solid	Silver-white solid
Molecular formula	O_2	S_2, S_4, S_6, S_8	Se_2, Se_8	$Te_2, (Te_8?)$
Melting point, °C	-218.8	119.0	217	450
Boiling point, °C	-183	444.6	685	1087
Ionization energy, ev	13.55 (312 kcal)	10.36 (239 kcal)	9.75 (225 kcal)	9.01 (208 kcal)
Radius of atom, A	0.66	1.04	1.14	1.32
Radius of ion (E^{2-}), A	1.40	1.84	1.98	2.21
Electronic structure	2,6	2,8,6	2,8,18,6	2,8,18,18,6
Electronegativity	3.5	2.5	2.4	2.1

freezing the hot liquid elements. When solid sulfur is formed very rapidly (for example, by pouring the boiling liquid into cold water), the sulfur molecules do not have time to orient themselves so as to form a well-developed crystal. As a result, the solid is a mass of tiny crystallites that have no over-all pattern; that is, they are amorphous.

It is not uncommon for pure substances to exist in more than one crystal pattern, depending on the temperature and pressure, and in some cases on the method of preparation, Sulfur is a good example of a substance that behaves in this manner. When a solution of sulfur in toluene is evaporated to dryness, the sulfur will crystallize in a rhombic lattice if the temperature is below 95.5°C, but it will crystallize in a monoclinic lattice if the temperature is above this. A **transition temperature** between crystal forms is usually just as definite a temperature as a boiling or freezing point. Furthermore, such a change in form involves a change in energy. In sulfur the change from rhombic to monoclinic is endothermic, because the molecules are not packed together so closely in the monoclinic patterns as in the rhombic:

$$S_8 \text{ (rhombic)} \rightarrow S_8 \text{ (monoclinic)} \quad \Delta H \text{ is positive}$$

When solid sulfur in the rhombic form is raised to a temperature above 95.5°C, the crystalline pattern slowly changes to monoclinic. The complete transformation may take days, because the molecules in the solid cannot move easily enough to reorient themselves quickly into the new crystal pattern.

The sulfur molecule, consisting of a ring of eight atoms, is pictured in Fig. 21-3. This is the unit particle commonly present in the solid and liquid states, although the rings tend to break up and form chains when the sulfur is heated.

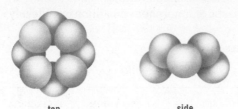

top   side

Fig. 21-3. Top and side views of a model of the sulfur molecule, S_8.

The existence of these chains explains the peculiar changes in the viscosity of liquid sulfur. Instead of decreasing regularly between melting and boiling temperatures, the viscosity of sulfur decreases from 119°C to a minimum at about 160°C, after which the viscosity begins to increase drastically, reaching a maximum at about 200°C; above 200°C the viscosity decreases till the boiling point of 444.6°C is reached. When sulfur first melts, it can be poured easily from a

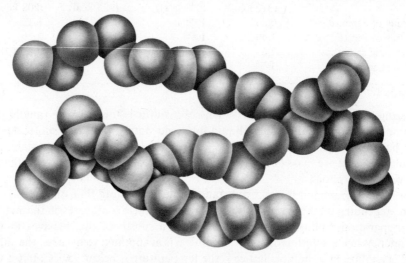

Fig. 21-4. Models representing the chains of sulfur atoms in viscous, molten sulfur; the chains may contain as many as 10,000 atoms each.

beaker or test tube, but at about 200°C it has the consistency of thick tar. Evidently the sulfur molecules still exist as S_8 rings just above the melting point, and these rings slip and roll over one another easily (not viscous). But when the liquid sulfur is heated, the rings are broken into chains that then join, forming very long molecules (see Fig. 21-4); these long molecules become entangled with one another (very viscous).[1] Above about 200°C these chains break up more and more, and sulfur begins to act again like a typical liquid; that is, the viscosity begins to decrease with further rise in temperature.

CHEMICAL PROPERTIES

A distinctive feature of the elements in the sulfur family is the fact that their atoms have four p electrons in the outside energy level. They therefore frequently react as oxidizing agents, achieving an oxidation state of -2. From the relative sizes of the atoms (Fig. 21-1), it would be expected that oxygen would be the strongest oxidizing agent and tellurium the weakest. This is indeed the case, as shown by the electronegativity data in Table 21-1. Sulfur, selenium, and tellurium can be oxidized by strong oxidizing agents—for example, oxygen or some of the halogens. When oxidized, the elements tend to be in the $+4$ or $+6$ oxidation state, as in SO_2, SeO_2, TeO_2, and SO_3, SeO_3, TeO_3. However, other oxidation states are known.

The most common inorganic compounds of sulfur are the sulfides and sulfates, containing S^{2-} and SO_4^{2-} ions, respectively. The sulfites contain SO_3^{2-} and are well known, as are a number of other combining forms:

Oxidation state:	-2	0	$+4$	$+6$
Examples:	H_2S	S	SO_2	SO_3
			H_2SO_3	H_2SO_4
	H_2Se	Se	SeO_2	SeO_3
			H_2SeO_3	H_2SeO_4
	H_2Te	Te	TeO_2	TeO_3
			H_2TeO_3	H_2TeO_4

To metals sulfur acts as an electron acceptor; to most nonmetals as a donor. The fact that it is both an oxidizing and a reducing agent accounts for its combining with all the elements save gold, platinum, and the noble gases. Yet sulfur is not very reactive unless heated above its melting point.

Selenium and tellurium have similar chemical properties but are somewhat less reactive than sulfur. Selenium compounds are so poisonous that in some regions the small amount of selenium taken from the soil by growing plants is enough to poison livestock (blind staggers).

Reactions with Metals. The equations below represent the type of reaction that occurs between a metal and members of the sulfur family:

$$Fe + S \rightarrow FeS$$

$$Fe + Se \rightarrow FeSe$$

$$Fe + Te \rightarrow FeTe$$

---

[1] This modern explanation is not unlike that given in the first century B.C. by the Roman poet Lucretius (as translated by R. E. Latham): "We see that wine flows through a strainer as fast as it is poured in; but sluggish oil loiters. This, no doubt, is either because oil consists of larger atoms, or because these are more hooked and intertangled and, therefore, cannot separate as rapidly, so as to trickle through the holes one by one."

As a rule, sulfur reacts more energetically then selenium and tellurium, and less energetically than oxygen. The heat evolved when a powdered metal is oxidized by powdered sulfur may be great enough to make the products red-hot. Some metals—among them copper, silver, and mercury—show a greater chemical affinity for sulfur than for oxygen. This is the opposite of what would be predicted on the basis of the electronegativities: O, 3.5; S, 2.5. The explanation is that a stronger crystal is formed when the larger sulfide ion is distorted or *polarized* by these small positive metal ions.

Reactions with Hydrogen. When hydrogen gas is bubbled through molten sulfur, the two elements react to form the gas hydrogen sulfide:

$$H_2 + S \rightarrow H_2S$$

The compounds H_2Se and H_2Te can also be prepared by direct combination of the elements.

Reactions with Certain Nonmetals. Sulfur reacts with hot carbon to form the colorless liquid carbon disulfide, CS_2, and with hot boron to form the solid boron trisulfide, B_2S_3. The bonds in these two compounds are covalent, the electron pair being relatively closer to the sulfur atom.

Sulfur burns in oxygen and in fluorine to form gaseous oxides, SO_2 and SO_3, and a gaseous fluoride, SF_6, respectively. It reacts less violently with chlorine to form the liquid sulfur monochloride, S_2Cl_2. The bonds in all these compounds are covalent, but the electron pairs are drawn away from the sulfur atom so that sulfur is relatively positive.

Note that, when the sulfur atom is relatively negative (oxidation state of -2), its symbol is placed second in the formula and the compound is called a *sulfide*; when the sulfur atom is relatively positive (oxidation state of $+1$, $+4$, or $+6$), its symbol is placed first and the compound is called an *oxide* or a *halide*, as the case may be.

OCCURRENCE

About the same amount of sulfur as of fluorine is present in the earth's crust, approximately 0.1 percent by weight. Selenium and tellurium are rare; the relative abundance of the two, taken together, is less than a millionth that of sulfur.

All three occur naturally, both in the free form and in compounds. Although most sulfur is found combined with metals in compounds, such as those listed in Table 21-2, large deposits of elemental sulfur have been found. Elemental selenium is often found mixed with the sulfur. When sulfur occurs as the element, it is usually mixed with rocks and earth; such a mixture is easily separated by heating it until the sulfur melts and runs out. The element was known to the ancients as the stone that burns, and later named *brimstone*.

PRODUCTION AND USES

Because sulfur is one of the most useful of all nature's raw materials, the United States is fortunate in having some of the world's richest deposits of

TABLE 21-2.

Abundant Sulfur-Containing Minerals

FORMULA	CHEMICAL NAME	MINERAL NAME
S	Sulfur	Sulfurite
FeS_2	Iron(II) disulfide	Pyrite
HgS	Mercury(II) sulfide	Cinnabar
PbS	Lead sulfide	Galena
ZnS	Zinc sulfide	Sphalerite
$CuFeS_2$	Copper(II) iron(II) sulfide	Chalcopyrite
$BaSO_4$	Barium sulfate	Barite
$CaSO_4 \cdot 2H_2O$	Calcium sulfate dihydrate	Gypsum

elemental sulfur. Located on the Gulf Coast in Texas and Louisiana are huge beds of rock that are shot through with veins of sulfur. These sulfur-rock beds, roughly circular in shape and about half a mile across, are often found in the cap rock that covers great subterranean salt domes of sodium chloride. The fact that the sulfur is itself covered by several hundred feet of quicksand permeated by poisonous gases often makes shaft-mining impractical.

Till early in this century, most sulfur was obtained from volcanic sources; then an American engineer, Herman Frasch, devised the ingenious method now used for mining sulfur. A hole or well is drilled down into the sulfur-containing stratum, and three concentric pipes are lowered into it. The sulfur is melted in the rock bed by superheated water (the water is kept under sufficient pressure so that it can be heated to about 180°C). Compressed air forces the water-molten sulfur mixture up to the surface of the ground, where it is separated. (See Fig. 21-5.) The sulfur is then pumped into a storage vat where it cools and freezes (at 119°C), becoming part of a huge block of sulfur.

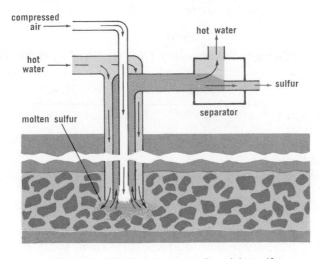

Fig. 21–5. The Frasch process for mining sulfur.

We can gain an idea of the immense scale of operations from the fact that such a block may be 400 ft long, 200 ft wide, and 100 ft high. When some of the sulfur is to be shipped, the required quantity is blasted off with dynamite; the broken sulfur is scooped up by a steam shovel and loaded onto freight cars on a railroad track laid right beside the storage block, or onto conveyor belts to barges in the Gulf. Sulfur is even shipped in the molten form via ocean tankers.

In spite of the fact that little attention is given to purifying it or protecting it during storage and shipping, sulfur is one of the purest (99.5 percent pure) of all the substances that are produced on a really large scale.

Elemental sulfur can be prepared by the oxidation of sulfides or the reduction of sulfites and sulfates. Selenium and tellurium can be prepared by similar reactions. In the case of sulfur, these reactions are not of primary industrial importance in this country, because the element itself can be mined so cheaply. Sulfur is also a by-product in the refining of certain sulfur-containing crude oils and in the treatment of some metal sulfides.

More than 8,500,000 tons of sulfur are produced annually in this country. A large portion of this is used in the manufacture of sulfuric acid, H_2SO_4. Considerable amounts are required for the vulcanization of rubber (Chap. 29), as a fuel in gunpowder and matches, for use in insecticides and soil conditioners, and for the manufacture of carbon disulfide (a solvent) and many other compounds.

Selenium is used in light-sensitive devices, for special vulcanizing processes for rubber, for ruby glass (red-colored glass), and in certain copper and steel alloys. Tellurium is used as an additive in some steel and lead alloys.

SULFIDES, SELENIDES, AND TELLURIDES

Compounds of Hydrogen. All five hydrogen compounds of the sulfur family are known: hydrogen oxide, H_2O, hydrogen sulfide, H_2S, hydrogen selenide, H_2Se, hydrogen telluride, H_2Te, and hydrogen polonide, H_2Po. Table 21-3 lists some of the important properties of the first four.

TABLE 21-3

Properties of Hydrogen Compounds of Elements (E) of the Sulfur Family

	H_2O	H_2S	H_2Se	H_2Te
Melting point, °C	0	−85.6	−60.4	−51
Boiling point, °C	100	−60.8	−41.5	− 1.8
$\Delta H°$ per mole for $H_2 + E \rightarrow H_2E$, kcal	−68.3	− 4.8	+20.5	+36.9
First ionization constant, $H_2E \rightleftarrows H^+ + HE^-$		1.2×10^{-7}	1.9×10^{-4}	2.3×10^{-3}

The trend in heat of formation is especially important, because it indicates the relative stability of the compounds. H_2O is by far the most stable, H_2S next, then H_2Se, and last, H_2Te. Hydrogen selenide and hydrogen telluride are actually endothermic compounds. Such compounds are usually unstable and will decompose with the evolution of energy if sufficient activation energy becomes available. This is the stability trend we would predict on the basis of the sizes of the atoms. The large tellurium atom has the least attraction for the electron pair joining it to the hydrogen atom, the selenium has more attraction, the sulfur atom still more, and the oxygen atom the most.

Note also the trend in the first ionization constant, K_i, of the water solutions of these compounds:

$$K_i = \frac{[H^+][HE^-]}{[H_2E]}$$

The small value for H_2S indicates that this proton donor is holding on to its protons rather than letting them go to the H_2O molecules. The trend from H_2S to H_2Te reveals that the smaller sulfur atom holds on to hydrogen more tightly than the larger tellurium atom does. The water solutions of H_2S, H_2Se, and H_2Te are acidic. Hydrosulfuric acid, the weakest acid of the three, is used in qualitative analysis for providing a low concentration of sulfide ion.

Hydrogen sulfide is an important chemical compound from the standpoint of general laboratory utility. It can be readily prepared by the action of acids on metallic sulfides. It is a poisonous gas whose offensive smell (rotten eggs) is well remembered by all who encounter it in the laboratory. Hydrogen sulfide is present in minute amounts in the fumes given off during the burning of sulfur-containing fuel oil and coal. It corrodes metals; for example, in air containing traces of H_2S silverware tarnishes, owing to the formation of a thin film of silver sulfide. The dark silver sulfide remains on the surface until buffed off.

Hydrogen sulfide dissolves in water and makes a slightly acidic solution, hydrosulfuric acid, which is widely used in qualitative analysis.

The salts of hydrosulfuric acid are the *sulfides* and the *acid sulfides*. Examples are sodium sulfide, Na_2S, sodium acid sulfide, NaHS, calcium sulfide, CaS, and calcium acid sulfide, $Ca(HS)_2$. Solutions of the soluble sulfides are used to dehair hides in the tanning industry. Some sulfide cosmetic preparations available in the stores have the same aim.

Compounds of Metals. The metallic sulfides are important well-known compounds, because they occur widely as minerals and are often encountered in laboratory and industrial chemistry. The selenides and tellurides have similar formulas and structures but are relatively rare.

Except for the metals of very low electronegativity (metals in groups IA and IIA), the sulfides tend to be drawn by ionic-covalent bonds into tight structures that have rather high melting points and are very insoluble in water.

The metal sulfides are formed (1) by direct union of the finely ground elements or (2) by reaction between the ions in water solution. Direct union occurs in the following reactions:

$$Ni + S \rightarrow NiS$$

$$2Na + S \rightarrow Na_2S$$

Ionic reactions in solution include

$$Cu^{2+} + S^{2-} \rightarrow CuS$$
$$Hg^{2+} + S^{2-} \rightarrow HgS$$

Aside from the metals in groups IA and IIA, the other metallic sulfides are characterized by their extremely slight solubility. From the solubility product of mercuric sulfide, $K_{sp} = 3.5 \times 10^{-52}$, it can be calculated that only one pair of ions (Hg^{2+}, S^{2-}) of mercuric sulfide is dissolved in 90 liters of the saturated solution. For many years the analytical chemist has taken advantage of the fact that the sulfides of many metals are so slightly soluble. By adding S^{2-} ions to a solution, a number of positive ions can be precipitated.

Nonmetallic sulfides are also well known, though as a rule they hydrolyze in contact with moist air. An exception to this rule is carbon disulfide, CS_2, a colorless liquid that neither reacts with nor mixes with water. Produced in the United States to the extent of about 820 million lb annually, carbon disulfide is used chiefly in the manufacture of rayon and cellophane (see Chap. 29) and of carbon tetrachloride.

METAL POLYSULFIDES. Not only do sulfur atoms tend to bond together in elemental sulfur (as rings and chains), but they may attach themselves to the sulfide ion. Solutions of certain metal sulfides will dissolve sulfur. For example,

$$Ba^{2+} + S^{2-} + 2S \text{ (solid)} \rightarrow Ba^{2+} + S_3^{2-}$$
$$S^{2-} + 2S \rightarrow S_3^{2-}$$

On evaporation of the solution, the salt, BaS_3, precipitates.

The polysulfide ions range in size from S_2^{2-} to S_6^{2-}. Large crystals of the most famous polysulfide, the common iron ore *pyrite*, FeS_2, look so much like gold that they are called *fool's gold*. Figure 21-6 shows how FeS_2 crystallizes in a modified NaCl pattern, with S-S pairs occupying the Cl^- positions and Fe^{2+} ions occupying the Na^+ positions.

A common laboratory reagent is ammonium polysulfide, $(NH_4)_2S_x$.[2] It is

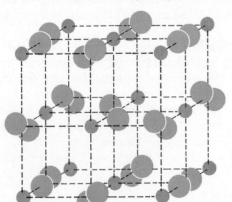

Fig. 21-6. Schematic representation of the structure of iron pyrite, FeS_2. (Redrawn by permission from A. F. Wells, *Structural Inorganic Chemistry*, The Clarendon Press, Oxford, 1962.)

---

[2] The properties of NH_4^+ salts are so similar to those of certain monovalent metal salts, especially K^+, that they are often studied with those metal salts.

usually made by dissolving an indefinite amount of sulfur in $(NH_4)_2S$ solution; the x indicates an unknown number between 2 and 5. Like sulfur, selenium and tellurium also form polyatomic anions, Se_x^{2-} and Te_x^{2-}, respectively.

The polysulfide ion is decomposed by acid solutions because of the tendency of S^{2-} to pick up two H^+ particles to form H_2S.

$$S_3^{2-} + 2H^+ \rightarrow H_2S \uparrow + 2S \downarrow$$

When an acid solution is added to a polysulfide solution, finely divided sulfur precipitates and H_2S is evolved:

$$K_2S_5 + 2HCl \rightarrow H_2S \uparrow + 4S \downarrow + 2KCl$$
$$S_5^{2-} + 2H^+ \rightarrow H_2S \uparrow + 4S \downarrow$$

OXIDES AND OXY-ACIDS

A number of oxides of the sulfur family are known. Five of sulfur—SO, S_2O_3, SO_2, SO_3, and SO_4—two of selenium—SeO_2 and SeO_3—and three of tellurium—TeO, TeO_2, and TeO_3—have been shown to exist. These many formulas are cited to emphasize the fact that atoms of elements combine in many ratios. In every case, the atoms tend to arrange themselves in the most stable way possible under the conditions prevailing at the moment of combination.

Of the sulfur oxides, SO and S_2O_3 are very unstable and decompose at ordinary room temperature. Of all the sulfur family oxides, only SO_3 and SO_2 have any great use. Both dissolve in water, the former yielding sulfuric acid, H_2SO_4, and the latter sulfurous acid, H_2SO_3. Note that in the -*ic* acid the oxidation state is $+6$, whereas in the -*ous* acid it is $+4$.

Sulfur Dioxide. Sulfur dioxide is the colorless choking gas formed when sulfur burns in air:

$$S + O_2 \rightarrow SO_2$$

and when certain metal sulfides are decomposed by roasting in air:

$$4FeS_2 + 11O_2 \rightarrow 2Fe_2O_3 + 8SO_2$$

Sulfur dioxide is often prepared for laboratory use by the action of hydrochloric or sulfuric acid on sulfites and bisulfites:

$$SO_3^{2-} + H^+ \rightarrow HSO_3^-$$
$$HSO_3^- + H^+ \rightarrow H_2SO_3$$
$$H_2SO_3 \rightarrow H_2O + SO_2$$

The gas is used as a fungicide, a fumigant (in the form of sulfur candles), and a food preservative. A person who is sensitive to sulfur dioxide may experience a mild choking sensation after eating uncooked dried apricots or peaches that have been treated with the gas. Most coal and fuel oil contain small amounts of sulfur compounds, which form sulfur dioxide upon combustion. This gas is one of the chief undesirable constituents of the smog afflicting our metropolitan areas. Some oil companies are now removing at least a part of the sulfur.

The structure of the SO_2 molecule is diagrammed in Fig. 21-7. A typical nonmetal oxide, sulfur dioxide dissolves in and reacts with water to make an acidic

SO_2 Fig. 21-7. A model of the sulfur dioxide molecule.

solution, sulfurous acid. The compound hydrogen sulfite, H_2SO_3, is known only in solutions, because it is too unstable to exist as the pure substance:

$$SO_2 + H_2O \overset{\text{water}}{\rightleftarrows} H_2SO_3$$

Sulfurous Acid. Sulfurous acid is a diprotic acid with a moderate tendency to lose one H^+ ion and a very slight tendency to lose two in water. A solution of the acid contains SO_2, H_2SO_3, HSO_3^-, and SO_3^{2-}, as well as the ions of water and water molecules. The equilibria involved in such a solution are

$$SO_2 + H_2O \rightleftarrows H_2SO_3 \rightleftarrows H^+ + HSO_3^-$$
$$\updownarrow$$
$$H^+ + SO_3^{2-}$$

When a solution of sulfurous acid is heated, the sulfur dioxide is driven from the solution. As a result, the reactions toward the left in the various equilibria are favored. Sulfurous acid is used to bleach delicate textiles, straws, and other natural fibers which chlorine would ruin.

When we compare SO_3^{2-} and SO_4^{2-} we find that, as in the case of the halogen oxy-acids, the ion that is the more symmetrical is the more stable. As shown in Fig. 21-8, the sulfite ion is an unsymmetrical pyramid with the sulfur exposed

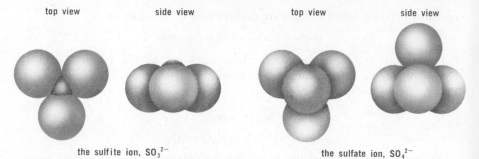

the sulfite ion, SO_3^{2-} the sulfate ion, SO_4^{2-}

Fig. 21-8. Models of the sulfite ion and the sulfate ion.

at the apex, whereas the sulfate ion is a regular tetrahedron with the sulfur atom nestled inside.

Sulfur Trioxide. Sulfur trioxide is formed when sulfur dioxide is heated with oxygen in the presence of a catalyst (platinum powder or vanadium oxide,

Group VIA: The Sulfur Family

V_2O_5, have long been used):

$$2SO_2 + O_2 \xrightarrow{\text{Pt or } V_2O_5} 2SO_3$$

This is the key reaction in the commercial production of sulfuric acid by the *contact process*. At room temperature SO_3 is a colorless liquid. It has a great affinity for water and combines with it in a highly exothermic reaction:

$$SO_3 + H_2O \rightarrow H_2SO_4 \quad \Delta H \text{ is negative}$$

In the battle against air pollution, a process has been developed for passing waste flue gases containing SO_2 and O_2 over catalytic V_2O_5 in order to form SO_3. The SO_3 can then be used to make H_2SO_4. In this way, instead of spewing forth objectionable waste, a company can recover a useful by-product.

Sulfuric Acid. The production of sulfuric acid is the largest of all the chemical industries in terms of total output. This useful acid has been manufactured in large quantities for many years; even prior to 1800 it was sold by the ton. About 29,000,000 tons per year are now manufactured in the United States.

One-third of the H_2SO_4 produced is used in the manufacture of fertilizer, 17 percent in the manufacture of other chemicals, 10 percent in the refining of petroleum, and considerable amounts in the manufacture of steel, paints, and cellulose products such as rayon.

Sulfuric acid is most commonly available in the laboratory in a highly concentrated solution that contains about 98 percent hydrogen sulfate and 2 percent water. This concentrated acid is a colorless oily liquid with a density of 1.8. When one lifts a large bottle of it, he readily notices that a gallon of it is much heavier than a gallon of water. Although sulfuric acid is very stable toward heat, a dilute solution gradually loses water as it boils; on the other hand, the pure compound loses some sulfur trioxide on boiling. Both dilute and 100 percent acid change in composition on boiling till they reach the equilibrium constant-boiling mixture of 98.5 percent H_2SO_4 and 1.5 percent H_2O.

Anhydrous H_2SO_4 can be prepared by adding the proper amount of SO_3 to 98 percent acid. The pure hydrogen sulfate will combine with more sulfur trioxide to form *pyrosulfuric acid* (or *oleum*, as it is called commercially):

$$\underset{\text{hydrogen sulfate}}{H_2SO_4} + SO_3 \rightarrow \underset{\text{pyrosulfuric acid}}{H_2S_2O_7}$$

Although it is decomposed at high temperatures, the stability of H_2SO_4 is in sharp contrast with the lack of stability of H_2SO_3. Two things should be mentioned in this connection: (1) the reaction of SO_3 with water is much more exothermic than that of SO_2 with water and (2) the H_2SO_4 molecule is more symmetrical in structure than the H_2SO_3 molecule.

IMPORTANT PROPERTIES OF SULFURIC ACID. The fact that such huge quantities of sulfuric acid are used is of course based on the physical and chemical characteristics of the acid and its low price.

It is a *strong acid*, because it reacts easily with water to form the hydronium ion. Because it is a diprotic acid, it loses a second H^+ after the first is lost.

$$H_2SO_4 + H_2O \rightleftarrows H_3O^+ + \underset{\text{acid sulfate ion}}{HSO_4^-}$$

$$HSO_4^- + H_2O \rightleftarrows H_3O^+ + SO_4^{2-}$$

In dilute solution very little of the acid is present as molecular H_2SO_4. Aside from the water, the most abundant entity present is H_3O^+, and there are relatively large numbers of SO_4^{2-} and HSO_4^- ions. Sulfuric acid is the cheapest acid available for dissolving metals and metal oxides, neutralizing bases, and cleaning corroded metal surfaces.

It is an *oxidizing agent*, especially when hot and concentrated. Even such inactive substances as copper and carbon are oxidized when the sulfur in sulfuric acid acquires two electrons and changes its oxidation state from +6 to +4:

$$Cu + 2H_2SO_4 \xrightarrow{heat} CuSO_4 + SO_2 + 2H_2O$$

Each copper atom loses two electrons, so that only one sulfate is reduced for each atom of copper.

In the case of carbon, two sulfates per carbon atom must react, because the oxidation state of carbon is changing from 0 to +4:

$$C + 2H_2SO_4 \xrightarrow{heat} CO_2 + 2SO_2 + 2H_2O$$

Sulfuric acid is an effective *dehydrating agent* because of its strong affinity for water. The reaction is extremely exothermic:

$$H_2SO_4 + xH_2O \rightarrow H_2SO_4 \cdot xH_2O \qquad \Delta H \text{ is negative}$$

In mixing the concentrated acid with water to make a dilute acid, the acid should always be added slowly to the water, not vice versa. When a little water is added to a large amount of acid, the heat evolved may be sufficient to turn some of the water to steam, causing the hot solution to pop and sputter. If the concentrated acid and water are mixed in an ordinary glass container, the solution can get hot enough so that the vessel cracks; hence it is safer to use a Pyrex glass vessel.

The great amount of heat energy released in this mixing of two liquids indicates that a chemical change is taking place. That this is the case is shown by the fact that definite hydrates have been identified. They have formulas of the type $H_2SO_4 \cdot xH_2O$, where x is 1, 2, 3, or 4.

Because of this affinity for water, concentrated sulfuric acid can be used to remove water from other substances, and even to remove hydrogen and oxygen from molecules that do not contain H_2O as such. It will remove most of the water vapor from a wet gas, such as humid air, and will decompose some molecules that contain firmly bound hydrogen and oxygen. In the case of ordinary sugar (sucrose), the residue appears charred, only black carbon being left:

$$C_{12}H_{22}O_{11} + 11H_2SO_4 \rightarrow 12C + 11H_2SO_4 \cdot H_2O$$

The *high boiling point* of sulfuric acid is one of its very important properties, because it makes possible the formation of more volatile acids when sulfuric acid is heated with certain salts:

$$2NaCl + H_2SO_4 \xrightarrow{heat} Na_2SO_4 + 2HCl \uparrow$$

$$CaF_2 + H_2SO_4 \xrightarrow{heat} CaSO_4 + 2HF \uparrow$$

$$2NaNO_3 + H_2SO_4 \xrightarrow{heat} Na_2SO_4 + 2HNO_3 \uparrow$$

Group VIA: The Sulfur Family

Pure HBr and HI, though volatile, cannot be made in this way. The hot concentrated sulfuric acid oxidizes iodide ion to elemental iodine, I_2, and, to a lesser extent, the bromide ion to elemental bromine, Br_2.

SULFATES, SELENATES, AND TELLURATES

In compounds where the oxidation state is $+6$, only the sulfates are common, but selenates and tellurates are known. Because H_2SO_4, H_2SeO_4, and H_2TeO_4 are diprotic acids, there are normal salts—for example, K_2SO_4 and K_2SeO_4—and acid salts—for example, $RbHSO_4$ and $RbHTeO_4$.

Two important mineral sulfates are listed in Table 21-2. Although most sulfates are soluble in water, those of group IIA (Ca, Sr, and Ba) are only slightly soluble. Barium sulfate, $BaSO_4$, is used as a filler to give glazed-paper a brilliant whiteness and the desired body. Calcium sulfate forms two familiar hydrates: gypsum, $CaSO_4 \cdot 2H_2O$, and plaster of paris, $(CaSO_4)_2 \cdot H_2O$. Gypsum is mined in large quantities; a spectacular natural occurrence is the 175,000-acre White Sands National Monument in New Mexico ("Sands" being a misnomer chemically). Plaster of paris, the fundamental ingredient of plaster and stucco, is made by heating gypsum:

$$2CaSO_4 \cdot 2H_2O \xrightarrow{\text{heat}} (CaSO_4)_2 \cdot H_2O + 3H_2O$$
$$\text{gypsum} \qquad\qquad \text{plaster of paris}$$

When the plaster of paris is mixed with water at room temperature, it slowly picks up enough water of crystallization to turn it back into gypsum. This is the process involved in the setting of plaster of paris.

Among the other useful sulfates are $CuSO_4$, Na_2SO_4, $MgSO_4$, and $Al_2(SO_4)_3$. Magnesium sulfate heptahydrate, $MgSO_4 \cdot 7H_2O$, is the medicine Epsom salts.

EXERCISES

1. (a) Although oxygen is in group VIA, it is not usually thought of as a typical member of the sulfur family. Why?
   (b) What are some ways in which oxygen does resemble members of the sulfur family?
2. Describe three allotropic forms of sulfur. Starting with monoclinic sulfur, how would you prepare two other allotropic forms? How would you show that each allotropic form is elemental sulfur and not a compound of sulfur?
3. Give three uses of elemental sulfur in everyday life.
4. Describe what happens when a test tube containing sulfur is heated slowly.
5. Account for the differences in the observed ionization energies of sulfur and tellurium on the basis of their atomic structures.
6. Write balanced equations for the reactions between (a) magnesium and selenium; (b) sulfur and potassium; (c) chlorine and sulfur; (d) hydrogen and polonium.
7. Give equations to show how sulfur dioxide might be prepared from (a) sulfur; (b) sodium sulfide; (c) calcium sulfite; (d) sulfuric acid.

8. Given a supply of sulfur dioxide and sodium oxide, state what other materials would be required to prepare each of the following; then write equations for the reactions: (a) sodium sulfite; (b) sulfuric acid; (c) sodium hydrogen sulfate.
9. Compare the physical properties of H_2O and H_2S; relate the differences to their molecular structures.
10. Compare the chemical properties of the disulfide ion, S_2^{2-}, and the peroxide ion, O_2^{2-}.
11. (a) Name two sulfur-containing minerals in which sulfur is in a reduced state and two in which it is in an oxidized state.
    (b) Name elements that occur in nature (1) only in the reduced state, (2) only in the oxidized state, and (3) usually in the elemental state. Compare sulfur with these elements on the basis of chemical reactivity.
12. Calculate the weight of FeS_2 required to produce 50 lb of 98 percent sulfuric acid. Assume 95 percent efficiency for the process.
13. Write equations to show how sulfuric acid is prepared from sulfur. State three properties of sulfuric acid which account for its economic importance.
14. Write equations for the ionization of hydroselenic acid. Write the expression for K_{i_1} and K_{i_2}. Would you expect K_{i_1} to be larger than K_{i_2}? Why? Would you expect K_{i_1} for H_2Se to be larger or smaller than K_{i_1} for H_2S? Why?
15. Calculate the pH of 0.1 N H_2S.
16. What weight of 98.5 percent sulfuric acid is required to neutralize 250 ml of a 4.8 M sodium hydroxide solution? What weight of pure NaOH is involved?
17. Write equations showing sulfuric acid reacting as (a) an acid; (b) an oxidizing agent; (c) a dehydrating agent.
18. Why might sulfurous acid be preferable to chlorine as a bleach?
19. "The reason your plaster of paris did not set was because it was too wet." Comment on this opinion.
20. Baking powders may consist of bicarbonate of soda, $NaHCO_3$, and an acid component such as $NaAl(SO_4)_2 \cdot 12H_2O$ (an alum). Explain how this substance is able to act as an acid when added to water and how it liberates carbon dioxide from bicarbonate of soda. Write equations.
21. (a) A technique for removing tarnish from silver involves heating silverware in water in an aluminum pan. Write a possible equation for the reaction.
    (b) The removal of the tarnish is speeded greatly by adding salt, NaCl, to the water. Explain.
22. Explain the part that selenium plays in the reproduction of printed material by a Xerox machine.
23. Why is it advisable to add sulfuric acid to water rather than water to sulfuric acid when diluting the concentrated acid?
24. Show with equations how H_2SO_4 could be used to prepare HNO_3 from one of its salts. (Boiling point of HNO_3 is 83°C.) Should concentrated or dilute H_2SO_4 be used? How could HNO_3 be separated from the reaction mixture?
25. Give two reasons for not using a strong base, e.g., NaOH, to neutralize H_2SO_4 on the skin. How would you neutralize it on the skin? In the eye?

SUGGESTED READING

Allen, C. W., "The Chemistry of Tetrasulfur Tetranitride," *J. Chem. Educ.*, **44**, 38 (1967).

Brasted, R. C., "Hydrogen Sulfide Under Any Other Name Still Smells," *J. Chem. Educ.*, **47**, 574 (1970).

Moews, P. C., Jr., and R. H. Petrucci, "The Oxidation of Iodide Ion by Persulfate Ion," *J. Chem. Educ.*, **41**, 549 (1964).

22

GROUP VA: THE NITROGEN FAMILY

The elements in group VA of the periodic table constitute the nitrogen family. These elements, each of which has five electrons (s^2p^3) in its outside main energy level, are nitrogen, phosphorus, arsenic, antimony, and bismuth. They are similar to one another in some respects but are more noted for their differences. The stepwise change from nonmetallic to metallic character within a group is more clearly evident in the nitrogen family than in any other.

In group VIA it was pointed out that oxygen differed greatly in physical and chemical properties from the other elements. Similarly, nitrogen is unlike the other group VA elements.

PHYSICAL PROPERTIES

A few of the important physical properties of the nitrogen family are listed in Table 22-1. The trend from nonmetal to metal is quite evident. (1) Both antimony and bismuth have the luster of metals on freshly broken surfaces; (2) the ionization energy values reveal that nitrogen holds onto its outside electrons most strongly, bismuth least strongly; (3) the electronegativities show that nitrogen has a high electron affinity and that the affinity decreases down the family to antimony (and presumably bismuth). Measurements of the electrical conductivity (not included in the table) show that it increases from nitrogen to bismuth.

The regular trend in size is apparent in the group, the size increasing steadily as the atomic number increases. An interesting comparison of the radii of atoms in different combinations can be made from Table 22-1. Note how large the

Group VA: The Nitrogen Family

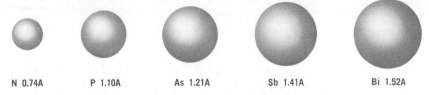

N 0.74A   P 1.10A   As 1.21A   Sb 1.41A   Bi 1.52A

Fig. 22–1. The relative sizes of atoms in the nitrogen family.

radius is for a particle in the -3 oxidation state and how much smaller it is for the $+5$ oxidation state. The ideal covalent radius is intermediate in value (an electron pair is equally shared with another atom). Figure 22-1 shows diagrams of the five atoms drawn to scale on the basis of their covalent radii.

All the elements except nitrogen are solids. The melting and boiling points of these elements are worthy of special attention; they are plotted in Fig. 22-2. Note the slight difference in the melting and boiling points for nitrogen. It has a liquid range of only 14 degrees, whereas bismuth has a liquid range of about 1200 degrees. The melting point for bismuth (271°C) is relatively low for a metal.

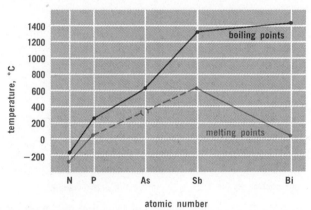

Fig. 22–2. Differences between the melting and boiling points for the members of the nitrogen family. (The melting point for arsenic is not plotted, since arsenic sublimes rather than melts when heated at atmospheric pressure.)

Note also, in Table 22-1, the tremendous difference in the melting points for the white and red forms of phosphorus. Because both are pure phosphorus, it is evident that this difference is due to the structures of the two. White phosphorus is made up of discrete P_4 molecules attracted to one another by weak van der Waals forces, whereas the red (or violet) form is crystallized in layers of tightly bound atoms (Fig. 22-3). In the P_4 molecule the bonds arise from the interaction of $3p$ orbitals, with the bonding orbitals bent outward from the line joining the centers of each pair of atoms. White phosphorus is very active chemically, is soluble in carbon disulfide, and is poisonous; red phosphorus has none of these properties.

TABLE 22-1.

Physical Properties of the Nitrogen Family

	NITROGEN N	PHOSPHORUS P	ARSENIC As	ANTIMONY Sb	BISMUTH Bi
Appearance at room temperature	Colorless gas	Waxy white, red (violet), or black solid	Steel-gray solid	Blue-white solid, metallic luster	Pinkish white solid, metall. luster
Common molecular formula	N_2	P_4	As_4	Sb	Bi
Melting point, °C	−210	44[a] 592[b] (43 atm)	814 (36 atm)	630	271
Boiling point, °C	−196	280	610 (sublimes)	1380	1470
Ionization energy, ev	14.5 (334 kcal)	11.0 (254 kcal)	9.81 (226 kcal)	8.64 (199 kcal)	7.29 (168 kcal)
Covalent radius, A	0.74	1.10	1.21	1.41	1.52
Crystal radius (X^{3-}), A	1.71	2.12	2.22	2.45	
Crystal radius (X^{5+}), A	0.11	0.34	0.47	0.62	0.74
Electronic structure	2,5	2,8,5	2,8,18,5	2,8,18,18,5	2,8,18,32,18,5
Electronegativity	3.0	2.1	2.0	1.8	

[a] White phosphorus.
[b] Red phosphorus.

Actually, solid phosphorus, arsenic, and antimony each have at least two allotropic (or polymorphic) modifications—a nonmetallic form of low density

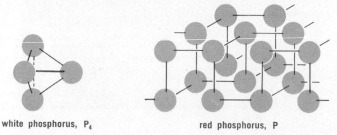

white phosphorus, P_4 red phosphorus, P

Fig. 22-3. The molecular structure of two forms of phosphorus. (Redrawn by permission from A. F. Wells, *Structural Inorganic Chemistry*, The Clarendon Press, Oxford, 1962.)

and a dense, more closely packed metallic form. Solid nitrogen has only a nonmetallic structure, and solid bismuth has only a metallic one.

The behavior of arsenic is typical of substances (for example, CO_2) that cannot exist as liquids at atmospheric pressure. When heated, arsenic does not melt; instead, it passes directly from the solid state into the vapor state (sublimes). If heated under pressure, it can exist in the liquid state.

The molecular formulas N_2, P_4, As_4, Sb, and Bi are informative. Of all the elements, only the nonmetals tend to form simple polyatomic molecules.[1]

---

[1] There are three solid nonmetals that do not form simple polyatomic molecules under ordinary conditions, but rather crystallize in indefinitely large polyatomic crystals. Their symbols indicate the absence of simple polyatomic molecules—boron, B, carbon, C, and silicon, Si.

The symbols Sb and Bi, lacking subscripts, indicate that these elements crystallize like metals, with the single atom acting as an independent unit.

CHEMICAL PROPERTIES

Reactivity. Perhaps the most striking chemical property of the nitrogen family (indeed one of the most striking chemical properties of any of the elements) is the inactivity of elemental nitrogen. With an electronegativity of 3.0 (equal to chlorine, exceeded only by oxygen and fluorine), nitrogen should be one of the most active of all elements. Actually, it resists combination with other atoms as a result of the great affinity that one nitrogen atom has for another. In molecules of elemental nitrogen, N_2, the two nitrogen atoms share three pairs of electrons. Such a bond is called a **triple covalent bond,** or more simply, a **triple bond** (see Fig. 5-7 for a molecular orbital representation):

$$:N:::N: \quad \text{or} \quad N \equiv N$$

Molecular nitrogen can be decomposed by a very high-voltage electric discharge:

$$N_2 \rightarrow 2N \quad \Delta H = 226 \text{ kcal electrical energy per mole } N_2$$

The same amount of energy is liberated when the atoms recombine:

$$2N \rightarrow N_2 \quad \Delta H = -226 \text{ kcal heat energy per mole } N_2$$

These two reactions are the basis for an atomic nitrogen torch, which is used to generate a very high temperature flame. A stream of nitrogen is directed between the poles of an electric arc that dissociates the N_2 molecules into atoms. As the gaseous atoms immediately recombine to form molecules, heat energy is released in the form of radiant energy and hot gas. There is also the atomic hydrogen torch that operates on the same principle. Such flames are useful because of their high temperatures and because they permit use of oxygen-free flames.

The inactivity of nitrogen is apparent in many common processes. In the changes involved in combustion, fermentation, decay, and the respiration of animals it is the oxygen of the air, not the nitrogen, that participates. When air is being used by a furnace or an automobile engine, almost all the nitrogen (78 percent of the air) passes unchanged through the furnace or engine and is returned to the atmosphere. (In automobile exhaust fumes there can be slight traces of nitrogen oxides. These acidic oxides contribute to air pollution, especially in crowded cities.)

Under extreme temperatures and pressures or in the presence of catalysts, nitrogen does react with other elements. For example, nitrogen and oxygen combine when a high-voltage spark passes through a mixture of the two:

$$N_2 + O_2 \rightarrow 2NO$$

The spark dissociates some O_2 and N_2 molecules into atoms to initiate the reaction. However, the reaction does not progress beyond the spark, probably because of the high dissociation energy of N_2. In a similar fashion, some of the nitrogen and oxygen in the path of a bolt of lightning during thunderstorms is converted into nitric oxide (NO). The nitric oxide then reacts with more oxygen of the air to form nitrogen dioxide:

$$2NO + O_2 \rightarrow 2NO_2$$

The nitrogen dioxide produced during lightning storms dissolves in rain water, forming a very dilute solution of nitric acid. In this way, a considerable amount of the inactive elemental nitrogen of the air is converted into nitrogen compounds and deposited in the soil for use by plants as food. (See Fig. 22-4.) This is one of nature's **nitrogen fixation processes**.[2] It is estimated that a region with moderate rainfall receives 5 to 7 lb of nitrogen (as HNO_3) per acre per year.

With some of the very active metals, nitrogen reacts and forms *ionic nitrides*:

$$3Ca + N_2 \rightarrow \underset{\substack{\text{calcium}\\\text{nitride}}}{Ca_3N_2}$$

Similar reactions occur with such metals as magnesium, lithium, and aluminum.

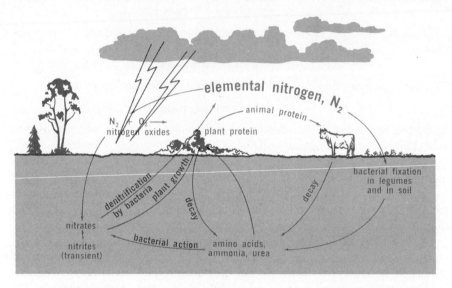

Fig 22-4. The nitrogen cycle in nature.

In contrast to nitrogen with its slight activity, phosphorus is very active. It burns readily in air, forming either the phosphorus(III) or the phosphorus(V) oxide,[3] depending on the availability of oxygen.

---

[2] Nitrogen fixation is any process in which elemental nitrogen reacts to form a compound, for example, a fertilizer. Artificial fixation of nitrogen by the direct union of nitrogen and oxygen has been commercially developed in some areas where electric power is cheap. Nitrogen and oxygen are blown through an electric arc; the resulting NO is oxidized to NO_2 and then combined with water to make nitric acid. The method is now obsolete.

[3] The names "trioxide" and "pentoxide" are still applied to these compounds. They are holdovers from the days when the formulas for these compounds were written as P_2O_3 and P_2O_5, respectively. The modern names phosphorus(III) oxide and phosphorus(V) oxide are used more and more in texts, but the old names, like those of many other compounds, are slow to disappear.

$$P_4 + 3O_2 \rightarrow P_4O_6$$
phosphorus(III) oxide

$$P_4 + 5O_2 \rightarrow P_4O_{10}$$
phosphorus(V) oxide

The kindling temperature of white phosphorus is about room temperature; hence it tends to ignite spontaneously when exposed to air. The kindling temperature of red phosphorus is much higher. Phosphorus reacts vigorously with the halogens to form such compounds as PCl_3, PBr_3, and PBr_5.

Arsenic, antimony, and bismuth are not affected by oxygen at ordinary temperatures. At elevated temperatures, however, each burns to the trioxide. For simplicity, P and As are usually used in equations, rather than P_4 and As_4:

$$4As + 3O_2 \rightarrow 2As_2O_3$$
$$4Sb + 3O_2 \rightarrow 2Sb_2O_3$$
$$4Bi + 3O_2 \rightarrow 2Bi_2O_3$$

Halogens combine directly with the VA elements, except nitrogen, to form either the pentahalides or the trihalides. Examples are

$$2P + 5F_2 \rightarrow 2PF_5$$
$$2P + 3I_2 \rightarrow 2PI_3$$
$$2Sb + 3Cl_2 \rightarrow 2SbCl_3$$

Oxidation State. Compounds are known in which one or more of the nitrogen family have oxidation states of $-3, -2, -1, +1, +2, +3, +4$, and $+5$, but the common ones are $-3, +3$, and $+5$. The three common states are explained nicely on the basis of the use in bonding of the three *p* electrons or the use of both the three *p* electrons and the two *s* electrons. Compounds of the various oxidation states are listed in Tables 22-3 and 22-4.

Trend from Nonmetals to Metals. We have observed that the physical properties of the nitrogen family change from those characteristic of nonmetals to

TABLE 22-2.

Hydrogen Compounds of Nitrogen Family

NAME	FORMULA	ΔH PER MOLE FOR $E + 3/2 H_2 \rightarrow EH_3$, KCAL	REMARKS
Ammonia	NH_3	-11.0	Stable in air; decomposes when strongly heated
Phosphine	PH_3	-1.9	Easily decomposed by heat
Arsine	AsH_3	$+36.7$	Easily decomposed by heat
Stibine	SbH_3	$+34$ (?)	Decomposes explosively when heated
Bismuthine	BiH_3	(?)	Decomposes spontaneously at room temperature

those characteristic of metals as one proceeds down the family. The chemical properties also show this important trend.

STABILITY OF HYDROGEN COMPOUNDS. To acquire electrons and become negative ions, or to be in a negative oxidation state, is characteristic of nonmetals. In the nitrogen family, nitrogen has the greatest tendency to have an oxidation state of -3, but bismuth does not form stable compounds in which its oxidation state is -3. The hydrogen compounds illustrate this trend very well, as shown by the data in Table 22-2. Only the nitrogen compound, NH_3, is stable enough to be made by the direct union of the elements. Note also that arsine, AsH_3, and stibine, SbH_3, are endothermic compounds, almost a sure sign of instability.

The same trend from nonmetal to metal behavior was indicated in group VIA by the decrease in stability of the compounds H_2O, H_2S, H_2Se, and H_2Te.

RELATIVE ACID-BASE CHARACTER OF THE OXIDES. The oxides of nitrogen family elements exhibit properties that range from those of the acidic oxides of the nonmetal nitrogen to those of the basic oxides of the metal bismuth. It is interesting to compare the compounds in which the parent element has an oxidation state of $+3$.

Nitrogen forms nitrous acid, HONO, usually written as HNO_2. This acidic compound reacts as though composed of the ions H^+ and NO_2^-.

Bismuth forms bismuthyl hydroxide, HOBiO, usually written as BiOOH. This basic compound reacts as though composed of the ions BiO^+ and OH^-.

Antimony forms a compound that has the formula HOSbO. When a base is present, HOSbO reacts as though composed of the ions $H^+ + SbO_2^-$. But when an acid is present, the compound reacts as though composed of OH^- and SbO^+. A compound such as HOSbO, which can react as either an acid or a base, is said to be amphoteric (see Chap. 16).

TABLE 22-3.

Minerals of the Nitrogen Family

NAME	FORMULA	OXIDATION STATE	MINERAL NAME
Sodium nitrate	$NaNO_3$	5	Chilean nitrate
Calcium phosphate	$Ca_3(PO_4)_2$	5	Phosphorite
Calcium chlorofluophosphate	$Ca_5(Cl,F)(PO_4)_3$	5	Apatite[a]
Iron arsenide sulfide	FeAsS	?	Arsenical pyrites
Arsenic(III) sulfide	As_2S_3	3	Orpiment
Arsenic(II) sulfide	As_2S_2	2	Realgar
Antimony sulfide	Sb_2S_3	3	Stibnite
Antimony oxide	Sb_2O_3	3	Senarmonite
Bismuth sulfide	Bi_2S_3	3	Bismuthite
Bismuth oxide monohydrate	$Bi_2O_3 \cdot H_2O$	3	Bismite

[a] In apatite, Cl^- and F^- ions replace each other in all proportions by isomorphic substitution. The fact that teeth are composed in part of a similar substance is the basis for the fluoridation of water. The proper amount of F^- ions in the water apparently promotes the formation of teeth that are especially resistant to decay by increasing the F^- content and decreasing the Cl^- content of the enamel.

OCCURRENCE

The most abundant of the elements in the nitrogen family is phosphorus; it ranks tenth among the elements in the earth's crust, 0.12 percent by weight. The other elements are not abundant on a weight basis, but they are not rare by any means. Elemental nitrogen, N_2, makes up 78 percent of the atmosphere, and nitrogen compounds (especially proteins) are constituents of all living things. Arsenic, antimony, and bismuth are found in both the elemental and combined form in nature and are easily prepared from their compounds. Only phosphorus is so active that it is found in nature only in compounds.

Common minerals of the family are listed in Table 22-3. Sodium nitrate, $NaNO_3$, and calcium phosphate, $Ca_3(PO_4)_2$, are both used in the manufacture of fertilizers. Phosphorus, like nitrogen, is essential to the growth of living things; phosphorus compounds are found in the protoplasm of cells. Note that the most common minerals of the elements of a metallic character, arsenic, antimony, and bismuth, are the *sulfides* and *oxides*.

PREPARATION AND USES

Nitrogen. Nitrogen is produced most easily be separating it from the atmosphere. Air is compressed and cooled till it is liquid and then allowed to boil; the nitrogen is separated by fractional distillation. The nitrogen produced in this way is not absolutely pure, for it contains small amounts of other gases such as oxygen and the noble elements.

Very pure nitrogen can be prepared by the thermal decomposition of an unstable nitrogen compound. Many nitrogen compounds are easily broken down, because the bonds that hold the molecule together are not so strong as the $N\equiv N$ bond, and therefore the nitrogen atoms tend to break away to form nitrogen molecules, N_2.

Ammonium nitrite is often used as a laboratory source of nitrogen gas; this compound is so unstable that it is not stored but is made as needed from two more stable compounds:

Step 1
$$NaNO_2 + NH_4Cl \xrightarrow[\text{water}]{\text{in}} NH_4NO_2 + NaCl$$

Step 2
$$NH_4NO_2 \xrightarrow[\text{solution}]{\text{gently heat}} 2H_2O + N_2$$

The decomposition of such compounds is the result of the tendency for atoms to group themselves together in the most stable structures possible. Under ordinary conditions N_2 is the most stable structure nitrogen atoms can attain.

Nitrogen gas is used in huge quantities for the production of ammonia, NH_3. Smaller amounts are used in the hardening of steel and in chambers where an inactive atmosphere is desired. Such chambers range from large rooms, to laboratory hoods, to electric light bulbs. Liquid nitrogen in a Dewar vessel can be used as a refrigerant, because it boils at $-196°C$.

Phosphorus. Phosphorus is produced on a large scale by a reaction between a phosphate mineral, coke, and sand. A number of complex reactions undoubtedly occur in steps, but the process can be summarized as follows:

$$2Ca_3(PO_4)_2 + 6SiO_2 \xrightarrow[\text{hot}]{\text{very}} 6CaSiO_3 + P_4O_{10}$$
calcium phosphate / silica (sand) / calcium silicate / phosphorus(V) oxide

then

$$P_4O_{10} + 10C \xrightarrow[\text{hot}]{\text{very}} 10CO + P_4$$

As the P_4O_{10} vapor rises through the mixture, it comes in contact with hot carbon. The second reaction tends to occur because the carbon forms a stronger bond with oxygen than phosphorus does. Carbon acts as a reducing agent here as phosphorus vapor is formed. The P_4 is separated from the gaseous CO by running the mixture through cool pipes in which the liquid phosphorus condenses. (Great care must be taken to protect personnel against any leakage of CO and P_4 vapors.)

Elemental phosphorus is used in the manufacture of certain widely used bronze alloys, in the production of phosphoric acid, H_3PO_4, and in the manufacture of matches (Fig. 22-5), smoke bomb mixes, tracer bullets, and pest poisons.

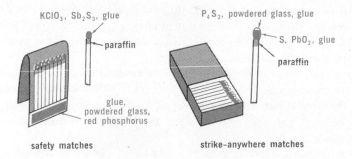

Fig. 22-5. The compositions of two types of matches.

Other Elements. *Arsenic, antimony,* and *bismuth* are often produced by reducing the appropriate oxide with carbon. This type of reaction is something of a classic, for it has been used for centuries to separate various metals from their ores. If M stands for As, Sb, or Bi, the reaction proceeds thus:

$$M_2O_3 + 3C \rightarrow 2M + 3CO$$

These three nitrogen family elements are used mainly in alloys to which they contribute certain desirable properties. A trace of arsenic is added to lead shot to make the lead harder and its surface tension higher.[4] Antimony is used in alloys to dilute more expensive metals, such as lead and tin. Antimony alloys do not have great tensile strength, but they can be used for cheap castings such as white

---

[4] Lead shot are formed by allowing tiny drops of molten lead to fall in air; the greater the surface tension of the liquid, the more nearly spherical the droplet before it solidifies.

metal (small toys and decorations). Babbitt, an alloy of antimony, is used in bearings.

Bismuth and antimony alloys tend to expand when they solidify;[5] hence they are useful in making good castings. The expanding metal fills even the finest details of the mold. Bismuth is a component of some alloys with very low melting points that are used as safety plugs in steam boilers and in automatic fire-control sprinkler systems.

REPRESENTATIVE COMPOUNDS OF THE NITROGEN FAMILY

Nitrogen forms a number of simple compounds in which its oxidation states range from -3 to $+5$ (see Table 22-4).

Ammonia. Ammonia, NH_3, is a common, extremely useful compound that is produced commercially in huge quantities by the direct union of nitrogen and hydrogen. It is made in nature by the decay of the protein in the bodies of plants and animals (see the nitrogen cycle, Fig. 22-4). Ammonia is used as a fertilizer, a refrigeration gas, and a solvent. It is a raw material in the manufacture of ammonium sulfate, $(NH_4)_2SO_4$, nitric acid, HNO_3, and ammonium nitrate, NH_4NO_3.

TABLE 22-4.

Oxidation States of Nitrogen

NAME	FORMULA	OXIDATION STATE	NAME	FORMULA	OXIDATION STATE
Ammonia	NH_3	-3	Nitric oxide	NO	$+2$
Hydrazine	N_2H_4	-2	Nitrogen trioxide	N_2O_3	$+3$
Hydroxylamine	NH_2OH	-1	Nitrogen dioxide	NO_2	$+4$
Nitrogen	N_2	0	Nitrogen pentoxide	N_2O_5	$+5$
Nitrous oxide	N_2O	$+1$			

Dissolved in water it forms a solution known as *ammonium hydroxide* or *ammonia water*. This weakly basic solution is used as a fertilizer, as a neutralizer of acid solutions, and as a cleansing agent (household ammonia). The odor of ammonia rising from the solution[6] reveals that the ammonia water is not very stable; indeed, the ammonia can be completely driven out of the water by boiling.

The great solubility of ammonia in water is unusual, about 700 liters of the gas dissolving in 1 liter of water under room conditions. The attraction of water for ammonia is so great that, if a little water is introduced into a chamber filled with the gas, a partial vacuum will result. This is the basis of a demonstration known

---

[5] The water-ice transformation is the most familiar example of this rare phenomenon.
[6] A whiff of ammonia water is sometimes used to revive a person who has fainted.

as the ammonia "fountain," similar to the hydrogen chloride "fountain" shown in Fig. 19-6. (Because ammonia is a base, the color change is opposite to that shown in the figure for the acidic hydrogen chloride.)

The NH_3 molecule is a pyramid with the nitrogen at the apex, as shown in Fig. 22-6. The molecule can be considered an incomplete tetrahedron (N at the center) with the two unused electrons held on the side away from the three hydrogen atoms. Like the H : Ö : molecule, the H : N̈ : H molecule is a polar molecule,
                                  H   H

because of the unsymmetrical distribution of positive and negative charges. In addition to the vibration and rotation motions that we would expect, the NH_3 molecule has the unusual and fascinating motion of *inversion*, as pictured in Fig. 22-6.

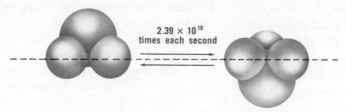

Fig. 22–6. Inversion motion of the ammonia molecule; the nitrogen oscillates back and forth through the plane of the three hydrogens.

COMMERCIAL PRODUCTION OF AMMONIA. As pointed out earlier, the most striking chemical property of nitrogen is its inactivity. Shortly before World War I broke out in 1914, the German chemist Fritz Haber solved the problem of commercial nitrogen fixation by inventing a process for making ammonia directly from nitrogen and hydrogen.[7] The key to the problem of how nitrogen could be made to react directly with hydrogen lay in finding a catalyst for the reaction. Haber found that iron oxide containing traces of other metal oxides would catalyze the reaction. Although other catalysts have been developed, iron or iron oxide is usually the main ingredient in the modern mixed catalysts.

The equilibrium reaction

$$N_2 + 3H_2 \rightleftarrows 2NH_3 \qquad \Delta H = -22 \text{ kcal}$$

was discussed from several standpoints in Chap. 15.

Considerable ammonia and ammonium compounds are by-products in the production of coke from coal, but over 90 percent of our total ammonia production of approximately 14,000,000 tons annually is made by the Haber process. Almost half of this ammonia is converted into ammonium nitrate for use as a

---

[7] Some historians have pointed out that the stockpile of Chilean nitrate in Germany in 1914 was not large enough to provide explosives for a long war. When it became evident that the war was not going to be won quickly, the Germans turned to the new Haber process. After months of feverish activity, it was developed to the point where it provided enough ammonia to supply the munitions industry. Thus it is possible that the new ammonia industry enabled the Germans to prolong the war by two or three years.

fertilizer:

$$NH_3 + HNO_3 \to NH_4NO_3$$

A large amount of ammonia is converted into nitric acid; some is absorbed in water to make aqua ammonia (household ammonia). Liquid ammonia, under a pressure of 8 to 10 atm, is shipped in iron cylinders and in tank cars.

Ammonium Compounds. The NH_4^+ ion is a perfect tetrahedron in which all the hydrogen atoms are identical; the positive charge is uniformly associated with the whole ion instead of being more on one atom than on another. In its compounds the NH_4^+ ion acts very much like the K^+ and Rb^+ ions, which have the same charge and are of about the same size (radii: K^+, 1.33A; Rb^+, 1.48A; NH_4^+, about 1.52A).

Ammonium compounds as a class are noted for their great solubility in water. Some of the important compounds are listed in Table 22-5.

TABLE 22-5.

Ammonium Compounds

NAME	FORMULA	USES
Ammonium chloride	NH_4Cl	As a "flux" for cleaning iron before galvanizing and for cleaning metals before soldering; used in dry cells and in preparing fabrics for dyeing
Ammonium fluoride	NH_4F	For etching glass
Ammonium sulfate	$(NH_4)_2SO_4$	Fertilizer; except for NH_3, cheapest source of NH_4^+ ions
Ammonium sulfide	$(NH_4)_2S$	Common reagent for qualitative analysis; used in making polysulfides
Ammonium carbonate	$(NH_4)_2CO_3$	Smelling salts
Ammonium nitrate	NH_4NO_3	Fertilizer; explosive

Most ammonium compounds are sufficiently stable to exist indefinitely at room temperature, but they decompose when strongly heated; NH_3 gas is usually one of the products:

$$\underset{\text{ammonium chloride}}{NH_4Cl} \xrightarrow{\text{heat}} \underset{\text{ammonia}}{NH_3} + \underset{\text{hydrogen chloride}}{HCl}$$

The action of heat on a solution of ammonium nitrite, NH_4NO_2, was described earlier in this chapter.

In water solution the ammonium ion reacts with strong bases in this way:

$$NH_4^+ + Cl^- + Na^+ + OH^- \to NH_3 \uparrow + H_2O + Na^+ + Cl^-$$

or

$$NH_4^+ + OH^- \to NH_3 \uparrow + H_2O$$

The presence of the NH_4^+ ion in solution can be demonstrated by adding a strong base, for example, sodium hydroxide, and heating. The ammonia gas that vaporizes from the solution can be detected by its odor or by holding a piece of moist red litmus paper in the fumes (the litmus turns blue).

Ammonium nitrate, NH_4NO_3, commonly used as a fertilizer, has also been used as an explosive in bombs and shells. An intimate mixture of ammonium nitrate and TNT (trinitrotoluene) is almost as powerful as the same weight of pure TNT and is cheaper to use. Although normally very difficult to detonate, ammonium nitrate has been responsible for some disastrous peacetime explosions.

It is interesting that all common explosives are nitrogen-containing compounds. Other examples, in addition to those just mentioned, include nitroglycerin and cellulose nitrate. Such compounds tend to decompose to form nitrogen and other gases in violent exothermic reactions. One solid propellant for rockets is reported to be a mixture of ammonium perchlorate and aluminum held in a plastic mold of nitroglycerin and cellulose nitrate. Fatal explosions have occurred in plants where these propellants are manufactured.

Nitric Acid. Nitric acid is the water solution of the colorless liquid covalent compound, hydrogen nitrate, HNO_3. Hydrogen nitrate dissolves in and reacts with water to form a very strong acid solution:

$$HNO_3 + H_2O \rightarrow H_3O^+ + NO_3^-$$

or

$$HNO_3 \rightarrow H^+ + NO_3^-$$

Along with sulfuric and hydrochloric it is one of the most useful and common acids known. Nitric acid is noted as a strong oxidizing agent. When it reacts in this capacity, nitrogen changes from a $+5$ oxidation state to a lower one. Because nitrogen can exist in any oxidation state from $+5$ to -3, the low oxidation state actually attained—i.e., the reduction product of nitric acid—depends on several factors, including activity of the reducing agent, concentration of the nitric acid, and temperature. The more common steps can be summarized as follows:

$$\overset{+5}{HNO_3} \rightarrow \overset{+4}{NO_2} \rightarrow \overset{+3}{HNO_2} \rightarrow \overset{+2}{NO} \rightarrow \overset{+1}{N_2O} \rightarrow \overset{0}{N_2} \rightarrow \overset{-3}{NH_3}$$

The most common reduction products evolved when nitric acid reacts with a metal are the gases *nitrogen dioxide* and *nitric oxide*. Because NO_2 is brown and NO is colorless, they are easily distinguished.[8] The more concentrated the nitric acid, the greater the tendency for NO_2 to be formed. Consider copper, a metal that is not attacked by hydrochloric or cold sulfuric acid, because it is not oxidized by hydrogen ions. With concentrated nitric acid the products are NO_2, $Cu(NO_3)_2$, and H_2O:

$$Cu + 4HNO_3 \rightarrow Cu(NO_3)_2 + 2NO_2 \uparrow + 2H_2O$$

When dilute acid is used, the products are NO, $Cu(NO_3)_2$, and H_2O:

$$3Cu + 8HNO_3 \rightarrow 3Cu(NO_3)_2 + 2NO \uparrow + 4H_2O$$

---

[8] But remember that NO reacts with the oxygen in the air to yield NO_2.

Metals above hydrogen in the activity series are readily oxidized by hydrogen ions. Magnesium reacts with hydrochloric acid as follows:

$$Mg + 2HCl \rightarrow MgCl_2 + H_2 \uparrow$$

Oxidation equation:

$$Mg \rightarrow Mg^{2+} + 2e^-$$

Reduction equation:

$$2H^+ + 2e^- \rightarrow H_2$$

When such a metal reacts with nitric acid, we might expect two reduction products, one from the reduction of hydrogen ions (to elemental hydrogen) and the other from the reduction of $+5$ nitrogen (to NO, N_2O, N_2, NH_3, etc.); this is often the case. Usually, however, hydrogen does not make up an appreciable amount of the final reduction products. Rather, they are largely those which arise from the reduction of the $+5$ nitrogen. With dilute nitric acid and with active metals, such as magnesium and zinc, it is possible for the reduction of $+5$ nitrogen to proceed to the formation of ammonia (-3 nitrogen). The ammonia (a base) then reacts with excess nitric acid to form the salt ammonium nitrate, NH_4NO_3:

$$Mg + HNO_3 \rightarrow Mg(NO_3)_2 + H_2O + NH_4NO_3$$

Oxidation equation:

$$4(Mg \rightarrow Mg^{2+} + 2e^-)$$

Reduction equation:

$$NO_3^- + 10H^+ + 8e^- \rightarrow NH_4^+ + 3H_2O$$

Balanced over-all equation:

$$4Mg + 10HNO_3 \rightarrow 4Mg(NO_3)_2 + NH_4NO_3 + 3H_2O$$

It must be understood that writing NO_3^- as the oxidizing agent in nitric acid is a simplification. Solutions of such salts as potassium nitrate, KNO_3, calcium nitrate, $Ca(NO_3)_2$, and iron(II) nitrate, $Fe(NO_3)_2$, have nitrate ions in them, but they are not strong oxidizing agents as are nitric acid solutions, in which the NO_3^- and H^+ ions are both present.

Aqua regia, a mixture of concentrated nitric acid (1 volume) and concentrated hydrochloric acid (3 volumes), is often more effective in dissolving metals and minerals than nitric acid alone. The mixture dissolves gold and platinum, whereas neither of the acids will do so when used separately. The efficiency of this mixture of acids is thought to depend on (1) the strong oxidizing action of nitric acid and (2) the tendency of chloride ions to form stable complex ions.

Phosphoric Acid. Phosphoric acid is the water solution of hydrogen phosphate, H_3PO_4. The anhydrous compound (a viscous, colorless liquid at room temperature) is seldom prepared. The acid solution of any concentration can be made easily by dissolving phosphoric oxide, P_4O_{10}, in water:

$$P_4O_{10} + 6H_2O \xrightarrow{\text{in water}} 4H_3PO_4$$

Soluble phosphates hydrolyze in water to make basic solutions, because the phosphate ion has sufficient proton affinity to take a hydrogen ion away from water:

$$PO_4^{3-} + H_2O \rightleftarrows HPO_4^{2-} + OH^-$$

For this reason, solutions of the soluble phosphates lower the surface tension of water and have a "soapy feel" like sodium hydroxide or potassium hydroxide solutions. Trisodium phosphate, long used as an additive in many soap and detergent powders, is an important plant nutrient and contributes to the *eutrophication* of lakes. Its use in soaps and detergents is now restricted.

EXERCISES

1. By a consideration of both chemical and physical properties of the elements in the nitrogen family, cite evidence to support the statement that the change from nonmetallic to metallic characteristics is quite clear in group VA.
2. On the basis of their properties, account for the fact that nitrogen occurs in nature as the free element while phosphorus does not.
3. Ammonia is formed by the hydrolysis of aluminum nitride, AlN. Write the equation for the reaction.
4. Give the oxidation state of nitrogen in each of the following compounds: HNO_3, KNO_2, N_2O, NH_4Br, N_2O_4, N_2H_4, N_2O_5, Ca_3N_2.
5. Which of the nitrogen family elements most readily forms E^{3-} ions? Why? Which most readily forms E^{3+} ions? Why?
6. Nitrous oxide can be formed by careful thermal decomposition of ammonium nitrate at about 200°C. Write a hypothetical equation for the reaction.
7. Pure ammonia is shipped commercially as a liquid, but hydrogen is shipped as a gas. Give a possible reason for this.
8. (a) Write equations to show how methane, air, and steam could react to give a mixture of N_2, H_2, and CO_2.
   (b) How could the carbon dioxide be separated from the mixture described in (a), leaving nitrogen and hydrogen?
9. Both nitric oxide and nitrogen dioxide are paramagnetic. What does this indicate about their electron arrangements? Propose electron-dot formulas for their molecules.
10. In the "striking" of a safety match (see Fig. 22-5), what is the function of each of the following: (a) powdered glass; (b) red phosphorus; (c) $KClO_3$; (d) paraffin?
11. Write equations for the following reactions:
    (a) combustion of bismuth;
    (b) antimony and chlorine;
    (c) lithium and nitrogen;
    (d) zinc and nitric acid;
    (e) lead and nitric acid (yields NO);
    (f) silver and hot, concentrated nitric acid (yields NO_2).

12. Outline the chemical steps in the production of elemental phosphorus.
13. Ammonium nitrate, although difficult to detonate, is an explosive compound associated with some of the most disastrous peacetime explosions in history. Write a hypothetical equation for the decomposition reaction and suggest an explanation for its exothermic character.
14. What is meant by nitrogen fixation? How does it occur naturally? How does man fix nitrogen?
15. Copper reacts with concentrated nitric acid to produce NO_2 (as one of the products) and with dilute acid to produce NO. Write oxidation and reduction ionic equations for both of these cases.
16. Describe how you could show experimentally that white phosphorus consists of molecules having four atoms.
17. About how many cubic feet of air at 20°C and 740 mm would be required to make 100 lb of ammonia?
18. The dissociation energies of nitrogen, oxygen, and nitric oxide are:

$$N_2 \rightarrow 2N \quad \Delta H = 226 \text{ kcal}$$
$$O_2 \rightarrow 2O \quad \Delta H = 118 \text{ kcal}$$
$$NO \rightarrow N + O \quad \Delta H = 151 \text{ kcal}$$

Calculate the heat energy required to produce 1 lb of NO from the elements.
19. Phosphate rock is not suitable as such as a plant food because of its insolubility in water. One method for overcoming this difficulty is to treat the pulverized rock with sufficient sulfuric acid to convert the insoluble tricalcium phosphate into the soluble monocalcium phosphate. Write the equation for the reaction.
20. What weight of sulfuric acid is required to produce 1 ton of monocalcium phosphate from phosphate rock (see Exercise 19)?
21. Solutions of household ammonia or trisodium phosphate are commonly used in cleaning. Write ionic equations for each to show that there is one ion common to the two solutions.
22. What volume of 0.25 N NaOH would be required to convert 30 ml of 0.125 N arsenic acid to trisodium arsenate?
23. Outline a method of preparing arsenic acid from As_2S_3.
24. What weights of $(NH_4)_2SO_4$, $Ca(H_2PO_4)_2$, and KCl are required to make 50 lb of fertilizer containing 6 percent nitrogen, 8 percent phosphorus pentoxide, and 6 percent potassium oxide or "potash," K_2O? This mixture of salts is commonly called "6-8-6" fertilizer. (Note that P and K percentages are expressed in terms of the oxides, although the oxides are not present as such.)

SUGGESTED READING

Bissey, J. E., "Some Aspects of d-Orbital Participation in Phosphorus and Silicon Chemistry," *J. Chem. Educ.*, **44**, 95 (1967).

Cowley, A. H., "The Structures and Reactions of the Phosphorus Sulfides," *J. Chem. Educ.*, **41**, 530 (1964).

Hirsch, P. R., "Effect of Liquid NH_3 on Wood," *J. Chem. Educ.*, **41**, 605 (1964).

Holmes, R. R., "Ionic and Molecular Halides of the Phosphorus Family," *J. Chem. Educ.*, **40**, 125 (1963).

Lagowski, J. J., "Liquid Ammonia, A Unique Solvent," *Chemistry*, **41** (4), 10 (1968).

23

CARBON AND SILICON

We have studied four nonmetal families of the periodic table in detail—groups VA, VIA, VIIA, and VIIIA. In three of these families there are trends in chemical differences that can be related to differences in the electronegativities of the members. The chemistry of the noble gases is so limited at present that we will exclude them from this discussion of trends. There is a great difference in electronegativity between fluorine and iodine, but still all the group VIIA elements are classified as nonmetals. In the next group, the trend from oxygen to polonium goes from nonmetals to a metalloid element. In group VA the change is even greater—from the nonmetals nitrogen and phosphorus, through the metalloid elements arsenic and antimony, to the metallic bismuth.

In group IVA we find that the elements carbon, C, and silicon, Si, differ so much from the other family members—germanium, Ge, tin, Sn, and lead, Pb—that it is not satisfactory to study the five elements collectively. In fact, the metallic elements in group IVA (Ge, Sn, and Pb) resemble the elements in group IVB (Ti, Zr, and Hf) more than they resemble the nonmetallic carbon and silicon. There is a greater similarity between the A and B families in group IV than in any other group. The trends in the A families are summarized below:

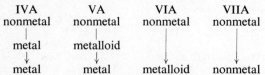

In this chapter we shall take up the two nonmetals carbon and silicon from group IVA. Boron, in group IIIA, is the remaining important element on the right side of the zigzag line that we can draw on a periodic table to divide the nonmetals from the metals. We will mention boron only in passing, but it should be noted that its chemistry resembles in many ways that of silicon. In this respect it is similar to some of the other elements in period 2, which chemically resemble elements in adjacent groups—lithium resembles magnesium, and beryllium is similar to aluminum.

PHYSICAL PROPERTIES

The very high melting and boiling points of carbon and silicon distinguish them from the other nonmetals (see Table 23-1). Both are rigid solids that may

TABLE 23-1.

Physical Properties of Carbon and Silicon

	CARBON C	SILICON Si
Melting point, °C		1410
Boiling point, °C	4347 (sublimes)	2677
Electron distribution	2,4	2,8,4
Ionization energy, ev	11.2	8.1
	(260 kcal)	(190 kcal)
Covalent radius, A	0.77	1.17
Crystal radius, A	0.15 (C^{4+})	0.41 (Si^{4+})
Electronegativity	2.5	1.8

be thought of as giant molecules consisting of a huge number of atoms. Silicon can be made in only one crystalline form, whereas carbon occurs in two well-defined crystalline forms. Both elements can be obtained in one or more amorphous modifications. The common amorphous forms of carbon are charcoal, coke, carbon black, and boneblack.[1]

The crystalline forms of carbon are famous for their physical differences. One, *graphite*, is a soft black substance that actually feels greasy; as a powder it is used as a lubricant, especially for locks. The other, *diamond*, a colorless solid capable of being cut into brilliant crystals, is the hardest, most abrasive mineral known. Yet both these substances consist of only carbon atoms.

In the case of graphite, the atoms crystallize in a pattern of layers (Fig. 23-1). In contrast, the carbon atoms in the diamond structure have strong bonds with

---

[1] Although these forms are often described as amorphous, it has been shown that some of them are microcrystalline (graphite structure).

neighbors in three dimensions, each atom being bound by equally strong covalent bonds to atoms on all sides (Fig. 23-2).

Silicon has an ionization energy and electronegativity that mark it as a borderline element, but the higher values for carbon show it to be a true nonmetal. Both elements are relatively poor conductors of heat and electricity,

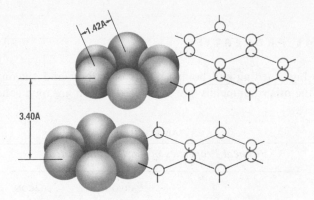

Fig. 23-1. In graphite, carbon atoms crystallize in layers with hexagonal symmetry. The atoms are much closer to their neighbors in the same layer than to atoms in an adjacent layer.

although the graphite form of carbon conducts electricity better than most other nonmetals. To account for the low electrical conductivity of the diamond structure, it is theorized that the electron pairs are held strongly. In the graphite structure the electrons in the bonds between the layers are not held so tightly and hence are freer to move through the crystal and conduct electricity. Because it adheres well to many materials and is a conductor, graphite is often brushed

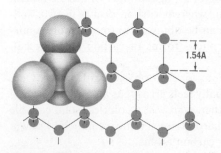

Fig. 23-2. The diamond structure. Carbon atoms crystallize with tetragonal symmetry. Each atom has four nearest neighbors.

over nonconducting surfaces, such as leather or plastic, that are to be electroplated.

The atoms of carbon and silicon are very small compared with other atoms. The ordinary crystal radii of these elements are even smaller, because the atoms are usually in positive oxidation states.

Carbon and Silicon

CHEMICAL PROPERTIES

Carbon and silicon are quite unreactive at ordinary temperatures, especially as large crystals. When they do react, there is no tendency for their atoms to lose outer electrons completely and form simple cations, such as C^{4+} or Si^{4+}. Small ions of this type would have such high charge densities that they would attract electrons from most other particles nearby. Instead of forming simple positive or negative ions, carbon and silicon usually react with other elements by sharing electrons to form covalent bonds.

Reactions with Halogens. Carbon reacts directly with fluorine. Silicon reacts with the halogens, even burning in gaseous fluorine (X refers to a halogen atom):

$$C + 2F_2 \rightarrow CF_4 \qquad \Delta H \text{ is negative}$$

$$Si + 2X_2 \rightarrow SiX_4 \qquad \Delta H \text{ is negative}$$

Common Oxy-Acids. When heated in air, these elements react with the oxygen in highly exothermic combustion reactions.

$$C + O_2 \rightarrow \underset{\text{carbon dioxide}}{CO_2} \qquad \Delta H \text{ is negative}$$

$$Si + O_2 \rightarrow \underset{\substack{\text{silicon}\\\text{dioxide}\\\text{(silica)}}}{SiO_2} \qquad \Delta H \text{ is negative}$$

Both of these oxides are *acidic*. Carbon dioxide reacts with water to give very weak acid solutions. This reaction gives carbonated drinks their slightly sour taste (see Fig. 16-1).

$$CO_2 + H_2O \rightleftarrows \underset{\substack{\text{carbonic}\\\text{acid}}}{H_2CO_3}$$

Silicic acid, H_4SiO_4, also well known, is not formed in appreciable concentration by the direct union of SiO_2 and H_2O. The acidity of a silicic acid solution is hardly measurable because of the extremely low solubility of the acid. However, the solid acid reacts with a base.

$$H_4SiO_4 + 4NaOH \rightarrow \underset{\text{sodium silicate}}{Na_4SiO_4} + 4H_2O$$

When partially dried, silicic acid is called silica gel (a material that looks something like rock salt). In this form it has great adsorbent capacity for vapors of water, sulfur dioxide, nitric acid, benzene, and other substances. It is widely used as a dehumidifier.

Salts of the Oxy-Acids. The salts of the above acids are well known. Carbonic acid, as a typical diprotic acid, reacts with bases to give such carbonates and bicarbonates as

K_2CO_3	Potassium carbonate
$KHCO_3$	Potassium bicarbonate
$MgCO_3$	Magnesium carbonate
$Mg(HCO_3)_2$	Magnesium bicarbonate

Two simple silicic acids are orthosilicic acid, H_4SiO_4, and metasilicic acid H_2SiO_3. Salts of these two acids include:

Na_2SiO_3	Sodium metasilicate
Na_4SiO_4	Sodium orthosilicate
Mg_2SiO_4	Magnesium orthosilicate
$LiAl(SiO_3)_2$	Lithium aluminum metasilicate

All the silicates except those of Na^+, K^+, Rb^+, Cs^+, and NH_4^+ are practically insoluble in water.

All the soluble carbonates and silicates form basic solutions when dissolved in water. The CO_3^{2-} and SiO_3^{2-} ions act as bases to remove protons from water (hydrolysis; see Chap. 16).

$$CO_3^{2-} + H_2O \rightleftarrows HCO_3^- + OH^-$$

A solution of sodium carbonate or sodium silicate turns red litmus blue because of the formation of OH^- ions by reactions similar to the one above.

Formation of Large Molecules. A most important chemical property of carbon and silicon is their tendency to form huge molecules. We should note two differences between silicon and carbon. The first is that carbon atoms join with each other to form a limitless variety of chains or rings of atoms. The second is that carbon atoms tend to form single, double, and triple covalent bonds, whereas silicon tends to form only single bonds. Beginning with Chap. 26, several chapters are devoted to the subject of carbon compounds.

Silicon forms giant molecules and ions in which oxygen atoms occupy alternate positions. These structures are built up in this fashion:

$$-\underset{|}{\overset{|}{Si}}-O-\underset{|}{\overset{|}{Si}}-O-\underset{|}{\overset{|}{Si}}-O-$$

Silicon-containing rocks and minerals are commonly high-melting, hard, brittle

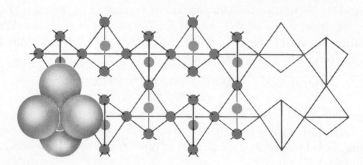

Fig. 23-3. The linking of (SiO_4) tetrahedra in one form of silica, SiO_2.

solids, each piece of which is a continuous lattice of tightly bound atoms. An example of such a solid is silicon dioxide, which occurs in nature in the form of quartz, agate, sand, etc. One SiO_2 crystal structure is shown in Fig. 23-3.

OCCURRENCE, PREPARATION, AND USES

Carbon. Carbon occurs in the earth's crust in both the free and combined states. The most common material that contains elemental carbon is anthracite coal; however, there is also combined carbon in it. Bituminous or soft coal contains some elemental carbon, but it is largely a mixture of complex and ill-defined carbon compounds. Both of the crystalline forms of carbon are found in nature; graphite is found in considerable quantities in deposits in Siberia, Ceylon, Canada, and the United States.

The principal natural compounds of carbon are the organic substances formed in the tissues of living things, both plant and animal, and in materials derived from living things, such as coal and petroleum. Among the common inorganic carbon compounds are carbon dioxide and the carbonate rocks, particularly calcium carbonate, $CaCO_3$. Carbonates of other group IIA elements are well known as minerals: magnesium, strontium, and barium carbonates—$MgCO_3$, $SrCO_3$, and $BaCO_3$, respectively.

Elemental carbon has many uses. *Graphite*, either natural or man-made, is used as the black constituent in ordinary pencil leads, as the pigment in black paints, in the manufacture of crucibles and electrodes to be used at extremely high temperatures, and as a dry lubricant.

Both of the polymorphic forms of carbon vaporize appreciably when heated to about 3500°C. On cooling, the vapors condense in the form of graphite. This is the process for the commercial production of graphite from anthracite coal and from coke. Diamonds could be converted into graphite in this way, a project in which there is little interest.

Diamonds, particularly discolored and small ones, are used in industry for making the hardest abrasive powders, for grinding wheels and the tips of drills and saws. Culminating 125 years of world-wide experimentation, application by General Electric scientists in 1954 of thermodynamic theories led to the commercial production of man-made diamonds. Pressures greater than 1,500,000 psi and temperatures above 2700°C were required. Man-made diamonds are not so large or so brilliantly white as the fine stones found in nature, but in most other characteristics they are identical with natural diamonds. (See Fig. 23-4.)

Carbon black, the sooty product formed when natural gas is incompletely burned, is mainly carbon. Light and fluffy, its extremely small particles are used in printing inks, carbon paper, shoe polish, and paint. When mixed with rubber, it increases the elasticity and durability of rubber so necessary in automobile tires. From 600 to 950 lb of carbon black is required per ton of rubber in tires. Carbon black is made by burning natural gas in a limited amount of air or by the thermal decomposition of natural gas. In the channel process, the smoky flame is directed against a water-cooled rotating iron cylinder, and the carbon that collects is scraped off. In the furnace process, the carbon resulting from an incomplete combustion is separated from the combustion gases in Cottrell precipitators.

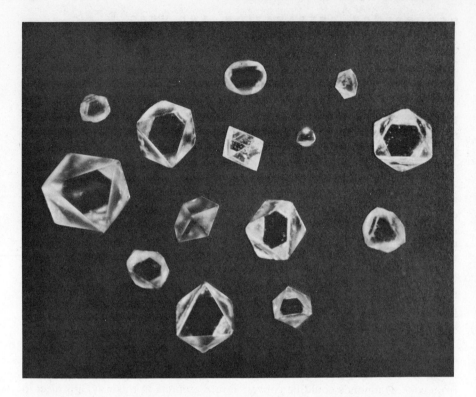

Fig. 23-4. Synthetic diamonds, highly magnified. The average diameter of synthetic diamonds is about 1/10 mm. These diamonds, used primarily in grinding wheels, cost nearly $6000 per pound. (Courtesy of H. Tracy Hall, Brigham Young University.)

Boneblack is made by the destructive distillation[2] of bones. It is composed chiefly of calcium phosphate, $Ca_3(PO_4)_2$, its active ingredient being finely divided carbon that has a great adsorptive capacity. It can be used in filters to adsorb coloring matter (for example, to decolorize crude sugar or molasses) and in gas masks to remove molecules of poisonous gases from the air.

Charcoal and *coke* are made by the destructive distillation of wood and coal,[3] respectively. Both charcoal and coke are used as high-quality solid fuels; also charcoal can be made more useful by "activating" by means of special treatment. Activated charcoal has a greatly increased adsorptive capacity and can be used for the same purposes as boneblack.

Silicon. In spite of its relative inactivity, silicon is not found free in nature. It occurs only in oxy-compounds, such as silica and the silicates. The silicon-oxygen compounds are the most abundant of all the compounds in the earth's crust.

---

[2] Destructive distillation refers to the process wherein a material is heated out of contact with air. As decomposition takes place, volatile products are boiled off, condensed, and collected. Nonvolatile products are left behind as a residue.

[3] Destructive distillation of these two materials produces valuable products other than charcoal and coke. See methyl alcohol in Chap. 27 and coal tar in Chap. 26 for lists of these products.

Most rocks and minerals[4] are silicates with an

$$-\underset{|}{\overset{|}{Si}}-O-\underset{|}{\overset{|}{Si}}-O-$$

lattice. This lattice can be thought of as derived from SiO_2 but with atoms of other elements sometimes attached to the silicon and oxygen atoms and sometimes substituted for these atoms. The formulas and names of some of the more abundant groups of minerals containing silicon are listed in Table 23-2.

TABLE 23-2.

Abundant Silicon-Containing Minerals

MINERAL GROUP	PERCENTAGE OF MINERALS IN EARTH'S CRUST	CHARACTERISTIC STRUCTURE	REPRESENTATIVE FORMULAS AND COMMON NAMES
Feldspars	49	Crystal large in three dimensions (boxlike)	$KAlSi_3O_8$, orthoclase $NaAlSi_3O_8$, albite $CaAl_2Si_2O_8$, anorthite $Na_4Al_3Si_3O_{12}Cl$, sodalite
Quartz	21	Same as above	SiO_2, silica
Amphi-boles, or pyroxenes	15	Crystal large in one dimension (chainlike)	$CaSiO_3$, wollastonite $NaAlSi_2O_6$, jadeite $Ca_2Mg_5Si_8O_{22}(OH)_2$, tremolite (an asbestos)
Mica	8	Crystal large in two dimensions (layerlike)	$KAl_2Si_3AlO_{10}(OH)_2$, muscovite $K_2Li_3Al_4Si_7O_{21}(OH,F)_3$, lepidolite

NOTE: The minerals listed here include only silica and the main silicates; yet they account for 93 percent of the total minerals in the earth's crust. The other 7 percent is comprised of some of the minor silicates and the myriad non-silicon-containing minerals, such as the carbonates, sulfates, sulfides, oxides.

Elemental silicon can be prepared from silica by reduction with magnesium or aluminum.

$$SiO_2 + 2Mg \xrightarrow{heat} 2MgO + Si$$

In its crystalline form silicon is gray or black. It has had few uses; however, it is being used as a component in transistors and in certain of the new solar batteries.

---

[4] The term *mineral* is restricted to naturally occurring, relatively homogeneous inorganic material with characteristic properties and a composition that is within certain limits. *Rocks* may be made of one or more minerals and may be homogeneous or heterogeneous.

COMMON INORGANIC COMPOUNDS OF CARBON

Carbon Monoxide. When carbon-containing fuels (for example, wood, coal, gasoline) are burned in the presence of a great deal of air, practically all the carbon combines with oxygen to make carbon dioxide, CO_2, but a very little *carbon monoxide*, CO, is formed. The latter, however, is present in small amounts in the fumes from practically all fires. The less air (or oxygen) there is available, the greater the relative amount of carbon monoxide formed. Also, at higher temperatures carbon dioxide tends to react with hot carbon:

$$CO_2 + C \xrightarrow{1000°C} 2CO$$

Extremely poisonous, carbon monoxide is especially dangerous because it is colorless and odorless. Even 1 part of the gas in 2000 parts of air causes death in about four hours.

Carbon monoxide forms a compound with the iron in the hemoglobin molecule in the blood. Normally, oxygen is removed from the air in the lungs by chemical combination with the hemoglobin in the red blood cells. The oxyhemoglobin, a bright red compound, is then carried though the arteries to the cells, where the oxygen is utilized in oxidizing food substances to carbon dioxide, water, and other products. The hemoglobin is then carried through the veins back to the lungs, where it again combines with oxygen. This cycle is broken by the presence of carbon monoxide, because the CO-hemoglobin bond is stronger than the O_2-hemoglobin bond. CO-hemoglobin therefore accumulates in the blood stream and less oxyhemoglobin is present to carry on its normal function.

In garages and long tunnels the amount of carbon monoxide may become dangerously high. One of the efforts to reduce air pollution has been the attempt to develop an automobile exhaust pipe device containing a catalyst for the reaction:

$$2CO + O_2 \rightarrow 2CO_2$$

Commercially, carbon monoxide has a number of uses. Mixtures of gases containing it have long been used as fuels; actually, more heat is liberated when carbon monoxide burns to carbon dioxide than when carbon burns to carbon monoxide:

$$C + \tfrac{1}{2}O_2 \rightarrow CO \qquad \Delta H = -26.4 \text{ kcal/mole}$$
$$CO + \tfrac{1}{2}O_2 \rightarrow CO_2 \qquad \Delta H = -67.6 \text{ kcal/mole}$$

Carbon monoxide reduces many metal oxides to their elements:

$$MO + CO \rightarrow M + CO_2$$

This reaction will be mentioned in Chap. 25.

Carbon Dioxide. Carbon dioxide is the common product of the combustion of carbon-containing fuels:

$$\underset{\substack{\text{carbon} \\ \text{(anthracite} \\ \text{coal)}}}{C} + O_2 \rightarrow CO_2 \qquad \Delta H \text{ is negative}$$

$$C_6H_{10}O_5 + 6O_2 \rightarrow 6CO_2 + 5H_2O \qquad \Delta H \text{ is negative}$$
cellulose
(wood)

$$CH_4 + 2O_2 \rightarrow CO_2 + 2H_2O \qquad \Delta H \text{ is negative}$$
methane
(in natural gas)

$$C_7H_{16} + 11O_2 \rightarrow 7CO_2 + 8H_2O \qquad \Delta H \text{ is negative}$$
heptane
(in gasoline)

Carbon dioxide is likewise a component of the breath exhaled by animals, because it results from the oxidation of food in the body. Present in the atmosphere to the extent of about 0.03 percent, its concentration may rise to 1 percent in a crowded room.

Carbon dioxide is not poisonous, but too high a concentration in the air (10 to 20 percent) is unhealthy, because it lowers the oxygen concentration and has harmful physiological effects (unconsciousness, failure of certain respiratory muscles, and a change in the pH of the blood).

There can also be too little carbon dioxide in a person's system. Fear or excitement may make him breathe so rapidly that carbon dioxide is removed from his blood faster than his cells can replenish it. Convulsions or loss of consciousness results if the acidity of the blood is sufficiently altered. The patient may be revived by having a paper bag put over his head. (A triumph of medical science via Dalton's and Henry's laws: As one breathes the same air over and over, the partial pressure of carbon dioxide increases, thereby increasing the amount dissolved in the blood.)

Large amounts of the gas are found in the water of some geysers and mineral springs. The atmosphere in a cave or valley where carbon dioxide seeps out of fissures in the ground may be dangerous. Because the gas has such high density (44 g/mole compared to 29 g/mole for air), it tends to stay in low places. It can be poured from one vessel to another like a liquid; it even remains in an open beaker for a short time before it diffuses out into the air.

Constantly entering the atmosphere in a variety of ways, carbon dioxide is constantly removed by photosynthesis in plants, the formation of carbonate rocks, and the formation of the shells of marine animals. (See Fig. 30-6.)

The most useful properties of carbon dioxide are described below.

It reacts with water to make a dilute solution of carbonic acid, H_2CO_3. Carbonated water—water saturated with carbon dioxide at 3 or 4 atm pressure—is the foundation of our huge soda-water beverage industry. Removing the cap from a bottle of carbonated drink releases the pressure, and some of the dissolved gas *effervesces*, or leaves the solution in the form of tiny bubbles (Fig. 16-1). The solubility of carbon dioxide in water follows Henry's law up to about 5 atm pressure (Chap. 12).

Carbon dioxide does not support combustion. Cylinders of gaseous or liquid carbon dioxide under about 60 atm pressure are used as fire extinguishers. A nozzle attached to the cylinder delivers a stream of the gas or liquid that cools itself by expansion. (A little of it is cooled enough to freeze into tiny crystals, producing a cloud of "snow.") Because the gas is so dense, it settles around the base of a fire and smothers it.

The solid sublimes at −78.5°C under a pressure of 1 atm. Solid carbon dioxide is called dry ice because it vaporizes without first melting. It is a convenient, clean refrigerant, especially useful when subzero temperatures are necessary. Crushed solid carbon dioxide in a Dewar flask of alcohol or acetone makes an excellent cold bath for laboratory work (see Fig. 23-5).

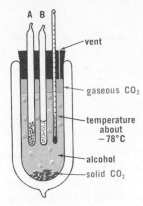

Fig. 23–5. A low-temperature bath of alcohol cooled with solid carbon dioxide. The sealed glass tube A contains solid ammonia which freezes at −77.7°C. Tube B contains liquid propane which condenses at −42.7° but freezes at about −190°C.

The change of state of carbon dioxide with temperature is shown in Fig. 23-6. Below a pressure of 5.3 atm (4030 mm) liquid carbon dioxide does not exist. This behavior can be compared with that of water, which does not exist as a liquid if the pressure is less than 4.6 mm. If some liquid H_2O were thrown into a

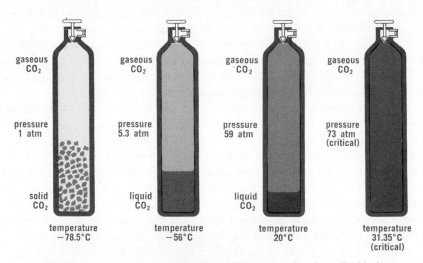

Fig. 23–6. Schematic representation of the state of carbon dioxide in four sealed cylinders at different temperatures.

vacuum chamber in which the pressure was maintained at 4 mm, some of the liquid would evaporate instantly, thereby cooling and freezing the remaining liquid to ice. The ice would then sublime at about 0°C. Strictly speaking, this would be dry ice.

The third tube in Fig. 23-6 shows a typical condition inside a liquid carbon dioxide fire extinguisher at room temperature. Such an extinguisher is shown in use in Fig. 23-7.

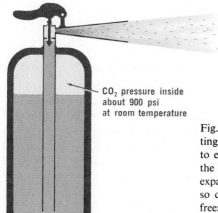

Fig. 23–7. Operation of a carbon dioxide fire extinguisher. When some of the liquid CO_2 is allowed to escape to the low pressure of the atmosphere, the liquid rapidly evaporates and the gas formed expands. These two endothermic changes occur so quickly that some of the CO_2 loses heat and freezes (an exothermic change) to form the small CO_2 crystals visible momentarily as a white fog.

Carbonates and Bicarbonates. As the most abundant inorganic carbon compounds, the carbonates and bicarbonates are both useful and well-known substances. Most carbonates are only slightly soluble in water—for example, calcium carbonate, $CaCO_3$, barium carbonate, $BaCO_3$, magnesium carbonate, $MgCO_3$, and lead carbonate, $PbCO_3$. Many of the bicarbonates are stable only in water solution. Examples are calcium bicarbonate, $Ca(HCO_3)_2$, and magnesium bicarbonate, $Mg(HCO_3)_2$. All group IA metals, except lithium, form soluble salts of carbonic acid, the most inexpensive and useful ones being sodium bicarbonate, $NaHCO_3$ (baking soda), and sodium carbonate, Na_2CO_3.

Because carbonic acid is such a weak acid and is unstable, the carbonates and bicarbonates react with most acids to give CO_2. The reaction is quite rapid, and the gas is evolved readily. For instance, barium carbonate reacts with hydrobromic acid in this way:

$$BaCO_3 + 2HBr \rightarrow BaBr_2 + H_2CO_3$$
$$\downarrow$$
$$H_2O + CO_2 \uparrow$$

A reaction that can be tested in any kitchen occurs when sodium bicarbonate (baking soda) and acetic acid (vinegar) are combined:

$$NaHCO_3 + HC_2H_3O_2 \rightarrow \underset{\text{sodium acetate}}{NaC_2H_3O_2} + H_2CO_3$$
$$\downarrow$$
$$H_2O + CO_2 \uparrow$$

The "fizz" is due to escaping carbon dioxide. Cake rises as a result of being inflated with bubbles of carbon dioxide when the baking soda, $NaHCO_3$, in the baking powder reacts with an acid ingredient in the baking powder. The acid component may be calcium acid phosphate, $Ca(H_2PO_4)_2$, sodium aluminum sulfate, $NaAl(SO_4)_2$, or potassium acid tartrate, $KHC_4H_4O_6$. The sodium bicarbonate and the acid do not react appreciably while in the dry state; but once the baking powder is in the batter, the two dissolve in the water solution and react.

All the reactions above of carbonates and bicarbonates with acids may be summarized by these equations:

$$CO_3^{2-} + H^+ \rightarrow HCO_3^-$$
$$HCO_3^- + H^+ \rightarrow H_2CO_3 \rightarrow H_2O + CO_2 \uparrow$$

As pointed out in Chap. 16, bicarbonates are amphoteric substances that can be used as buffers for solutions, because they react with both acids and bases. Bicarbonates are unstable. When heated, they decompose to form carbonates.

Calcium carbonate is one of the most widely distributed nonsiliceous minerals.[5] The chemistry of this simple compound is both interesting and instructive.

If carbon dioxide is passed into a solution of a cation that forms an insoluble carbonate (for example, Ca^{2+}, Ba^{2+}, Mg^{2+}, or Pb^{2+}), a white precipitate will form. Calcium carbonate precipitates when carbon dioxide is bubbled through a solution of lime water, $Ca(OH)_2$:

$$CO_2 + H_2O \rightleftarrows H_2CO_3 \rightleftarrows 2H^+ + CO_3^{2-}$$
$$Ca^{2+} + CO_3^{2-} \rightarrow CaCO_3 \downarrow$$

Surprisingly enough, if we continue to add carbon dioxide, the precipitate will dissolve. This is contrary to the expectation that if a little carbon dioxide causes a precipitate to form, a lot of carbon dioxide should result in even more precipitate. Actually the solid carbonate dissolves because it is reacting with carbonic acid and forming the *more soluble bicarbonate*.

$$CaCO_3 + H_2CO_3 \rightarrow \underset{\substack{\text{calcium} \\ \text{bicarbonate}}}{Ca(HCO_3)_2}$$

If the solution of $Ca(HCO_3)_2$ is heated, the bicarbonate decomposes and the precipitate reappears:

$$Ca(HCO_3)_2 \xrightarrow{\text{heat}} CaCO_3 \downarrow + H_2O + CO_2 \uparrow$$

Or if the bicarbonate solution is simply allowed to stand in the open air, the calcium carbonate will reappear as the water evaporates:

$$Ca(HCO_3)_2 \xrightarrow{\text{on drying}} CaCO_3 \downarrow + H_2O \uparrow + CO_2 \uparrow$$

CAVE FORMATION. The spectacular limestone caves (such as Mammoth Cave and Carlsbad and Luray Caverns) have been dissolved out of solid rock by the gentle action of the carbonic acid formed by the solution of carbon dioxide in rain water. An original fissure or weakness in the rock probably allowed the first trickles of carbonic acid to seep through. As more and more of the carbon-

---

[5] Although we are not mainly concerned with mineral names, it is of interest to note that a mineral is not named simply on the basis of its chemical composition. For example, there are calcite, aragonite, limestone, marble, travertine, and chalk. Each of these natural substances is chiefly $CaCO_3$, but differences in mode of formation have caused the six to be quite different in appearance and usefulness.

ate was converted into the soluble bicarbonate, the hole grew in size, till there was room for an underground stream or river. Thereafter erosion by the moving water aided the dissolving action of the carbonic acid.

When the cave became open enough to allow a current of air to circulate, the most beautiful changes began to take place. Ground water saturated with calcium bicarbonate seeped through from above to the ceiling of the cave and evaporated in the circulating air. As a result of this evaporation, calcium carbonate was precipitated on the ceiling (*stalactites*). In the places where the concentrated solution dripped to the floor of the cave before evaporating, the carbonate was precipitated on the ground (*stalagmites*). (See Fig. 23-8.)

Fig. 23-8. The broken end of a stalactite that shows rings of impurities precipitated along with the calcium carbonate.

BOILER SCALE. There are other less attractive deposits of calcium carbonate that cost us hundreds of thousands of dollars each year in heating and plumbing bills.

One of the substances usually present in hard water is calcium bicarbonate. When such water is heated in boilers or hot-water tanks, the precipitation of the calcium carbonate forms a scale in the boiler or hot-water pipes. Magnesium carbonate ($MgCO_3$) also forms if magnesium bicarbonate is present in the water. Heat transfer becomes less efficient as this scale thickens. Eventually, a pipe or a boiler may become almost completely blocked with hard, rocklike carbonate scale.

COMPOUNDS OF SILICON

Silicon Dioxide. Silicon dioxide, or silica, is one of the most common chemical compounds. Pure SiO_2 crystals are found in nature in three polymorphic forms, the most common of which is *quartz*. Sand, agate, onyx, opal, amethyst, and flint are silicon dioxide with traces of impurities.

Fused quartz is used to make crucibles and other laboratory vessels that must be heated to extremely high temperatures. Not only does it have a high softening point (about 1500°C) but another equally valuable property is its very low coefficient of thermal expansion. A substance with a high coefficient of thermal expansion expands and contracts a great deal when heated and then cooled. Quartz expands very little when heated; therefore, it is not likely to crack even if it is cooled rapidly and unevenly.

The forms of silica are some of the truly important crystal structures, not only because silica itself is such an abundant and useful substance, but because the (SiO_4) structure is the fundamental unit in most minerals. As is evident in Fig. 23-3, SiO_2 crystals have two main features: (1) each silicon atom is at the center of a tetrahedron of four oxygen atoms, and (2) each oxygen atom is midway between two silicon atoms. One way of looking at the quartz structure is to

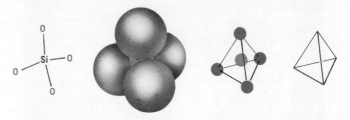

Fig. 23-9. Four ways of representing (SiO_4) tetrahedra in silica and silicates.

picture the tetrahedron (SiO_4) as in Fig. 23-9 and then picture each of the oxygens as a member of another tetrahedron. Billions upon billions of such tetrahedra are tightly linked, in three dimensions, in each grain of sand on the beach or in the desert.

Attention is often called to the great differences between SiO_2 and CO_2 in physical properties. The former does not soften till heated to about 1500°C; the latter sublimes at −78°C. This and other differences can be correlated with the types of molecules, which in turn depend on the types of bonds, Si—O single bonds or C=O double bonds.

Carbon unites with oxygen by forming two covalent bonds with each oxygen, O=C=O. Carbon dioxide is made up of tiny triatomic molecules and is a gas at room temperature. In contrast, the endless

$$-O-\underset{|}{\overset{|}{Si}}-O-$$

structure in silicon dioxide, Fig. 23-3, is a single molecule, whether a grain of sand or a magnificent quartz crystal larger than a man's head.

Silicon dioxide is an acidic oxide. Though it is practically insoluble in water, it does dissolve in such bases as sodium hydroxide. One of the few other compounds that attacks it is hydrogen fluoride, which reacts to produce the volatile substance SiF_4.

FAMILIAR SILICON-CONTAINING MATERIALS

Two of man's oldest industries are pottery and brick making. Sometime later he learned to make mortar, plaster, and cement. In spite of having mastered the art of cement and ceramics long ago, we still have much to learn about the science of these important materials. However, the knowledge we do have indicates that the

$$-\overset{|}{\underset{|}{Si}}-O-\overset{|}{\underset{|}{Si}}-O-$$

bonds are probably responsible for the strength of both cement and ceramics.

Ceramics. Ceramic products are made of mixtures of various finely divided minerals and rocks that will form a strong rocklike mass when heated to a high temperature. The most important ingredient is *clay*, a naturally occurring material that is formed by the action of the weather upon certain feldspars. Although clay is not a pure substance, it is chiefly *kaolinite*, $Al_2Si_2O_5(OH)_4$, a soft, easily pulverized mineral. Kaolinite is a hydrated aluminum silicate.

Cement. Cement is a material whose rocklike strength is due to silicon-oxygen bonds. It becomes hard as the result of a low temperature reaction, instead of the high temperature used in the ceramic kiln.

Portland cement, first made in the early 1800s in England (and named for its similarity to a natural rock mined on the Isle of Portland), can be made from limestone, clay, and gypsum. Other materials, such as blast furnace slag or iron ore, may be used, depending partly on what is available in the locality of the cement plant.

Glass. The glassy state is something between the true crystalline solid and the liquid state. Unlike a crystalline material, glass does not have a sharp melting point but begins to soften far below the temperature at which it flows like a liquid.

At temperatures slightly above the softening point, glass can be bent easily into almost any desired shape. At somewhat greater temperatures, when it begins to behave like a very viscous liquid, small pieces of the molten glass are drawn into spherical drops by surface tension, and two pieces of molten glass will join and become a single piece. In this melted condition it can be blown, molded, or rolled into sheets.

Figure 23-10 compares a hypothetical crystalline and a glassy structure. The building units are roughly the same; but in the crystal they are arranged with perfect symmetry, whereas in the glass they have a random arrangement and are distorted.

Although it appears to be rigid and brittle, glass will bend at room temperature if put under strain and will eventually be deformed, taking a new shape till the stress is relieved. Filaments of glass bend so readily that they can be spun into cloth; encased in plastic, they are used to make the important new structural material fiberglass.

Glass is so inexpensive and common that it is surprising to learn that the glassy state is really quite rare. Only one element (selenium), a few oxides (such as B_2O_3, SiO_2, GeO_2, and P_2O_5), and a few oxy-salts (such as the borates, silicates, and phosphates) exist in a glassy form. Neither simple ionic nor simple molecular substances form glasses.

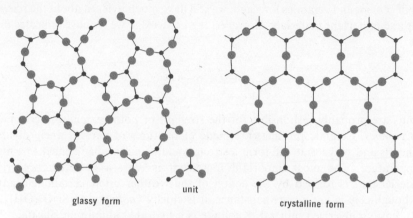

Fig. 23-10. A schematic diagram showing how the same structural units can join in a random way, forming glass, or how they can join in perfect symmetry, forming a crystal. (Redrawn by permission from A. F. Wells, *Structural Inorganic Chemistry*, The Clarendon Press, Oxford, 1945.)

Lime glass, exceeding all other types in tons produced, is used as window glass, plate glass, and bottle glass. It is essentially a mixture of sodium and calcium silicates, with an excess of silica.

Pyrex is a borosilicate glass with a very low coefficient of thermal expansion, hence its wide use for cooking and serving utensils. Moreover, it can be shaped and blown by amateur glassblowers. For this reason, Pyrex is a great boon to research chemists and physicists who often need to make special glass equipment.

Hundreds of different glasses are made: *lead glass* for fine cut glassware; *colored glass* for beauty (for example, selenium or gold added to make it red; cobalt compounds for blue; chromium or copper compounds for green); *tempered glass* for great strength.

Silicones. Some of the most interesting synthetic compounds, unlike anything found in nature, are the *silicones*. These substances are chainlike molecules of Si, O, C, and H atoms. Methyl silicone,[6] shown in the formula below, is an example:

$$\begin{array}{ccc} CH_3 & CH_3 & CH_3 \\ | & | & | \\ -Si-O-Si-O-Si-O- \\ | & | & | \\ CH_3 & CH_3 & CH_3 \end{array}$$

---

[6] The methyl group, —CH_3, is the simplest carbon-hydrogen group that can be added to the side of the Si atom. For the formulas of more complex groups, such as —C_2H_5, ethyl —C_3H_7, propyl, see Chap. 26.

As the molecular weight changes, the properties change. A silicone made of short chain molecules is an oily liquid; silicones with medium-length chains behave as viscous oils, jellies, and greases; those with very long chains have a rubberlike consistency.

Other useful properties of the silicones are their resistance to chemical attack and their water-repellent nature (i.e., they are *hydrophobic*). Textiles and wood and metal surfaces can be waterproofed with a thin silicone coating; furthermore, it is tough and long-lasting.

Zeolites. The *zeolites* are a class of silicate minerals in which a pair of ions—for example, Al^{3+} and K^+—has been substituted for one Si^{4+}. The trivalent Al^{3+} takes the place of the Si^{4+} in the center of an SiO_4^{4-} tetrahedron, and the monovalent K^+ fits nearby in a hole in the crystal structure. Figure 23-11 shows

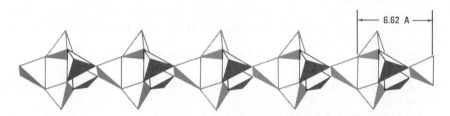

Fig. 23–11. A possible arrangement for the naturally occurring zeolite natrolite, $Na_2(Al_2Si_3O_{10}) \cdot 2H_2O$. Each colored tetrahedron has an Al atom at the center, whereas each gray one has an Si atom. The order in the natural crystal is not necessarily as regular as here, but the ratio is 3:2. Chains like this are crosslinked with others in three dimensions; Na^+ ions and water molecules fit into holes in the lattice.

the structure of *natrolite*, a zeolite in which Na^+ ions fit in the holes; the formula can be written $Na_2(Al_2Si_3O_{10}) \cdot 2H_2O$ to emphasize the freedom and ionic character of the sodium ions. The zeolites are characterized by a porous structure through which water can circulate rather freely.

The great use of the zeolites is in water softening. Certain ions, such as calcium and magnesium, are objectionable in water, because they form insoluble precipitates (curd or scum) with soaps. When water is allowed to trickle slowly through a bed of crushed zeolite, the Ca^{2+} and Mg^{2+} ions in solution tend to be attracted to the mineral; K^+ or Na^+ ions leave the zeolite and take the place of the divalent ions in solution. (One Ca^{2+} ion will displace two K^+ ions.) Such a process is called **ion exchange.** In this way Ca^{2+} and Mg^{2+} ions are exchanged for the unobjectionable Na^+ and K^+ ions.

EXERCISES

1. Describe the complete sublevel electronic structure of carbon and silicon atoms. Why do they not tend to form simple cations?
2. Carbon and silicon have very high melting and boiling points which distinguish them from other nonmetals. Account for these high values for carbon and silicon.

3. Compare the physical properties of diamond and graphite. Account for their differences on the basis of structure.
4. Outline a possible series of natural processes by which carbon in atmospheric carbon dioxide could eventually be converted to diamond.
5. What are some structural similarities between diamond and silica?
6. Name the following acids or salts: $Al_4(SiO_4)_3$, Na_2SiO_3, H_3BO_3, $MgCO_3$, $Ca(HCO_3)_2$, H_4SiO_4, HBO_2, $NaPO_3$.
7. What is the composition of baking powders? Discuss how they work.
8. Complete and balance the following equations:
   (a) $Si + Cl_2 \rightarrow$
   (b) $CO_2 + C \rightarrow$
   (c) $CaCO_3 + H_2CO_3 \rightarrow$
   (d) $H_4SiO_4 + Ca(OH)_2 \rightarrow$
   (e) $K_2CO_3 + HCl \rightarrow$
   (f) $SiO_2 + Al \rightarrow$
9. Fossil fuels (petroleum and coal) contain appreciable amounts of sulfur which upon combustion produces sulfur dioxide. It has been suggested that this component of the various air pollutants leads to a rapid deterioration of limestone and marble statues, particularly in metropolitan areas. Write equations to account for this corrosive action.
10. When liquid carbon dioxide issues from a fire extinguisher, some of it turns to gas, some to solid. Explain.
11. In localities where soil tends to be acidic, crushed limestone is often applied regularly to "sweeten" the soil.
    (a) Why might the soil be acidic?
    (b) Why would limestone lessen this acidity?
    (c) How is the term "sweeten" appropriate for this treatment?
12. If one blows his breath into lime water through a straw, the liquid becomes milky and turbid. If the blowing is continued for some time, the liquid becomes a clear solution again. Explain with equations.
13. Suppose a home hot-water tank is heated with an electrical immersion rod. Is a scale of calcium carbonate more likely to form in the water entry pipe, on the rod, on the walls of the tank, or in the water exit pipe? Explain.
14. (a) Show with at least four specific balanced equations the amphoteric character of sodium bicarbonate.
    (b) On the basis of the equations you have just written, explain why a solution of sodium bicarbonate is often kept handy in chemical laboratories as a safety precaution.
15. (a) Suppose a company has as waste product a large volume of dilute sodium hydroxide solution it wishes to neutralize before discharging it into a nearby river. Would it be possible to neutralize the solution by bubbling through it exhaust gases formed by burning coal? Explain with equations.
    (b) Repeat (a) for a waste product of calcium hydroxide solution.
16. Explain, using a crude diagram, the mechanism whereby carbon monoxide poisons a person. Cyanide poisoning is very similar to carbon monoxide poisoning. Show with electron-dot formulas that CO and CN^- have similar structures.
17. The enthalpy change for the process $C(s) + O_2 \rightarrow CO_2$ is different for graphite and diamond. Account for this fact. Which releases the larger amount of heat, graphite or diamond?
18. Describe in detail how man's activities tend to increase the amount of carbon dioxide in the atmosphere. Although there is evidence that the carbon dioxide content of the air is increasing (see Chap. 30), the increase is extremely slow. Why? Does Henry's law have a bearing on this discussion? Explain.

19. Show how the use of ammonium nitrate as a fertilizer in mica-containing soil could increase the supply of available potassium.
20. An unskilled glass worker may be able to heat and successfully reshape a tumbler made of Pyrex glass, but probably not one of ordinary glass. Why?
21. Calculate the amount of heat energy liberated by the combustion of (a) 100 lb of carbon to carbon dioxide; (b) 100 lb of carbon monoxide to carbon dioxide.
22. How could one analyze a sample of red glass to determine if it contained gold or selenium?
23. It has been claimed that cigarette smoking can interfere with physical and mental activity by decreasing the amount of oxygen available to the cells. Can you suggest possible chemical reactions consistent with this claim?
24. Calculate the heat energy released by the combustion of 10.0 liters of a 50 : 50 mixture of carbon monoxide and carbon dioxide.
25. Buttermilk contains lactic acid, $HC_3H_5O_3$. Write the equation for the reaction that occurs when buttermilk and baking soda are used in baking.

SUGGESTED READING

Brill, R. H., "Ancient Glass," *Sci. Amer.*, **209** (5), 120 (1963).

Brunauer, S., and L. E. Copeland, "The Chemistry of Concrete," *Sci. Amer.*, **210** (4), 80 (1964).

Dalton, R. H., "Recent Developments in Glass," *J. Chem. Educ.*, **40**, 99 (1963).

Dingledy, D., "Flow of Glass under Its Own Weight," *J. Chem. Educ.*, **39**, 84 (1962).

Encke, F. L., "The Chemistry and Manufacturing of the Lead Pencil," *J. Chem. Educ.*, **47**, 575 (1970).

Krishnamurty, K. V., "Carbon Trioxide," *J. Chem. Educ.*, **44**, 594 (1967).

Massey, A. G., "Boron," *Sci. Amer.*, **210** (1), 88 (1964).

Payne, D. A., and F. H. Fink, "Electronegativities and Group IVA Chemistry," *J. Chem. Educ.*, **43**, 654 (1966).

Rainey, R. G., "Making Baking Powder Biscuits," *J. Chem. Educ.*, **39**, 363 (1962).

Tee, P. A. H., and B. L. Tonge, "The Physical and Chemical Character of Graphite," *J. Chem. Educ.*, **40**, 117 (1963).

24

THE TRANSITION METALS AND THEIR NEIGHBORS

Of the slightly over 100 known elements, about 80 are classified as metals. Twelve of these, comprising the IA and IIA families were discussed in Chap. 18. Among the remaining elements are the well-known metals aluminum, iron, chromium, nickel, lead, tin, copper, silver, gold, mercury, and platinum. These, as well as the less familiar ones, are not nearly as active chemically as the IA and IIA metals, so that they may be used as elemental metals in contact with water, oxygen, and carbon dioxide without a rapid chemical change occurring. Indeed, some do not react at all with these substances under normal conditions.

As was pointed out in Chap. 18, in each of the long periods beginning with Group III there is a series of elements called the *transition* elements. These elements, and some of their neighbors just to the right in the periodic table, are the metals whose chemical and physical properties are to be discussed. We will find that their chemistry is more complex than that of the metals in Groups IA and IIA.

CLASSIFICATION

On the basis of their position in the periodic table, the metals under discussion are often grouped into families as follows:

Transition and Inner Transition Metals. The *scandium family*: scandium, Sc, yttrium, Y, lanthanum, La (and the lanthanide series), and actinium, Ac (and the actinide series).

The *titanium family:* titanium, Ti, zirconium, Zr, and hafnium, Hf.
The *vanadium family:* vanadium, V, niobium, Nb, and tantalum, Ta.
The *chromium family:* chromium, Cr, molybdenum, Mo, and tungsten, W.
The *manganese family:* manganese, Mn, technetium, Tc, and rhenium, Re.
The *iron family:* iron, Fe, cobalt, Co, and nickel, Ni.
The *platinum family:* ruthenium, Ru, rhodium, Rh, palladium, Pd, osmium, Os, iridium, Ir, and platinum, Pt.

Neighbors of Transition Metals. The *copper family:* copper, Cu, silver, Ag, and gold, Au.

The *zinc family:* zinc, Zn, cadmium, Cd, and mercury, Hg.

The *aluminum family:* aluminum, Al, gallium, Ga, indium, In, and thallium, Tl.

The *germanium family:* germanium, Ge, tin, Sn, and lead, Pb.

Antimony, Sb, and bismuth, Bi, although metallic in character, are included in the nitrogen family (Chap. 22), and polonium, Po, is classified in the sulfur family (Chap. 21).

Metallo-Acid Elements. The titanium, vanadium, chromium, and manganese families have several common characteristics, one of which is that they form metallic oxides that are acidic:

$$CrO_3 + H_2O \rightarrow H_2CrO_4$$
chromic oxide → chromic acid

Because this discussion can be simplified by grouping similar metals together, we shall use the term *metallo-acid elements*, a name based on their acid-forming characteristics, to refer to the 12 metals in these four families.

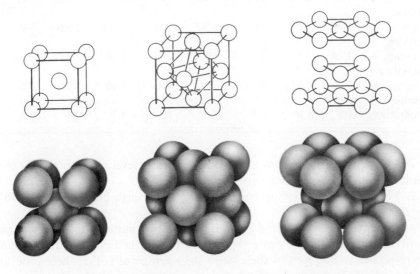

Fig. 24-1. The three common types of metal structures. From left to right: body-centered cubic, face-centered cubic, and hexagonal close-packed.

PHYSICAL PROPERTIES

The transition elements and their neighbors exhibit typical metallic properties. In general, they are *malleable, ductile,* and *good conductors of heat and electricity,* and they show a *metallic luster.*

The structures of these metals are of the close-packed types. The atoms are stacked like balls into a body-centered, or face-centered, or hexagonal close-packed structure (see Fig. 24-1). Except for copper and gold, which have characteristic colors, all the metals are similar in appearance, resembling tin or iron to some extent.

The properties of the elements in the first transition series are summarized in Table 24-1. A number of trends that are obvious in this table are typical of transition elements:

TABLE 24-1.

Physical Properties of Elements of First Transition Series

	Sc	Ti	V	Cr	Mn	Fe	Co	Ni
Electron configuration	2,8, 9,2	2,8, 10,2	2,8, 11,2	2,8, 13,1	2,8, 13,2	2,8, 14,2	2,8, 15,2	2,8, 16,2
Density, g/ml	3.1	4.43	6.07	7.19	7.21	7.87	8.70	8.90
Melting point, °C	1400	1812	1730	1900	1244	1535	1493	1455
Boiling point, °C	3900	3277	3377	2642	2087	2800	3100	2800
Covalent radius, Å	1.44	1.32	1.22	1.17	1.17	1.16	1.16	1.15
Ionization energy, ev	6.56	6.83	6.74	6.76	7.43	7.90	7.86	7.63
Electronegativity	1.3	1.6	1.7	1.6	1.5	1.7	1.7	1.8

1. The number of electrons in the outside energy level tends to remain constant (although there is some variation).

2. The melting and boiling points are uniformly high but do not vary in a regular way.

3. The covalent radius of the atoms is remarkably constant. In the second and third series these values increase slightly, starting at rhodium, Rh, and iridium, Ir, respectively. But even in these cases the main point is that these radii are very nearly the same (see Fig. 18-1).

4. The ionization energy tends to increase slightly as the positive charge on the nucleus (atomic number) increases.

These four points reveal strong "horizontal similarities" between adjacent transition elements.

There are even greater similarities among the members of the first inner transition series, the lanthanides. Some of the lanthanides, especially the adjacent ones, are so nearly identical in chemical and physical properties that it is difficult to tell their compounds apart. It is extremely tedious to separate the individual elements from the mixtures of compounds found in nature, because they react in similar ways to all ordinary physical or chemical forces. Undoubtedly this similarity is due to the presence of the same number of electrons in both the outside and the first inner energy levels.

The melting points, boiling points, and densities for the second and third transition series have the same general trends as those noted above for the first series. However, the melting points and densities of the second transition series are generally higher than those for corresponding members of the first series; those for the third series are the highest of all. For example, osmium in the third series has the greatest density (22.6 g/ml) of all elements—or, for that matter, of all known substances. Tungsten and rhenium in the same series have the highest boiling points of all the elements (W, 5680°C; Re, 5630°C). Their melting points are exceeded only by that of carbon.

TABLE 24-2.

Neighbors of transition elements

	ELECTRONS	DENSITY, G/ML	MELTING POINT, °C	BOILING POINT, °C
Family IB:				
Copper	2,8,18,1	8.9	1087	2582
Silver	2,8,18,18,1	10.5	961	2193
Gold	2,8,18,32,18,1	19.3	1063	2710
Family IIB:				
Zinc	2,8,18,2	7.1	420	907
Cadmium	2,8,18,18,2	8.6	321	767
Mercury	2,8,18,32,18,2	13.5	−39	357
Family IIIA:				
Aluminum	2,8,3	2.7	660	2327
Gallium	2,8,18,3	5.9	30	1983
Indium	2,8,18,13,3	7.3	157	2000
Thallium	2,8,18,32,18,3	11.8	304	1457
Family IVA:				
Germanium	2,8,18,4	5.3	960	2830
Tin	2,8,18,18,4	7.3	232	2340
Lead	2,8,18,32,18,4	11.4	328	1744

Table 24-2 shows certain properties of the neighbors of the transition metals. The densities here range from that of the light metal, aluminum (2.7), to that of one of the denser metals, gold (19.3). The melting points range from those of the lowest melting metals, mercury and gallium, to those of metals with intermediate values, copper, silver, and gold.

CHEMICAL PROPERTIES

Chemical Activity. The general activity of the metals in the middle of the periodic table varies greatly but is always less than that of the alkali and alkaline earth metals. For example, aluminum and zinc are quite active and iron and lead moderately so. However, silver, gold, and the platinum metals are inactive to the point of semi-inertness and so are referred to as *noble metals*. The relative activity of metals can be estimated by comparing their electrode potentials (Table 20-2):

1. The group IIIB elements and the members of the lanthanide and actinide series are very active, with oxidation potentials usually below that of sodium (2.71 volts) but above that of aluminum (1.66 volts).

2. The metallo-acid elements are moderately active, with oxidation potentials usually below aluminum (1.66 volts) but above hydrogen (0.00 volt).

3. The three members of the iron family are also moderately active. Comparable values of oxidation potentials are iron (0.44 volt), cobalt (0.28), and nickel (0.25).

4. The platinum metals are chiefly notable for their chemical inactivity, and their oxidation potentials are in line with this behavior. All are below copper in the activity series, with oxidation potentials ranging from ruthenium (-0.45 volt) down to iridium (-1.15 volts).

5. The neighbors of the transition metals range widely in activity from aluminum to gold. A number of them are listed in Table 20-2: aluminum (1.66 volts), zinc, tin, lead, copper, mercury, silver, and gold (-1.50 volts).

Oxidation States. A noteworthy characteristic of the transition metals and their neighbors is that most of them tend to show several oxidation states. This is in contrast to the alkali and alkaline earth metals (groups IA and IIA), which form cations only of $+1$ and $+2$ charge, respectively. The most common oxidation states for the members of the first transition series (Sc through Ni) are indicated in Fig. 6-1. There are many possible oxidation states not indicated; for instance, the $+6$ state of iron, or the $+2$, $+3$, and $+4$ states of vanadium.

Some important generalizations applicable to all three transition series can be made. Reference to Fig. 6-1 will show how these apply to the first series.

1. None of these elements has a common oxidation state of less than $+2$.

2. Each of the elements in groups IIIB to VIIB can exhibit the appropriate maximum oxidation state for its group. Examples of compounds include titanium dioxide, TiO_2, chromium trioxide, CrO_3, and rhenium heptoxide, Re_2O_7.

3. Most of the elements in group VIIIB have $+4$ as their maximum oxidation state. But two of the nine elements have an oxidation state of $+8$, the highest shown by any element (osmium oxide is OsO_4).

By a suitable choice of oxidizing agents and of concentration and temperature, an element can be made to assume any of its possible oxidation states. We must emphasize that for oxidation states of +4 and over, the ions are not simple. Indeed, no discrete ions may be formed; rather, the small highly charged particle will form covalent bonds with other molecules or ions if they are available. Examples include osmium oxide, OsO_4, wherein the osmium has a calculated state of +8, even though the Os^{8+} ion does not exist as a separate particle. A similar case is potassium dichromate, $K_2Cr_2O_7$, in which chromium has a calculated state of +6.

The neighbors of the transition elements exhibit positive oxidation numbers corresponding to their group numbers in the periodic table. However, they exhibit additional oxidation numbers, particularly those toward the end of a group, as is seen in Table 24-3. Note that the lower members of Groups IIIA

TABLE 24-3.

Common Oxidation States of the Neighbors of the Transition Elements

IB	IIB	IIIA	IVA	VA	VIA
		Al +3			
Cu +1 +2	Zn +2	Ga +3	Ge +2 +4		
Ag +1 +2	Cd +2	In +3	Sn +2 +4	Sb −3 +3 +5	
Au +1 +3	Hg +1 +2	Tl +1 +3	Pb +2 +4	Bi −3 +3 +5	Po −2 +2 +4 +6

through VA have oxidation numbers that are two less than the group number. It has been suggested that this is due to a lesser tendency of the outermost s electrons to become involved in bond formation. To illustrate this point, consider $PbCl_2$ and $PbCl_4$. The outermost main energy level of lead contains four electrons, $6s^2$ and $6p^2$. It is thought that in $PbCl_2$ only the $6p^2$ electrons are involved, while in $PbCl_4$, both $6s^2$ and $6p^2$ electrons are involved in bond formation.

Note also in Table 24-3 that Sb, Bi, and Po can have negative oxidation states. These metals lie along the zigzag line in the periodic table that separates the metals and nonmetals, hence, they have certain nonmetallic characteristics, one of which is the tendency to exhibit negative oxidation states in compounds that contain more electropositive elements.

Corrosion. Closely related to the chemical activity of the metals is the phenomenon of **corrosion,** the chemical attack on a metal by its environment. The most common type of corrosion is due to the action of the atmosphere in conjunction with water and various substances dissolved in the water. Essentially, corrosion is a reaction in which a metal is oxidized. In most cases the reaction is definitely electrochemical in nature and involves the type of electron transfer that is characteristic of voltaic cells (batteries).

Moisture is necessary for the atmospheric corrosion of all but the most active metals. If salts are dissolved in the moisture, the corrosion is speeded up, presumably because a solution of electrolytes is an ideal medium for electrochemical oxidation-reduction. The presence of any substance that dissolves and forms an acid solution—for example, sulfur dioxide or carbon dioxide—usually increases the rate of corrosion. Oxygen that is dissolved in water is an important corrosion agent.

In general, metals high in the electrochemical series, that is, those which have large positive oxidation potentials, corrode easily. In this category are the group IIIB elements and the lanthanides (and also the alkali and alkaline earths, which corrode rapidly even in dry air). Metals low in the series—the noble metals—do not corrode easily.

The position of a metal in the electrochemical series is not the only factor that determines the extent or rate of corrosion. Just as important is the type of film or coating that is formed on the surface of the metal. Aluminum and magnesium actually react quickly when exposed to air; but the thin, closely packed film of oxide that forms on the surface protects the underlying metal from further corrosion. These two metals corrode less completely in air than the less active metal iron. The rust formed on the surface of iron is so flaky and porous that the corroding chemicals can pass through it easily and attack the underlying metal.

The formation of a protective film explains the paradox of galvanized iron. Iron is protected from corrosion by being coated with the more active metal zinc. Zinc reacts readily with moisture, oxygen, and carbon dioxide, but it forms a tough film of basic carbonate, $Zn(OH)_2 \cdot ZnCO_3$, that resists further attack.

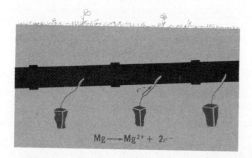

Fig. 24–2. Cathodic protection of an underground pipeline.

The rusting of underground pipes may necessitate costly repairs or even replacement. One ingenious method of preventing corrosion of iron pipes is *cathodic protection.* Pieces of an active metal, such as magnesium, are buried in the ground near the pipe and connected to it by a wire. Instead of the iron losing its own electrons directly to the oxidizing agents (corrosion agents) that attack it, it merely relays, via the wire, electrons from the more active metal. The slug of active metal corrodes away; but the costly pipeline is protected. A cathodic protection system is diagramed in Fig. 24-2.

Color of Compounds. One of the most characteristic properties of compounds is their color. Color is not a trivial property, but is related to the electronic structures of the molecules or ions present in the compound.

The transition elements are noted for the fact that they form colored compounds. This behavior is often associated with the excitation of d electrons in partially filled sublevels. Radiation in the visible range, between about 4000 A and 7500 A, is of the proper energy to excite d electrons to higher energy levels. Another factor related to color is the coexistence of two oxidation states of an element. Some of the most intensely colored materials are those containing different ions of the same element. Blueprint dye, for example, contains Fe^{2+} and Fe^{3+} in complex ions. Having ions that can easily exchange electrons evidently increases the number of electronic excitations that absorb wavelengths of visible radiation.

The colors of some of the ions of the transition elements and other elements are listed in Table 24-4. Note that the color is often different for different oxidation

TABLE 24-4.

Colors of Ions in Dilute Water Solutions

Colorless	Na^+, Mg^{2+}, Sc^{3+}, Ti^{4+}, Cu^+, Zn^{2+}, Al^{3+}
Pink	Co^{2+}, Mn^{2+}
Yellow	Fe^{3+}, CrO_4^{2-}, Au^{3+}, Au^+
Green	Fe^{2+}, V^{3+}, Ni^{2+}
Blue	Cr^{2+}, Cu^{2+}
Purple	MnO_4^-, Ti^{3+}
Violet	Mn^{3+}, Cr^{3+}, V^{2+}

states of the same element. The color is also modified by the formation of hydrates, ammoniates, and other complex ions. An interesting example is afforded by Co^{2+} complex ions. A dilute aqueous solution of cobalt(II) chloride has the pale pink color characteristic of the complex ion $Co(H_2O)_6^{2+}$. This solution is sometimes used as an "invisible" ink. Upon warming, the writing becomes visible because of the formation of anhydrous $CoCl_2$, which is deep blue. The addition of considerable hydrochloric acid to pink cobalt chloride solutions changes the color to blue because of the formation of the complex chloride, $CoCl_4^{2-}$:

$$\underset{\text{pink}}{Co(H_2O)_6^{2+}} + 4Cl^- \rightarrow \underset{\text{blue}}{CoCl_4^{2-}} + 6H_2O$$

Another common example involves Cu^{2+}. Anhydrous copper sulfate, $CuSO_4$, is colorless. When it is dissolved in water, the characteristic blue color of hydrated copper ions, $Cu(H_2O)_4^{2+}$, appears. If excess chloride ions (from HCl or other very soluble chlorides) are dissolved in the solution, the color becomes green. The green color is attributed to an equilibrium mixture of blue and yellow ions:

$$Cu(H_2O)_4^{2+} + 4Cl^- \rightleftarrows CuCl_4^{2-} + 4H_2O$$
$$\text{blue} \qquad\qquad\qquad \text{yellow}$$

COORDINATION COMPOUNDS

Metals form practically all the simple compounds with nonmetals that we expect, with the metals existing in all their varied oxidation states. In addition,

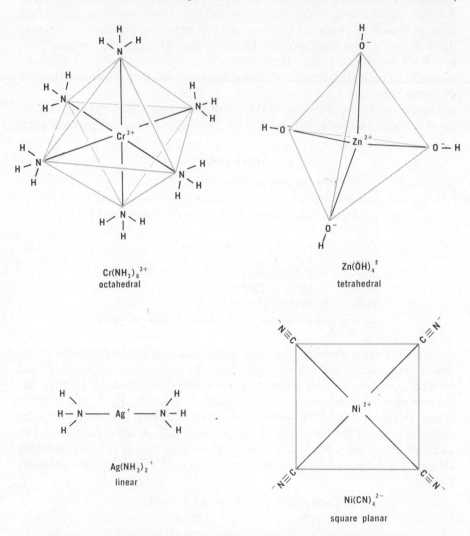

Fig. 24-3. Examples of complex ions. The ligand-metal bonds are denoted by lines and the over-all geometry of the complex is shown in color.

many transition metal atoms or ions form molecules or complex ions by combining via covalent bonds with species that donate both the electrons. The atoms, ions, or groups attached to the central metal are called **ligands** (Latin *ligare*, to bind). Examples are given in Fig. 24-3. Compounds containing such molecules or complex ions are called **coordination compounds.** Few compounds of the transition metals contain a simple positive ion. Many anhydrous compounds that are usually referred to by simple empirical formulas, such as $FeCl_3$ and $Cu(NO_3)_2$, are best viewed as coordination compounds in which the metal ions are covalently bound to the negative species (see Fig. 24-4). Elements other than

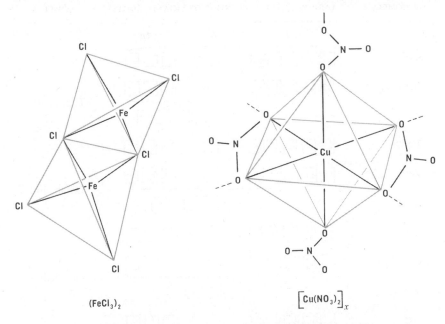

Fig. 24-4. Representation of the structure of two anhydrous transition-metal compounds. The structures show that these compounds do not contain simple positive and negative ions.

the transition elements can form coordination compounds, particularly if ions of these elements have high charge densities, for example, Be^{2+} and Mg^{2+}.

The reaction that produces the coordinated molecule or ion is an example of a Lewis acid-base reaction; the central atom acts as the Lewis acid (electron pair acceptor), and the ligand acts as the base (electron pair donor). The ligand may be an atom, ion, or molecule having a pair of unshared electrons that it can donate. Commonly, the element in the ligand that has an electron pair to donate is oxygen, sulfur, or nitrogen. The central atom or ion that accepts the electron pair must have a vacant orbital of relatively low energy available for bonding.

In the case of coordination compounds of the transition elements, *d* orbitals are usually involved in the bonding. Available *d* orbitals join with *s* and *p* orbitals of comparable energy to form *hybrid orbitals* similar to the sp^3 hybrid described in Chap. 5. A few specific examples are shown in Table 24-5. Note that for $Co(CN)_6^{3-}$ and $Ni(CN)_4^{2-}$ all electrons are shown paired, whereas for $(CoCl_4)^{2-}$ and $Fe(CN)_6^{3-}$ there are 3 and 1 unpaired electrons, respectively. The number of unpaired electrons in a compound is determined experimentally by making magnetic measurements.

TABLE 24-5.

Hybrid Orbitals in Coordination Compounds

SIMPLE ION OR COMPLEX	ELECTRONS OF ION OR CENTRAL ATOM (ARROWS) AND ELECTRONS OF LIGANDS (DOTS)	GEOMETRY OF COMPLEX
Co^{2+}	3d: ↑↓ ↑↓ ↑ ↑ ↑ ; 4s: ☐ ; 4p: ☐☐☐	
$(CoCl_4)^{2-}$ sp^3	3d: ↑↓ ↑↓ ↑ ↑ ↑ ; 4s: ·· ; 4p: ·· ·· ·· (bracketed as sp^3: ·· ·· ·· ··)	Tetrahedral
Fe^{3+}	3d: ↑ ↑ ↑ ↑ ↑ ; 4s: ☐ ; 4p: ☐☐☐	
$Fe(CN)_6^{3-}$ d^2sp^3	3d: ↑↓ ↑↓ ↑ ·· ·· ; 4s: ·· ; 4p: ·· ·· ·· (bracketed d^2sp^3: ·· ·· ·· ·· ·· ··)	Octahedral
Co^{3+}	3d: ↑↓ ↑ ↑ ↑ ↑ ; 4s: ☐ ; 4p: ☐☐☐	
$Co(CN)_6^{3-}$ d^2sp^3	3d: ↑↓ ↑↓ ↑↓ ·· ·· ; 4s: ·· ; 4p: ·· ·· ·· (bracketed d^2sp^3: ·· ·· ·· ·· ·· ··)	Octahedral
Ni^{2+}	3d: ↑↓ ↑↓ ↑↓ ↑ ↑ ; 4s: ☐ ; 4p: ☐☐☐	
$Ni(CN)_4^{2-}$ dsp^2	3d: ↑↓ ↑↓ ↑↓ ↑↓ ·· ; 4s: ·· ; 4p: ·· ·· ☐ (bracketed dsp^2: ·· ·· ·· ··)	Square planar

Lability of Ligands. The cyanide ion, CN^-, is an example of a ligand that is usually strongly held in a coordination compound. Although a solution of KCN is extremely poisonous, owing to free CN^- ions, a solution of an ion, such as $[Fe(CN)_6]^{3-}$, is relatively harmless, because the CN^- ligands are so strongly bound to the central atom. If a ligand in a complex ion is not strongly held, it is said to be *labile*, or free to move. In a water solution of Fe^{3+} and Cl^- ions, the complex $(FeCl_6)^{3-}$ might be expected to form. However, the Cl^- ion is a labile ligand that can exchange places in aqueous solution with a labile H_2O ligand. Complexes such as $[FeCl_2(H_2O)_4]^+$ and $[FeCl_3(H_2O)_3]$ exist, with the proportions of chloride ion and water molecules in the complexes depending on the concentration of chloride ion in solution.

Chelates. A ligand that has two or more points of attachment to the central atom is called a **chelate** (Greek *chele*, claw). An example of a chelate ligand is the compound $H_2N-CH_2-CH_2-NH_2$, ethylenediamine. Each nitrogen has a pair of electrons to donate, so that both ends of the molecule can be bound to a central atom. We shall use the symbol (en) for an ethylenediamine molecule.

One ethylenediamine molecule can replace two singly bound ligands, as shown in the reaction with the octahedral $(CoCl_6)^{3-}$ ion:

$$(CoCl_6)^{3-} + (en) \rightarrow [Co(en)Cl_4]^- + 2Cl^-$$

As might be expected, a chelating ligand is held by a central atom more strongly than singly held ligands bound by comparable bonds.

The compound ethylenediamine is called a *bidentate* (two teeth) ligand, because it has two points of attachment to a central atom; there are also *ter-*, *quadri-*, and even *sexadentate* chelates. Two quadridentate compounds of great interest are chlorophyll and heme. There are truly remarkable similarities in structure between these two life-essential compounds, one in the plant and one in the animal kingdom.

Hydrates and Ammoniates. Because most reactions are carried out in water solution, the H_2O molecule is often of interest as a ligand. The NH_3 molecule is a similarly held neutral molecule ligand. In most cases these ligands are labile, but in some compounds they are very tightly held.

Consider the addition of water to anhydrous copper(II) sulfate to form the blue pentahydrate:

$$\underset{\text{white}}{CuSO_4} + 5H_2O \rightarrow \underset{\text{blue}}{CuSO_4 \cdot 5H_2O}$$

X-Ray studies show that there are four water molecules close to each copper(II) ion and that the fifth water molecule is close to a sulfate ion. When ammonia gas is continuously passed over copper(II) sulfate pentahydrate, it is found that only four of the water molecules can be replaced:

$$CuSO_4 \cdot 5H_2O + 4NH_3 \rightarrow CuSO_4 \cdot H_2O \cdot 4NH_3 + 4H_2O$$

The water molecules attached to the Cu^{2+} ion are labile, but the fifth molecule is not.

The copper(II) ion in water solution actually has the formula $Cu(H_2O)_4^{2+}$. Ammonia can be substituted for water in solution:

$$\underset{\text{blue}}{Cu(H_2O)_4^{2+}} + 4NH_3 \rightarrow \underset{\text{deeper blue}}{Cu(NH_3)_4^{2+}} + 4H_2O$$

This reaction takes place when ammonium hydroxide (ammonia water) is added to water solutions of the copper(II) ion. The $Cu(NH_3)_4^{2+}$ ion has a deeper blue color than does the hydrated ion.

The dissolving of a silver chloride precipitate by ammonium hydroxide solution is a result of the formation of complex ions. Even though AgCl is practically insoluble in pure water, ammonia molecules attach so strongly to Ag^+ ions that they cause solution to occur:

$$AgCl + 2NH_3 \rightarrow Ag(NH_3)_2^+ + Cl^-$$

Although it is common to write M^{y+} as a simple formula for a metal ion, in a water solution such an ion often exists as $M(H_2O)_x^{y+}$, a hydrated ion; x usually has a value from 2 to 6. Even the metals in groups IA and IIA are hydrated in solution, but they do not tend to form ions with definite numbers of coordinated water molecules as readily as the transition elements do.

The following are representative examples of stable complex ions. Although all of them involve water and ammonia molecules, molecules other than can these also form stable complexes:

$Al(H_2O)_6^{3+}$ $Ni(H_2O)_6^{2+}$
$Co(H_2O)_6^{3+}$ $Ni(NH_3)_6^{2+}$
$Co(H_2O)_6^{2+}$ $Zn(NH_3)_4^{2+}$
$Co(NH_3)_6^{3+}$ $Hg(NH_3)_2^{2+}$
$Co(NH_3)_6^{2+}$ $Hg(NH_3)_4^{2+}$

Uses of Coordination Compounds. A major use of coordinating and chelating ligands is to influence the available concentration of ions in solution. Phosphate chelating agents are used in water softeners to keep calcium ions in solution; chelated iron is used in fertilizers to provide iron that will not react with soils to form insoluble iron compounds; and chelated metal ions can be formed that are more soluble in oil than in water, the reverse of the usual behavior. New dyes, especially for some of the difficult to tint synthetic fabrics, have been made from coordination compounds.

Chelating agents are used to remove unwanted impurities from drinking water. They have also been used in separating the mixtures of metal ions found in the fission products of nuclear reactors.

Not only recent discoveries, but many classical chemical reactions are interpreted today in terms of coordination chemistry. The long-known ability of *aqua regia*, a mixture of concentrated HNO_3 and HCl, to dissolve gold and platinum, depends in large part on the formation of the complex ions $AuCl_4^-$ and $PtCl_6^{2-}$, respectively.

EXERCISES

1. Explain in terms of electronic structure why "horizontal similarities" are more pronounced among the transition elements than among A family elements.
2. (a) It is stated in the text that the electron configurations are very nearly constant in the outer two main levels of elements in the lanthanide and actinide series. Check this statement and discuss its validity.

The Transition Metals and Their Neighbors

(b) In the light of your discussion, what would you expect about the extent of chemical similarities among the lanthanides as compared with the actinides?

3. Cite evidence that indicates that half-filled d or f sublevels have a special stability. (See inside front cover for electron configurations.)
4. Suggest reasons why group VIIIB of the periodic table has three vertical columns of elements while all other groups have only one.
5. On the basis of electron configurations, name for each of the following elements two elements that should be very similar in chemical properties: lutetium, Lu; thorium, Th; europium, Eu; element 105; indium, In.
6. Although element 120 is unknown we can postulate a position for it in the periodic table, based upon a probable electronic structure. We can then predict its properties from its location. Work out a probable electronic structure for element 120 and then predict several of its properties.
7. Which would you expect to have the more acidic properties, uranium dioxide, UO_2, or uranium trioxide, UO_3? Explain.
8. The maximum oxidation state of the elements in the series Sc through Mn progresses from $+3$ for scandium up to $+7$ for manganese. Account for this progression in terms of the electron distributions of the elements.
9. The standard oxidation potential of silver is -0.80 volts. What information does this convey about the properties of silver?
10. Compare the corrosion resistance of aluminum with that of silver.
11. Using electrochemical equations, explain why the iron in a "tin can" will corrode more rapidly than the iron in a galvanized can when the thin tin or zinc coatings are broken.
12. Write equations for the reaction of sodium hydroxide with chromic acid, permanganic acid, and vanadic acid. Name the products. Repeat for ammonium hydroxide and calcium hydroxide.
13. A water solution of zinc sulfate is colorless, while one of copper(II) sulfate is blue. Account for this observation on the basis of the electron structure of the two cations.
14. Give the sublevel electron structure for each of the following atoms or ions: Ni^{2+}, Sn, Sn^{2+}, Nb, Nb^{5+}.
15. What characteristic of tungsten makes it suitable for filaments in incandescent bulbs?
16. Work out orbital diagrams of the type in Table 24-5 for the following complexes. The number of unpaired electrons and the type of hybrid orbitals for each are as follows:

Complex	Unpaired e^-	Type of hybrid orbitals
$NiCl_4^{2-}$	2	sp^3
$Fe(H_2O)_6^{3+}$	5	sp^3d^2
$Cu(NH_3)_4^{2+}$	1	dsp^2

17. Explain or define each of the following: cathodic protection; bidentate ligand; coordination compound; Lewis acid-base reaction.
18. It is reported that freezing-point depression measurements show that $Co(NH_3)_4Cl_3$ exists in solution in the form of $[Co(NH_3)_4Cl_2]^+$ and Cl^- ions. Explain in terms of a hypothetical numerical example the sort of data that might have been obtained.
19. Name four coinage metals? Which one would an industrial nation miss most if it could get no more of it? Justify your choice.

20. Consider the relative positions of the nine elements in the small square area of the periodic table defined by the atomic numbers 19, 20, 21, 37, 38, 39, 55, 56, and 57. Assuming that the trends in family IIIB are similar to those in IA and IIA, predict (a) which of the nine elements has the smallest atom, which the largest; (b) which has the smallest electronegativity, which the greatest; (c) which forms the ion with the greatest charge density, which the least; (d) which has the greatest ionization energy, which the least; (e) which would have the greatest metallic character; (f) which would be the most dense; (g) which would have the highest melting point. Check your predictions against data tabulated in the text, where possible, and discuss any discrepancies.

SUGGESTED READING

Burns, R. M., "Chemical Reactions in the Corrosion of Metals," *J. Chem. Educ.*, **30**, 318 (1953).

Gavin, R. M., "Simplified Molecular Orbital Approach to Inorganic Stereochemistry," *J. Chem. Educ.*, **46**, 413 (1969).

Manch, W., and W. C. Fernelius, "The Structure and Spectra of Nickel(II) and Copper(II) Complexes," *J. Chem. Educ.*, **38**, 192 (1961).

Markowitz, M. M., "Alkali Metal–Water Reactions," *J. Chem. Educ.*, **40**, 633 (1963).

Petrocelli, A. W., and D. L. Kraus, "The Inorganic Superoxides," *J. Chem. Educ.*, **40**, 146 (1963).

Schlessinger, G. G., "A Summer Short Course in Coordination Chemistry. I, II, and III," *Chemistry*, **39** (6, 7, and 8), 8 13, and 15 (1966).

Schubert, J., "Chelation in Medicine," *Sci. Amer.*, **214**(5), 40 (1966).

Swann, P. R., "Stress-Corrosion Failure," *Sci. Amer.*, **214**(2), 72 (1966).

Vavoulis, A., "A Spot Test Scheme for the Identification of Metal Ions," *J. Chem. Educ.*, **39**, 395 (1962).

25

METALS: THEIR PRODUCTION AND USE

Whether used in a relatively pure form or alloyed with one another, the elemental metals have been the materials used to make the tools and machines of civilization. Man's mastery of his environment became efficient only when he learned how to obtain metals from minerals. Today, the demand for metals threatens to exhaust our natural ore supplies. As the earth's richer ores are depleted, methods will have to be devised for separating elements from poorer mineral sources such as the silicates. Although the sea is a vast storehouse of metallic compounds, the dilution makes it uneconomical at present to produce many metals from sea water; however, we may be able to look to the sea as a source of metals in the future.

SOURCES OF METALS

Except for magnesium from sea water and some manganese from the floor of the oceans, all metals are produced commercially from the earth's crust, a region discussed in detail in Chap. 30. The natural materials that are found in the earth's crust are called **minerals**. Minerals that can be used as a source for the commercial production of materials are called **ores**.

The most common ores of the metals are *oxides, sulfides, halides, silicates, carbonates,* and *sulfates*. The silicates, although the most abundant minerals, are of relatively little value as ores of metals, because they are so difficult to decompose chemically and because there are still available deposits of simpler ores that are easier and cheaper to process.

Alkali and Alkaline Earth Elements. The elements in groups IA and IIA never occur in nature as free elements. Four of them are extremely common: calcium, sodium, potassium, and magnesium ranking respectively as the fifth, sixth, seventh, and eighth most abundant elements in the world. Data on their relative abundances are given in Fig. 30-3.

The compounds of the alkali elements are so soluble that they tend to be leached out of the soil by rain water and thence carried by streams and rivers to the sea. (The solubilities of some common metal compounds are given in Table 4 in the Appendix.) Sea water contains about 3 percent by weight of alkali compounds, calculated as the chlorides. Large quantities of alkali salts are found around the world in salt lakes and salt flats, and in underground beds of sedimentary rock left by the evaporation of seas in times long past. There are huge underground domes of almost pure sodium chloride near the Gulf Coast shores of Louisiana and Texas.

The alkaline earth compounds are not quite so soluble as the alkali compounds; indeed, some of the common ones—the carbonates and certain of the sulfates—are insoluble enough to resist the weathering action of rain water strongly. The magnesium ion, however, is one of the very common ions in sea water, and a continuing supply of calcium ion is available in the sea for building the carbonate shells of crustaceous sea animals.

Transition Elements and Their Neighbors. Of the transition metals and their neighbors, the four most abundant in the earth's crust are aluminum, iron, titanium, and manganese (see Fig. 30-3). The relative scarcity of some of the familiar metals is noteworthy; copper, lead, and tin are not so plentiful as zirconium and titanium, two metals unfamiliar to most people.

Many of the relatively rare, precious metals (silver, gold, and the members of the platinum family) are so highly prized because of their strength, beauty, or resistance to chemical attack that they are searched for all over the world. An important characteristic of elements that are rare, yet well known, is that they are relatively easily separated from the minerals in which they occur naturally. By contrast, aluminum, the earth's third most abundant element, was discovered centuries later than gold because of the difficulty of decomposing its natural compounds. Methods of producing titanium, the tenth most common element, are only now being perfected to the point where it can be produced cheaply. Both titanium and aluminum form such strong bonds with oxygen (in oxides, silicates, and similar compounds) that special chemical methods must be resorted to in separating these elements from their natural compounds.

Perhaps the most useful types of ore of the transition elements and their neighbors are the oxides and sulfides. As we would expect, the very inactive metals are sometimes found in nature in their elemental or free state.

METALLURGY

The subject of **metallurgy** is concerned with the various processes for obtaining elemental metals from natural ores and preparing them for use as either pure metals or as alloys. This is in part an art and in part a science; many of the most useful techniques have been developed over the centuries by trial and error, many more have been discovered more recently by applying advanced theories and the latest knowledge.

The general problem of metallurgy is to decompose a compound in which a metal exists in a positive oxidation state, often as positive ions, and to transform the positive ions of the metal into atoms, that is, into the element. A general simplified equation, x being 1, 2, 3, etc., is

$$M^{x+} + xe^- \rightarrow M$$

The typical reaction is a *reduction reaction*. The metal in its positive oxidation state is not necessarily a simple ion; indeed, it is often found in an anion with oxygen, e.g., MO_4^{x-}, or in a silicate.

In rare cases, the desired metal is in the elemental state, and hence needs only to be separated from impurities.

The production of most metals has three main parts: (1) concentration of the ore, (2) chemical reduction to the element, and (3) refining, purifying, and perhaps alloying. In some cases these activities overlap; for example, chemical reduction can be part of the concentration process or of the refining operation.

Concentration of the Ore. The ore that is mined usually contains some worthless rock called *gangue*. The ore is usually crushed and ground until the particles of the mineral are broken apart from the gangue. If possible, these particles are separated by physical means, such as washing, flotation, or magnetic attraction.

Washing with a turbulent stream of water often washes the lighter gangue away from the desired mineral.

Flotation involves agitating the ore in a vessel with a detergent or foaming agent. The more valuable denser mineral may stick to the bubbles of foam and float off with it, leaving the gangue behind, or the gangue may be attracted to the foamy layer and float off with it (see Fig. 25-1).

With an *electromagnet* some minerals can be drawn out of their crushed ores. An example is magnetite, Fe_3O_4. Also, certain minerals can be charged electrically and then attracted to a charged plate, leaving the gangue behind.

If the ore cannot be sufficiently concentrated by physical means, chemical processes are used, such as those described in the following paragraphs.

In many cases the ore is *roasted* to drive off volatile impurities and to burn off organic matter. Roasting in air usually converts sulfides and carbonates to oxides. For example,

$$2ZnS + 3O_2 \rightarrow 2ZnO + 2SO_2$$

$$CaCO_3 \rightarrow CaO + CO_2$$

Ores generally contain considerable gangue, even after the most careful physical separation. To remove the last of the gangue, a *flux* is added. When the mixture is heated in a furnace, the flux combines with the gangue and makes a molten material called *slag*. At high temperatures, the slag is a liquid that is insoluble in the molten metal so the two can be separated. If the gangue is an acidic oxide such as silica, SiO_2, a cheap basic oxide like lime, CaO, will be used

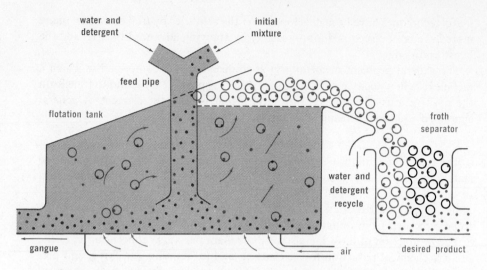

Fig. 25-1. Diagram of a flotation tank showing the separation of a desired material (colored). A rotating paddle at the bottom of the cell distributes the feed and also sweeps the bubbles around the central pipe.

for the flux. These two react in a furnace to form the low-melting compound calcium silicate, the slag.

$$SiO_2 + CaO \rightarrow CaSiO_3$$

If the gangue is basic, e.g., calcium or magnesium carbonate, the flux will be a cheap acidic oxide.

Acids or bases may be used to *dissolve* part of the ore. Sometimes a compound of the desired metal is precipitated from the solution; sometimes impurities are precipitated.

Reduction to the Element. Several chemical methods can be used to reduce a given metal from its oxidation state in the ore to the elemental state. If reduction is difficult for a particular metal, powerful reducing processes will be necessary.

REDUCTION BY HEAT IN AIR. Compounds of the precious metals in Groups VIII and IB are reduced easily. Platinum, gold, and sometimes silver are found in their elemental form and have only to be heated to melt them out of the gangue. Since many of the oxides of the less active metals are decomposed by extreme heat, roasting in air is all that is needed for reduction. For example, roasting the sulfide ore of mercury forms the metal rather than the metallic oxide.

$$HgS + O_2 \rightarrow Hg + SO_2$$

REDUCTION WITH CARBON. Oxides of many moderately active metals can be reduced by carbon. The reaction for zinc oxide is

$$ZnO + C \xrightarrow{heat} Zn + CO$$
$$ZnO + CO \longrightarrow Zn + CO_2$$

This reduction method is suitable for metals of the iron family and for some others such as lead and tin. Note that the carbon may be oxidized to carbon monoxide, CO, or carbon dioxide, CO_2. In many cases, as in the production of zinc or iron, the carbon monoxide formed acts as a reducing agent.

Carbon tends to form carbides with certain metals such as chromium and manganese, so it cannot be employed for the reduction of all oxide ores. But it is used when possible, because it is both cheap and convenient.

REDUCTION WITH AN ACTIVE METAL. If compounds of metals are not satisfactorily reduced by carbon, a more active metal can be used as the reducing agent. Aluminum, magnesium, sodium, and calcium are active enough to be good reducing agents. Uranium(IV) fluoride is reduced by calcium. Titanium chloride is reduced by magnesium or sodium.

$$TiCl_4 + 2Mg \rightarrow 2MgCl_2 + Ti$$

REDUCTION BY HYDROGEN. Reduction by hydrogen is more expensive than reduction by carbon and is used only when carbon is not suitable. Tungsten oxide is reduced in this way:

$$WO_3 + 3H_2 \rightarrow W + 3H_2O$$

REDUCTION BY ELECTROLYSIS. It was pointed out previously that very active metals, such as the alkali metals and the alkaline earth elements, are most efficiently produced by the electrolysis of anhydrous fused salts. The Group III elements and the lanthanide series are also usually prepared by electrolysis of fused salts. The chlorides are commonly used for this purpose.

Refining and Purifying. The metal produced commercially by any of the above methods usually contains considerable amounts of impurities.

Metals with low boiling points—e.g., mercury, bismuth, and tin—can be separated from most impurities simply by melting the metal and pouring it off, or by distilling it. Like salts, metals can also be refined by fractional crystallization. Probably the most widely used refining process is the electrolytic process.

INDUSTRIAL PRODUCTION OF SEVERAL METALS

To illustrate some general principles of metallurgy, we will consider the production of four different elements. Three of these, sodium, aluminum, and magnesium, are very active chemically. Certain aspects of their preparation, such as electrolysis, are typical of the production of *active metals* in general. Metallurgical operations used for *moderately active metals* are described in the section on iron.

Sodium and Magnesium. By far the most practical and widely used method for the reduction of active metals is the *electrolysis* of fused hydroxides or chlorides. The chloride is generally employed in the United States. Large amounts of elemental sodium, magnesium, and calcium are produced in this way.

In the case of sodium chloride the reaction is

$$2NaCl \xrightarrow{\text{electrolysis}} 2Na + Cl_2$$
(molten)

Chlorine gas is a valuable by-product of this process.

So far as the alkali metals are concerned, there are two reasons why sodium is the only one produced on a very large scale: (1) the raw material (NaCl) required for its production is available in enormous high-purity deposits at many places in the earth's crust, and (2) sodium serves as well for most industrial purposes as would one of the more expensive alkali metals.

The first step in the production of magnesium by the electrolysis of molten magnesium chloride is the preparation of the chloride, because of the lack of deposits of relatively pure magnesium chloride. One method starts with sea water, which contains a practically inexhaustible supply of magnesium ions. The magnesium is first precipitated as magnesium hydroxide by the addition of lime to the sea water:

$$CaO + HOH \rightarrow Ca^{2+} + 2OH^-$$

$$Mg^{2+} + 2OH^- \rightarrow Mg(OH)_2$$

The magnesium hydroxide is removed by filtration and then converted into the chloride by the addition of hydrochloric acid:

$$Mg(OH)_2 + 2HCl \rightarrow MgCl_2 + 2HOH$$

After the water is removed by evaporation, electrolysis of the magnesium chloride is carried out in a large steel pot that serves as the cathode; graphite bars immersed in the molten magnesium chloride serve as the anodes (Fig. 25-2). As

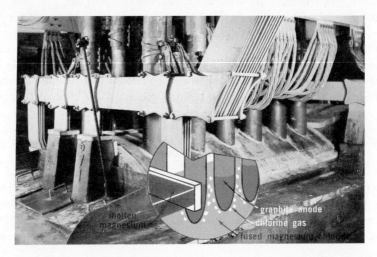

Fig. 25–2. The production of magnesium by the electrolysis of molten magnesium chloride. The molten magnesium, formed at the steel wall of the pot (cathode), rises at the sides and is separated from the chlorine by a metal barrier. (Photograph courtesy Dow Metal Products Company.)

the molten magnesium forms, it floats to the surface and is removed at suitable intervals:

$$MgCl_2 \text{ (molten)} \xrightarrow{\text{electrolysis}} Mg + Cl_2$$

The chlorine is recovered and converted into hydrochloric acid for use in a previous step of the process. It is interesting to note that the coastal region of Texas, one place where this process is carried out, has no deposits of limestone rock, the usual source of calcium carbonate for the production of lime. However, oyster shells, which are abundant and rich in calcium carbonate, are calcined to supply the lime:

$$CaCO_3 \xrightarrow{heat} CaO + CO_2$$

Production of magnesium in the United States is about 80,000 tons annually.

Aluminum. Aluminum, the most common metallic element in the earth's crust, is found in some of the earth's most abundant silicate minerals. But it was not until 1886 that the electrolytic process was invented for recovering aluminum from its ores. Because aluminum ions are more difficult to reduce than hydrogen ions, electrolysis cannot be carried out successfully in a water solution. The successful commercial process was discovered practically simultaneously by two investigators, Charles M. Hall in the United States and Paul Héroult in France.

The process devised by the twenty-two-year-old Hall is practically identical to that used commercially today. Anhydrous Al_2O_3 is dissolved in molten cryolite, Na_3AlF_6, where it is thought to be in the form of the ions Al^{3+} and O^{2-}. Electrolysis of the solution yields elemental aluminum at the cathode and oxygen at the anode.

A schematic diagram of electrolytic cells for the production of aluminum is shown in Fig. 25-3. An iron tank lined with carbon is the cathode of the cell, and large blocks of carbon serve as anodes. The cryolite is melted in the tank, and purified anhydrous aluminum oxide (alumina) is added to it.

When current is passed through the cell, molten aluminum forms at the walls and bottom of the tank (the cathode). Essentially all of the oxygen liberated at the anode attacks the carbon and forms carbon dioxide.

At the cathode: $\quad\quad\quad Al^{3+} + 3e^- \rightarrow Al$

At the anode: $\quad\quad\quad C + 2O^{2-} \rightarrow CO_2 + 4e^-$

The carbon anodes are continually used up and hence must be replaced from time to time. Under normal operating conditions, about 0.5 lb of the anode is consumed per pound of aluminum produced. The cost of the carefully made anodes is a major item of expense.

The passage of the electric current through the cryolite generates enough heat to keep it molten. The temperature, about 1000°C, is above the melting point of aluminum, so the metal produced collects in the bottom of the cell as a liquid. Periodically, the molten aluminum is siphoned off into molds, cooled, and formed into large ingots. The aluminum produced by the primary electrolysis is better than 99 percent pure; further electrolytic refining gives metal that is more than 99.9 percent pure.

The cost of the electrical power is of prime importance in producing aluminum. For this reason, aluminum plants are located near sources of cheap power, rather than either near the source of raw materials or near markets for the finished product. The first big customer of the Niagara Falls Power Company was an aluminum plant. Hydroelectric power plants in the Tennessee River Valley and on the Columbia and St. Lawrence Rivers also supply current to aluminum reduction plants.

Fig. 25-3. Alumina (Al_2O_3) electrolysis "pots." (Photograph courtesy Aluminum Company of America.)

Iron. Iron compounds are present in most rocks and soils; but iron ore, to be workable, must contain a high percentage of the iron compound, and consequently only small amounts of sand, clay, and other minerals.

The most important ore in the United States is *hematite*, Fe_2O_3. Some *limonite*, $2Fe_2O_3 \cdot 3H_2O$, and *magnetite*, Fe_3O_4, are also used as ores. These oxide deposits are located in northern Minnesota, the Birmingham (Alabama) district, and in Wyoming, Utah, and California. The red color of clay and sandstone usually is due to hematite. The deposits in Minnesota originally contained such a high percentage of iron oxide that the ore could be reduced in blast furnaces without preliminary treatment. These rich deposits have been largely used up. Huge amounts of leaner ores remain, but they must receive special physical and chemical treatment before reduction. The annual production of iron ore in the United States ranges from 140 to 150 million tons. This is supplemented by annual imports of from 30 to 40 million tons of ore from Canada, Chile, Sweden, and Liberia, among other countries.

THE BLAST FURNACE. The primary reduction of the iron oxide ore is carried out in a mammoth chimney called a *blast furnace*. In this fiery reaction chamber molten iron is formed, and silica, SiO_2, the chief impurity, is largely removed. Lime, CaO, added in the form of calcium carbonate, combines with the silica to form a slag of calcium silicate, $CaSiO_3$. As shown in Fig. 25-4, the molten slag and iron settle in separate layers at the bottom of the furnace.

A mixture of crushed iron ore, coke, and limestone is added by means of a hopper at the top of the furnace. A blast of hot air is blown up through this

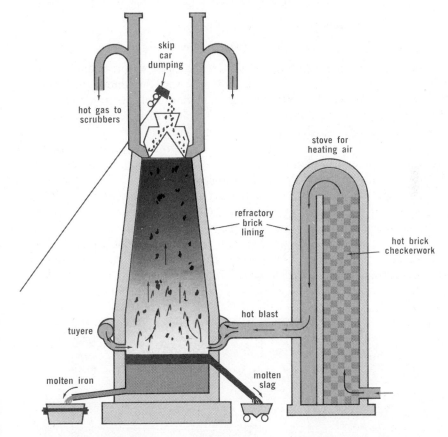

Fig. 25-4. Schematic diagram of a blast furnace. (Courtesy Bethlehem Steel Company.)

mixture from the bottom of the furnace. The complex series of reactions can be summarized as follows: Near the bottom of the furnace where the blast of hot air enters, coke burns furiously.

$$C + O_2 \rightarrow CO_2$$

As the carbon dioxide rises in the chimney, it is reduced almost immediately by hot coke.

$$CO_2 + C \rightarrow 2CO$$

The carbon monoxide reacts with the oxide to form metallic iron. The reaction takes place in a series of steps that occur in the middle and upper portions of the furnace and can be summarized by the over-all equation:

$$Fe_2O_3 + 3CO \rightarrow 2Fe + 3CO_2$$

Near the middle of the furnace, the limestone decomposes to form lime and carbon dioxide.

$$CaCO_3 \rightarrow CaO + CO_2$$

Farther down, the lime and silica react to form a slag.

$$CaO + SiO_2 \rightarrow CaSiO_3$$

The operation of the furnace is continuous; the mixture of reactants is fed into the top at regular intervals to begin its journey toward the white-hot lower levels. The temperature near the bottom is high enough to melt the iron and slag, and they collect as immiscible layers at the bottom. The furnace must be "tapped" about every six hours to drain off the molten iron.

For each ton of blast furnace iron, or pig iron,[1] produced, about 2 tons of iron ore, 1 ton of coke, 0.3 ton of limestone, and 4 tons of air are required. The main by-products are 0.6 ton of slag and 5.7 tons of flue gas.

STEEL. Pig iron is so brittle and so low in tensile strength that it is of little use. Its poor quality is due chiefly to the presence of several impurities: 3 to 5 percent carbon and smaller percentages of silicon, phosphorus, and sulfur.

Melting pig iron and high-grade scrap iron together produces *cast iron*. It has about 2 percent miscellaneous impurities, chiefly carbon, but is too brittle to be forged, rolled, or welded. Articles made from it are cast into the desired shape in a mold.

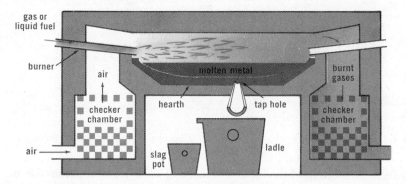

Fig. 25-5. Schematic diagram of an open-hearth furnace for steel production. (Courtesy Bethlehem Steel Company.)

Most iron is made into *carbon steel*, an alloy that contains up to 2.0 percent carbon and can be made tough as well as hard by special heat treatment. To make

---

[1] The molten blast furnace iron is sometimes run into molds where it hardens into small ingots called *pigs*. Increasingly in modern plants, the molten iron is converted directly into steel, instead of into ingots or pigs. The terms *pig iron* and *blast furnace iron* are synonymous.

steel from pig iron, some of the carbon and practically all the phosphorus, sulfur, and silicon must be removed.

The *open-hearth* method accounts for most[2] of the steel produced in this country. The average hearth (Fig. 25-5) is a shallow vessel 40 × 18 ft and 2 ft deep. Over it is a roof of arched fire brick against which hot fuel gases are burned. The materials charged into the hearth are pig iron, rusty scrap iron (Fe_2O_3 on the surface) or smaller amounts of iron ore, and other materials for special alloys.

The hearth is lined with either a basic or an acidic lining, depending on the type of pig iron being purified. In this country the ores usually have acid impurities (phosphorus or sulfur), so basic linings such as magnesium and calcium oxide are used. When the charge of 100 or more tons is melted in the hearth, the iron rust (or iron ore) may take part in the following typical reactions:

$$2Fe_2O_3 + 3S \rightarrow 4Fe + 3SO_2$$
$$2Fe_2O_3 + 3C \rightarrow 4Fe + 3CO_2$$
$$10Fe_2O_3 + 12P \rightarrow 20Fe + 3P_4O_{10}$$

The CO_2 gas bubbles out of the melt, and the sulfur and phosphorus oxides combine with the basic oxides of the lining to form a slag. The process is slow enough (about eight hours) so that chemical analyses can be made periodically to check the composition of the steel.

ALLOYS

Although there are only about 80 metal elements, there are thousands of different combinations of them, each with its own special properties. Metals have the ability to mix with and combine with one another to form a practically unlimited number of alloys. An **alloy** is the solid that results when two or more metals are melted together to form a homogeneous mixture and then allowed to cool.

The metals in an alloy may form as follows: (1) They may dissolve completely in one another and on cooling form a *solid solution* (homogeneous). (2) They may crystallize separately and be present in the alloy as a *mixture* of tiny crystals (heterogeneous). (3) Their atoms may combine in a definite ratio, forming an *intermetallic compound* (homogeneous).

Properties of Alloys. Alloys have properties of their own that often differ markedly from those of the elements of which they are made. The melting points of some alloys—for example, solder—are lower than that of any of their constituents. Other alloys have higher melting points than do any of their constituents.

Alloys are usually harder than the parent metals. Pure iron is quite soft and ductile in comparison with the steel alloys, which are composed mainly of iron.

---

[2] Other methods for producing steel include the Bessemer and the electric furnace processes, and the recently developed basic oxygen process.

Gold is too soft even for jewelry; it must be hardened by alloying with copper or some other metal. Alloys are generally poorer conductors of heat and electricity than pure metals.

One of the chief differences between pure metals and alloys is corrosion resistance. Alloys are usually more resistant to corrosion; hence they are more useful for materials that are exposed to the action of corrosive agents. Aluminum alloys are notable exceptions to this; they are sometimes covered with pure aluminum to improve their corrosion resistance.

Some of the more common alloys of copper, silver, and gold, as well as a few other nonferrous alloys, are described in Table 25-1.

TABLE 25-1.

Nonferrous Alloys

NAME	PERCENTAGE COMPOSITION (APPROXIMATE)	USE
Babbitt	Sn 85, Sb 10, Cu 5	Bearings
Brass, red	Cu 85, Zn 15	Hardware, radiator cores
Brass, yellow	Cu 67, Zn 33	Musical instruments, cartridges
Bronze, ordinary	Cu 90, Sn 10	Valves, rods
Bronze, aluminum	Cu 90, Al 10	Gilt paint
Bronze, phosphor	Cu 95, Sn 4.8, P 0.2	Spring metal
Gold, white	Au 75, Cu 3.5, Ni 16.5, Zn 5	Jewelry
Linotype metal	Pb 79, Sb 16, Sn 5	Printing
Nickel silver	Cu 64, Ni 18, Zn 18	Silverware, plating, resistance wire
Pewter	Sn 85, Cu 7, Bi 6, Sb 2	Metal dishes
René 41	Ni 55, Cr 19, Co 11, Mo 10, (Ti + Al) 5	Sheathing of space capsules, jet engines
Solder, soft	Sn 60, Pb 40	Solder
Sterling silver	Ag 92.5, Cu 7.5	Silverware
Wood's metal	Bi 50, Cd 12.5, Pb 25, Sn 12.5	Low-melting alloy, fuses

Alloy Steels. Hard carbon steels are much too brittle for structural use or for the manufacture of machine parts. Alloying steel with nickel, chromium, manganese (in excess of 1.65 percent), tungsten, vanadium, and other metals produces *alloy steels*. These steels range in tensile strength as high as 110 tons/in.^{2}, can be made uniformly hard, are not brittle, and have other desirable properties. A vast number of such steels are on the market today; in general, however, they are more expensive and more difficult to work than carbon steels. Alloy steels are used in such diverse objects as armor plate, drills and cutting tools, rock crushers, piston rings, gears, springs, safes, kitchen ware, automobile frames, and stainless cutlery.

EXERCISES

1. List several metals that are found in the earth's crust as atoms; as cations; as, or in, anions.
2. Distinguish between an ore and a mineral.
3. Exposed minerals containing aluminum are found widely scattered in the earth's crust, while minerals of sodium are very localized and are nearly always found underground or in desert regions. Why?
4. How could you probably distinguish, on the basis of a simple test with hydrochloric acid, between a sulfide, a carbonate, and a silicate ore?
5. Copper has been used by man much longer than chromium, although the latter also resists corrosion and is more common in the earth's crust. Why?
6. In the development of man's history, the iron age was preceded by the bronze age. Why?
7. List several oxide ores for which carbon might be used as a reducing agent and several for which it could not.
8. Name a cheap basic oxide that might be used as a flux to remove SiO_2.
9. Write equations for the reactions when each of the following is roasted in air: PbS, HgO, Ag_2S, NiS, $SrCO_3$.
10. Outline the various methods for concentrating ores.
11. Suggest a means of reducing each of the following to the element. Write an equation for each reduction: (a) $SrCl_2$; (b) Cr_2O_3; (c) ZnO; (d) PtO; (e) Fe_3O_4.
12. Consider a small particle of limestone dropped into a blast furnace. Write equations for all possible reactions that it or its products undergo while in the furnace.
13. In the production of blast furnace iron, which raw material is used in the greatest amount by weight?
14. An ore contains 3.2 percent calcium carbonate after flotation treatment. What is the theoretical weight of sand needed per ton of ore for slag formation? What weight of slag would be formed?
15. What weight of carbon from the anode is used in the production of 1 ton of aluminum?
16. Could aluminum ingots be purified in a manner similar to the removal of sulfur and phosphorus from pig iron? Explain using equations and a discussion.
17. Describe some of the differences between cast iron and steel.
18. What properties would be desirable in an alloy used to sheath capsules that are to carry men into space? To what laboratory tests might alloys be subjected to evaluate them for this use?
19. Discuss the economic location of a magnesium plant in terms of raw materials and other requirements (a) if the magnesium is obtained from the ocean and (b) if the magnesium is obtained from a deposit of $MgCO_3$.
20. Explain how it is possible for the geologist to find valuable ores by examining plant life.
21. For what uses might a pure metal be preferred to an alloy? What are some properties of alloys that might make them more useful than the pure metal?
22. If a recently discovered Canadian iron ore deposit (69 percent iron) is practically a pure compound, what might it be? What would be another way of accounting for the percentage composition only?
23. Calculate the volume of magnesium that has the same weight as 1 cu in. of gold. (See Tables 18-2 and 24-2.)

24. (a) Calculate the weight of oyster shells required to precipitate the magnesium from 1 cu mile of seawater, based on the following data: oyster shells, 95 percent $CaCO_3$; seawater, 0.13 percent Mg^{2+} by weight; density of seawater, 1.0 g/ml.
(b) What weight of magnesium can be obtained if the recovery is 50 percent efficient?

SUGGESTED READING

Bachmann, H. G., "The Origin of Ores," *Sci. Amer.*, **202**(6), 146 (1960).

Blumenthal, W. B., "Zirconium Chemistry in Industry," *J. Chem. Educ.*, **39**, 604 (1962).

Cook, N. C., "Me.alliding," *Sci. Amer.*, **221** (2), 38 (1969).

Cottrell, A. H., "The Nature of Metals," *Sci. Amer.*, **217** (3), 90 (1967).

Feiss, J. W., "Minerals," *Sci. Amer.*, **209**(3), 128 (1963).

Fleischer, M., "The Abundance and Distribution of the Chemical Elements in the Earth's Crust," *J. Chem. Educ.*, **31**, 446 (1954).

Fyfe, W. S., "Modern Geochemistry," *J. Chem. Educ.*, **40**, 330 (1963).

Garrett, A. B., "Aluminum Metallurgy: Charles Martin Hall," *J. Chem. Educ.*, **39**, 415 (1962).

Giannini, G., "Electrical Propulsion in Space," *Sci. Amer.*, **204**(3), 57 (1961).

Kaufman, N., "High Temperature Alloys for Small Gas Turbines," *J. Chem. Educ.*, **39**, 158 (1962).

Stone, J. H., "Oxygen in Steelmaking," *Sci. Amer.*, **218** (4), 24 (1968).

Zackay, V. F., "The Strength of Steel," *Sci. Amer.*, **209**(2), 72 (1963).

26

ORGANIC CHEMISTRY I: HYDROCARBONS

Organic Chemistry. The early chemist drew a sharp distinction between the compounds that form the rocks and soils and the compounds that originate during the growth of plants and animals. In his view, the latter came into being only through some vital force associated with life processes; their synthesis in the laboratory was impossible. Accordingly, he referred to them as **organic** compounds to distinguish them from the earthy compounds, which he called **inorganic** compounds.

The development of laboratory procedures for analyzing and synthesizing compounds during the first half of the nineteenth century enabled chemists to learn a great deal about organic compounds. They were able to synthesize the simple ones in the absence of a "vital force"[1] and to show that the same laws and principles apply to both types of compounds.

Since the first organic compound was synthesized in the laboratory, a tremendous effort has been made to isolate, identify, and synthesize all the compounds that originate in plants and animals. This has been, and is, a most important type of research, for organic compounds constitute all or a major portion of petroleum, coal, proteins, fats, carbohydrates, vitamins, hormones, cellulose, anesthetics, antiseptics, antibiotics, enzymes, and a host of other useful products.

Although perhaps most of the natural organic compounds have been isolated and characterized, man has not been able to synthesize some of the more complex ones. An interesting and important outcome of this work has been the synthesis of a large number of compounds that do not originate in living organisms but are

---

[1] In 1828, the German chemist Wöhler synthesized urea by heating ammonium cyanate. Previously (1824) he had synthesized oxalic acid from cyanogen. Because of this work, Wöhler is usually given credit for initiating the downfall of the "vital force" idea.

very similar in composition and properties to those which do. Inasmuch as both the natural and synthetic compounds contain carbon, we now define **organic compounds** as the *compounds of carbon*.[2] No distinction is made between those of natural origin and those of synthetic origin. **Organic chemistry** is the science that deals with carbon compounds.

Number of Organic Compounds. The number of organic compounds has been estimated recently at well over a million. This is many times more than the number of known compounds that do not contain carbon. The large number of carbon compounds is explained on the basis of two characteristics of carbon atoms: (1) Carbon atoms unite with one another by sharing one or more pairs of electrons to form chain or ring molecules. (2) Carbon atoms, with four valence electrons, can form four covalent bonds. This means that the carbon atoms are able to form rings and chains and still have valence electrons left over that can be used to form bonds with atoms of other elements.

The other group IVA elements—silicon, germanium, tin, and lead—also have four valence electrons. However, their outer-shell electrons are located progressively farther from the nucleus with increasing atomic number and are less strongly held. Hence, these elements are more metallic than carbon and are less likely to form chain or ring molecules by means of covalent bonds. Silicon, which is next to carbon in the family, does form compounds in which silicon atoms are joined in chains, but these chains are neither long nor very stable.

FOUR SERIES OF HYDROCARBONS

Hydrocarbons are defined as compounds that contain only carbon and hydrogen. Hydrocarbons occur abundantly in nature, largely as petroleum, natural gas, and coal. Thousands are known; the number theoretically possible is virtually without limit. They are of special importance to the student of chemistry, because they are the basis for the systematic naming and classification of organic compounds.

We shall confine our study largely to four classes or series: the **alkanes**, the **alkenes**, the **alkynes**, and the **benzene (aromatic) hydrocarbons**. The first three series are frequently referred to collectively as the **aliphatic hydrocarbons**.[3]

Alkanes. The alkane series is also called the *methane* hydrocarbons or the *paraffins*. The first few members are listed in Table 26-1.

The number of carbon atoms, beginning with the pentanes, is indicated by the Greek prefix, except that occasionally the Latin is used, for example, *nona* or *non* for nine.

Note that the molecules of each member differ from those of the preceding and succeeding members by a constant number of atoms (1 carbon and 2 hydro-

---

[2] As previously pointed out, there are certain rocklike or earthy carbon-containing substances that are usually classed as inorganic compounds, for example, the carbonates, carbides, and cyanides.

[3] The molecules of aliphatic hydrocarbons are characterized by carbon chains (an important exception is the cycloalkane group), whereas those of the aromatic hydrocarbons are characterized by carbon rings and a bond system similar to that of benzene.

gen atoms). Such a series is called a **homologous series,** and each member is a **homolog** of the series. Furthermore, a general formula can be assigned to the

TABLE 26-1.

Alkanes (C_nH_{2n+2}, General Formula)

CH_4	Methane	C_5H_{12}	Pentanes (3)	C_9H_{20}	Nonanes (35)
C_2H_6	Ethane	C_6H_{14}	Hexanes (5)	$C_{10}H_{22}$	Decanes (75)
C_3H_8	Propane	C_7H_{16}	Heptanes (9)	$C_{15}H_{32}$	Pentadecanes
C_4H_{10}	Butanes (2)[a]	C_8H_{18}	Octanes (18)		(4347)

[a] The numbers in parentheses are the calculated numbers of possible isomers for the individual molecular formulas. In the cases of the larger molecules no attempts have been made to isolate all the isomers. For example, $C_{20}H_{42}$ has 366,319 possible different structural arrangements.

series that will represent any member. For the alkanes, the general formula is C_nH_{2n+2}, where n is the number of carbon atoms. For example, if the molecule of a homolog contains 100 carbon atoms, n is 100 and $2n + 2$ is 202. The formula is $C_{100}H_{202}$.

Alkenes. The alkene series is also referred to as the *ethylene* series. The first few members are

C_nH_{2n}, general formula

C_2H_4	ethene (ethylene)	C_4H_8	butenes (4)
C_3H_6	propene (propylene)	C_5H_{10}	pentenes (6)

Alkynes. The alkynes are also referred to as the *acetylene* series. The first few members are

C_nH_{2n-2}, general formula

C_2H_2	ethyne (acetylene)	C_4H_6	butynes (2)
C_3H_4	propyne	C_5H_8	pentynes (3)

Aromatic Hydrocarbons. There are many homologous series of aromatic hydrocarbons, the most important ones being derived from benzene, C_6H_6. We shall give most of our attention to the series that can be represented by the general formula C_nH_{2n-6}:

C_6H_6	benzene	C_8H_{10}	xylenes (3) and ethyl benzene
C_7H_8	toluene	C_9H_{12}	8 isomers

STRUCTURE OF HYDROCARBON MOLECULES

In accounting for the large number of organic compounds, it was pointed out that a carbon atom could form four bonds with other atoms, including other carbon atoms. Another distinctive feature of carbon atoms is their capacity for combining with one another via different types of bond hybridization. For example, the carbon atoms in alkane molecules are joined by sp^3 bonds (see

Chap. 5). Some of the carbon atoms in molecules of the other series, except for the first member of each, may also be joined by sp^3 bonds, but each molecule of an alkene, alkyne, or benzene hydrocarbon has at least two carbon atoms joined by bonds that represent a different kind of orbital hybridization and that endow the hydrocarbons of that series with the properties characteristic of the series.

Structure of Alkane Molecules. A striking characteristic of the bonds in alkane hydrocarbons is their *equivalence*. For example, the four carbon-hydrogen bonds in methane, CH_4, are identical in reactivity and stability. Moreover, the valence forces are uniformly directed from the carbon atom toward the corners of an imaginary tetrahedron at angles of 109°28'. It is difficult to account for these characteristics in terms of valence orbitals of the most stable state of an isolated carbon atom, a state in which it is thought that two of the four valence electrons occupy the $2s$ orbital as paired electrons and that each of the remaining two occupies separate $2p$ orbitals. Instead, as pointed out in Chap. 5, it is believed that, in compound formation of the alkane type, one of the $2s$ electrons is elevated into a vacant, higher-energy $2p$ orbital to provide four orbitals with four unpaired electrons (see Fig. 5-6). In the process, the resulting orbitals become identical in all respects and are directed outward from the nucleus at 109°28' angles. Because they result from the blending or hybridization of one s and three p orbitals, they are referred to as sp^3 orbitals. The energy required to promote an s electron to a p orbital is thought to be more than balanced by the energy liberated in forming four equivalent bonding orbitals in the resulting alkane molecule.

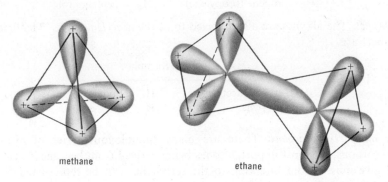

methane   ethane

Fig. 26-1. A schematic representation of the bonding orbitals in methane and ethane.

In the formation of methane, each sp^3 orbital is thought to interpenetrate the s orbital of a hydrogen atom; that, is, each hydrogen atom is bonded to carbon by a sigma bond involving an s orbital of hydrogen and an sp^3 hybrid orbital of carbon (see Fig. 26-1). This results in a tetrahedrally shaped molecule with the four hydrogen nuclei at the four corners and the carbon nucleus at the center. Because carbon and hydrogen have electronegativities of 2.5 and 2.1, respectively, we might expect the carbon-hydrogen bond to be somewhat polar, with the electron cloud slightly displaced toward carbon. The properties of the alkanes

Organic Chemistry I: Hydrocarbons

show that this is correct. However, the symmetrical shapes of most alkane molecules have the effect of balancing the partial charges, so that the molecules as a whole are nonpolar. For example, the dipole moments of methane and ethane are zero.

In an ethane molecule, C_2H_6, two carbon atoms are joined by a sigma bond formed by the overlap of an sp^3 orbital from each carbon atom (Fig. 26-1). The six hydrogen atoms, three to each carbon, are bonded to carbon by sigma bonds in the same manner as in methane.

In propane molecules, C_3H_8, three carbon atoms are joined by sigma bonds formed by the overlap of sp^3 orbitals. Because sp^3 bonding orbitals lie along axes that form 109°28' bond angles with one another, the lines that connect the three carbon nuclei form the same angle.

The carbon atoms in molecules of butane, pentane, hexane, and, as a matter of fact, all alkanes, are thought to be joined by sigma bonds of the same type shown for ethane in Fig. 26-1. The hydrogen atoms are also bonded by sigma bonds in the same manner as described for methane.

SIMPLE STRUCTURAL FORMULAS. Molecular formulas, i.e., ones that show only the number and kind of atoms in molecules, such as C_5H_{12} and C_6H_{14}, are generally inadequate for the following reasons. (1) Unless the molecule is small, the molecular formula represents more than one compound. The numbers in parentheses in Table 26-1 show the possible number of hydrocarbons that may have the particular molecular formula. (2) The molecular formula is of little value in correlating and classifying properties, and (3) it offers no clue as to how the compound can be synthesized.

The very detailed structural formulas shown in Fig. 26-1 require too much space and are too difficult to reproduce for the representation of the structures of a million-odd organic molecules. Long before such detailed information was available, as far back as about 1850, the organic chemist had worked out experimentally the relative positions of atoms in certain molecules. Gradually, a simple type of structural formula was developed for showing this kind of information, a graphic representation in which each symbol represents an atom of the element, and one short line represents a single pair of shared electrons (a covalent bond). In the following structural formulas, note that four valence bonds are formed by each carbon, whereas each hydrogen atom forms only one bond:

$$\begin{array}{cc} \text{H} \\ | \\ \text{H}-\text{C}-\text{H} \\ | \\ \text{H} \\ \text{methane, CH}_4 \end{array} \qquad \begin{array}{cc} \text{H} \quad \text{H} \\ | \quad | \\ \text{H}-\text{C}-\text{C}-\text{H} \\ | \quad | \\ \text{H} \quad \text{H} \\ \text{ethane, C}_2\text{H}_6 \end{array}$$

The structure may also be shown by using dots to represent a pair of shared electrons, as in earlier chapters:

$$\begin{array}{c} \text{H} \\ \text{H} : \ddot{\text{C}} : \text{H} \\ \ddot{\text{H}} \\ \text{methane} \end{array} \qquad \begin{array}{c} \text{H} \quad \text{H} \\ \text{H} : \ddot{\text{C}} : \ddot{\text{C}} : \text{H} \\ \ddot{\text{H}} \quad \ddot{\text{H}} \\ \text{ethane} \end{array}$$

The structural formula for propane is

$$\begin{array}{c} \text{H} \quad \text{H} \quad \text{H} \\ | \quad\ | \quad\ | \\ \text{H}-\text{C}-\text{C}-\text{C}-\text{H} \\ | \quad\ | \quad\ | \\ \text{H} \quad \text{H} \quad \text{H} \end{array}$$

LIMITATIONS OF STRUCTURAL FORMULAS. The formulas above do not show the correct positions of the atoms in space. For example, they seem to indicate that all the atoms are in one plane and that the bond angles are either 90° or 180°. In spite of these drawbacks, simple structural formulas that employ lines for valence bonds and symbols for atoms are universally used to represent organic molecules. When the occasion demands, we interpret them in terms of spatial relationship of atoms, bond angles, and bonding orbitals.

When it is desirable to represent the spatial arrangement of atoms in molecules, three-dimensional diagrams and models, such as are shown in Figs. 26-2 and 26-3, are often used.

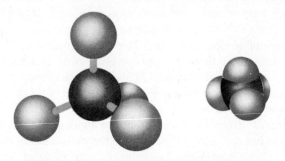

Fig. 26–2. Two methods of representing the spatial arrangement of atoms in a molecule of methane. Left, a ball-and-stick model; right, a model which shows the relative sizes of the atoms as well as their positions in space.

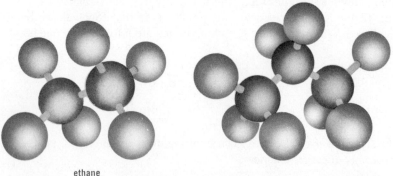

ethane

Fig. 26–3. Ball-and-stick models of ethane and propane. Each stick represents one covalent bond and each ball an atom (black for carbon and red for hydrogen).

Structure of Alkene Molecules. Alkene molecules (C_nH_{2n}) have two hydrogen atoms less than the corresponding alkane molecules (C_nH_{2n+2}), except that an alkene corresponding to methane and having the formula CH_2 does not exist. Apparently, at least two carbon atoms must be present in an alkene molecule to allow a special kind of bond hybridization that brings stability to the molecule. Let us examine proposed structures for the first two members, ethene and propene, to see the type of bonding that is believed to exist.

Measurements show that the bond angles in ethene are not 109°28' but approximately 120°, as shown in Fig. 26-4a, and that the molecule is planar. Moreover, the distance C—C is 1.34 A as compared with 1.54 A in ethane. These facts suggest that all the bonds in ethene are different from the sp^3 bonds of ethane. It is thought that one s and only two p orbitals hybridize to form three equivalent sp^2 bonding orbitals, which extend toward the corners of an equilateral triangle, with the carbon nucleus at the center (the remaining p orbital with its one electron will be dealt with in the next paragraph). Two hydrogen atoms are then bonded to a carbon atom, each by a sigma bond resulting from the overlap of a hydrogen s orbital and a carbon sp^2 orbital. The two carbon atoms are bonded (in part) by the overlap of the remaining sp^2 orbital of each. This much of the structure of ethene is shown in Fig. 26-4b.

The remaining p orbital of each carbon atom contains one electron and lies above and below the plane of the sp^2 bond orbitals with the axis of its lobes perpendicular to this plane; the two axes are parallel to one another. Because of the nearness of the two carbon atoms, the p orbitals overlap laterally (along the edges rather than head-on) to form a second bond, called a **pi** bond, between the two carbon atoms (Fig. 26-4c). Thus, the carbon atoms in ethene can be thought of as being joined by a *double* bond consisting of a sigma bond and a pi bond (Fig. 26-4c). Thermochemical measurements of bond energies reveal that the energy of the double bond in ethene is not twice that of the sigma C—C bond in ethane, a fact that indicates that the pi bond is a weaker bond than the sigma bond.

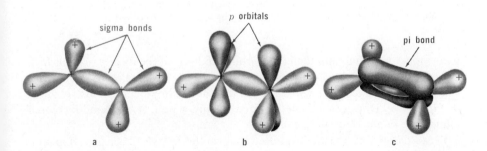

Fig. 26-4. Structure of ethene. (a) The six nuclei of C and H, shown with + signs, lie in a single plane. Only the sigma bonding orbitals are shown. (b) The two electrons not involved in sp^2 sigma bond formation are shown as occupying p atomic orbitals. (c) The p orbitals shown in (b) are now represented as having interpenetrated laterally to form the pi bond.

The bonding system that we have discussed for ethene is thought to exist for one pair of carbon atoms in the molecules of all ethene homologs, the remaining atoms being bonded as in the alkanes. The double bond is usually shown in structural formulas by a pair of parallel lines that join the two carbons bonded by the pi and the sigma bonds as shown by the following formulas:

$$\begin{array}{c} H\ H \\ |\ \ | \\ C{=}C \\ |\ \ | \\ H\ H \end{array} \quad \text{or} \quad CH_2{=}CH_2 \qquad \begin{array}{c} H \\ | \\ H{-}C{-}C{=}C{-}H \\ |\ \ |\ \ | \\ H\ H\ H \end{array} \quad \text{or} \quad CH_3CH{=}CH_2$$

ethene propene

Bond angles or positions of atoms in space are not shown by these formulas.

Structure of Alkyne Molecules. It can be shown by chemical reactions and by physical methods, such as spectroscopic methods, that any alkyne molecule contains a pair of carbon atoms that behave differently from any carbon atoms in alkane or alkene molecules. Let us consider the first member of the series, ethyne (acetylene, C_2H_2), to see what is characteristic of the structure of alkyne molecules.

Ethyne molecules are linear; that is, all bond angles are 180°, as shown in Fig. 26-5a. It is thought that one of the s electrons in a carbon atom is promoted

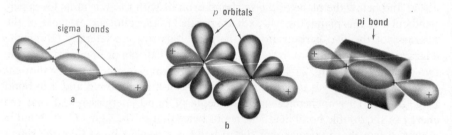

Fig. 26–5. Structure of ethyne. (*a*) Sigma bonds join the C and H nuclei, shown with + signs, at bond angles of 180°. (*b*) The *p* atomic orbitals not involved in *sp* hybridization are shown as they might exist in isolated carbon atoms. (*c*) The *p* orbitals shown in (*b*) are now represented as having interpenetrated to form an electron cloud that surrounds the carbon nuclei. This electron cloud includes 4 electrons and constitutes 2 pi bonds.

to an empty *p* orbital, and then the *s* orbital and one of the three *p* orbitals form two *sp* equivalent hybrid orbitals whose axes lie at 180° angles to each other. The two carbon and hydrogen atoms are then bonded by sigma bonds, as shown in Fig. 26-5a. However, the sigma bond is only a part of the bonding system joining the two carbon atoms.

Because only one *p* orbital was used to form the *sp* hybrid, each carbon atom possesses two ordinary *p* orbitals, each orbital containing one electron (Fig. 26-5b). The carbon atoms are close enough to each other for the *p* orbitals to overlap laterally to form two pi bonds. Furthermore, the two pi bonds are thought to interact with each other to form an electron cloud that is cylindrically symmetrical around the line joining the two carbon nuclei, as shown in Fig. 26-5c.

A simple representation of the structures of the alkyne molecules shows the three bonds (*triple* bond) with three parallel lines:

$$H-C\equiv C-H \qquad H-\underset{H}{\overset{H}{C}}-C\equiv C-H$$

ethyne propyne

Structure of Benzene Hydrocarbons. Each benzene hydrocarbon molecule contains a group of six carbon atoms linked together in the form of a hexagonal ring and lying in a single plane. In benzene itself, a hydrogen atom is attached to each carbon atom. In toluene, the next member of the series, a hydrogen atom has been replaced by a methyl group. These ideas were first incorporated into structural formulas by the German chemist F. A. Kekulé in 1865. In order to allow for the tetravalence of carbon, he postulated alternate double and single bonds between the carbon atoms, as shown in the following formulas:

benzene   toluene   benzene toluene
(abbreviated formulas)

Because of the presence of the three double bonds, the Kekulé formula suggests that benzene hydrocarbons have chemical properties very much like the alkene hydrocarbons. They do not. Also, the Kekulé formula suggests that the distances from carbon to carbon in the ring are alternately the distance between two adjacent carbon atoms in alkane molecules (1.54 A) and the distance between two carbon atoms joined by a double bond in alkene molecules (1.34 A). However, the distances, measured by X-ray diffraction, are uniform around the ring and are 1.39 A. Today the benzene molecule is considered to be planar, with all carbon-carbon bonds the same length and all bond angles 120°, as shown in Fig. 26-6a.

The 120° bond angle indicates sp^2 hybridization of the type described in alkene molecules. The bonds shown in Fig. 26-6a are sigma bonds, resulting from the overlap of an sp^2 orbital of a carbon atom with an sp^2 orbital of a second carbon or with an s orbital of hydrogen.

The fourth valence electron of carbon is not involved in sp^2 hybridization and may be thought of for the moment as occupying a normal p orbital that lies above and below the plane containing the carbon nuclei (Fig. 26-6b). Because there are six carbon atoms in the ring, there are six of these orbitals, all parallel to one another and close enough for lateral overlap, i.e., pi bond formation. However, each p orbital overlaps two neighboring p orbitals, one on either side, thus resulting in two doughnut-shaped electron clouds. One such doughnut lies above and one lies below the plane through the carbon nuclei (Fig. 26-6c).

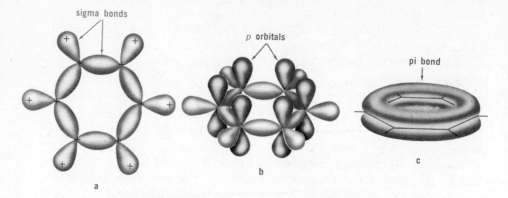

Fig. 26–6. Structure of benzene. (a) Sigma bonds involving sp^2 and s orbitals are shown. The C and H nuclei, indicated with + signs, lie in the plane of the paper. (b) The plane containing the C and H nuclei is now shown almost perpendicular to the plane of the paper. The atomic p orbitals not involved in sigma bond formation are indicated above and below this plane. (c) The p orbitals shown in (b) are shown here as having interpenetrated laterally to form doughnut-shaped electron clouds above and below the plane containing the C and H nuclei.

Since the discovery of benzene by Michael Faraday in 1825, several graphic representations have been proposed. Some are

Kekulé 1865 Claus 1867 Dewar 1867 Armstrong-Baeyer 1887 a

The graphic representation a has been proposed recently by several chemists and has had considerable acceptance. In this representation, the hexagon refers to that part of the benzene molecule shown in Fig. 26-6a and the circle refers to the delocalized pi electrons shown in Fig. 26-6c. Many chemists, especially organic chemists, continue to use the Kekulé representation. In this text, we shall use the circle formula unless it is important to center attention on specific bonds joining pairs of carbons, in which case we shall use the Kekulé formula.

SATURATED AND UNSATURATED HYDROCARBONS. The carbon atoms in alkane molecules (and cycloalkanes to be discussed later in the chapter) are joined by single bonds only. Such hydrocarbons are called **saturated hydrocarbons.** On the other hand, at least one pair of carbon atoms in alkene, alkyne, and benzene hydrocarbon molecules are joined by multiple bonds, a condition represented in structural formulas with double and triple bonds for alkenes and alkynes, or with a circle inside a hexagon for benzene hydrocarbons. Hydrocarbons with one or more pairs of carbon atoms joined by multiple bonds are called **unsaturated hydrocarbons.**

Organic Chemistry I: Hydrocarbons

ISOMERISM

Compounds that have the same molecular formula but different structural formulas are *isomers* (Greek, *isos*, equal, and *meros*, part). The phenomenon is known as **isomerism**. Isomeric compounds are not possible among the alkanes till there are enough carbon atoms to permit more than one arrangement of the carbon chain. For each of the compounds CH_4, C_2H_6, and C_3H_8, there is only one possible arrangement of the carbon chain. For C_4H_{10}, only two arrangements of the atoms are possible, and only two compounds with the formula C_4H_{10} have been discovered. For C_5H_{12} three arrangements are possible, and only three have been discovered. As the molecules become more complex, the numbers of isomers increase markedly (see Table 26-1).

Isomeric compounds differ both chemically and physically from one another. They can be identified experimentally by differences in melting point, boiling point, refractive index, and chemical activity. The structural formulas and boiling points for the two isomeric butanes and the three isomeric pentanes are as follows:

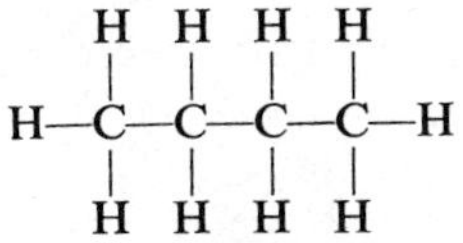

butane
(normal butane)
bp 0.5°

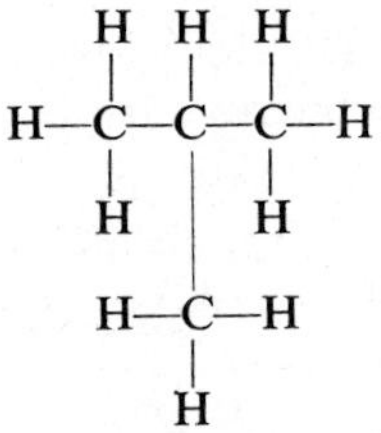

methylpropane
(isobutane)
bp −11.7°

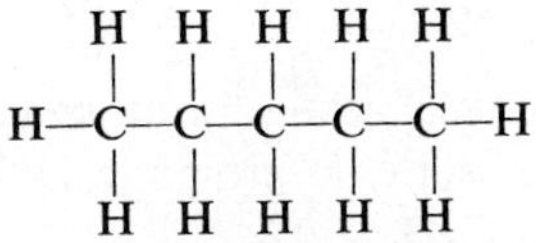

pentane
(normal pentane)
bp 36.5°

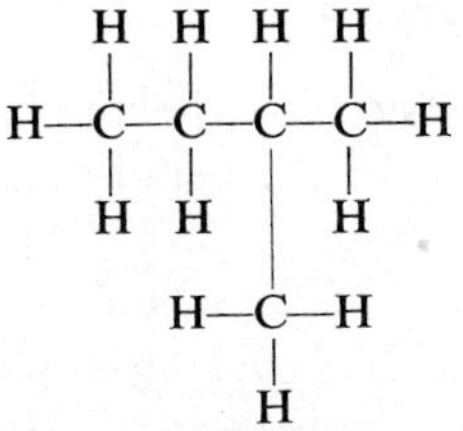

methylbutane
(isopentane)
bp 27.9°

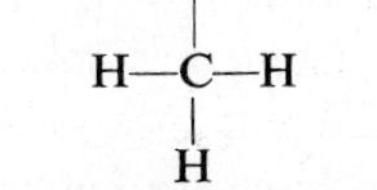

dimethylpropane
(neopentane)
bp 9.5°

In the examples of isomerism just cited, the difference in structure arises from a different arrangement of the carbon atoms. That is, in normal butane, the four carbon atoms are joined in a continuous or unbranched chain, but in isobutane, three carbon atoms are in a continuous chain arrangement, and the fourth carbon is joined to the middle carbon as a branch. Note that the isomer that has a continuous carbon chain without carbon branches is named as the normal isomer. The prefixes *iso-* and *neo-* are sometimes used to indicate branched-chain isomers.

There are many other ways by which isomeric compounds can result. For example, among the alkenes and alkynes, isomeric compounds can result when double and triple bonds occur at different places in the molecules. To illustrate this, let us consider the number of isomeric butenes, C_4H_8, theoretically possible.

If we use ball and stick models to represent possible molecules, we find that there are three different ways of organizing the four carbon atoms and the double bond into chain-type molecules, as shown by the structural formulas below. (One of these, 2-butene, represents two isomers, for there are two spatial arrangements of the groups attached to the chain, see page 505.)

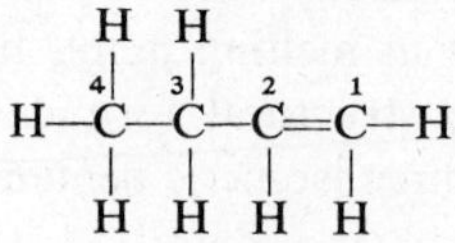

1-butene
bp −5°

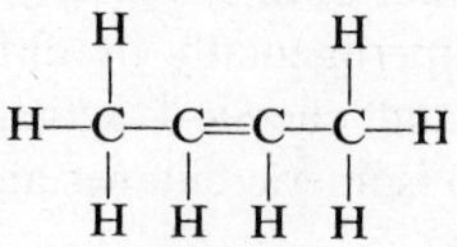

2-butene
bp 2.5°

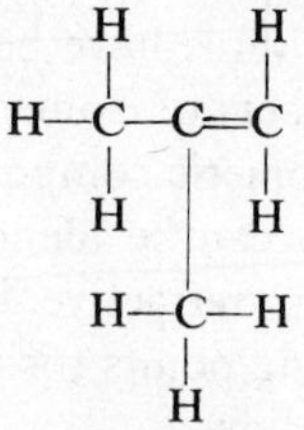

methylpropene
bp −6°

Again using balls and sticks, we find that we can construct only two model molecules that contain a triple bond and have the composition C_4H_6 (butyne). Instead of four isomers, as was the case for the butenes, C_4H_8, only two isomeric butynes, C_4H_6, are predicted, and only two have ever been discovered. Their structural formulas are given below:

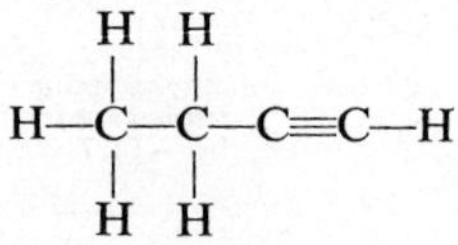

1-butyne

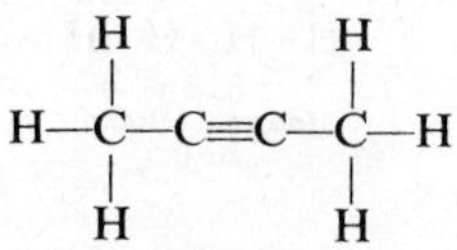

2-butyne

It is a fact of great importance that the number of isomers predicted by putting together models is precisely the number of isomers found by the most painstaking laboratory experimentations.

Isomeric compounds result if certain atoms or groups of atoms are attached at different locations in molecules. For example, four isomeric aromatic hydrocarbons are known that have the molecular formula C_8H_{10}:

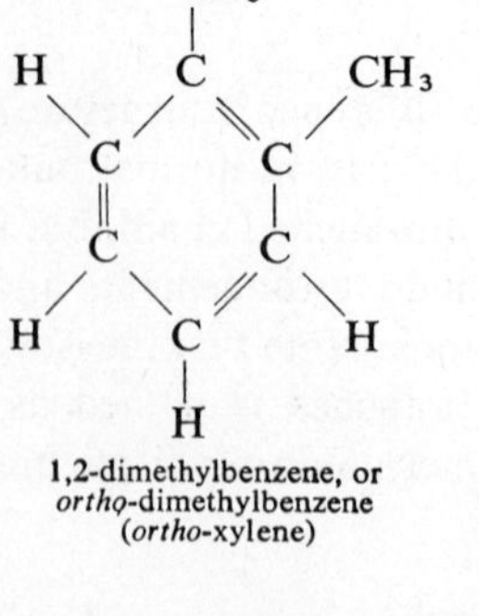

1,2-dimethylbenzene, or
ortho-dimethylbenzene
(*ortho*-xylene)

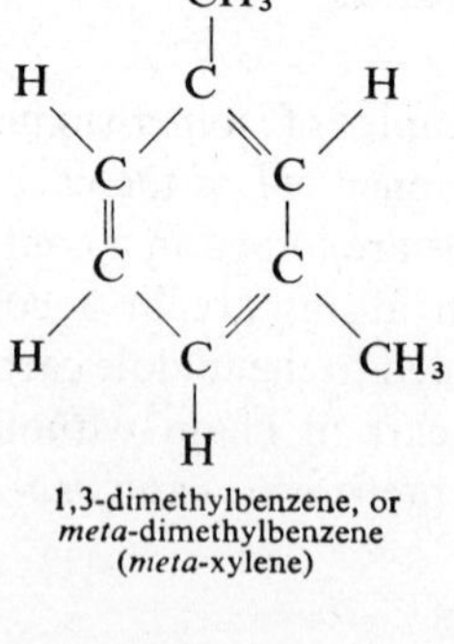

1,3-dimethylbenzene, or
meta-dimethylbenzene
(*meta*-xylene)

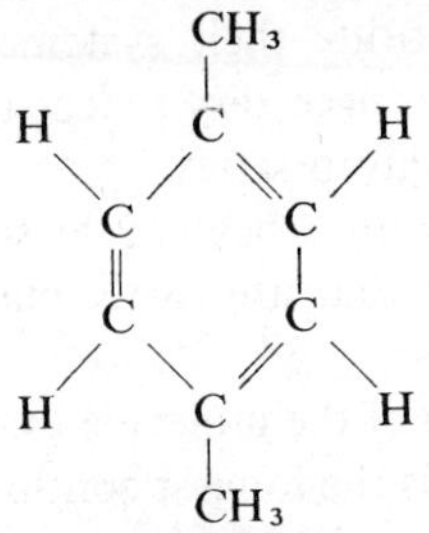

Organic Chemistry I: Hydrocarbons

[Structure of 1,4-dimethylbenzene (para-dimethylbenzene, para-xylene) with CH₃ groups at positions 1 and 4 of benzene ring]

1,4-dimethylbenzene, or
para-dimethylbenzene
(para-xylene)

[Structure of ethylbenzene with CH₂CH₃ group on benzene ring]

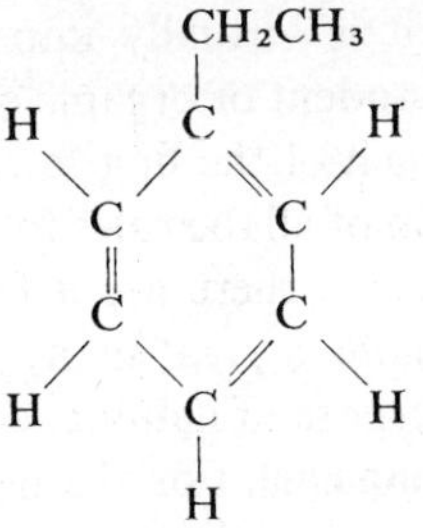

ethylbenzene

Other differences in structure that can give rise to isomerism are discussed in later chapters.

ABBREVIATED STRUCTURAL FORMULAS. The structural formulas written thus far for isomers are unnecessarily cumbersome and can be condensed, once we understand clearly what ideas they convey. For example, structural formulas for the four isomers of C_8H_{10} can be condensed to

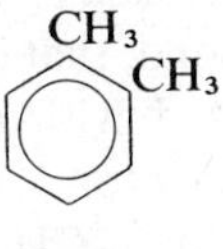

[Four condensed structures of benzene rings with substituents:]

ortho-
dimethylbenzene

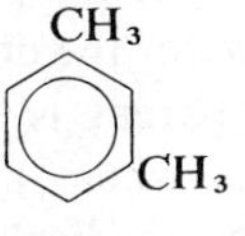

meta-
dimethylbenzene

para-
dimethylbenzene

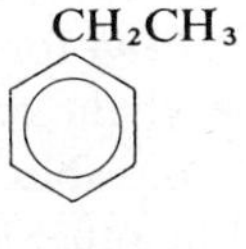

ethylbenzene

The isomeric butenes shown in the preceding section can be represented thus:

$CH_3CH_2CH{=}CH_2$ $CH_3CH{=}CHCH_3$ $CH_3\underset{|}{\overset{CH_3}{C}}{=}CH_2$

1-butene   2-butene   methylpropene

In order to save space, we shall frequently use condensed structural formulas, but the student should be able to translate these semistructural formulas into more complete versions.

SYSTEMATIC NOMENCLATURE

The systematic naming of a million-odd compounds is a vast problem indeed. Before the turn of the century the collection of known substances became so huge that a special international meeting, the Geneva Conference of 1892, was called to establish a systematic scheme of naming compounds. The rules of naming agreed on at this conference[4] are known as the Geneva System. One of the key rules establishes similar names for members of homologous series, for example, the alkanes.

To complicate matters, however, the first few members of a given series have often been known for a long time, and their names do not agree with the systematic names. These common names have become so firmly fixed in the chemical

---

[4] The rules have been kept up to date by later conferences sponsored by the International Union of Pure and Applied Chemistry.

literature that it is virtually impossible to substitute their systematic names. Therefore, the student of organic chemistry finds it necessary to learn more than one name for each of the first few members of a given series.

The discussion of all the rules for systematic naming is beyond the scope of this book. Nevertheless, there are a few simple rules that are easily mastered and enable one to name a great many compounds:

1. For the purpose of naming, choose a portion of the molecule that can serve as the **parent compound.** For alkanes, this portion is the longest continuous chain of carbon atoms; for alkenes and alkynes, it is usually the longest continuous chain containing the double or triple bond; for benzene hydrocarbons, it is usually the benzene ring.

2. For aliphatic hydrocarbons, the number of carbon atoms in the parent compound is denoted by the proper prefix, that is, meth-, eth-, prop-, but-, pent-, etc., for 1, 2, 3, 4, 5, etc., carbon atoms, respectively. The suffix -ane, -ene, or -yne is added to denote the hydrocarbon class. That is, if the longest chain of carbon atoms contains seven carbon atoms joined as a continuous chain and contains a double bond, the name of the parent compound is made by combining the prefix *hept-* for seven and the suffix *-ene* for the double bond, that is, *heptene*.

3. The chain or benzene ring of the parent is numbered with the lowest numbers possible to locate the positions where hydrogen atoms have been replaced by other atoms or groups of atoms (radicals), such as —CH_3 (methyl), —CH_2CH_3 (ethyl), —$CH_2CH_2CH_3$ (normal propyl), —$\underset{\underset{CH_3}{|}}{CHCH_3}$

(isopropyl), etc. *Normal* and *iso* may be abbreviated *n-* and *i-*, respectively. The positions 1, 2-; 1, 3-; and 1, 4- in the benzene ring may also be designated by *ortho*, *meta*, and *para* (abbreviated *o*, *m*, and *p*), respectively.

4. The complete name of the compound is a combination of (*a*) the name of the parent portion, (*b*) the names of the attached radicals, and (*c*) the numbers that show where the radicals are attached. (These same rules may also be applied to the naming of certain compounds that are not hydrocarbons.)

Let us apply these rules to the naming of an isomer of nonane that has the structure shown below:

$$\underset{\underset{\underset{\underset{H}{|}}{\underset{H-C-H}{|}}}{\underset{H-C-H}{|}}}{\overset{H}{\underset{|}{H}}\overset{H}{\underset{|}{-}}\overset{H}{\underset{|}{C}}\overset{H}{\underset{|}{-}}\overset{H}{\underset{|}{C}}\overset{H}{\underset{}{-}}\overset{H}{\underset{|}{C}}\overset{H}{\underset{|}{-}}\overset{H}{\underset{}{C}}\overset{H}{\underset{|}{-}}\overset{H}{\underset{|}{C}}\overset{H}{\underset{|}{-}}\overset{H}{\underset{|}{C}}-H} \quad \text{or} \quad \overset{6}{CH_3}\overset{5}{CH_2}\overset{4}{\underset{\underset{CH_3}{|}}{\underset{CH_2}{|}}}\overset{3}{CH}\overset{2}{\underset{\underset{}{|}}{\underset{CH_3}{|}}}\overset{1}{CH_3}$$

a. The longest continuous chain of carbon atoms contains six atoms, all joined by single bonds. This part is to be considered the parent hydrocarbon and is named *hexane* (*hex-* for six and *-ane* for single bonds).

Organic Chemistry I: Hydrocarbons

b. The hexane chain is numbered from right to left so as to locate with the lowest possible numbers the positions where hydrogen atoms of the parent hydrocarbon have been replaced. These numbers, 2- and 4-, are used to designate the positions in the molecule of the —CH$_3$ and —CH$_2$CH$_3$ radicals. Note that, if the chain is numbered in the opposite direction, the numbers 3- and 5- would have to be used to locate these groups. Because we wish to use the lowest numbers possible in the name, we choose to number this chain from right to left.

c. The complete name is 2-methyl-4-ethylhexane.

The structural formulas and names for several compounds, some of which are not hydrocarbons, follow. Study these to see how the names are derived from the rules that we have stated. The student is also advised to study the names given in earlier pages of this chapter in the light of these rules:

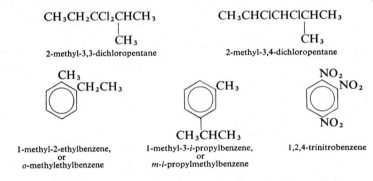

Double and triple bonds are located by mentioning the lower of the numbers of the two carbon atoms joined by the multiple bond:

$$\overset{1}{C}H_3\overset{2}{C}H=\overset{3}{C}H\overset{4}{C}H_2\overset{5}{C}H_3$$

2-pentene

$$\overset{6}{C}H_3\overset{5}{C}H_2\overset{4}{C}H_2\overset{3}{C}HCH_2CH_3$$
$$\underset{2\ \ \ 1}{C\equiv CH}$$

3-ethyl-1-hexyne

PROPERTIES OF HYDROCARBONS

Physical Properties. The alkanes, alkenes, alkynes, and benzene hydrocarbons are very much alike physically. All are colorless compounds, insoluble or only slightly soluble in water, but quite soluble in nonpolar solvents. The hydrocarbons with low molecular weights, C$_1$ to about C$_5$, are gases; those with intermediate weights are liquids; and those with high molecular weights are solids (compare Table 26-2). The actual melting and boiling points vary for molecules with the same number of carbon atoms, depending on the presence or absence of double and triple bonds and the number and kind of chain branches. For the unbranched-chain members of a given aliphatic series, the melting and boiling points increase fairly regularly as the molecular weight increases (Fig. 26-7). The boiling points of the branched-chain hydrocarbons are generally lower than those of their unbranched isomers. The melting and boiling points of benzene hydrocarbons are usually higher than the aliphatic hydrocarbons with the same number of carbon atoms; for example, the boiling points of hexane, 1-hexene, 1-hexyne, and benzene are 69, 64, 72, and 80°C, respectively.

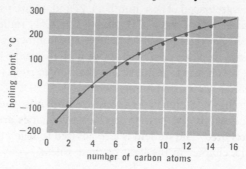

Fig. 26-7. Boiling points of the first 15 normal alkanes.

Chemical Properties. COMBUSTION. All hydrocarbons have one important chemical property in common. They are all combustible, burning in oxygen or air to carbon dioxide and water.

$$C_2H_6 + 3\tfrac{1}{2}O_2 \rightarrow 2CO_2 + 3H_2O$$
ethane

$$C_2H_4 + 3O_2 \rightarrow 2CO_2 + 2H_2O$$
ethene

$$C_2H_2 + 2\tfrac{1}{2}O_2 \rightarrow 2CO_2 + H_2O$$
ethyne

$$C_6H_6 + 7\tfrac{1}{2}O_2 \rightarrow 6CO_2 + 3H_2O$$
benzene

The combustion of hydrocarbons in a gasoline engine, a diesel engine, an oil furnace, or a gas stove does not take place so completely as the above equations imply. Some carbon monoxide, carbon (soot), and even hydrocarbons may constitute an appreciable fraction of the combustion products, depending on how the combustion is carried out. When natural gas, consisting mainly of methane with some ethane and propane, is burned in a laboratory burner or a gas stove, the ratio of air to gas and the mixing of the two can be controlled so as to give very little carbon monoxide, carbon, and other undesired combustion products. On the other hand, when diesel fuel ($C_{14}H_{30}$, $C_{15}H_{32}$, etc.) is injected into the combustion chamber as a spray, the mixing with air is not nearly so good, and the combustion products contain an undesirable amount of unburned products. The same is true, usually to a lesser extent, of combustion of gasoline in an automobile engine.

ADDITION REACTIONS. In alkenes, alkynes, and benzene hydrocarbons, two or more carbon atoms are joined by both sigma and pi bonds. Under the proper conditions, atoms or groups of atoms react with these carbons by *addition*. During the addition, pi bonds change to sp^3 orbitals and then form sigma bonds with the atoms being added to the molecule. The net result is that addition occurs without cleavage of the sigma bond between the pair of carbon atoms. Reactions of this type are called **addition reactions.** Examples follow.

Organic Chemistry I: Hydrocarbons

$$H-\underset{\underset{H}{|}}{\overset{\overset{H}{|}}{C}}=\underset{\underset{H}{|}}{\overset{\overset{H}{|}}{C}}-H + H_2 \xrightarrow{Ni} H-\underset{\underset{H}{|}}{\overset{\overset{H}{|}}{C}}-\underset{\underset{H}{|}}{\overset{\overset{H}{|}}{C}}-H$$

ethene → ethane

$$CH_3C{\equiv}C-H + 2Cl_2 \longrightarrow CH_3\underset{\underset{Cl}{|}}{\overset{\overset{Cl}{|}}{C}}-\underset{\underset{Cl}{|}}{\overset{\overset{Cl}{|}}{C}}-H$$

propyne → 1,1,2,2-tetrachloropropane

benzene + $3H_2 \xrightarrow{Pt}$ cyclohexane

The first and third of the foregoing reactions are examples of *hydrogenation* reactions, i.e., the addition of hydrogen to a multiple bond.

SUBSTITUTION REACTIONS OF ALKANES. The alkanes are inactive toward common laboratory reagents, such as acids, bases, and oxidizing agents, unless the temperature is unusually high. And they do not participate in addition reactions; they are *saturated compounds*.

At elevated temperatures or in the presence of ultraviolet light, one or more hydrogen atoms in an alkane molecule may be replaced by atoms of chlorine, bromine, or other similar groups:

$$CH_4 + Cl_2 \rightarrow CH_3Cl + HCl$$

$$CH_3Cl + Cl_2 \rightarrow CH_2Cl_2 + HCl$$

$$C_2H_6 + Br_2 \rightarrow CH_3CH_2Br + HBr$$

SOURCES OF HYDROCARBONS

Coal. Coal has long been an indirect source of many aromatic compounds. On destructively distilling coal to produce coke, some of the coal is converted into a black, viscous liquid called *coal tar*. This tar-like liquid is separated into pure compounds by fractional distillation, together with crystallization processes, centrifuging, and solvent extraction. One ton of coal yields about 140 lb of tar, along with about 1500 lb of coke, 10,000 cu ft of coal gas (mostly hydrogen and methane), and about 25 lb of ammonium salts. The 140 lb of coal tar in turn produces from about 14 to 18 lb of naphthalene, 1 lb of benzene, 3 lb of toluene and the xylenes, varying amounts of many other aromatic compounds, and about 70 lb of pitch.

Natural Gas. Natural gas, often encountered in drilling wells for petroleum, contains from 50 to 94 percent methane. Some ethane, propane, and other paraffin hydrocarbons of low molecular weight are also present. Natural gas is thought to have originated as petroleum did, i.e., from plant and animal materials.

Petroleum. Petroleum is a complex liquid solution of gaseous, liquid, and solid hydrocarbons. Also present are small amounts of nitrogen and sulfur compounds. The low-distilling hydrocarbons that make up petroleum ether, gasoline, and kerosene belong mostly to the methane series. The crude oil remaining after the volatile hydrocarbons have been removed may be composed mainly of alkanes (paraffin base) or of cycloparaffins (asphalt base), both with high molecular weights. The present U.S. rate of production of crude petroleum is about 8 million barrels per day.

THE REFINING OF PETROLEUM. The first step in refining petroleum consists of separating petroleum by **fractional distillation** into batches with different boiling ranges (Table 26-2). Crude oil is pumped continuously through a furnace where it is heated. The hot liquid then passes into a "flash chamber" where, by

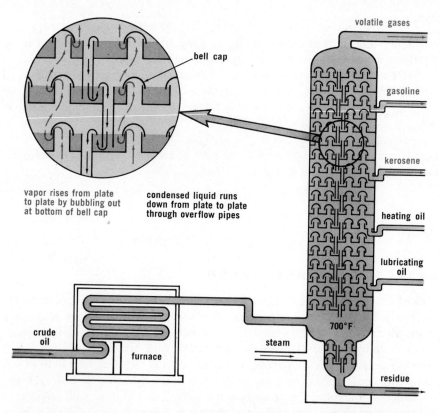

Fig. 26–8. Diagram of a fractionating tower for petroleum distillation. The cutaway view shows how the vapor and liquid phases are kept in contact with each other so that condensation and distillation occur throughout the column.

lowering the pressure, the lower-boiling components are vaporized. The vapors pass into a fractionating tower (see Fig. 26-8). Here the higher-boiling components are condensed in the lower portion of the tower and flow downward from plate to plate. The ascending vapors bubble through the liquids that have condensed on the plates; hence the more volatile lower-boiling components collect in the upper portion of the tower. Distillation carried out in this manner affords an excellent separation into portions called **fractions**. Each fraction consists of a mixture of compounds, boils in a certain range of temperature, and is of rather definite composition. Practically speaking, each plate represents a separate distilling unit, and the whole process amounts to several distillations.

TABLE 26-2.

Hydrocarbon Fractions Obtained from Petroleum

FRACTION	SIZE RANGE OF MOLECULES	BOILING-POINT RANGE (°C)	USES
Gas	C_1–C_5	$-164°$ to $30°$	Gaseous fuel; production of carbon black, hydrogen, or gasoline (by polymerization)
Petroleum ether (ligroin)	C_5–C_7	30° to 90°	Solvent; dry cleaning
Gasoline (straight-run)	C_5–C_{12}	30° to 200°	Motor fuel (gasoline)
Kerosene	C_{12}–C_{16}	175° to 275°	Illuminant; fuel
Gas oil, fuel oil, and Diesel oil	C_{15}–C_{18}	250° to 400°	Furnace fuel; fuel for Diesel engines; cracking
Lubricating oils, greases, vaseline	C_{16} up	350° up	Lubrication
Paraffin (wax)	C_{20} up	melts 52°–57°	Candles; waterproofing fabrics; matches; home canning
Pitch and tar		residue	Artificial asphalt
Petroleum coke		residue	Fuel; electrodes

EXERCISES

1. List the objects you can see from where you are sitting that are probably organic.
2. Suggest a test that one might use to determine if a newly discovered compound is organic or inorganic.
3. What discoveries aided in discrediting the vital force theory? Over what period of time did this occur and what was the effect on the development of the field of organic chemistry?
4. Write the general formulas for the four series of hydrocarbons. Give name and structural formula for the simplest member of each series.

5. Could a homologous series of hydrocarbons exist in which each member differs from a preceding member by one carbon and three hydrogen atoms? Explain.
6. Write the molecular formula of a homolog of the alkane series containing 21 carbon atoms; of a homolog of the alkyne series containing 21 carbon atoms.
7. Why is it necessary to show a different kind of bond hybridization for ethene molecules from that used for ethane molecules and still another kind for ethyne molecules?
8. Show with a sketch similar to Fig. 26-1 the sigma bonds in the two isomeric butanes, C_4H_{10}.
9. Show with a sketch similar to Fig. 26-5 the bonding orbitals in propyne.
10. What distinguishes a sigma bond from a pi bond?
11. Write structural formulas for all the isomers of C_6H_{14}. Name each.
12. Write structural formulas for 1-pentene, 3-methyl-1-hexyne, 1,3-dichloro-2-methylbenzene.
13. Write condensed molecular formulas for the compounds mentioned in Exercises 11 and 12.
14. What determines if a compound is classed as an aliphatic hydrocarbon or as an aromatic hydrocarbon?
15. Write condensed structural formulas for all the benzene hydrocarbons with molecular formula C_9H_{12}. Name each.
16. What is the source of most aliphatic hydrocarbons? How are they separated from each other? What is the source of most aromatic hydrocarbons?
17. Distinguish between addition reactions and substitution reactions. Will alkanes undergo addition reactions? Why or why not?
18. Write the equation for the reaction of ethane with chlorine to produce monochloroethane. The substitution of a second chlorine atom into the molecule produces two compounds. What are they?
19. How does kerosene differ chemically from gasoline?
20. Starting with 1,2,3-trichlorobenzene, replacement of a hydrogen atom with another chlorine would give rise to how many isomeric tetrachlorobenzenes? Starting with 1,2,4-trichlorobenzene, how many isomeric tetrachlorobenzenes are possible?
21. Write an equation for each of the reactions indicated below. Name each product:
    (a) the complete addition of hydrogen to toluene;
    (b) the addition of nitric acid to benzene;
    (c) the complete combustion of gasoline.
22. Write equations for the action of each of the following on benzene: (a) hydrogen; (b) oxygen at high temperatures; (c) chlorine.

SUGGESTED READING

Allinger, N. L., "Energy Relations in Teaching Organic Chemistry," *J. Chem. Educ.*, **40**, 201 (1963).

Benfey, O. T., "August Kekulé and the Birth of the Structural Theory of Organic Chemistry in 1858," *J. Chem. Educ.*, **35**, 21 (1958).

Bykov, G. V., "The Origin of the Theory of Chemical Structure," *J. Chem. Educ.*, **39**, 220 (1962).

Kimberlin, C. N., Jr., "Chemistry in the Manufacture of Modern Gasoline," *J. Chem. Educ.*, **34**, 569 (1957).

Mills, G. A., "Ubiquitous Hydrocarbons," *Chemistry*, **44** (2), 8 (1971).

Orchin, M., "Determining the Number of Isomers from a Structural Formula," *Chemistry*, **42** (5), 8 (1969).

27

ORGANIC CHEMISTRY II: DERIVATIVES OF THE HYDROCARBONS

The hydrocarbons serve as the basis for the classification and systematic naming of all organic compounds. In this classification, a chain or ring type of non-hydrocarbon compound is considered as being derived from the hydrocarbon that contains the same carbon chain or ring. For example, 1-chloropropene is considered to be a derivative of propene.

```
      H                            H
      |                            |
  H—C—C=C—Cl                   H—C—C=C—H
    | |                          | | |
    H H H                        H H H
   1-chloropropene              propene
                             (parent hydrocarbon)
```

Functional Groups. Replacement of one or more hydrogen atoms of a hydrocarbon molecule with oxygen, chlorine, hydroxyl, nitrate, or other groups of atoms produces chemically active centers in an otherwise less active molecule. These groups are able to function in a variety of reactions and hence are referred to as **functional groups**. The compounds in which hydrogen atoms have been replaced in this way include alcohols, ethers, aldehydes, etc.

DERIVATIVES OF ALKANE HYDROCARBONS

ALCOHOLS

Derivatives whose molecules contain one or more hydroxyl (—OH) groups in place of hydrogen atoms are known as **alcohols**. Hundreds of alcohols, derived from the alkanes, alkenes, alkynes, and other hydrocarbons, are known. The simplest alcohols are derived from the alkanes and contain only one hydroxyl group per molecule. These have the general formula ROH, where R is an alkyl radical of the composition —C_nH_{2n+1}. The first few members are as follows:

CH_3OH	Methanol (methyl alcohol)	C_3H_7OH	The propanols
C_2H_5OH	Ethanol (ethyl alcohol)	C_4H_9OH	The butanols

Since all the hydrogen atoms in molecules of methane or in molecules of ethane are equivalent, there are no isomeric methanols or ethanols. The structural formulas for these two compounds are shown below:

A ball-and-stick model of a molecule of ethanol is shown in Fig. 27-1.

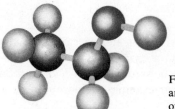

Fig. 27-1. Ball-and-stick model of a molecule of ethanol: carbon atoms, black; hydrogen atoms, light red; oxygen atom, dark red.

In the case of propane, the hydrogen atoms attached to the middle carbon are not equivalent to the hydrogen atoms attached to the end carbons. Replacement of a hydrogen atom on an end carbon with a hydroxyl group forms an alcohol that is different from the alcohol formed by replacement of a hydrogen on the middle carbon. Both of these alcohols are well known.

```
    H H H                           H
    | | |                           |
                                H O H
                                | | |
H—C—C—C—O—H               H—C—C—C—H
    | | |                       | | |
    H H H                       H H H
   1-propanol                  2-propanol
(normal propyl alcohol)     (isopropyl alcohol)
     b.p. 97°                   b.p. 82°
```

Organic Chemistry II: Derivatives of the Hydrocarbons

Nomenclature. The name of the hydrocarbon radical along with the word *alcohol* suffices for the first few members. Methyl alcohol, ethyl alcohol, and normal propyl alcohol are examples. Unfortunately, people are prone to consider the word *alcohol* as the only significant part of the name; many have lost their lives by drinking alcohol that was not ethyl alcohol. Consequently most manufacturers now use the systematic names instead of common ones, i.e., methanol instead of methyl alcohol, ethanol instead of ethyl alcohol.

In the systematic naming of simple and complex alcohols alike, the following principles are followed:

1. The longest carbon chain that contains the hydroxyl group is considered as the parent compound.
2. The *-e* in which the name of this carbon chain ends is changed to *-ol*.
3. The location of the hydroxyl and any other groups is shown by the smallest possible numbers.

These principles are illustrated in the preceding section in the case of methanol, ethanol, and the isomeric propanols.

Physical Properties. In contrast to methane (boiling point $-161°C$), methyl alcohol boils at 65°C; and in general, other alcohols boil at considerably higher temperatures than their parent hydrocarbons. This is thought to be due to the *association* of alcohol molecules through hydrogen bonding. (The dashed line indicates the hydrogen bond.)

$$H:\overset{H}{\underset{H}{\ddot{C}}}:\ddot{O}:H + H:\overset{H}{\underset{H}{\ddot{C}}}:\ddot{O}:H \rightarrow H:\overset{H}{\underset{H}{\ddot{C}}}:\ddot{O}:H - :\overset{H:\overset{H}{\underset{H}{\ddot{C}}}:H}{\ddot{O}}:$$

It will be remembered that water has a relatively high boiling point for the same reason.

Also, unlike the parent hydrocarbons, the alcohols with low molecular weights

TABLE 27-1.

Saturated Aliphatic Alcohols

ALCOHOL	FORMULA	MELTING POINT	BOILING POINT	DENSITY G/ML	SOLUBILITY IN 100 G OF WATER
Methanol	CH_3OH	$-98°C$	65°C	0.793	Completely miscible
Ethanol	CH_3CH_2OH	$-117°$	78°	0.789	Completely miscible
1-Propanol	$CH_3CH_2CH_2OH$	$-127°$	97°	0.804	Completely miscible
1-Butanol	$CH_3CH_2CH_2CH_2OH$	$-89°$	118°	0.810	9 g at 15°
1-Pentanol	$CH_3CH_2CH_2CH_2CH_2OH$	$-78°$	138°	0.817	2.7 g at 22°
1-Hexanol	$CH_3CH_2CH_2CH_2CH_2CH_2OH$	$-52°$	156°	0.820	Slight

are very soluble in water. This, too, is accounted for on the basis of hydrogen bonding between the hydroxyl group of the alcohol and the water molecules. However, as the molecular weight increases, the van der Waals forces between the hydrocarbon portions of the alcohol molecules become more effective in attracting alcohol molecules to each other, thereby offsetting the effect of hydrogen bonding. For this reason, methyl alcohol is soluble in water in all proportions, whereas 1-decanol, $C_{10}H_{21}OH$, is insoluble.

The location of the hydroxyl group on the carbon chain and the amount of branching have an important bearing on the melting and boiling points and on the solubility in water.

The properties of certain alcohols are summarized in Table 27-1.

Chemical Properties. In contrast to the paraffin hydrocarbons, the alcohols are quite active chemically because they enter into a number of reactions that involve the hydroxyl group. The hydroxyl group, therefore, is the *functional* group that characterizes alcohols. Only two of the characteristic properties will be mentioned here: (1) dehydration to form ethers and unsaturated hydrocarbons, and (2) partial oxidation to form aldehydes, ketones, and acids. Because of these properties, the alcohols are starting materials in the synthesis of many useful substances. Specific examples are given in the sections that follow.

Some Common Alcohols. METHYL ALCOHOL. Methyl alcohol (methanol, wood alcohol) was once made largely by the destructive distillation of hardwoods such as birch, beech, oak, and maple. One cord of wood yielded about 225 gallons of an aqueous distillate containing up to 6 percent methyl alcohol and 10 percent acetic acid. Today about 99 percent of our output is produced by the catalytic hydrogenation of carbon monoxide.

$$CO + 2H_2 \rightarrow CH_3OH$$

A catalyst (ZnO, Cr_2O_3) and elevated temperature and pressure are required for the success of the reaction.

Our annual production of methyl alcohol is about 4.6 billion pounds. It is used chiefly in the manufacture of formaldehyde, as a radiator antifreeze, and a solvent for shellac and varnishes.

ETHYL ALCOHOL. Ethyl alcohol (ethanol, grain alcohol) is the physiologically active ingredient of beer, wine, and whiskey. It has been produced for centuries by the fermentation of *carbohydrates*. In the fermentation processes, organic compounds are broken down into simpler compounds by the action of *enzymes*. Enzymes are complex organic compounds that originate in living organisms. The production of ethyl alcohol from *starches* (corn, potatoes, rice, etc.) involves first the enzymatic conversion of the starch into sugar (glucose). The sugar is then converted into alcohol and carbon dioxide by the action of *zymase*, an enzyme produced by living yeast cells. The fermentation must be carried out in dilute water solutions, because the yeast cells cannot live and multiply in concentrated sugar or alcohol solutions. The dilute alcohol solutions thus produced are distilled if a more concentrated product is desired. Alcohol produced from starches is used largely in beverages. The reactions follow.

Organic Chemistry II: Derivatives of the Hydrocarbons

$$(C_6H_{10}O_5)_x + xH_2O \rightarrow xC_6H_{12}O_6$$
$$\text{starch} \qquad\qquad\qquad \text{glucose}$$

$$C_6H_{12}O_6 \rightarrow 2C_2H_5OH + 2CO_2$$
$$\text{glucose}$$

Industrial ethyl alcohol (annual production about 2.4 billion pounds) is mostly prepared by two methods: (1) fermentation of black-strap molasses imported from islands in the Caribbean Sea, and (2) the indirect addition of water to ethene, a by-product of the cracking of hydrocarbons to obtain gasoline.

$$CH_2\!\!=\!\!CH_2 + H_2SO_4 \rightarrow CH_3CH_2HSO_4$$

$$CH_3CH_2HSO_4 + H_2O \rightarrow CH_3CH_2OH + H_2SO_4$$

DENATURED ALCOHOL. Denatured alcohol is ethyl alcohol that contains an ingredient that renders it unfit for drinking. Methyl alcohol is sometimes used for this purpose. It is interesting to note that tax-free alcohol sells at about 50 cents per gallon.

Di- and Trihydroxy Alcohols. The simplest and most important alcohol that contains two hydroxy groups is **ethylene glycol**. It is prepared commercially from ethene.

```
        H   H
        |   |
        O   O
        |   |
    H—C—C—H
        |   |
        H   H
```
ethylene glycol, or 1,2-dihydroxyethane

It is extensively used as a permanent antifreeze for automobile radiators.

The most important trihydroxy alcohol is **glycerol,** a derivative of propane.

```
       H   H   H
       |   |   |
       O   O   O
       |   |   |
   H—C—C—C—H
       |   |   |
       H   H   H
```
glycerol, or 1,2,3-trihydroxypropane

Glycerol is a by-product in the manufacture of soap (see Chap. 29) and is also synthesized from propane.

ETHERS

The ethers are a class of compounds that contain two similar or dissimilar hydrocarbon radicals in combination with an atom of oxygen. The general formula is ROR'. Although hundreds of ethers are known, we shall center our attention on a few that are derived from the alkane hydrocarbons, i.e., those in which R is C_nH_{2n+1}. Below are the formulas and names of the first few in the series.

CH$_3$OCH$_3$ Dimethyl ether

CH$_3$CH$_2$OCH$_2$CH$_3$ Diethyl ether

CH$_3$OCH$_2$CH$_3$ Methyl ethyl ether

CH$_3$CH$_2$CH$_2$OCH$_2$CH$_2$CH$_3$ Di-*n*-propyl ether

Unlike alcohols, the ethers cannot form associated molecules by means of hydrogen bonds. Hence the boiling point of an ether is considerably below the boiling point of the alcohol with the same number of carbon atoms. For example, ethyl alcohol is isomeric with dimethyl ether (both have the formula C$_2$H$_6$O), yet dimethyl ether is a gas, and ethyl alcohol is a liquid at room temperature. Furthermore, the ethers are either sparingly soluble or insoluble in water.

Common ether, the most widely used general anesthetic, is diethyl ether with the structural formula:

$$\begin{array}{c} \text{H H} \quad\quad \text{H H} \\ | \; | \quad\quad\; | \; | \\ \text{H—C—C—O—C—C—H} \\ | \; | \quad\quad\; | \; | \\ \text{H H} \quad\quad \text{H H} \end{array}$$

A model of a diethyl ether molecule is shown in Fig. 27-2.

Fig. 27-2. Ball-and-stick model of a molecule of diethyl ether: carbon atoms, black; hydrogen atoms, light purple; oxygen atom, dark purple.

Ethers are generally prepared by the dehydration of the proper alcohol. Thus, diethyl ether is prepared by heating ethyl alcohol with concentrated sulfuric acid to a temperature of about 140°C and maintaining that temperature until the reaction is complete. The reaction may be summarized by the following equation:

$$\text{CH}_3\text{CH}_2\underline{\text{OH} + \text{H}}\text{OCH}_2\text{CH}_3 \xrightarrow[\text{heat}]{\text{H}_2\text{SO}_4} \text{CH}_3\text{CH}_2\text{—O—CH}_2\text{CH}_3 + \text{HOH}$$

In addition to its use as an anesthetic, diethyl ether is widely employed as a solvent for fats, waxes, and other substances insoluble in water. It must be used with caution, however, because it is highly flammable.

The functional group that is characteristic of ethers, the $\begin{array}{c} | \quad | \\ \text{—C—O—C—} \\ | \quad | \end{array}$ group, is not very active chemically; hence ethers as a class are not very active.

ALDEHYDES

The aldehydes are a group of compounds with the general formula R—C(=O)—H. In the first member of the aldehydes derived from the paraffin series, R is hydrogen; in the other members R is the usual alkyl radical. It is obvious from the general formula that aldehyde molecules differ from the parent hydrocarbon molecules in that two hydrogen atoms on an end carbon have been replaced with a single oxygen atom. The first few members derived from the methane series are:

$$\text{H—C(=O)—H} \quad \text{Methanal (formaldehyde)}$$

$$\text{CH}_3\text{—C(=O)—H} \quad \text{Ethanal (acetaldehyde)}$$

$$\text{CH}_3\text{CH}_2\text{—C(=O)—H} \quad \text{Propanal}$$

Systematic naming involves changing the end of the name of the parent hydrocarbon from *-e* to *-al*. The common names of the first two members, *formaldehyde* and *acetaldehyde*, are usually used by chemist and layman alike, instead of the systematic names.

Aldehydes are prepared by the mild oxidation of alcohols. Inasmuch as the aldehyde group is always located at the end of a carbon chain, the hydroxy group to be oxidized must also be located at the end of a carbon chain.[1]

Controlled Oxidation. The cheapest and most widely used oxidizing agent is atmospheric oxygen. The majority of organic substances will not undergo atmospheric oxidation in the absence of a catalyst unless the temperature is very high. For example, alcohols, ethers, and the hydrocarbons in gasoline, natural gas, and other fuels show little tendency to oxidize at room temperature. Once the temperature is raised sufficiently, oxidation is violent and more or less complete, with carbon dioxide and water the main oxidation products.

In order to bring about *partial* oxidation of an organic molecule—for example, the oxidation of an alcohol to an aldehyde—special procedures must be employed to control the oxidation. Either a special oxidant or atmospheric oxygen with a catalyst may be used; in each case the temperature must be fairly low.

SPECIAL OXIDANTS. Certain compounds of oxygen such as H_2O_2, $KMnO_4$, $K_2Cr_2O_7$, and HNO_3 are used as the oxidizing agents. The oxidation is usually carried out in solution so that the temperature cannot rise above the boiling point of the mixture. Further, the solution may be jacketed with an ice bath or other cooling medium to keep the temperature as low as necessary. Under these

---

[1] Alcohols with the —OH group at the end of a carbon chain, —CH_2OH, are called primary alcohols.

conditions, oxidation of the organic molecule will occur only at reactive points in the molecule. With the alcohols, only the portion of the molecule that is joined to the hydroxyl group is susceptible to oxidation. The following equation shows the over-all oxidation, the brackets indicating that the oxygen is derived from an oxydizing agent.

$$\underset{\text{ethyl alcohol}}{H-\underset{\underset{H}{|}}{\overset{\overset{H}{|}}{C}}-\underset{\underset{H}{|}}{\overset{\overset{H}{|}}{C}}-O-H} + [O] \rightarrow \underset{\text{acetaldehyde}}{H-\underset{\underset{H}{|}}{\overset{\overset{H}{|}}{C}}-\overset{\overset{H}{|}}{C}=O} + H_2O$$

Use of this method of controlled oxidation is limited to research work and to the manufacture of fine chemicals because of the high cost of the oxidizing agents.

ATMOSPHERIC OXYGEN AND A CATALYST. At a moderate temperature and with the proper catalyst, oxygen attacks the portion of the molecule most susceptible to chemical change without reacting with the remainder of the molecule. This procedure is frequently used in the large-scale production of chemicals by oxidation. For example, formaldehyde is prepared commercially by passing a mixture of air and methyl alcohol at 40° to 50°C over a copper catalyst.

Properties and Uses of Aldehydes. Formaldehyde, a gas with a boiling point of $-21°C$, is marketed as a 37 percent water solution called *formalin* for use as an embalming agent, a preservative for biological specimens, and a fumigant. The important plastic known as Bakelite is produced by the reaction of formaldehyde with phenol. Production of formaldehyde in the United States is around 1.8 billion pounds annually.

Acetaldehyde is used chiefly as an intermediate in the production of other organic chemicals, especially acetic acid, *n*-butyl alcohol, and ethyl acetate.

The aldehydes as a class are very reactive compounds. They are easily converted to the corresponding acid by controlled oxidation.

$$R-\underset{H}{\overset{|}{C}}=O + [O] \rightarrow R-\underset{OH}{\overset{|}{C}}=O$$

Aldehydes also undergo a great variety of addition and condensation reactions involving the aldehyde group.

KETONES

The alkane ketones are a class of compounds with general formula $RR'C=O$. A few examples are

$$CH_3-\underset{\underset{O}{\|}}{C}-CH_3 \quad \text{Dimethyl ketone (acetone)}$$

$$CH_3-\underset{\underset{O}{\|}}{C}-CH_2CH_3 \quad \text{Methyl ethyl ketone}$$

$$\text{CH}_3\text{CH}_2-\underset{\underset{\text{O}}{\|}}{\text{C}}-\text{CH}_2\text{CH}_3 \quad \text{Diethyl ketone}$$

Notice that a ketone is very similar to an aldehyde, except that the =O is not bound to an end carbon.

Ketones are prepared by the partial oxidation of *secondary alcohols*, alcohols that have the structure

$$\begin{array}{c}\text{H}\\|\\-\text{C}-\text{C}-\text{C}-\\|\\\text{O}\\|\\\text{H}\end{array}$$

The oxidation of 2-propanol gives acetone:

$$\text{CH}_3\text{CHOHCH}_3 + [\text{O}] \rightarrow \text{CH}_3\text{COCH}_3 + \text{H}_2\text{O}$$

Acetone, the most important ketone commercially, is widely used as a solvent for waxes, plastics, and lacquers. More than 1.6 billion lb is produced annually.

CARBOXYLIC ACIDS

The simple carboxylic acids have the general formula $\text{R}-\underset{\underset{\text{O}}{\|}}{\text{C}}-\text{OH}$. For those derived from the methane series, R represents H for the first member and C_nH_{2n+1} for the remaining members. Structurally, a carboxylic acid is considered to be derived by replacing the three hydrogen atoms on an end carbon with an atom of oxygen and a hydroxyl group.

$$\begin{array}{cc}\text{H H}\\||\\\text{H}-\text{C}-\text{C}-\text{H}\\||\\\text{H H}\\\text{ethane}\end{array} \qquad \begin{array}{c}\text{H}\\|\\\text{H}-\text{C}-\text{C}-\text{O}-\text{H}\\|\|\\\text{H}\text{O}\\\text{ethanoic acid}\\\text{(acetic acid)}\end{array}$$

The $-\underset{\underset{\text{O}}{\|}}{\text{C}}-\text{O}-\text{H}$ group is called the **carboxyl** group.

Nomenclature. Many of the carboxylic acids were isolated from natural sources long before systematic naming was introduced. They were given names indicative of their source, and these names have persisted. For example, the acid derived from methane was first obtained by the destructive distillation of ants and was called *formic acid* from the Latin, *formica*, meaning ant. Similarly, the name *acetic acid* was derived from the Latin word *acetum*, vinegar; *butyric* is from the Latin word meaning butter. Systematic naming involves replacing the final *-e* of the name of the hydrocarbon with *-oic*, and using the word *acid*. Thus, the systematic names for formic and acetic acid are *methanoic acid* and *ethanoic acid*, respectively. Some of the important members that are derived from the methane series are listed in Table 27-2. They are frequently referred to as the fatty acids.

TABLE 27-2.

Aliphatic Carboxylic Acids

FORMULA	COMMON NAME	OCCURRENCE	FREEZING POINT	BOILING POINT
HCOOH	Formic	Ants, pine needles	8.4°C	100.7°C
CH_3COOH	Acetic	Soured fruit juices, vinegar	16.6°	118.2°
CH_3CH_2COOH	Propionic		−20.8°	141.4°
$CH_3CH_2CH_2COOH$	Butyric	Rancid butter, limburger cheese	− 5.5°	164.1°
$CH_3(CH_2)_{14}COOH$	Palmitic	Fats (as esters)	62.8°	
$CH_3(CH_2)_{16}COOH$	Stearic	Fats (as esters)	69.6°	

Preparation of Carboxylic Acids. There are numerous methods for preparing the RCOOH type of acid. The controlled oxidation of alcohols and aldehydes is widely used. In laboratory work, potassium dichromate and potassium permanganate are often used as the oxidizing agents. Atmospheric oxygen with a suitable catalyst is sometimes employed in commercial preparations.

$$RCH_2OH + 2[O] \rightarrow RCOOH + H_2O$$

$$RCHO + [O] \rightarrow RCOOH$$

Acetic acid is the most important carboxylic acid commercially. Production in the United States amounts to about 1.6 billion lb annually, exclusive of that in vinegar. The acid is sold in the pure form as *glacial acetic acid*, so called because it freezes to an ice-like solid on cold days. The melting point is 16.6°C. The acid is used in the manufacture of cellulose acetate, white lead for paints, dyes, and medicines.

Acetic acid is usually prepared by the oxidation of ethyl alcohol or acetaldehyde, with atmospheric oxygen as the oxidizing agent.

ENZYME-CATALYZED OXIDATION OF ETHYL ALCOHOL. The alcohol in fermented fruit juices and fermented malt (cider, wines, beer) undergoes low-temperature oxidation to acetic acid in the presence of the proper enzyme and air.

$$CH_3CH_2OH + O_2 \xrightarrow[\text{ferment}]{\text{mother of vinegar}} CH_3COOH + H_2O$$

Vinegar produced in this manner is from 4 to 5 percent acetic acid and is used almost exclusively as a condiment and food preservative. Its flavor is enhanced by the flavors in the cider, wine, or malt.

OXIDATION OF ACETALDEHYDE. In the presence of cobalt or manganese(II) acetate, acetaldehyde rapidly reacts with oxygen from the air, yielding acetic acid.

Organic Chemistry II: Derivatives of the Hydrocarbons

$$2CH_3CHO + O_2 \xrightarrow{catalyst} 2CH_3COOH$$

Most glacial acetic acid is now synthesized by this method.

Properties of the Carboxylic Acids. The first four members derived from the methane series are completely miscible with water. As the molecular weight increases, however, the solubility in water decreases. Those that are soluble in water affect indicators as do other acids and give a sour taste to the solution, but they are weak acids with low ionization constants.

$$CH_3COOH + HOH \rightleftarrows CH_3COO^- + H_3O^+$$
$$ \text{acetate} \ \text{hydronium}$$
$$ \text{ion} \ \ \text{ion}$$

The lower members have an acrid odor; those from butanoic (C_4) through heptanoic (C_7) have disagreeable odors. The odor of rancid butter and strong cheese is due to acids in this group.

Many of the acids that originate in plant and animal materials contain two or more functional groups. The following are familiar examples:

		COOH		COOH
	CH$_3$	CHOH	CH$_2$COOH	CH$_2$
COOH	CHOH	CHOH	HO—C—COOH	CHOH
COOH	COOH	COOH	CH$_2$COOH	COOH
oxalic acid (rhubarb, sorrel)	lactic acid (buttermilk)	tartaric acid (grapes)	citric acid (citrus fruits)	malic acid (unripe fruits)

ESTERS

Esters are compounds that may be considered as being derived from acids by the replacement of the ionizable hydrogen with hydrocarbon radicals. If the source acid is a carboxylic acid, the general formula for the ester is RCOOR'.

There are many methods for the preparation of these compounds. In one method, which is widely used and which indicates the structure of ester molecules, a molecule of water is lost when a molecule of an alcohol reacts with a molecule of a carboxylic acid. Sulfuric acid is used as the dehydrating agent in this reaction. The over-all reaction is as follows:

$$\underset{\text{acid}}{R-\underset{\underset{O}{\|}}{C}-\boxed{O-H + H-}O-R'} \xrightarrow{H_2SO_4} \underset{\text{ester}}{R-\underset{\underset{O}{\|}}{C}-O-R'} + \underset{\text{water}}{H_2O}$$

Each carboxylic acid gives rise to a homologous series of esters. Therefore the name of an individual ester is designed to show the acid and the hydrocarbon radical.

From *formic acid* (HCOOH):

HCOOCH$_3$	Methyl formate
HCOOCH$_2$CH$_3$	Ethyl formate
HCOOCH$_2$CH$_2$CH$_3$	*n*-Propyl formate

From *acetic acid* (CH_3COOH):

CH_3COOCH_3 Methyl acetate

$CH_3COOCH_2CH_3$ Ethyl acetate

$CH_3COOCH_2CH_2CH_3$ *n*-Propyl acetate

The esters with low molecular weights are volatile liquids with pleasant fruity odors and are often found in fruits and flowers. The odor of banana is due to isoamyl acetate:

$$CH_3COOCH_2CH_2CHCH_3$$
$$|$$
$$CH_3$$

Hydrolysis converts an ester into the alcohol and the acid from which it is derived. Thus, the hydrolysis of ethyl acetate produces acetic acid and ethyl alcohol.

$$CH_3COOC_2H_5 + HOH \rightleftarrows CH_3COOH + C_2H_5OH$$

The hydrolysis goes to completion when the ester is boiled with dilute bases.

$$CH_3COOC_2H_5 + NaOH \rightarrow CH_3COONa + C_2H_5OH$$
ethyl acetate   sodium acetate

Up to this point the discussion of esters has involved only those derived from carboxylic acids. Several common and important esters are derived from inorganic acids. A noteworthy example is glyceryl trinitrate (popularly called nitroglycerin), an ester obtained by the action of nitric acid on glycerol in the presence of sulfuric acid.

$$\begin{array}{c} H \\ | \\ H-C-O-H \\ | \\ H-C-O-H + 3HONO_2 \rightarrow \\ | \\ H-C-O-H \\ | \\ H \end{array} \quad \begin{array}{c} H \\ | \\ H-C-ONO_2 \\ | \\ H-C-ONO_2 + 3HOH \\ | \\ H-C-ONO_2 \\ | \\ H \end{array}$$

Nitroglycerin, an oily liquid, is a powerful explosive. When absorbed in wood flour or some other absorbent material it is called dynamite. Because it is fairly safe to handle in this form, it is the staple explosive for road building, mining, and many other peacetime operations.

The explosion of nitroglycerin involves the rapid oxidation of the carbon and hydrogen to carbon dioxide and water at the expense of the nitrate groups, which are reduced to nitrogen and to oxides of nitrogen. *All the products are gases.* As in all combustions, great heat is evolved. The hot gases formed exert terrific pressure in all directions, pushing aside obstacles in their path until their force of expansion is dissipated. Approximately 10,000 ml of hot gases is produced from 1 ml of liquid nitroglycerin.

DERIVATIVES OF ALKENES AND ALKYNES

Replacements of hydrogen atoms in molecules of alkenes and alkynes with the appropriate functional group usually give rise to the expected compounds. For example, alcohols, or aldehydes, or ketones, or ethers, or carboxylic acids may be formed that are comparable to the corresponding compounds derived from the alkanes.

However, the chemical properties of these derivatives of unsaturated compounds are complicated by the fact that there are two active places in the molecule; the functional group and the double or triple bond. When the functional group (e.g., alcohol group, or carboxyl group) is separated from the double or triple bond by two or more carbon atoms, the functional group exhibits its usual chemical properties as does the double or triple bond. On the other hand, if the functional group is near the double or triple bond, the characteristics of each may be modified by the presence of the other. In some cases the tendency of the two groups to interact may lead to an entirely different compound. For example, ethenol (vinyl alcohol) is unknown; attempts to prepare this alcohol usually result in the formation of the isomeric acetaldehyde:

$$\begin{array}{c} \text{H} \\ | \\ \text{H}-\text{C}=\text{C}-\text{O}-\text{H} \\ | \quad | \\ \text{H} \quad \text{H} \\ \text{vinyl alcohol} \\ \text{(unknown)} \end{array} \rightarrow \begin{array}{c} \text{H} \quad \text{H} \\ | \quad | \\ \text{H}-\text{C}-\text{C}=\text{O} \\ | \\ \text{H} \\ \text{acetaldehyde} \end{array}$$

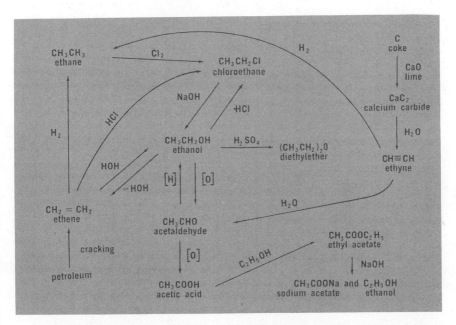

Fig. 27-3. A flow diagram showing some interconversions of organic compounds whose molecules are built of two carbon units. All reactions shown here will take place under proper conditions of temperature, concentration, and catalysis.

Figure 27-3 illustrates a method by which derivatives may be obtained from parent compounds.

COMPOUNDS DERIVED FROM AROMATIC HYDROCARBONS

The aromatic hydrocarbons form the same types of derivatives as the aliphatic hydrocarbons do. Some of these, derived from benzene, are as follows:

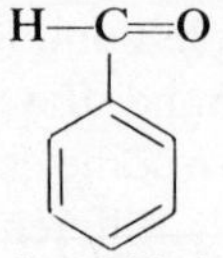

benzaldehyde (an aldehyde)

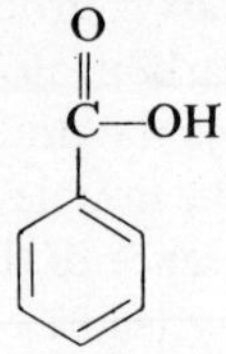

benzoic acid (an acid)

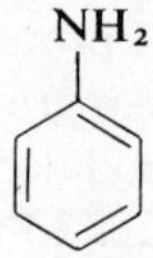

aniline (an amine)

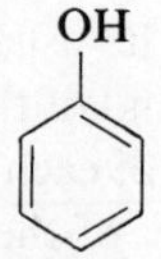

phenol (a phenol)

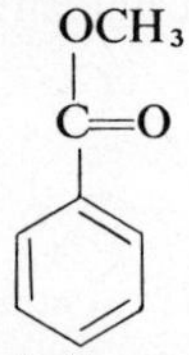

methyl benzoate (an ester)

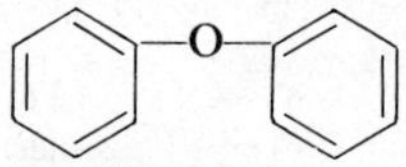

diphenyl ether (an ether)

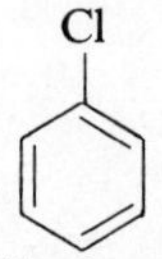

chlorobenzene (a halogen derivative)

Note that the combining form *phenyl-* is used to indicate a substituted benzene ring in a molecule.

The methods of preparing aromatic derivatives in some cases parallel those for preparing the corresponding aliphatic compounds. For example, methyl benzoate can be prepared by the method used to prepare aliphatic esters.

$$CH_3COOH + HOCH_3 \xrightarrow{H_2SO_4} CH_3COOCH_3 + HOH$$
acetic acid — methyl alcohol — methyl acetate

$$C_6H_5COOH + HOCH_3 \xrightarrow{H_2SO_4} C_6H_5COOCH_3 + HOH$$
benzoic acid — methyl alcohol — methyl benzoate

The properties of a functional group depend to some extent on the type of carbon ring or chain to which it is attached. The hydroxyl group provides excellent examples of this statement. When attached to the carbon of an aliphatic hydrocarbon, the hydrogen of the hydroxide (—OH) group does not ionize, and water solutions of alcohols therefore are neutral. When attached to a carbon of the benzene ring, the hydrogen of the hydroxide group does ionize to a small extent. Hence water solutions of such aromatic derivatives contain an appreciable concentration of the hydronium ion and are weakly acidic.

$$C_6H_5OH + H_2O \rightleftarrows C_6H_5O^- + H_3O^+$$
phenol

For this reason, the hydroxy derivatives of aromatic hydrocarbons are called **phenols** rather than alcohols.

Phenol is widely used in the synthesis of certain dyes, drugs, and plastics. For example, it is used in making salicylic acid.

$$\underset{}{\text{C}_6\text{H}_4(\text{OH})(\text{COOH})}$$

This in turn serves as an intermediate in the synthesis of wintergreen (methyl salicylate) and aspirin (acetylsalicylic acid). At the current bulk price, the acetylsalicylic acid in a 5-grain aspirin tablet costs about 0·05 cent.

methyl salicylate   aspirin

The annual production of phenol in the United States is almost 2 billion lb. Of this, about 35 million lb is obtained from coal tar as such. The remainder is synthesized from benzene by a variety of methods.

The *cresols* are the hydroxy derivatives of toluene. Three isomers exist; their formulas are as follows:

ortho-cresol   *meta*-cresol   *para*-cresol

Creosote oil, a coal-tar fraction, contains a large proportion of cresols; it is widely used as a wood preservative.

EXERCISES

1. Write structural formulas and names for all the isomeric alcohols that have the molecular formula $C_5H_{11}OH$.
2. What is the purpose of sulfuric acid in the conversion of ethanol into diethyl ether?
3. Write structural formulas and names for all the isomeric alcohols and ethers with the molecular formula $C_4H_{10}O$.
4. Dimethyl ether and ethanol, both with a molecular formula of C_2H_6O, have boiling points of $-23°C$ and $78°C$, respectively. Account for this difference on the basis of their molecular structures. Predict which of the two compounds is more soluble in water.
5. Write equations for the partial oxidation of 1-butanol; for the partial oxidation of 2-butanol. Name the products in each case.
6. (a) Calculate the freezing point of a solution containing 100 g of ethylene glycol in 400 g of water.

(b) What weight of methanol would be required in 400 g of water to give a solution that freezes at the same temperature as the solution in part (a)?
7. Starting with propene, show how the following compounds can be prepared: 2-propanol, acetone, 1,2-dibromopropane, di-i-propyl ether.
8. Write balanced equations for the processes indicated in Fig. 27-3.
9. Why are the hydroxy derivatives of the aromatic hydrocarbons called phenols rather than alcohols?
10. Identify the class of organic compound indicated by each of the following general formulas: ROR′, ROH, RCOR′, RCOOR′, RCHO, RCOOH.
11. Indicate with equations how each of the following syntheses could be carried out; the reagents available in addition to those indicated in the problem are inorganic chemicals:
    (a) Glucose → ethyl acetate
    (b) CO → methanoic acid
    (c) Apple juice → vinegar
    (d) Salicylic acid and ethanol → aspirin
12. What is the conjugate base of sodium phenolate? K_i for phenol is 1.3×10^{-10}. Would the pH of a solution of sodium phenolate be 7, less than 7, or greater than 7? Why?
13. Name the following compounds:
    (a) $CH_3CH_2CH_2COCH_2CH_3$
    (b) $CH_3CHOHCH_2CH_2CH_2CH_3$
    (c) $CH_3CH_2COOCH_2CH_2CH_3$
    (d) $CH_3CH_2CH_2OCH_2CH_2CH_3$
    (e) $CH_3CH_2CH_2CH_2CH_2COOH$
14. Write the names and structural formulas for the following compounds:
    (a) an alcohol obtained by the destructive distillation of wood;
    (b) a ketone widely used as a solvent;
    (c) a compound found in citrus fruit;
    (d) an ester of an inorganic acid;
    (e) a wood preservative;
    (f) a diprotic organic acid.
15. Write equations for each reaction indicated below:
    (a) the explosion of nitroglycerin;
    (b) the base hydrolysis of propyl benzoate;
    (c) the reaction of sodium and ethanol.
16. Consider propane, 2-propanol, propanal, dimethyl ketone, propanoic acid, and methyl ethyl ether, all of which contain three carbon atoms. Draw structural formulas for these compounds and predict which ones are soluble in water. Which has the highest vapor pressure? Which the lowest?

28

BIOCHEMISTRY

Biochemistry may be defined as the chemistry of living things. This simple statement covers a vast amount of subject matter that is of the utmost importance to our comfort and well-being.

First, biochemistry is concerned with the isolation and structural identification of the different substances that make up plant and animal organisms. A living organism is composed of more than just fats, carbohydrates, and proteins. Hundreds of other substances are necessary to the proper functioning of the organism. Many of these are very complex compounds, and some exist in such minute amounts that merely detecting them is extremely difficult.

Second, biochemistry is concerned with all the chemical changes that take place in the cells to provide for energy, growth, reproduction, and aging. Once these reactions are understood, their control can be undertaken.

THE CELL

Cells are the structural units of plants and animals just as atoms, ions, and molecules are the structural units of inanimate matter. While cells may vary in size, form, and function, with few exceptions all have the same intracellular organization, consisting of a *nucleus* surrounded by the nuclear membrane and imbedded in the *cytoplasm*, which in turn is surrounded by the cell membrane. There are several smaller organizations in these two major components of the cell, such as hereditary material, nucleolus, mitochondria, ribosomes, etc. From a chemical point of view, the cell is made of many different substances, water being the most abundant single chemical species, proteins perhaps next, and lesser amounts of fats, carbohydrates, and other types of organic compounds. The ashing of cellular material leaves a residue amounting to about 1 percent

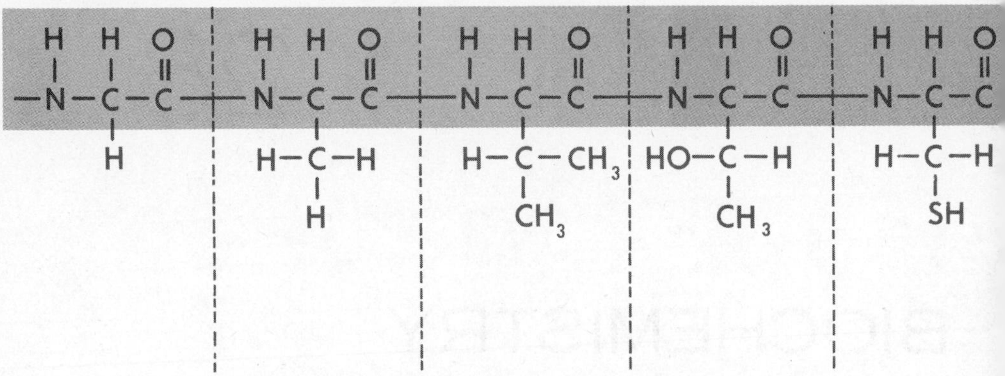

```
  H  H  O     H  H  O     H  H  O     H  H  O     H  H  O
  |  |  ‖     |  |  ‖     |  |  ‖     |  |  ‖     |  |  ‖
—N—C—C——N—C—C——N—C—C——N—C—C——N—C—C—
  |  |        |  |        |  |        |  |        |  |
  H  H      H—C—H      H—C—CH₃   HO—C—H      H—C—H
              |            |            |            |
              H           CH₃          CH₃          SH
```

Fig. 28–1. A portion of a protein molecule, showing amino acid residues joined by peptide bonds.

of the total weight. This ash is composed of carbonates, phosphates, chlorides, and sulfates of sodium, potassium, calcium, and magnesium, with traces of salts of copper, manganese, zinc, silicon, and iodine.

The living cell is no more merely an aggregation of chemical substances than a molecule is merely an aggregation of atoms. The atoms of the elements in both cells and molecules are organized according to some plan, but the plan of organization in a living cell is exceedingly complex and not completely understood. This organization of the components of a cell gives rise to the characteristics of life, characteristics that are different from the properties of known combinations of elements present in inanimate, non-living matter.

What are the distinguishing features of living matter? One of the most significant characteristics is the power of *reproduction*. All living organisms—either plants or animals—have this ability. Other characteristics of living matter include growth, aging, and the eventual loss of these capacities.

Interesting—but not unusual, for we often observe borderline examples for every classification we attempt—is the fact that the distinction between living and non-living matter is not always clear cut. Tobacco mosaic, a virus that infests the tobacco plant, is a white, crystalline compound that does not appear to possess the qualities of life. Yet in the cells of the tobacco plant it is capable of reproduction. Whether viruses are living organisms or not is a moot question.

PROTEINS

Proteins are among the most complex chemical substances known; they play a primary part in the life processes of all living organisms. Lean meat, skin, hair, tendons, nails and hoofs, vital organs, many hormones, enzymes, and antibodies are composed largely of proteins. Reproduction, growth, and the relaying of inherited characteristics involve protein synthesis.

Biochemistry

[Structural diagram showing a protein chain with five linked amino acid residues, each with the backbone pattern H-N-C(H)-C(=O)- and different side chains: CH₂-CH₂-S-CH₃ (methionine); COOH; phenyl ring; CH₂-CH₂-NH₂; and an imidazole ring.]

Protein molecules are very large, ranging in molecular weight from 10,000 to several million awu. All contain nitrogen in combination with oxygen, carbon, and hydrogen. Many also contain sulfur or phosphorus, and some contain iron, manganese, copper, or iodine.

Amino Acids. The large molecules of simple proteins[1] are readily broken up (in digestion or by boiling with dilute acids) into small molecules called *amino acids*. The simplest amino acid is glycine (aminoacetic acid). Its formula is as follows:

$$\underset{\text{amino group}}{H-N}\underset{\text{acid group}}{\overset{H}{\underset{H}{-C-}}\overset{H}{\underset{\parallel}{-C-OH}}}$$

The natural amino acids can be represented by the general formula

$$H-N\overset{H}{\underset{Z}{-C-}}\overset{H}{\underset{O}{-C-OH}}$$

in which Z represents hydrogen (as in glycine) or some alkyl radical such as methyl, isopropyl, etc. Z can also be a carbon chain that contains sulfur atoms, or cyclic, or other groups.

Only 24 amino acids have been isolated and identified from plant and animal proteins. Thus, it appears that all protein materials found in plants and animals are built from 24 simple compounds.

---

[1] Simple proteins give only amino acids on hydrolysis. (There are a few exceptions.) Conjugated proteins are made up of simple protein molecules linked to nonprotein molecules. The hemoglobin of blood is an example. The protein is combined with *heme*, a complex red compound of iron. Casein in milk and vitellin in egg yolk are also conjugated proteins in which the protein is combined with phosphoric acid. In other conjugated proteins, the additional group may contain carbohydrates, or compounds of nitrogen, magnesium, copper, manganese, cobalt, or other substances.

Experiments have shown that eight, or perhaps nine, of the 24 amino acids are *essential* in man's diet. The human body is able to synthesize the *nonessential* amino acids as needed from nitrogen compounds and other nutrients. A balanced or good protein food contains proteins that hydrolyze (digest) to form all the essential amino acids. Examples include milk, eggs, cheese, and lean meat. The digestion products of an unbalanced or poor protein food is lacking in one or more of the essential amino acids. Examples include gelatin, beans, and peas.

Simple protein molecules are long-chain molecules that are formed by the union of hundreds or even thousands of amino acid molecules. The union is due to valences that can be considered to originate from the elimination of a hydrogen from the —NH_2 group and an —OH from the —COOH group. This important linkage is called the **peptide bond.** Figure 28-1 shows how amino acid molecules join to make large protein molecules.

Amino Acid Sequence. Look again at Fig. 28-1. If we number the amino acid residues from left to right as 1, 2, 3, etc., we note that the sequence of amino acids in this portion of the protein molecule is –1–2–3–4–5–6–7–8–9–10–. We would expect to find this sequential portion in every molecule of this particular protein. In a different protein we find a different sequence of amino acid residues. For example, the sequence in a portion of some other protein molecule might be –2–16–4–6–7–13–11–13–16–2–. That is, not only is the order different, but some of the amino acids present in the first protein are missing from the second one, and vice versa.

Experimental methods for identifying each amino acid along the protein chain have been developed recently. F. Sanger and co-workers, during the period 1945–1952, were the first to establish the amino acid sequence in a protein, the hormone insulin (from beef). Insulin is a relatively simple protein, each molecule consisting of only 51 amino acid residues. Since that time, the sequence has been determined for a number of simple proteins. It goes without saying that this type of research is extremely tedious, since the amino acid residues may range from a hundred or so in simple proteins found in certain hormones, viruses, and blood hemoglobin, up to several thousand in the more complex proteins comprising muscle, skin, and hair.

A second important aspect of protein structure is concerned with the way that

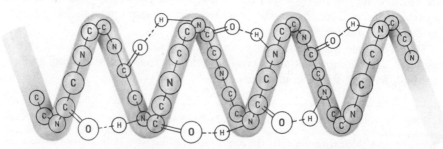

Fig. 28–2. The helical structure of proteins. The main strand consists of the –C–C–N– unit of amino acids. Hydrogen bonding, shown by dotted lines from O of one amino acid residue to H of another, stabilizes the coiled structure. (Only a few H bonds are shown; Z groups and most H atoms are omitted for clarity.)

the amino acid residues are organized in space along the polypeptide chain. (Note that Fig. 28-1 shows the kinds of atoms in the chain but does not show their relative positions in space.) According to Linus Pauling, these units are held in a coiled fashion, called the **alpha helix configuration**, by bonding forces other than those in the chain structure. These bonds are mainly hydrogen bonds (Fig. 28-2). In some proteins—hair, skin, and muscle meat, for example—the alpha helices are believed to be twisted about one another to form rope-like structures. In other proteins—enzymes and insulin, for example—the chains are folded to give a globular structure (see Fig. 28-6).

OPTICAL ISOMERISM. Consider for a moment the thumb, little finger, palm, and back of both your hands. These four components are arranged in space so that the two hands cannot be superposed to make the four components coincide, no matter how you turn your hands. Note also that one hand is the mirror image of the other.

There are also pairs of molecules that are mirror images of one another and that cannot be superposed. "Right-" and "left-handed" molecules are possible, provided a carbon atom is present to which four different groups are attached. Such a carbon atom is said to be **asymmetric.**

It will be seen from the general formula of natural amino acids that all these acids, except glycine, are capable of existing as right- and left-handed molecules, because each contains an asymmetric carbon atom:

$$NH_2-\underset{Z}{\overset{H}{\underset{|}{\overset{|}{C}}}}-COOH \quad \text{asymmetric carbon}$$

The arrangement in space of the two possible isomers of $NH_2CH(CH_3)COOH$ is shown in Fig. 28-3.

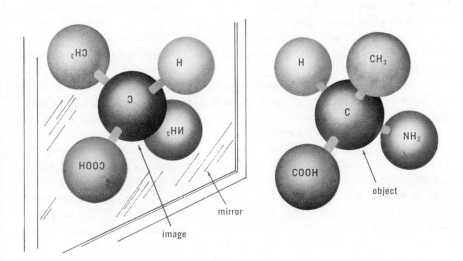

Fig. 28–3. Spatial arrangement of the optical isomers of an amino acid, $NH_2CH(CH_3)COOH$.

Isomerism such as this, which involves the spatial arrangement around an asymmetric carbon atom, is called **optical isomerism,** because the isomers rotate plane-polarized light.[2] One isomer rotates the plane of polarization to the right; the other isomer rotates the plane to the left.

Optical isomers behave alike so far as most chemical reactions are concerned. In biochemical reactions, however, a molecule with one spatial arrangement may participate freely, whereas its optical isomer will either react not at all, or only extremely slowly. In laboratory synthesis of such compounds, there is usually a random positioning of the four groups in space so that the number of molecules of right- and left-handed types are equal. But in the biosynthesis of proteins (and most other plant and animal materials), this randomness does not occur. Rather, the groups are positioned in proteins so that only the left-handed amino acids are present. It has been pointed out that if man does reach one of the other planets and finds life there, the carbohydrates and proteins will be worthless to him as foods unless their biosynthesis involves the same spatial arrangement about asymmetric carbon atoms as on this planet. We will see later that most of the chemical reactions that occur in plants and animals take place at a measurable rate only if catalytic substances—enzymes, for example—are present. For the catalytic action to be effective, the molecules involved must fit each other snugly at active sites. And, just as a right-handed glove does not properly fit a left hand, an enzyme with one type of shape will not fit a protein or carbohydrate molecule with the mirror-image shape.

PROTEIN SYNTHESIS

The problem of how cells utilize some 20 simple compounds—the amino acids—to build the very complex proteins, each with its unique amino acid sequence, and how this capacity is handed from parent to offspring is a problem that has fascinated biologists and biochemists for many decades. In the past twenty years great progress has been made toward its solution. The problem is so complex, involving most of the areas of chemistry and biology, and the amount of research literature that has accumulated is so vast that we can only touch a few points.

As a beginning, let us remember that many ordinary substances tend to react with each other when brought together under favorable conditions. Iron, oxygen, carbon dioxide, and water react to form iron rust; acetic acid and bicarbonate of soda react to form carbon dioxide, and so on. It is apparent that we have to go one step further in analyzing the chemical reactions involving the union of amino acids to form proteins. We must postulate some sort of guide or coding system to allow for the formation of all the different proteins from the same 20-odd amino acids. It is now quite clear that the coding systems in all cells, from bacteria to the cells of man, are embodied in molecules of **deoxyribonucleic acid (DNA).** (The term *genes* has long been used to refer to the

---

[2] When light is passed through certain substances, the emergent beam is vibrating in one plane only and is said to be **plane-polarized.**

Biochemistry

hereditary factors in a fertilized cell. A single gene is now thought to be a molecule of DNA.)

The molecular structure of DNA as proposed by J. D. Watson and F. H. C. Crick in 1953 consists of two parallel **polynucleotide chains** wound around a

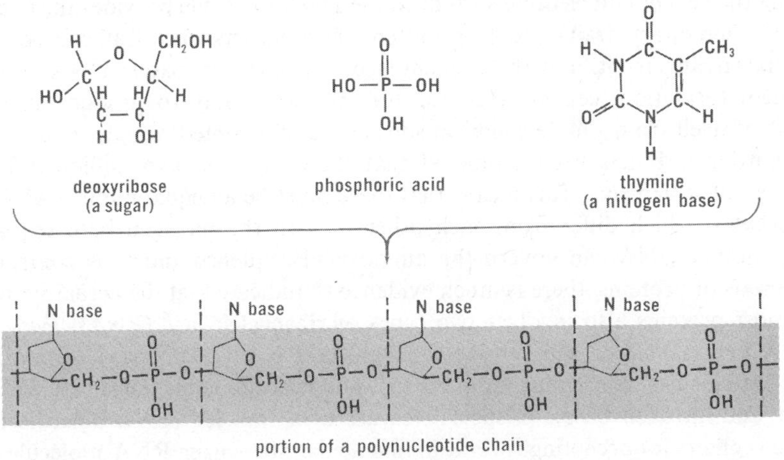

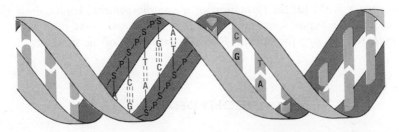

Fig. 28–4. A portion of a DNA molecule showing how the three structural units are joined to make a huge molecule. Note the use of an outline formula for deoxyribose in the DNA representation.

common axis into a helix and held together by hydrogen bonding between nitrogen bases spaced along the chain. The polynucleotide chain, or **nucleic acid**

Fig. 28–5. The DNA structure is believed to consist of two polynucleotide chains wound around each other in a double helix (P, phosphate, and S, sugar, in the figure). Along each chain are attached the nitrogen bases (A, T, G, and C in various pairings in the figure) which form hydrogen bonds with each other to hold the helix together. These features are shown in some detail in the central portion, schematically to right and left of this portion.

as it is commonly called, is formed by the repetitious union of the nucleotide unit, a unit formed by joining together a molecule of the pentose sugar *deoxy ribose*, a molecule of *phosphoric acid*, and a molecule of a *nitrogen base*. Figure 28-4 shows these structural units and how they are joined to form the polynucleotide chain. Figure 28-5 shows how two polynucleotide chains are wound around a common axis to form a molecule of DNA. The four nitrogen bases that are repeated in this structure project inward and form hydrogen bonds to stabilize the structure.

According to the Watson and Crick proposal, the sequence of the four bases along the two polynucleotide strands in the DNA molecule provides the code by which hereditary traits are transmitted. Through its chemical composition, spatial arrangement, and the sequences of the nitrogen bases, DNA controls protein formation, cell metabolism, and its own ability to produce an exact copy of itself from simple chemical substances that come through the cell walls as nutrients. It has been estimated that there are over two million different species of organisms. This means that there must be an equal number of DNA molecules, which differ from each other only in the nitrogen base sequence.

Although DNA can govern the amino acid sequence and thus control the synthesis of proteins, there is much evidence to indicate that the actual synthesis of most proteins also involves two types of **ribonucleic acid (RNA)**: *messenger* RNA and *soluble* or *transfer* RNA. Transfer RNA is composed of relatively small molecules, containing from 80 to 100 nucleotide units (sugar, phosphoric acid, and nitrogen base). A special feature of a transfer RNA molecule is its coded centers for accepting specific amino acids. Messenger RNA molecules are much larger, consisting of from 900 to 1200 nucleotide units. RNA differs from DNA in that the sugar unit in the latter has one less oxygen.

To summarize the protein synthesis process, it is now believed that DNA in the nucleus of the cell serves as the template for the formation of messenger RNA. Messenger RNA then moves into the cytoplasm where it lines up briefly with the surface of the ribosomal nucleoprotein of the cell. At this point the transfer RNA molecules with their attached amino acids line up as dictated by the code carried by messenger RNA for the formation of a specific protein. Under these conditions, peptide bonds form rapidly to join the amino acids into a protein molecule, thus freeing the transfer RNA for repetition of the synthesis.

CARBOHYDRATES

Carbohydrates are the simple sugars and the substances which, on hydrolysis, yield simple sugars. These compounds, which are produced in green plants from carbon dioxide and water, are a major source of energy for animal life. Chemically, carbohydrates are composed of carbon, hydrogen, and oxygen. A highly abridged classification follows.

1. Monosaccharides, such as $C_6H_{12}O_6$. Examples: glucose, fructose, galactose. Ribose, mentioned in connection with DNA and RNA, is a 5-carbon monosaccharide.

2. Disaccharides, such as $C_{12}H_{22}O_{11}$. Examples: sucrose, maltose, lactose.

3. Polysaccharides, such as $(C_6H_{10}O_5)_x$. Examples: starch, cellulose.

Biochemistry

MONOSACCHARIDES

The most important monosaccharides, aside from ribose, are those whose molecules contain six carbon atoms—i.e., the hexoses, $C_6H_{12}O_6$. These are crystalline, water-soluble sweet substances that occur in ripe fruits and honey. Their condensation products are cane sugar, malt sugar, milk sugar, starch, and cellulose.

Glucose. Glucose, $C_6H_{12}O_6$, one of the isomeric hexoses, is an important sugar in nature, both because of its widespread occurrence and because of the prominent part it plays in biological processes. It is the sugar into which all other carbohydrates are converted prior to oxidation in the body. Glucose occurs as such in all ripe fruits; it is especially abundant in grapes. Many other carbohydrates—e.g., maltose, sucrose, starch, and cellulose—yield it on hydrolysis. Other names for it are grape sugar, blood sugar, and dextrose.

Chemical reactions and analyses indicate that glucose molecules contain five hydroxyl groups and an aldehyde group joined to a 6-carbon chain. This structure can be represented by the formula

$$CH_2OH\overset{*}{C}HOH\overset{*}{C}HOH\overset{*}{C}HOH\overset{*}{C}HOHCHO$$

Note that there are 4 asymmetric carbon atoms (marked by asterisks) in a molecule of glucose. Since there is a right- and left-handed arrangement about each asymmetric carbon atom, sixteen optically active isomers are possible; i.e., ordinary glucose is one of sixteen sugars, all of which have the above structural formula. These sixteen sugars, referred to as the *aldohexoses*, have been isolated and identified.

Determining the spatial arrangement of the four groups about each asymmetric carbon is a complex process that will not be discussed here. However, the spatial formulas for *dextro*-glucose (common glucose) and its mirror-image isomer, *levo*-glucose, are as follows:

```
      H—C=O                H—C=O
        |                    |
      H—C—OH             HO—C—H
        |                    |
     HO—C—H               H—C—OH
        |                    |
      H—C—OH             HO—C—H
        |                    |
      H—C—OH             HO—C—H
        |                    |
       CH₂OH               CH₂OH
   dextro-glucose       levo-glucose
```

Glucose molecules are now believed to exist mostly in a closed-chain or cyclic form. It will be remembered that the valence angles between carbon atoms are 109° 28′, rather than the 180° angles that the structural formulas would seem to indicate. This means that the aldehyde group on the No. 1 carbon atom can be very close to the hydroxyl group on the No. 5 carbon atom if the chain twists around on itself. The aldehyde and hydroxyl parts can enter into an addition reaction with each other (dotted arrows in the formula below) to form a cyclic molecule.

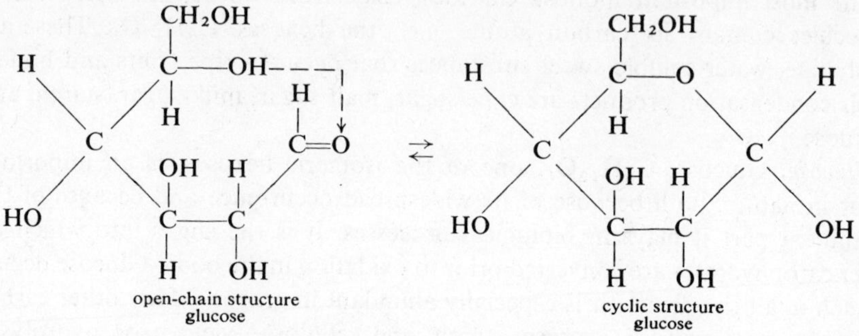

open-chain structure
glucose ⇌ cyclic structure
glucose

Formation of this ring changes the carbon atom of the aldehyde group into an asymmetric carbon atom. Postulation of this additional asymmetric carbon is necessary to account for the pair of optical isomers that can be derived from *dextro*-glucose. The existence of these isomers is the chief evidence in support of the ring structure. However, the open-chain structure cannot be completely abandoned, because glucose acts as an aldehyde in many of its reactions. Therefore an equilibrium between open-chain and ring molecules is proposed, with a majority of the molecules being in the closed structure at any one time.

Fructose. Fructose, $C_6H_{12}O_6$ (levulose, fruit sugar), is a crystalline sugar that occurs with glucose in honey and in fruits. It is considerably sweeter than either glucose or sucrose (table sugar). The open-chain structural formula is:

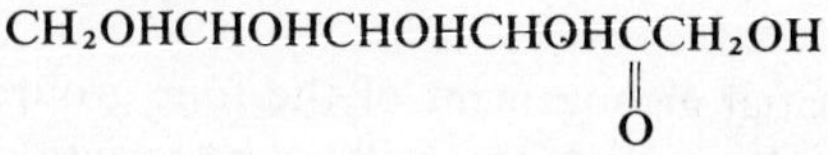

$$CH_2OHCHOHCHOHCHOHCCH_2OH$$
$$\underset{O}{\|}$$

DISACCHARIDES

Sucrose (table sugar), maltose (malt sugar), and lactose (milk sugar) are important members of the disaccharide group, $C_{12}H_{22}O_{11}$. As the group name indicates, each molecule of these sugars is composed of two monosaccharide units. The units are joined together through valences that result from the elimination of a molecule of water. For example, a sucrose molecule is composed of a glucose and a fructose unit joined as shown in the following formula:

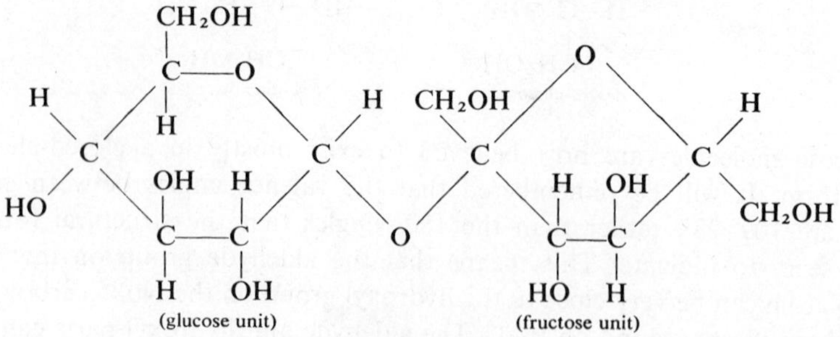

(glucose unit)   (fructose unit)
sucrose, $C_{12}H_{22}O_{11}$

A lactose molecule is also made up of two hexoses, a glucose unit and a galactose unit. Molecules of maltose are built up only of glucose units.

POLYSACCHARIDES

As the name indicates, molecules of polysaccharides are composed of many monosaccharide units. The ratio of atoms in a polysaccharide is slightly different from the ratio in the simple sugars from which the polysaccharide is derived. This is due to the fact that the two valence bonds necessary to link two monosaccharide units together can no longer be attached to the hydrogen atom and hydroxyl radical as in a simple sugar molecule.

The most abundant polysaccharides, $(C_6H_{10}O_5)_x$, are those in which the monosaccharide unit is glucose. *Starch* and *cellulose* are in this group. The glucose units are joined in starch molecules in the following way:

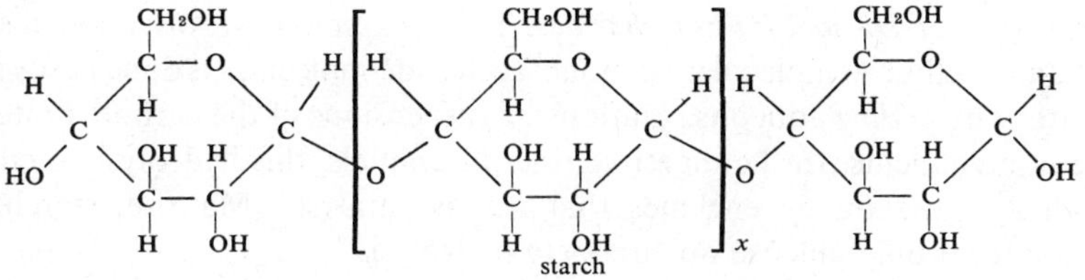

starch

Molecular-weight determinations show that starch molecules contain from 200 to 3000 glucose units per molecule. The weight of cellulose molecules is more difficult to ascertain. However, the best estimates indicate that the number of glucose units per molecule is of the order of several thousand.

There are at least two important differences in the structure of starch and cellulose molecules. In cellulose, the glucose units are attached end-to-end to form long, filament-like molecules. In starch, these units are generally joined in a branched-chain pattern, although the smaller unbranched molecules are present in varying amounts in most starches. For example, starch from potatoes contains about 20 percent of the unbranched type.

A second difference in the structure of starch and cellulose molecules concerns the spatial arrangement of the groups around the two asymmetric carbons by which two glucose units are joined. If we think of the arrangement around these carbons in starch as being "right-handed," the arrangement around the same carbons in cellulose is then "left-handed." A comparison of the formula below for cellulose with the one above for starch will illustrate this relationship.

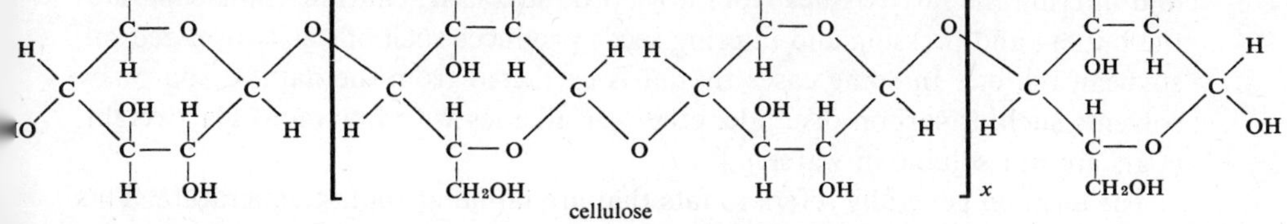

cellulose

Although this difference may appear trivial, it seems to account for the fact that man cannot digest cellulose but can digest starch. (See specificity of enzyme action, page 495.)

Occurrence of Starch and Cellulose. Starch is found in almost all plants. It is especially abundant in the *seeds* of common cereals (wheat, rye, oats, rice, and corn), constituting from 60 to 80 percent of the seed. Seeds from peas and beans are about 50 percent starch. Unripe *fruits* such as bananas and apples contain large proportions of starch, which becomes sugar when the fruit ripens. Grains of sweet corn, however, contain an abundance of sugar when the kernels are immature, and only a little sugar but much starch when the grains are ripe.

In the plant, the starch molecules are aggregated into larger particles called *granules*. Each species has granules with characteristic shape, size, and markings.

Cellulose occurs in the *woody* and *fibrous* parts of plants and is responsible for the structure of plants. Dry wood contains from 60 to 70 percent of carbohydrate material, about half of which is true cellulose. *Cotton* is essentially pure cellulose.

Hydrolysis of Di- and Polysaccharides. The breaking down of complex sugar, starch, and cellulose molecules to monosaccharide molecules is done easily in the laboratory by *boiling* aqueous solutions or suspensions of the carbohydrate with dilute mineral acids. In the digestive tract of animals, this hydrolysis is effected at *body temperature* by enzymes that act as catalysts. Maltose, starch, and cellulose form only glucose on complete hydrolysis.

$$\underset{\text{maltose}}{C_{12}H_{22}O_{11}} + H_2O \rightarrow 2\underset{\text{glucose}}{C_6H_{12}O_6}$$

$$\underset{\substack{\text{starch or}\\\text{cellulose}}}{(C_6H_{10}O_5)_x} + xH_2O \rightarrow x\underset{\text{glucose}}{C_6H_{12}O_6}$$

Sucrose yields equal amounts of glucose and fructose.

$$\underset{\text{sucrose}}{C_{12}H_{22}O_{11}} + H_2O \rightarrow \underset{\text{glucose}}{C_6H_{12}O_6} + \underset{\text{fructose}}{C_6H_{12}O_6}$$

FATS

Many plant and animal materials—for example, peanuts, coconuts, cottonseed, olives, soybeans, cream, butter, cheese, and certain meats—contain large percentages of compounds called *fats*. The cereals as a rule are comparatively poor in these compounds.

The fats are obtained from natural sources by a variety of methods. Heating and filtering the fatty tissues from hogs produces lard; churning milk separates the butter; and pressing and filtering seeds produces such oils as cottonseed oil, soybean oil, etc. In some cases the fat is extracted from the natural source by solvents such as carbon disulfide, ether, or alkanes with low molecular weight. (Fats are not soluble in water.)

The term *oil* generally refers to fats that are liquid at room temperature. This use of the word oil must not be confused with its use in such terms as kerosene oil, lubricating oils, and oil of wintergreen. The latter are not fats.

HYDROLYSIS OF FATS; STRUCTURE

When lard, tallow, and other fats are boiled in dilute sodium hydroxide or subjected to the action of specific enzymes (digestion), the fat molecules are completely hydrolyzed to form simpler molecules. *Glycerol*, $CH_2OHCHOHCH_2OH$, always constitutes a portion of the product, and acids with high molecular weights make up the remainder. The acids thus produced are frequently called *fatty acids*; the more common ones are:

$C_{11}H_{23}COOH$	Lauric acid
$C_{13}H_{27}COOH$	Myristic acid
$C_{15}H_{31}COOH$	Palmitic acid
$C_{17}H_{35}COOH$	Stearic acid
$C_{17}H_{33}COOH$	Oleic acid
$C_{17}H_{31}COOH$	Linoleic acid
$C_{17}H_{29}COOH$	Linolenic acid

Lauric, myristic, palmitic, and stearic are unbranched *saturated* acids and are homologs of formic and acetic acids, i.e., derivatives of the methane series. Oleic acid, $CH_3(CH_2)_7CH{=}CH(CH_2)_7COOH$, containing one double bond; linoleic, containing two; and linolenic, containing three, are *unsaturated* acids. Table 28-1 shows the relative amounts of fatty acids obtained by the hydrolysis of different fats.

TABLE 28-1.

Fatty Acids Obtained from Hydrolysis of Fats and Oils (*In percentages[a]*)

FAT OR OIL HYDROLYZED	SATURATED			UNSATURATED		
	LAURIC	PALMITIC	STEARIC	OLEIC	LINOLEIC	LINOLENIC
Animal fats						
Butter	2–3	23–26	10–13	30–40	4–5	
Lard		28–30	12–18	41–48	6–7	
Tallow		24–32	14–32	35–48	2–4	
Vegetable fats (oils)						
Coconut	45–51	4–10	1–5	2–10	0–2	
Olive		5–15	1–4	69–84	4–12	
Peanut		6–9	2–6	50–70	13–26	
Corn		7–11	3–4	43–49	34–42	
Cottonseed		19–24	1–2	23–33	40–48	
Soybean		6–10	2–4	21–29	50–59	4–8
Linseed		4–7	2–5	9–38	3–43	25–58
Tung		1–3	1–3	4–16	0–1	74–91[b]

[a] The percentages do not total 100 because all the acids are not listed.
[b] An isomer of linolenic.

In the fat itself, the glycerol portion and the fatty acid are joined by means of ester linkages. A **fat** may be defined as an ester of glycerol and long-chain carboxylic acids. A general structural formula is:

$$\begin{array}{l} R'\overset{O}{\overset{\|}{C}}-O-\overset{H}{\underset{|}{C}}-H \\ R''\overset{O}{\overset{\|}{C}}-O-\overset{|}{\underset{|}{C}}-H \\ R'''\overset{O}{\overset{\|}{C}}-O-\overset{|}{\underset{|}{C}}-H \\ \phantom{R'''\overset{O}{\overset{\|}{C}}-O-}\overset{|}{H} \end{array}$$

fatty acid part / glycerol part

$(RCOO)_3C_3H_5$ (abbreviated formula)

R′, R″, and R‴ may or may not be the same in the same molecule; they correspond to $C_{11}H_{23}$, $C_{13}H_{27}$, $C_{15}H_{31}$, $C_{17}H_{35}$, $C_{17}H_{33}$, $C_{17}H_{31}$, etc. It is interesting to note that in natural fats R is made up of an *odd* number of carbon atoms, and therefore the fatty acids derived from these fats have an *even* number of carbon atoms.

During digestion fats are hydrolyzed to fatty acids and glycerol.

$$\begin{array}{l} R'COO-CH_2 \\ | \\ R''COO-CH + 3HOH \xrightarrow{\text{enzymes}} \\ | \\ R'''COO-CH_2 \end{array} \begin{array}{l} R'COOH \\ + \\ R''COOH \\ + \\ R'''COOH \\ \text{fatty acids} \end{array} \begin{array}{l} CH_2OH \\ | \\ CHOH \\ | \\ CH_2OH \\ \text{glycerol} \end{array}$$

BIOCATALYSIS

ENZYMES

Few of the chemical reactions that take place in protoplasm have been carried out in the laboratory, and then only with considerable difficulty and under extreme reaction conditions. But these very complex chemical reactions occur readily in protoplasm at very mild temperatures, pressures, and *p*H. This difference in ease of reaction appears in many cases to be due to the presence of catalytic substances called **enzymes**.

Enzymes are definite chemical substances produced by living organisms. However, life itself is apparently not necessary to the action of enzymes, once they have been formed. For example, Buchner in 1896 ground up yeast cells with sand. From this mixture he pressed out a juice that caused sugar to ferment, although no yeast cells were present. Many enzymes have now been obtained in the crystalline state and in a high degree of purity. All of them are proteins.

A common method of naming enzymes is to add the ending *-ase* to part of the name of the substance that is changed catalytically by the enzyme. Thus the enzyme that catalyzes the hydrolysis of maltose is called *maltase*. Other enzymes are named by adding *-ase* to the word indicating the reaction that is catalyzed. Thus, an enzyme that influences an oxidation is an *oxidase*.

Nature of Enzyme Action. Enzymes are believed to influence the speed of biological reactions in somewhat the same way that catalysts influence the speed of ordinary chemical reactions. For example, the enzyme may react with the substance undergoing change (the **substrate**), producing a more reactive intermediate substance, which in turn reacts, forming end products and regenerating the enzyme.

Specificity of Enzyme Action. In digestion, enzymes that break down carbohydrates have no effect on fats and proteins. Nor do the enzymes that hydrolyze fats or proteins have any effect on the hydrolysis of carbohydrates. Thousands of different enzymes are required to catalyze the many chemical reactions that occur in the cells of plants and animals, with each enzyme being quite specific in its action. This specificity is due to the folded shape of the protein chain composing the large enzyme molecule and to the location of *active sites* on the enzyme surface into which parts of the substrate molecule must fit closely (see Fig. 28-6). Only specifically shaped molecules can gain access to these active sites,

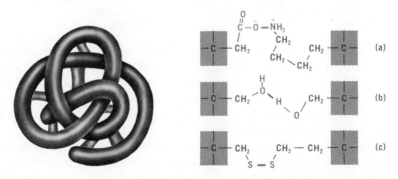

Fig. 28–6. A molecule of an enzyme, a globular protein, is shown schematically on the left. For a specific enzyme, not only is there a unique sequence of amino acids, but there is a unique shape of the folded molecule. The folds are stabilized by bonds through the Z groups along the peptide chain. The peptide chain, –C–C–N–C–C–N–, is shown in black; some of the possible cross-linking is shown in red. The nature of the cross-linking is shown on the right: (a) electrostatic attraction between ions; (b) hydrogen bonds; (c) disulfide links.

and, in many cases, the *fit* is due to the proper spatial arrangements of groups around an asymmetric carbon atom.

Enzyme Inhibition. Frequently a coenzyme is required in the process being catalyzed by an enzyme, and this too must fit properly into the reaction site. Vitamins are thought to act as coenzymes in certain biochemical reactions.

According to one theory, possibly only a portion of the coenzyme molecule is required to fit into the reaction site. One part of a second kind of molecule may resemble the coenzyme sufficiently to also fit into this site. However, another part of this second molecule may be so different from the coenzyme that the enzymatic action is inhibited or even stopped if this kind of molecule is present in a relatively high concentration. Thus, metabolic processes (see page 498) affecting growth, reproduction, etc., may be seriously hindered or brought to a complete stop. Substances that inhibit metabolic processes are called **antimetabolites.**

An example of antimetabolic activity is seen in the use of the *sulfa* drugs in blood stream infections, scarlet fever, gonorrhea, septic sore throat, and certain other infectious diseases. The organisms responsible for these diseases are believed to require the vitamin *p-aminobenzoic acid* for normal growth and reproduction. Like certain other vitamins, this vitamin probably reacts first with an enzyme and thus starts a sequence of reactions necessary for the organism's growth. The effectiveness of the sulfa drugs is thought to be due to the fact that their molecules so closely resemble those of *p*-aminobenzoic acid in shape and functional groups that they can fit in and react at the active centers of the large enzyme molecules. The formulas of the two follow:

$$\text{para-aminobenzoic acid} \qquad \text{sulfanilamide (a sulfa drug)}$$

Penicillin is thought to act by blocking cell wall constuction in microorganisms, thus preventing their multiplication. Also, common poisons often act as enzyme inhibitors. For example, the cyanides and carbon monoxide, by combining with iron-containing catalysts (cytochromes), block the transport of oxygen. The nerve poison diisopropylphosphofluoridate inhibits acetylcholine esterase, the enzyme that plays a major role in the functioning of nerves.

One type of research involving anticancer agents has as its aim the development of compounds that will act as inhibitors to the enzymes that control the metabolism of cancer cells. Obviously, if such antimetabolites are to be usable, they must not interfere with the metabolism of normal cells.

VITAMINS

Vitamins are organic compounds, minute quantities of which are necessary components of the diet. Systematic study of vitamins began in 1881 when it was shown that mice cannot live on a diet consisting of pure carbohydrates, fats, proteins, and mineral salts. In this early work, the chemical nature of vitamins was not yet known; hence their alphabetical designation, vitamin A, B, etc. In some cases a vitamin that was thought to be a single vitamin proved to be a mixture of *two* or *more*—e.g., vitamins B, D, and E. Most of the known vitamins have now been isolated in pure form, and their chemical structures have been determined. The more recently discovered vitamins have been given systematic chemical names. Some commercially synthesized vitamins are now available to supplement the vitamins derived from natural sources.

HORMONES

The active chemical substances in the secretions of ductless glands (endocrine organs) are called **hormones**; they are discharged into the circulating fluids (blood, etc.). *Hormones are necessary in small amounts for the proper functioning of the organs and systems in the body.*

Among the glands that produce hormones are the adrenal glands on top of each kidney, the pituitary gland at the base of the brain, the thyroid and parathyroid glands in the neck, glands in the stomach and intestines, the pancreas, and the sex organs (ovaries and testes).

The isolation, identification, and synthesis of every hormone are important objectives of medical research. Once the function of a particular hormone is understood and its chemical synthesis has been achieved, the disorders due to its undersecretion in the body may be corrected by supplying the synthesized hormone. Disorders resulting from its oversecretion may also be remedied. Whereas the molecules of some hormones are large and complex, the molecules of others are relatively simple. For example, *adrenalin* (epinephrine), secreted by the adrenal glands, has the following structure:

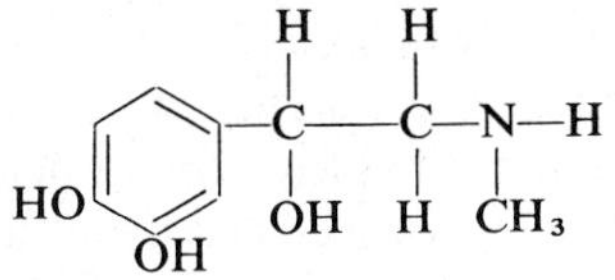

$$HO-C_6H_3(OH)-CH(OH)-CH_2-NH-CH_3$$

adrenalin

Adrenalin, a rather simple derivative of benzene, is now prepared synthetically. Its injection is followed by a rapid increase in blood pressure because of a more rapid pulse and the contraction of the small arteries. This hormone also brings about contraction of the iris, relaxation of the muscles controlling the bronchial tubes, and an increase in the amount of glucose in the blood stream.

The secretion of adrenalin is increased during a period of emotion (fear, rage, etc.). Sugar is poured into the blood stream and the heart beats more rapidly, thus setting up processes that supply the individual with energy that enables him to deal with the emergency.

DIGESTION AND METABOLISM

DIGESTION

Digestion includes all the changes that foods undergo in the alimentary canal which make possible their subsequent passage through the intestinal walls into the lymph and blood. For certain foods such as inorganic salts and monosaccharides, no chemical change is necessary because their small, water-soluble molecules readily diffuse through the intestinal membranes. In general, however, molecules of di- and polysaccharides, fats, and proteins are broken down into smaller molecules by hydrolysis during digestion. Thus, the starch in potatoes, bread, rice, and other starchy foods becomes glucose; proteins are changed to amino acids, and fats to fatty acids (or their salts) and glycerol. Since these hydrolytic reactions do not occur readily at room temperature, they must be catalyzed. Along with acidic and basic substances, specific enzymes and salts in saliva, gastric juice, pancreatic juice, intestinal secretions, and bile act as catalysts in hydrolyzing foods. Gastric juice has a pH which ranges from 1.0 to 2.5, due mainly to hydrochloric acid. The pH of bile and pancreatic juice is slightly above 7.

METABOLISM

The term **metabolism** refers to all the processes that digested foods go through to supply energy or to be made into tissue, bone, etc. The metabolism of food has two aspects. One involves exothermic oxidation reactions that produce carbon dioxide, water, and other waste products to supply energy; the other involves a building-up process that forms all the complex materials of bones, organs, muscles, and other body materials.

Recent research with radioactive isotopes of carbon and phosphorus, heavy hydrogen, and other tracer atoms has led to the conclusion that food substances continually become a part of some tissue and then leave it. In fact, living matter seems to be changing constantly. Atoms in a particular food molecule might participate in both aspects of metabolism; i.e., they might become part of some tissue and later be oxidized to provide heat and energy. It is estimated that in one year 98 percent of the atoms in the body are lost via excretion or respiration and are replaced via digestion or respiration. Even the atoms in bones and teeth are not permanently situated.

Metabolism of Carbohydrates and Fats. Carbohydrate metabolism begins when the products of the hydrolysis of carbohydrates—mostly glucose, fructose, and galactose—pass through the intestinal walls and are carried to the liver. There they are converted to glycogen (animal starch) and stored for future use; glycogen can also be stored in muscles to a limited extent. The glycogen is changed back to glucose as needed, thus enabling a rather constant concentration of glucose to be maintained in the blood.

When large quantities of sugars are eaten, the body is unable to change all of it to glycogen, and the concentration of sugar in the blood increases. When the sugar level approaches 160 mg per 100 ml of blood, sugar appears in the urine. The concentration at which this occurs is called the **sugar threshold.**

If more protein is eaten than is required, certain amino acids are converted into glucose. (About 60 percent of these acids are of the type that can be thus converted.) Glycerol, one of the products of the hydrolysis of fats, can also be changed to glucose.

The oxidation of glucose and fats to supply energy for work and body heat produces carbon dioxide and water as the end products. These oxidation processes occur at low temperature, but the total energy released is the same as when carbohydrates and fats are burned in a flame and the energy is released in a high-temperature combustion. The carbohydrates and fats undergo a long series of chemical changes, each one altering the molecule slightly and producing an intermediate substance. Each step is catalyzed by specific enzymes and coenzymes.

Although fats and carbohydrates are usually thought of as energy foods, it must be remembered that they also take part in the formation of cells and other body tissues and organs.

Protein Metabolism. The amino acids formed during digestion are carried to all parts of the body where they are used as needed in building new proteins in tissues or replacing old proteins. Aside from building proteins, amino acids may undergo many other chemical reactions in the course of their metabolism. The amino groups may be used to build nitrogen compounds from carbohydrate and fat; an amino acid may be oxidized to carbon dioxide, water, and ammonia. The ammonia is converted to *urea*, NH_2CONH_2, in the liver and excreted in urine.

VIRUSES

Many common ailments—smallpox, chicken pox, yellow fever, influenza, mumps, poliomyelitis, and common colds, to name a few—are caused by substances called *viruses*. In general, all viruses consist of two principal parts: a nucleic acid core and a protein outer shell. The nucleic acid may be either RNA or DNA and, of course, carries the code characteristic of the particular virus. The protein shell serves as a protective covering of the RNA or DNA core and, in some cases, acts to help the core attack and penetrate the walls of cells. Once the RNA or DNA of the virus is inside a cell, it reorganizes the cell's function to manufacture complete virus particles. When a sufficient number of virus particles have been formed, the cell ruptures, spilling out the virus particles to attack other cells.

Most viruses range in size from 100 A to 2000 A and, consequently, cannot be seen with optical microscopes (see Fig. 14-1). In the past few years, much has been learned about the structures of viruses by "viewing" them with electron microscopes. For example, the influenza virus has been revealed as an irregularly shaped sphere with regularly spaced surface projections and a coiled core (Fig. 28-7). Viruses possess hereditary traits coded in the DNA or RNA core, and are able to reproduce themselves, although the help of host cells is required. They are the smallest biological structures capable of reproduction.

The body appears to have two ways of defending itself against virus infections. One of these is the so-called *immune reaction*. The lymphatic cells manufacture a specific protein, an **antibody**, that prevents the virus from invading the cell. While the antibody may or may not be of much value in the initial onslaught, it sets up immunity against later attacks by the same virus. The other line of defense, discovered only recently, is **interferon**, a protein. The invasion of a cell by the nucleic acids of a virus appears to stimulate the cell to produce interferon in quantity. Interferon then blocks the reproduction of the virus. How this is done is not now known.

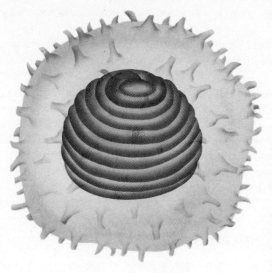

Fig. 28–7. The influenza virus has a protein covering that envelops the coiled DNA core. The protein protrusions are believed to aid the DNA in penetrating the host cell. Magnification about 500,000. (The drawing is based on electron micrographs and model supplied by Dr. R. W. Horne, University of Cambridge, Cambridge, England. See also Hoyle, L., Horne, R. W., and Waterson, A. P., *Virology*, **13**, 448 [1961]).

It is interesting that the first successful prevention of a virus disease by a synthetic chemical agent was reported in 1963. Methyl isatin-β-thiosemicarbazone has been shown to give effective protection against smallpox exposure. The formula is

methyl isatin-β-thiosemicarbazone

DRUGS

Many substances not naturally associated with cell metabolism are known that, in relatively small amounts, have a profound effect on cell metabolism including that of the brain and nervous system. Included among common types are stimulants, pain killers, tranquilizers, antibiotics, anesthetics, hallucinogens, and nerve poisons. Just how these substances produce their effects is well known for some and completely unknown for others.

In many cases the action of a drug is due to a specific inhibitory action on a particular enzyme or vitamin. For example, penicillin prevents wall construction

Biochemistry

in bacteria, presumably by interrupting enzyme and coenzyme reactions responsible for the construction. Sulfa drugs interfere with the uptake of a coenzyme essential to growth and cell division of certain types of organisms. Carbon monoxide and cyanides form compounds with the metal ions that are parts of enzyme systems in man, thereby blocking their functions. The nerve poison diisopropyl fluorophosphate inhibits acetylcholine esterase, an enzyme necessary to the functioning of nerves, by combining with an active site of the enzyme. The effectiveness of insecticides and herbicides is thought to be due to the inhibition of enzyme action in many cases.

However, the mechanisms by which many drugs produce their effects are not known. The hallucinogenic drug LSD (Fig. 28–8) has been the subject of extensive study; over 500 technical papers have been published about it, and its chemical structure and properties are well known, yet the mechanism by which it produces its effect is not known.

The brain has been studied experimentally by a great number of biochemists and physicians, and features such as oxygen utilization and rate of protein synthesis have been determined. The electrical nature of nerve impulses is thought to depend on a concentration gradient of Na^+ and K^+ ions inside and outside nerve cells. As an impulse passes, there is a small change in the voltage of the "electrochemical cells" due to a momentary change in the permeability of the cell walls to the passage of ions. These impulses bring about the release of acetylcholine and other chemicals that stimulate tissue at the nerve endings or stimulate other nearby nerve endings.

Tranquilizers, stimulants, pain killers, and hallucinogens are examples of drugs that act on the brain (Fig. 28-8). In some cases, the effective dose is a matter of a tiny fraction of a gram—0.000025 to 0.00005 g being an effective LSD dose in man.

LSD (lysergic acid diethylamide).

meprobamate, Equanil
a tranquilizer.

Fig. 28–8. Two chemical substances that act on the brain.

EXERCISES

1. Illustrate with an equation the formation of a peptide bond.
2. List four elements found in all proteins. On the basis of this list, suggest how proteins could be distinguished from carbohydrates and fats.

3. Show with structural formulas that seven or more isomeric molecules can be formed from the same four amino acids (choose the four from Fig. 28–1).
4. You are handed a carbohydrate in the laboratory and told to determine whether or not it is a monosaccharide. How would you proceed?
5. How many asymmetric carbon atoms are present in a molecule of each of the following: $CH_3CHOHCH_3$, $CH_2BrCHClCH_3$, $CH_3CHOHCHClCOOH$, $CH_3CHClCHClCHOHCH_3$?
6. How many asymmetric carbon atoms are shown in Fig. 28–1?
7. Consult the open-chain and closed-chain (cyclic) formulas for glucose. Compare the number of asymmetric carbon atoms in the two structures. Why is it necessary to postulate the cyclic structure?
8. Outline a method for converting wood into sugar. Approximately what weight of sugar could be obtained from a pound of wood?
9. Discuss the role of DNA in the synthesis of protein.
10. Write equations for the reactions that occur during the digestion of (a) sucrose; (b) starch; (c) simple proteins; (d) fats.
11. Suggest a reason why 1 g of fat releases more energy than 1 g of starch upon combustion.
12. Explain how reactions occur in metabolic processes in the body which in the laboratory require considerably higher temperatures.
13. Discuss the metabolism of fats.
14. Explain how a sulfa drug might function as an antimetabolite.
15. Distinguish between hormones and vitamins.
16. To what is the specific catalytic action of enzymes due?
17. Suggest how radioactive tracers such as carbon-14 or phosphorus-33 might be used to study metabolism in the body.

SUGGESTED READING

Bassham, J. A., "The Path of Carbon in Photosynthesis," *Sci. Amer.*, **206**(6), 88 (1962).

Bogue, J. Y., "Drugs of the Future," *J. Chem. Educ.*, **46**, 468 (1969).

Crick, F. H. C., "The Genetic Code: III," *Sci. Amer.*, **215**(4), 55 (1966).

Gates, M., "Analgesic Drugs," *Sci. Amer.*, **215**(5), 131 (1966).

Geyer, R. P., "Careers in Nutrition," *Chemistry*, **44**(1), 12 (1971).

Green, D. E., "The Synthesis of Fat," *Sci. Amer.*, **202**(2), 46 (1960).

Hanawalt, P. C., and R. H. Haynes, "The Repair of DNA," *Sci. Amer.*, **216**(2), 36 (1967).

Kellenberger, E., "The Genetic Control of the Shape of a Virus," *Sci. Amer.*, **215**(6), 32 (1966).

Nunes, F., "LSD—An Historical Reevaluation," *J. Chem. Educ.*, **45**, 688 (1969).

Phillips, D. C., "The Three-Dimensional Structure of an Enzyme Molecule," *Sci. Amer.*, **215**(5), 78 (1966).

Raw, I., "Enzymes—How They Operate," *Chemistry*, **40**(6), 8 (1967).

Williams, R. J., "The Functioning of Vitamins and Hormones," *J. Chem. Educ.*, **36**, 538 (1959).

29

APPLIED ORGANIC CHEMISTRY

Man's material comforts have been greatly enhanced through the development and application of organic chemistry. Clothing, dyes, medicines, foods, housing, insecticides, germicides, fuels for machines and for warmth, detergents, explosives—these and many other commonplace materials have undergone marked changes as man's knowledge of organic chemistry developed during the past century. Most of these synthetic products are made from the abundant natural materials wood, coal, petroleum, and natural gas.

Actually, from these abundant sources there are a few key compounds obtained that serve as the starting materials in the syntheses of most organic products; some of the more notable ones are ethylene (ethene), propylene (propene), acetylene (ethyne), methanol, ethanol, benzene, toluene, phenol, aniline, naphthalene, and cellulose. Figure 29-1 shows schematically a few of the ways one of these key compounds, propylene, serves as the starting point in the synthesis of other substances.

Most propylene is produced as a by-product of the production of gasoline; some comes from a process to produce ethylene. Of the total 24 billion pounds used annually in the U.S.A., only 5 to 6 billion pounds go into synthetic routes such as are shown in Fig. 29-1, the remainder is converted into gasoline or used directly as a fuel.

Often compounds that are important as intermediates in the production of useful end products are produced in more than one way in order to supply the demand. Phenol is an example. As we see from Fig. 29-1, phenol can be synthesized via the cumene route from propylene; it is also synthesized from benzene by two or three processes; and it is separated as such from coal tar.

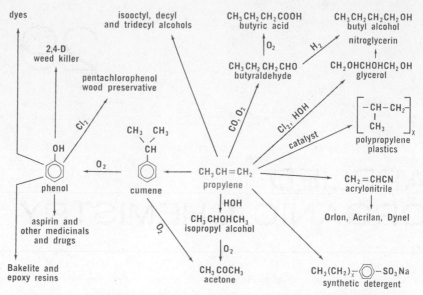

Fig. 29-1. Schematic representation of how one compound, available in large quantities from petroleum, serves as a key compound in the synthesis of many useful products.

Let us again point out that we could choose a dozen or so compounds and show with figures similar to Fig. 29-1 how each serves as the starting point in the syntheses of many useful products.

POLYMERS

Natural Polymers. Polymers are compounds whose large molecules are formed by the repetitious union of many small molecules. Proteins are natural polymers, the large molecules being formed by the linking together of the amino acid molecules. Starch and cellulose are also natural polymers; in them glucose units serve as the building blocks.

Polymers whose repeating units are bonded by valence bonds made available by the elimination of simple molecules such as HOH, HCl, and NH_3 are called *condensation polymers*. Proteins, starch, and cellulose (see Chap. 28) are in this category.

A second type of polymer is called an *addition polymer*. In forming this type, the building block molecules are unsaturated, i.e., they contain double or triple bonds. The unsaturated bonds rearrange in the polymerization reaction to form the valence bonds which link the smaller molecules together as a polymer. Rubber is a natural polymer that is sometimes classified as an addition polymer.[1]

---

[1] Actually, the mechanism by which natural rubber is formed in plants is not known. However, a synthetic rubber much like natural rubber can be formed by the addition polymerization of isoprene.

Applied Organic Chemistry

When its bark is cut, the rubber tree exudes a milky colloidal dispersion of rubber particles in water. The particles are coagulated as a pasty white material by the addition of an electrolyte, usually acetic acid.

The molecules of rubber are long and chain-like; they are thought to be built up of isoprene units. How these large molecules form in the rubber tree from carbon dioxide and water is unknown.

Isoprene itself is a low-boiling colorless liquid whose molecules are of the unsaturated hydrocarbon type:

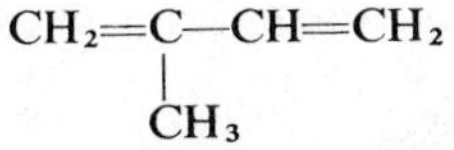

$$CH_2=C-CH=CH_2$$
$$|$$
$$CH_3$$

isoprene

The polymer obtained by the laboratory polymerization of isoprene is much like natural rubber, and the molecules of each are thought to be similar, i.e., built up of isoprene units in a linear fashion. If a catalyst is used, isoprene undergoes an addition type of polymerization, forming large molecules that contain up to 5000 isoprene units. In this addition process, the end carbons of the isoprene units become saturated and the middle carbons remain unsaturated. This is shown in the following formula for part of a molecule of synthetic rubber:

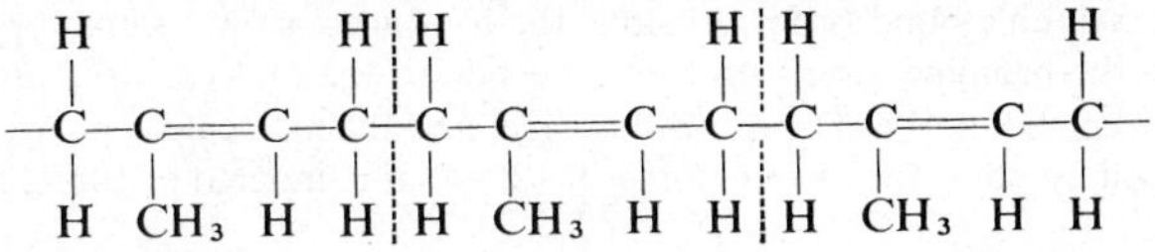

$$-\overset{H}{\underset{H}{C}}-\overset{H}{\underset{CH_3}{C}}=\overset{H}{\underset{H}{C}}-\overset{H}{\underset{H}{C}}-\overset{H}{\underset{H}{C}}-\overset{H}{\underset{CH_3}{C}}=\overset{H}{\underset{H}{C}}-\overset{H}{\underset{H}{C}}-\overset{H}{\underset{H}{C}}-\overset{H}{\underset{CH_3}{C}}=\overset{H}{\underset{H}{C}}-\overset{H}{\underset{H}{C}}-$$

Cis and Trans Isomerism. It was emphasized in Chap. 28 that chemical reactions that can proceed to form two or more isomers seldom do so when taking place in living cells. For example, we saw that molecules that could exist as either left-handed or right-handed molecules were always produced in cells as one or the other, but rarely, if ever, as a mixture of the two. Rubber affords another example of the biochemical production of molecules in which the parts occupy a single position in space rather than both of two possible positions. The

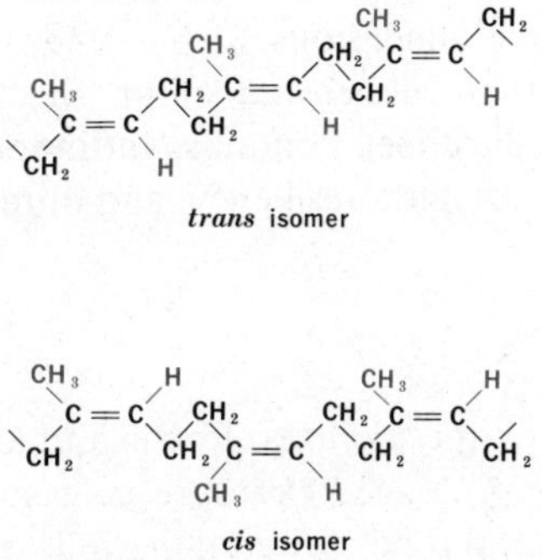

Fig. 29–2. Geometrical isomers of natural rubber: *trans* isomer, top; *cis* isomer, bottom.

type of space isomerism here does not involve asymmetric carbon atoms and optical activity; rather, it involves the positions in space of the four groups attached to a pair of carbon atoms joined by a double bond. Figure 29-2 shows two possible ways in which isoprene units can join to form polymeric molecules. Note that in the first formula, —CH₃ and —H groups are always attached on opposite sides (*trans*) of a double bond, while in the second formula they are always attached on the same side (*cis*) of a double bond. Rubber from the common rubber plants (*Hevea* family) is always made up of molecules with the *cis* configuration, while gutta-percha, a rubber-like material from trees that grow in the East Indies (*Sapotaceae* family) is composed of molecules of the *trans* configuration. This type of spatial isomerism is called geometrical isomerism. A simpler example of this type of isomerism is illustrated in Fig. 29-3.

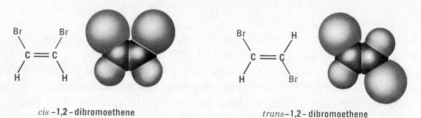

cis -1,2-dibromoethene   *trans*-1,2-dibromoethene

Fig. 29-3. *Cis* and *trans* isomers of CHBr=CHBr. Since free rotation of two carbon atoms about a double bond is not possible, the four groups must remain positioned as shown. With the bromine atoms on the same side of the molecule, the *cis* isomer has a greater concentration of electrons on that side, and the molecule is polar. This results in a higher boiling point for the *cis* isomer (112.5°C as compared to 108°C for the *trans* isomer).

VULCANIZATION. Raw rubber is soft and sticky at room temperature. To be useful, it must be heated with sulfur, a process called **vulcanization.** From 5 to 30 percent of sulfur is added, depending on the type of rubber desired. Small amounts are used for rubber for soft, easily stretchable articles; larger amounts are used for hard rubber. In vulcanization, the sulfur atoms combine, by addition, with the unsaturated carbon atoms of the rubber molecule. In this addition a sulfur atom may form a bridge between a carbon in one molecule and a carbon in a second molecule. Thus the sulfur atoms link the long-chain molecules together, forming molecules that are large in all three dimensions.

Most rubber articles contain substances other than rubber and sulfur. Black rubber, for example, contains carbon black; red rubber contains antimony sulfide. Such substances are added to increase the strength, resiliency, and durability.

SYNTHETIC POLYMERS

Natural polymers are not always satisfactory for specific uses. For example, natural rubber swells and loses its elasticity after prolonged exposure to gasoline or motor oil; silk and wool (proteins) are natural foods for certain kinds of bacteria and larvae of insects, as is also cellulose; and most natural polymers are hydrophilic, so that they readily absorb water. Moreover, the stability and melting points of natural polymers are such that they cannot be melted and then cast into desired shapes.

TABLE 29-1.

Synthetic Polymers

MONOMERS	POLYMERS	PRINCIPAL USES
Ethylene $CH_2=CH_2$	Polyethylene $-CH_2CH_2CH_2CH_2-$	Films and sheets, tubing, molded objects, electrical insulation
Vinyl chloride $CH_2=CHCl$	Poly(vinyl chloride) $-CH_2CHCH_2CH-$ with Cl, Cl on the CH carbons	Phonograph records, copolymer with vinyl acetate for floor coverings
Acrylonitrile $CH_2=CHCN$	Polyacrylonitrile $-CH_2CHCH_2CH-$ with CN, CN on the CH carbons	Fibers, e.g., Acrilan, Orlon
Tetrafluoroethylene $CF_2=CF_2$	Teflon $-CF_2CF_2CF_2CF_2-$	Objects resistant to chemical attack; nonstick kitchenware; metal coating against corrosion
Butadiene $CH_2=CH-CH=CH_2$ Styrene 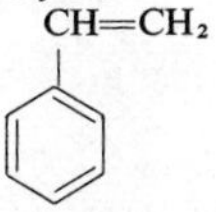 $CH=CH_2$ (with phenyl group)	Buna S and GR-S rubber 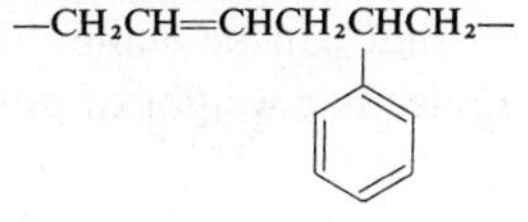 $-CH_2CH=CHCH_2CHCH_2-$ (with phenyl on one CH)	Synthetic rubber
Chloroprene $CH_2=CHC=CH_2$ with Cl	Polychloroprene $-CH_2CH=CCH_2CH_2CH=CCH_2-$ with Cl, Cl	Oil-resistant rubber, e.g., neoprene
Ethylene glycol $HOCH_2CH_2OH$ Terephthalic acid 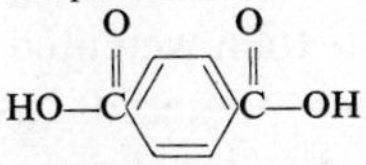 $HO-\underset{\underset{O}{\|\|}}{C}-\text{(phenyl)}-\underset{\underset{O}{\|\|}}{C}-OH$	Poly(ethylene terephthalate) 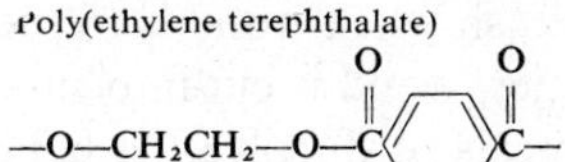 $-O-CH_2CH_2-O-\underset{\underset{O}{\|\|}}{C}-\text{(phenyl)}-\underset{\underset{O}{\|\|}}{C}-$	Fibers, e.g., Dacron; films
Hexamethylenediamine $NH_2CH_2CH_2CH_2CH_2CH_2CH_2NH_2$ Adipic acid $HO-\underset{\underset{O}{\|\|}}{C}-CH_2CH_2CH_2CH_2\underset{\underset{O}{\|\|}}{C}-OH$	Nylon 66 $-NH-(CH_2)_6-NH-\underset{\underset{O}{\|\|}}{C}-(CH_2)_4-\underset{\underset{O}{\|\|}}{C}-$	Fibers, molded objects
Phenol 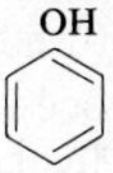 (phenyl-OH) Formaldehyde $HCHO$	Bakelite 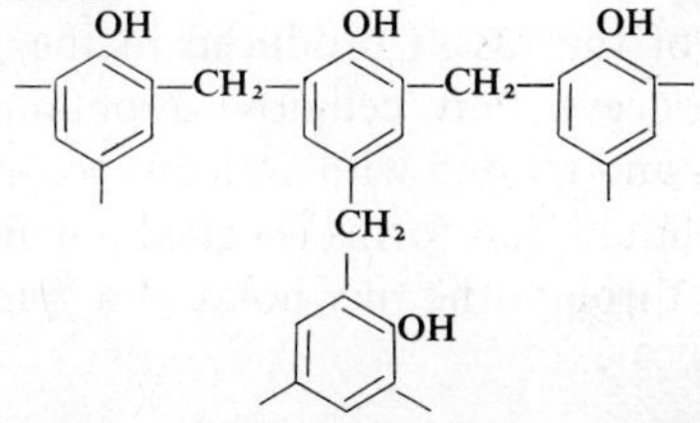 — phenol rings linked by $-CH_2-$ bridges with OH groups	Molded objects, varnishes, lacquers

During this century organic chemists have developed hundreds of synthetic polymers that have specific properties and can be cast into desired shapes including threads and sheets. In building these polymers, the chemist may choose simple molecules that can unite with each other by the elimination of small molecules, usually water, and form the condensation type of polymer; or he may choose unsaturated compounds that can join and form the addition type of polymer. Table 29-1 lists a number of common polymers. Note that the first six are of the addition type, the remainder of the condensation type.

A recent development in this field is the discovery of specific catalysts that can link the building blocks into ordered structures rather than random structures. As we have mentioned, in the natural polymers such as the proteins, viruses, starch, carbohydrates, and rubber, the structural units are always arranged in a particular sequence; and in many instances atoms or groups of atoms are arranged in specific relative positions in space. The fact that this same kind of positional order can now be achieved with some of the synthetic polymers makes it possible to build, from the same raw materials, polymers that have different properties. Figure 29-4 shows the ordered and random molecular structures resulting from the polymerization of propylene, $CH_2=CHCH_3$. The considerable variation in properties that can be achieved by varying both the structure (as to regularity) and the molecular weight of polypropylene molecules is shown in Fig. 29-5.

CELLULOSE PRODUCTS

Cellulose constitutes the woody portion of all plants and is present in all plant cells. In wood itself, the long molecules of cellulose are laid down in parallel rows to form the wood fibers; the fibers are bound together by a sticky organic substance called *lignin*. (Wood also contains about 8 percent mineral salts; these become the ash when wood is burned.)

Paper. In making paper, wood is cut into small pieces and cooked in calcium bisulfite or other chemicals to dissolve the lignin. The cellulose is removed by filtration, is bleached with chlorine or hydrogen peroxide, and is then weighted, sized, and rolled into sheets. In weighting and sizing, such materials as starch, glue, casein, rosin, aluminum silicate, and clay are added. *Glazed* paper is heavily weighted with minerals (e.g., barium sulfate) to reduce porosity; paper toweling contains little mineral additives. Filter paper is almost pure cellulose.

A large amount of the newsprint made in this country consists of a blend of ground wood (about 90 percent) and cellulose. Since the dry weight of wood is less than 50 percent cellulose, this method of producing paper makes for a more economical use of wood.

Rayon. Most of the rayon produced in the United States is manufactured by the *viscose* process. Pure cellulose is obtained from wood by the process described above, and treated with aqueous sodium hydroxide and carbon disulfide; the syrupy liquid that forms is called *viscose*. After aging and filtering, the viscose is forced through the tiny holes of a spinneret into a sulfuric acid bath.

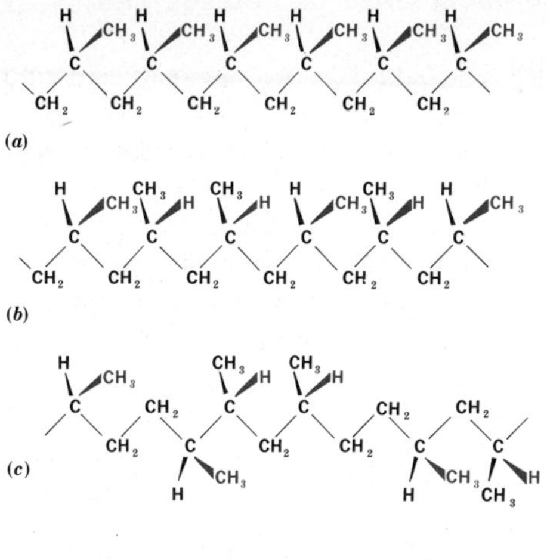

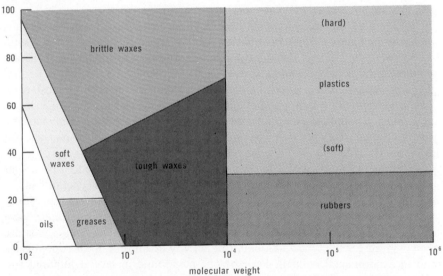

Fig. 29-4. Order and disorder in polypropylene. In (a), the unit

$$-\underset{\underset{CH_3}{|}}{CH}-CH_2-$$

is shown joined so that the —CH_3 group appears always on alternate carbons along the chain. Moreover, the —CH_3 groups are arranged in space in a regular way. This ordered arrangement allows the molecules to become organized in a regular fashion, and the polymer is said to be crystalline. In (b), the methyl group is also attached to every second carbon, but it occurs at random positions in space. In (c), the methyl groups appear in an irregular sequence along the carbon chain. Too, the polymer has been formed in a way so that the methyl groups appear on first one side of the chain and then on the other.

Fig. 29-5. The molecular weight and the regularity of the polymer molecules determine the nature of plastic material. (Courtesy Research Laboratories, Tennessee Eastman Co., Division of Eastman Kodak.)

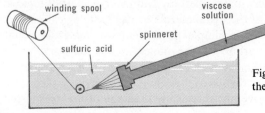

Fig. 29-6. Schematic representation of the manufacture of rayon.

This precipitates the cellulose as continuous threads that are gathered and twisted into rayon yarn (Fig. 29-6).

Cellophane is made by the same process as rayon, except that the precipitated cellulose is passed between a series of rollers to press it into sheets.

Cellulose Acetate. Each glucose unit in a cellulose molecule contains three hydroxyl groups. When cellulose is treated with acetic acid and sulfuric acid, or with acetic anyhdride, the hydroxyl groups are replaced by acetate groups, and *cellulose acetate* is formed. This plastic is widely used in fabrics and in photographic films.

CHEMOTHERAPY

The phase of chemistry that is concerned with the use of chemical substances to alleviate disease is called **chemotherapy.** The aim, of course, is to find specific compounds that will serve as a remedy for every malady known to man. The use of vitamins and hormones to control certain disorders is a phase of chemotherapy; the importance of these substances in the proper functioning of organs and systems in the body has already been mentioned.

Considerable success has also been achieved in using chemotherapeutic agents in the fight against infectious diseases. The problem here is to find chemical compounds that will destroy the microorganisms responsible for a disease without harming the patient. Originally such compounds were sought in natural products such as roots, leaves, and bark of plants, and in mineral substances. *Quinine*, a crystalline compound obtained from the bark of the cinchona tree, is an excellent example of a substance derived from a natural product that acts as a germicide against a specific microorganism, i.e., the organism responsible for malaria.

The modern approach to discovering substances that are toxic to a particular microorganism is through structural organic chemistry. Organic chemists in the laboratories of universities and colleges, the federal government, and drug companies are constantly synthesizing new compounds whose molecules have features that are known to impart physiological activity. The compounds are then assayed for the type and amount of activity in cultures of microorganisms and in experimental animals. Perhaps one compound in several hundred will offer enough promise to warrant clinical testing on human beings.

The Sulfa Drugs. The sulfa drugs provide a good example of the structural approach to the discovery of chemotherapeutic agents. In the early part of the twentieth century there was discovered a group of dyes that had a very complex chemical structure and also bactericidal properties. Efforts on the part of organic chemists to relate the structure of the dye molecule to its physiological activity led to the conclusion that a fairly simple portion of the molecule could account for this action. This portion of the molecule was very similar to *p*-aminobenzenesulfonamide (sulfanilamide), a well-known aromatic compound that had never been physiologically assayed. Clinical testing revealed that sulfanilamide was even more active physiologically than any of the dyes tested originally.

Applied Organic Chemistry

The next step in the structural approach was relatively straightforward. It involved making all manner of modifications in the sulfanilamide molecule and testing the physiological activity of these new compounds. The modification in which a hydrogen of the —SO_2NH_2 group was replaced with complex ring structures gave rise to the compounds that were best suited as medicines. Sulfapyridine and sulfathiazole are examples. The sulfa drugs were invaluable in the treatment of gangrene and pneumonia during World War II:

sulfanilamide   sulfapyridine   sulfathiazole

In this brief discussion of chemotherapeutic agents, it should be pointed out that many different kinds of physiologically active compounds that are not present in natural materials have been synthesized, and that others are constantly being sought. Of those now known, some are used as anesthetics, others as soporifics and analgesics, and still others for such purposes as lowering the body temperature, lowering or raising the blood pressure, relaxing the muscles, contracting the blood vessels, and acting on nerve centers. More recent triumphs include vaccines for poliomyelitis, oral antidiabetic drugs, and anticancer agents.

SOAPS; DETERGENTS

In soap making, fat is heated in huge iron kettles with aqueous sodium hydroxide until the fat is completely hydrolyzed. (Review fats, Chap. 28.) The equation for the hydrolysis is

$$(RCOO)_3C_3H_5 + 3NaOH \rightarrow 3RCOONa + C_3H_5(OH)_3$$
fat   soap   glycerin (glycerol)

Alkaline hydrolysis of a fat is also called **saponification.**

It is obvious from this equation that common soap may be a mixture of such compounds as sodium stearate, $C_{17}H_{35}COONa$, sodium palmitate, $C_{15}H_{31}COONa$, sodium oleate, $C_{17}H_{33}COONa$, and the sodium salts of other fatty acids. The soap is precipitated by the addition of salt. It is then removed by filtration, washed, and mixed with dyes, perfumes, or any other special ingredient. Next it is permitted to harden and is then cut and pressed into cakes. Scouring powders contain a high percentage of an abrasive such as volcanic ash or fine sand.

Soap functions in a variety of ways as a cleansing agent. It lowers the surface tension of water, thus enabling the water to moisten the material being washed more effectively; it acts as an emulsifying agent to bring about the dispersion of oil and grease (Fig. 14-5); and it adsorbs dirt.

The so-called *synthetic detergents* are of many types. In all of them a water-soluble salt-like group is attached to a long hydrocarbon chain. A typical example is $C_{17}H_{35}-OSO_3Na$. An important difference between ordinary soap and the synthetic detergents is that the latter in most cases do not form greasy precipitates with the polyvalent cations (e.g., Ca^{2+} and Mg^{2+}) that are normally present in unsoftened water:

$$2C_{17}H_{35}COONa + Ca(HCO_3)_2 \rightarrow (C_{17}H_{35}COO)_2Ca \downarrow + 2NaHCO_3$$
common soap — water-insoluble soap curd

$$2C_{17}H_{35}OSO_3Na + Ca(HCO_3)_2 \rightleftarrows (C_{17}H_{35}OSO_3)_2Ca + 2NaHCO_3$$
synthetic detergent — soluble

The use of synthetic products often is accompanied by unforeseen difficulties. Common soap is attacked by certain bacteria in sewage and is degraded into simpler compounds that reenter the life cycle, i.e., soap is biodegradable. The first synthetic detergents used on a large scale were resistant to bacterial attack and accumulated in lakes, rivers, etc.; most synthetic detergents now in use are biodegradable.

The widespread use of sodium phosphate (a necessary plant nutrient) in cleansing preparations has contributed to the *eutrophication* of lakes. That is, certain lakes that were once clear and largely free of plant life because of limited nutrients now are murky and have shore lines covered with a thick mat of vegetation and a bottom covered with a thick layer of decaying organic matter, due to phosphate and nitrate nutrients brought into them by sewage water and water from agricultural lands.

FOOD ADDITIVES

In many of today's foods, chemical substances are added to improve the color, flavor, texture, and nutritional value, and to maintain freshness. The chemical industry produces several hundred million lb of synthetic food additives annually, including dyes, flavors, vitamins, sweeteners, acidulants, antioxidants, emulsifiers, preservatives, stabilizers, thickeners, antistaling agents, leavening agents, mineral supplements, and amino acids. Actually, the food industry is so varied and complex that we can mention here only a few facets of this important area of chemistry.

Flavors. It is estimated that the total production of flavorings, including flavor enhancers, involves some 750 synthetic flavors and a total annual production of about 75 million lb. The trend to synthetic flavors is largely due to the lower cost, more uniform flavors, and longer shelf life of the synthetic compounds, as well as the limited availability of the natural flavors. For example, the world supply of natural vanilla would be insufficient to flavor all the vanilla ice cream consumed in the United States. Table 29-2 shows a few of the more common synthetic compounds used alone or in blends to produce specific flavors.

Certain compounds, called *flavor enhancers*, when used at low concentrations intensify or modify the natural flavors of foods. Among the synthetic flavor enhancers, monosodium glutamate, $HOOCCHNH_2CH_2CH_2COONa$, is the most widely used, with an annual production of about 45 million lb. It is used to intensify the flavor of canned meats, chicken, sea foods, spreads, soups, frozen

TABLE 29-2.

Synthetic Flavorings

NAME	FORMULA	ALONE OR IN BLENDS TO GIVE FLAVOR OF
Ethyl acetate	$CH_3COOC_2H_5$	Banana, strawberry, apple, mint
Isoamyl acetate	$CH_3COOCH_2CH_2CHCH_3$ $\|$ CH_3	Raspberry, strawberry, caramel
Ethyl butyrate	$CH_3CH_2CH_2COOC_2H_5$	Blueberry, pineapple, butter
Benzaldehyde	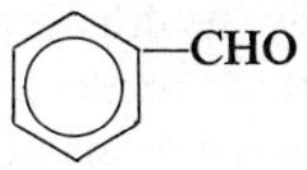 C₆H₅—CHO	Cherry, peach, coconut, almond
Methyl salicylate	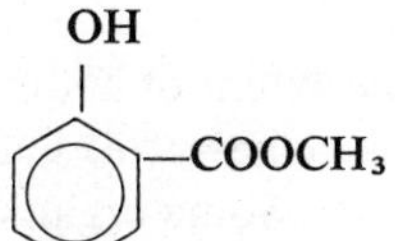 2-HO-C₆H₄—COOCH₃	Grape, wintergreen, mint, walnut
m-Methoxy-p-hydroxy-benzaldehyde	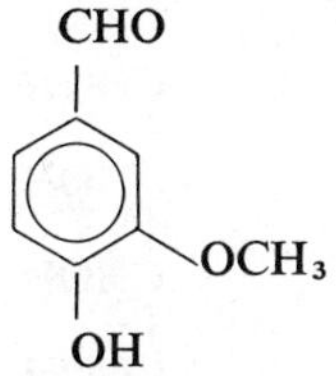 4-HO-3-CH₃O-C₆H₃—CHO	Vanilla

meats, sauces, and other high-protein foods. The compound contributes no flavor of its own. According to one theory, it stimulates the activity of the taste buds; according to another one, it increases the flow of saliva.

Other compounds are used to intensify or modify the flavors of foods high in carbohydrates such as in soft drinks, gelatin desserts, and jams. Maltol, a degradation product from carbohydrate materials, is an example:

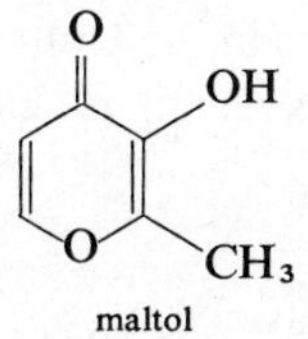

maltol

Artificial Sweeteners. Nonnutritive sweeteners, used mainly in low-calorie soft drinks and liquid diet foods such as Metrecal and Sego, are now being produced at the rate of millions of pounds annually. Since the ban of cyclamates in 1969, only saccharin is currently allowed in foods.

saccharin, about 300
times sweeter than sucrose

Nutritional Supplements. Many commercial foods are now enriched with vitamins or minerals, or both. Vitamin enrichment began on a large scale in 1941 with the addition of B vitamins to wheat flour, an action that resulted in practically eliminating pellagra in many impoverished areas. With the development of relatively cheap synthetic methods for the production of most vitamins, the addition of common vitamins to foods has become widespread. Various foods are also fortified with mineral supplements such as compounds of iodine, iron, calcium, magnesium, and copper. Recently, health foods heavily fortified with minerals and vitamins have come under attack on the grounds that most Americans eat natural foods that provide all the vitamins and minerals necessary to good health.

Miscellaneous Additives. Acids, to the extent of about 125 million lb annually, are added to soft drinks, candies, jams, jellies, fruit juices, gelatin desserts, and other foods to provide an appealing taste. Some organic acids used for this purpose are

| citric acid | adipic acid | tartaric acid | lactic acid |

Phosphoric acid, an inorganic acid, accounts for about 25 percent of the acidulents used in foods.

Fat-containing foods, such as cooking oils, potato chips, salted nuts, bacon, dehydrated soups, and a large number of other foods, undergo slow oxidation in air to produce compounds with rancid odors and flavors. For example, potato chips become rancid in about three days unless certain compounds, called *antioxidants*, are present to prevent the oxidation of fats and oils. The most widely used synthetic antioxidants in the United States are derivatives of phenol. With an antioxidant, potato chips remain rancid free for 20 days or longer. Two widely used antioxidants are

2,6-di-*tert*-butyl-*p*-cresol   propyl gallate

Chelating agents such as salts of ethylenediaminetetraacetic acid are added to certain foods to inactivate traces of iron, copper, or other metals that catalyze the oxidative spoilage of foods. Canned shrimp, mayonnaise, potato salads, and sandwich spreads are examples of commercial foods to which chelating agents may be added. Small amounts of iron compounds in beer tend to cause the sudden release of carbon dioxide and foaming when a beer container is opened. This can be avoided by using a chelating agent that forms a complex with the iron.

EXERCISES

1. Distinguish between condensation and addition polymers. Classify the polymers in Table 29-1 as condensation or addition polymers.
2. Write structural formulas for *cis* and *trans* isomers of 1-chloropropene; of 2,3-dichloro-2-butene.
3. Write a structural formula for 1-chloropropyne. Are two geometrical isomers possible for this compound? Why or why not?
4. Discuss the vulcanization of rubber.
5. Which of the following are properly classed as polymers: (a) fats; (b) proteins; (c) glyceryl trinitrate; (d) ethylene; (e) propylene.
6. Is a "crystalline" polymer crystalline in the same sense that ice is? Explain.
7. A polypropylene molecule might weigh as much as 500,000 awu. How many propene molecules would be needed to form such a molecule?
8. To what class of compounds do soaps belong? Describe how soaps can be prepared from fats. How do natural soaps differ from synthetic detergents?
9. Discuss how the structural approach is applied by research chemists who seek new drugs.
10. A shirt made of ordinary rayon was boiled in dilute HCl till it completely dissolved. After filtration to remove buttons, etc., the clear solution was neutralized with NaOH and then carefully evaporated to dryness. A white solid resulted. What is the probable composition of this solid?
11. Filter paper, on burning, leaves little ash; the paper in this book, if burned, would leave considerable ash. Why?
12. Chapter 29 deals with applications of organic chemistry. Can you think of important applications that have not been discussed?

SUGGESTED READING

Anderson, A. B., "The Composition and Structure of Wood," *J. Chem. Educ.*, **35**, 487 (1958).

Collier, H. O. J., "Aspirin," *Sci. Amer.*, **209**(5), 96 (1963).

Moncrief, R. W., "Linear Polymerization and Synthetic Fibers," *J. Chem. Educ.*, **31**, 233 (1954).

Price, C. C., "The Geometry of Giant Molecules," *J. Chem. Educ.*, **36**, 160 (1959).

Price, C. C., "The Effect of Structure on Chemical and Physical Properties of Polymers," *J. Chem. Educ.*, **42**, 13 (1965).

Roderick, W. R., "Structural Variety of Natural Products," *J. Chem. Educ.*, **39**, 2 (1962).

Thomas, B. B., "The Production of Chemical Cellulose from Wood," *J. Chem. Educ.*, **35**, 493 (1958).

30

CHEMICAL NATURE OF OUR WORLD

Our study of the composition of matter is based on the thesis that the chemical properties of matter can be explained in terms of electrons, protons, and neutrons. In our world, these particles are organized into specific units that form the atoms of more than 100 elements; the atoms of the elements are further organized into molecules and ions of compounds, which make up the great bulk of the nonliving part of our world. In the living part of our world, molecules are further organized, in an exceedingly complex manner, into what might be called the molecule of living matter—the cell. An important conclusion that has been appreciated only recently is that energy plays a dominant role in all these organizations of electrons, protons, and neutrons into the animate and inanimate forms of matter that compose the earth. Indeed, as brought out in Chap. 17, matter and energy are interconvertible.

In the present chapter we shall undertake a brief survey of the gross composition of our planet, and at the same time give attention to a few of the hypotheses with which we seek to account for the organization of the elements and compounds into the structural components of our world. We must realize that many conflicting theories have been advanced to explain how our physical world attained its present condition. Two or more hypotheses pertaining to the same single phase often appear equally plausible, and the scientist finds that he does not have enough data, at least for the moment, to help him choose between them. In general, the few hypotheses selected for presentation here have been advanced recently and have been well received by scientists.

Chemical Nature of Our World

THE UNIVERSE

Origin. The solar system, consisting of the sun, the planets and their satellites, the asteroids, the comets, and the meteors, is part of a galaxy that is composed of an estimated 100 billion stars (suns). Farther out in space are a large number of other galaxies of about the same size as ours. These systems of stars are scattered in space at tremendous distances from one another, the nearest one to the earth being about 1 million light-years away.

Light coming from the distant stars shows a displacement of the spectral lines toward the red portion of the spectrum (the Doppler effect, see Fig. 30-1). This

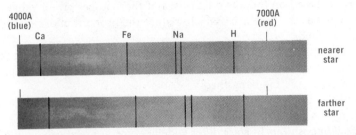

Fig. 30–1. The Doppler effect. Light coming to us from a star moving rapidly away from the earth has a slightly longer wavelength than it would have if the star were moving with the earth. Since all wavelengths are shifted by the same amount, the absorption lines (shown here by dark lines), as well as the continuous background of emitted light (colored portion), are shifted as a group. The greater the velocity of the star, the greater is the Doppler shift.

spectral shift is interpreted as meaning that all the galaxies, including our own, are moving outward and away from a common center, the more distant ones moving at greater velocities. This is the concept of the *expanding universe*.

According to one hypothesis that had wide acceptance till recently, all the matter in the known universe, at some time billions of years ago, was located in a comparatively small space under high internal pressure and temperature. With these conditions, elements did not exist; instead, all the matter was in a highly compressed form, consisting perhaps only of neutrons. Presumably, a violent expansion of this primordial matter began, accompanied by cooling as a result of the radiation of energy. The elements then formed from neutrons. (Remember that in certain radioactive changes, a neutron changes into a proton and an electron.) Condensation of the elements around centers of turbulence in the expanding material gave rise to clusters of matter that became stars, planets, etc. All this was supposed to have happened quite rapidly, perhaps within an hour or so. This theory, often called the *big-bang* theory, was pioneered by the astronomer-priest Georges Lemaitre and further developed by the physicist George Gamow.

One of the chief supports of the big-bang theory, until recently, was that the age of the elements as calculated from isotopic ratios was roughly the same as the calculated time since the chaotic explosion that sent the galaxies rushing apart. The recession velocities, as determined by the spectral red shifts, indicated that about 5 billion years had passed to allow the galaxies to move from a common origin to their present positions.

If it is assumed that the various isotopes of a given element were formed in equal amounts at the beginning of the violent expansion, the age of the elements can be calculated by measuring the relative amounts of radioactive species found remaining today. In the case of uranium, mass spectrographic analyses of ores reveal the relative abundance of the uranium-235 and uranium-238 isotopes to be in a 1 : 140 ratio. Assuming that the two were formed in equal amounts, the shorter half-life of uranium-235, 7.1×10^8 years as compared with 4.5×10^9 years for uranium-238, must be responsible for the present ratio of 1 : 140. Calculations based on these two half-lives of the isotopes give a value of about 6 billion years as the time necessary to reduce the proportion of uranium-235 to the present value.

Thus, two entirely different calculations indicated that 5 to 6 billion years had elapsed since the elements formed and the expansion started, and this agreement gave support to the big-bang theory.

Recently, enough evidence has accumulated to cast serious doubt on the big-bang theory. Among the new findings are:

1. The compositions of stars differ widely; some are relatively young and are still in the hydrogen fusion process; others are much older, having used all the original hydrogen. These stars probably did not originate at the same time.

2. A redetermination of the time elapsed for the observed recession, as gauged by the spectral shift, gives values of from 7 to 15 billion years instead of 5 billion years.

According to present-day theories, it is still quite possible that some element building did occur before differentiation into stars and planets occurred. However, it is now believed by many scientists that most of the building of elements occurs in the interior of stars and that this process can account for the natural abundance of elements. Other interesting postulates about our universe include (1) the universe is finite; (2) the universe is infinite; and (3) matter is being created continuously from energy, the so-called *steady-state theory*.

Composition. Knowledge of the chemical composition of the universe is obtained by the spectroscopic analysis of light from stars and planets, by the analysis of meteorites, and earth and moon materials, and from rockets that fly by or into planets and radio back information about their compositions.

The earth is made from fairly heavy elements, whereas in the universe as a whole hydrogen and helium are by far the most abundant. The stars are the massive heavenly bodies and so account for most of the mass in the universe. Our own star, the sun, is estimated to be about three-fourths hydrogen and one-fourth helium, with traces of many other elements.

THE EARTH

The planets can be divided into three classes. The small ones nearest the hot sun are composed of substances that condense easily. The second class is made up of substances that can condense if the temperature is low; such planets exist

Chemical Nature of Our World

outside the asteroid belt, much farther from the sun than the earth. The third class of planets consists mainly of gases.

Because only a small percentage of the earth (exposed surfaces, drill holes of a depth to about 5 miles, and mines) is available for direct analysis, we do not know much about the total composition of this planet. Data that have been helpful in formulating theories as to the composition of the earth's interior include (1) average density of the earth, (2) composition of meteorites, and (3) seismological data.

Density of the Earth. The mass of the earth as calculated from the force of gravity is 5.98×10^{27} g. The volume of a sphere is given by the formula $4\pi r^3/3$; the approximate volume (the earth is not a true sphere) is obtained by substituting its known radius, 6.37×10^8 cm, in the formula above and solving. This gives 1.08×10^{27} cc as the volume of the earth. The average density calculated from these data is

$$D = \frac{m}{v} = \frac{5.98 \times 10^{27} \text{ g}}{1.08 \times 10^{27} \text{ cc}} = 5.5 \text{ g/cc}$$

However, the average density of earth material in the accessible crust is only about 2.8 g/cc. We must assume, therefore, in order to account for an over-all density of 5.5 g/cc, that matter in the interior has a density that is considerably greater than 2.8 g/cc.

The high density of the earth's interior has been popularly accounted for in two ways: (1) the earth's interior contains a large percentage of dense elements, for example, iron; or (2) the gravitational compression of material similar to that in the crust gives rise to greater densities. (See also the footnote to Table 30-1.)

Composition of Meteorites. It has been estimated that perhaps 1000 tons of meteors and cosmic dust bombard the earth daily, much of which burns in the stratosphere or above. Although the many meteorites that have fallen to the earth vary considerably in composition and type, they may be classified roughly under two headings: (1) the metallic and (2) the stony iron and stone type.

The metallic type consists essentially of a nickel-iron alloy. The average composition of the alloy, obtained by analyzing many meteors collected around the world, is 90.8 percent Fe, 8.6 percent Ni, and 0.6 percent Co.

The stony iron and stone type consists of about 12 to 50 percent nickel-iron alloy, the remainder being made up mostly of metallic silicates of such common metals as Fe, Ni, Al, Ca, Na, Mg, Cr, and Mn. Oxides and sulfides may also be present.

Inasmuch as meteorites are thought to represent samples from different depths of one or more planets, their composition has been of great importance in formulating hypotheses as to the nature of the earth's interior. Available information (calculated densities and spectrographic data) indicates that the over-all composition of different planets varies; consequently too much reliance should not be placed on the composition of meteorites in deducing the composition of the inner portions of our planet.

Seismological Data. Earthquake waves travel at varying speeds through the earth, depending on the characteristics of the material through which they pass;

hence they are subject to refraction and reflection at the surface of discontinuities. By comparing the time at which the waves from an earthquake arrive at different seismological stations (recorded by an instrument called a *seismograph*), the velocities at different depths can be ascertained (see Fig. 30-2). Such

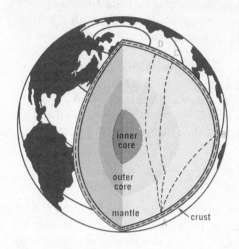

Fig. 30–2. Earthquake waves travel at different velocities through different earth materials. For this reason, the waves are refracted or bent as they pass from one type of material to another. By comparing the time required for shock waves originating at point A to reach recording stations at points B, C, D, etc., around the globe, three major discontinuities and several minor ones have been located.

data show clearly that there are three major discontinuities in the earth's interior. The earth is therefore divided into four major parts: the **crust** (lithosphere), which extends down for about 18 to 30 miles to the surface of the first discontinuity; the **mantle**, which extends for about 1800 miles below the crust; the **outer core**, which extends from the mantle to the inner core, a distance of about 1300 miles; and the **inner core**, which extends from the outer core to the center, about 4000 miles from the surface.

To complete the major divisions of our planet, three other zones must be added: the **hydrosphere**, the **biosphere**, and the **atmosphere**. Table 30-1 summarizes the chemical character of different zones. The data given there for the composition of the crust, mantle, and core account nicely for the calculated density of the earth and the different velocities of earthquake waves. It is also in accord with the composition of meteorites and with the composition of material brought up from the interior of the earth by volcanic action.

Many scientists think that the earth was formed by gravitational accumulation of cosmic dust. At the same time, other dust eddies condensed to form the sun, moon, and other planets. At the beginning of the formative stage, the temperature was relatively low, but owing to the heat released by the gravitational compression and radioactive changes the earth became molten and remained molten for about 2 billion years, till most of the original radioactive atoms present had changed to stable atoms. Some scientists think that the moon, like the earth, became molten and differentiated chemically at a very early period in its history. Some of the recently obtained facts and their correlating theories about the moon's composition and geological history are discussed at the end of the chapter.

During the molten period, great amounts of gaseous substances, such as

H_2, CH_4, NH_3, and the noble gases, escaped from the earth. This assumption is based on the fact that, at the high temperature of the molten earth, their molecules had sufficient kinetic energy to move out of the earth's gravitational field. The *escape velocity* of a hydrogen molecule is about 7 miles/sec, seven times its speed at sea level at present ordinary temperatures.

TABLE 30-1.

Composition of Major Zones of the Earth

NAME	CHEMICAL CHARACTER
Atmosphere	N_2, O_2, CO_2, H_2O, the noble gases, other gases
Hydrosphere	Salt and fresh water, ice, snow
Biosphere	Living matter, organic materials
Crust (lithosphere)	Igneous rocks, shale, sandstone, limestone
Mantle	Metal silicates, mostly magnesium and iron silicates, with probably considerable metallic sulfides, oxides, etc.
Core[a]	Iron-nickel alloy, the outer part probably molten

[a] According to one theory, the outer core is liquid and the inner core is solid, both being of the same approximate composition. Recently, consideration of the tremendous pressures that exist at the base of the mantle on through to the earth's center has given rise to a number of new and interesting speculations as to the chemical composition of the core. These pressures, about 1½ million atm at the base of the mantle and about 4 million atm at the center, are believed to cause matter to display properties that are very different from its normal characteristics. Consequently, one hypothesis holds that it is not necessary to consider the core made up of iron and nickel in order to account for the known mass of the earth. Substances of low densities could acquire higher densities under the enormous pressures in the core. Indeed, in 1941 Kuhn and Rittmann of Germany postulated that the earth's core is composed of highly compressed hydrogen. This speculation is contradicted by many arguments. Bullen of Australia considers that the outer core consists of high-density modifications of mantle material and uncombined iron and nickel, and that the inner core is mostly iron and nickel with perhaps some denser materials as well. He argues that this concept makes plausible the view that Mars, Venus, Mercury, and Earth were of the same primitive over-all composition.

The hot molten mass of the young earth was kept in a fairly homogeneous condition by convection currents. The abundant elements were iron, magnesium, silicon, and oxygen, with smaller amounts of calcium, nickel, sodium, and sulfur.

According to one hypothesis, as cooling continued and stable compounds formed, there was not enough oxygen and sulfur to convert all the abundant metals into oxides and sulfides. The excess elemental iron and nickel had higher densities than the compounds and so sank toward the center of the planet, forming the core. The mantle, consisting largely of magnesium and iron silicates, formed a liquid layer around the liquid core. Perhaps a third liquid phase of molten iron sulfide and other metallic sulfides formed as immiscible drops dispersed through the mantle and the core.

It appears likely that all the remaining elements and their compounds became scattered in these three liquid phases, not according to their densities but according to their chemical affinities for either the iron-nickel melt, or the silicate melt, or the metallic sulfide melt. For example, cobalt, the platinum metals, and gold are believed to have concentrated in the core, and uranium and thorium minerals in the crust. As cooling continued, the first substances to become solids were possibly iron and magnesium silicates, members of the *olivine* minerals.[1] Another hypothesis maintains that the first solids were largely metallic oxides and sulfides. All these crystalline compounds are believed to have first formed at the lowermost part of the mantle, thus forming a solid crust around the core that greatly reduced its further cooling. Indeed, the cooling of the core may have been retarded sufficiently so that at least part of it is still in the molten state. The refraction of earthquake waves by the core supports the hypothesis that at least its exterior is liquid.

As crystallization continued in the mantle, the liquid that still remained became enriched with compounds of Al, Ca, Si, O, K, and Na, and, to a lesser extent, compounds of certain other elements. As enrichment continued, these substances formed minerals, such as feldspar and quartz, that accumulated at the top of the molten liquid and eventually aggregated to form a *granitic*[2] type of crust over the mantle.

This granitic crust was probably too thin to resist the strong convection currents of the molten layer underneath it; hence it was broken and rebuilt many times. At some later time after it had thickened considerably, it apparently was broken again by the currents of the molten mass underneath it, with the parts being folded and piled in a way that formed the high and low areas that later became continents and oceans.

The primitive atmosphere was probably composed mostly of carbon dioxide and water. As long as the temperature remained high, only water vapor was present. But, as the temperature fell vast quantities of water condensed, and the crust was subjected to considerable chemical and physical erosion.

Clearly, continent-building processes have taken place during the long period of time that has elapsed since the crust finally solidified; otherwise the land would have been leveled to low plains by erosion. Earthquakes, upheavals, and other types of crust movement have continued throughout geologic time. Land at one time under water may now be several thousand feet above sea level, as evidenced by the presence of fossils and other sedimentary materials; or land that at one time must have been above sea level is now under the sea. And these massive displacements in the crust continue today.

Today the uppermost crust of all continental areas is a layer of granitic rock that is usually covered by a relatively thin layer of sedimentary material (soil, limestone, sandstone, etc.), material originally made available by erosion of parts of the primeval granitic layer. Because the deep parts of the Pacific Ocean have a basaltic bed and because volcanic extrusions are of the basaltic type, it is believed that beneath the granitic layer is a basaltic layer.

---

[1] The olivine minerals crystallize in the orthorhombic system and are orthosilicates of the bivalent metals, such as Mg_2SiO_4 (*forsterite*), Fe_2SiO_4 (*fayalite*), $(Mg,Fe)_2SiO_4$ (*olivine*), $CaMgSiO_4$ (*monticellite*), and $PbZnSiO_4$ (*larsenite*).

[2] Igneous rocks may be classed as *granitic* or *basaltic*, depending on whether they are acidic or basic. Granitic rocks (granite is a common example) are relatively poorer in the baseforming (metallic) elements and are acidic. Most of the igneous rocks close to the surface are granitic. Both types of igneous rocks contain complex compounds of O, Si, Al, Fe, Mn, Mg, Ca, Na, K, Ti, and P, as well as lesser amounts of other elements. Oxygen, aluminum, and silicon usually account for 75 percent or more of the rock.

Chemical Nature of Our World

THE CRUST

Composition. The upper 10 miles of the **crust** is estimated to be 95 percent **igneous rocks** (rocks formed by the solidification of molten mineral mixtures, for example, granites and basalts), 4 percent shale, 0.75 percent sandstone, and 0.25 percent limestone. In comparison with these, the total amount of other types of sedimentary and metamorphic matter is insignificant. Our knowledge about the average composition of the upper 10 miles of the crust comes from thousands of analyses of minerals and rocks sampled from all parts of the globe.

The percentages by weight of the 12 most abundant elements are shown in Table 30-2. It will be seen from the table that oxygen (combined in such minerals as silicates, carbonates, and oxides) makes up nearly half of the weight of the earth's crust, and that 12 elements (oxygen included) make up 99.5 percent of the weight.

TABLE 30-2.

Twelve Most Abundant Elements in the Earth's Crust (Percent by Weight)

Oxygen	49.5	Calcium	3.4	Hydrogen	0.9
Silicon	25.7	Sodium	2.6	Titanium	0.6
Aluminum	7.5	Potassium	2.4	Chlorine	0.2
Iron	4.7	Magnesium	1.9	Phosphorus	0.1
					99.5

Figure 30-3 shows the relative abundance of the elements in the 10-mile crust. Elements with atomic numbers 43, 61, 85, 87, 93, and greater occur in such meager amounts, if at all, that they are obtained by bombardment reactions, rather than from minerals. However, studies of radiations emitted from concentrates of natural radioactive materials have revealed the presence of very low concentrations of astatine, francium, and some others of the above.

Figure 30-3 reveals that (1) elements with low atomic numbers constitute most of the earth's crust; (2) with some exceptions, elements with even atomic numbers are more abundant than those with odd numbers. Note that an odd-number element is usually less abundant than the two adjacent elements with even numbers. See also Fig. 17-5.

From the standpoint of the second observation, helium, neon, argon, and krypton (even atomic numbers) are much less abundant than would be expected. It is possible that these gases were present in greater quantities early in the earth's history. Great quantities are thought to have escaped from the earth's gravitational field while the planet was in a molten condition. Spectroscopic analyses of light from the sun and other stars indicate they have much greater concentrations of these elements than has the earth.

Magnesium, chromium, sulfur, selenium, and tellurium (even atomic num-

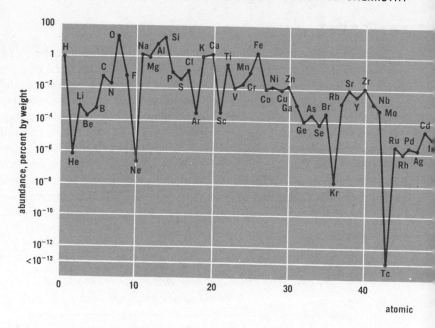

Fig. 30-3. Relative abundances of the elements in the earth's crust. (Data from Michael Fleischer, "The Abundance and Distribution of the Chemical Elements in the Earth's Crust," *J. Chem. Educ.*, **31**, 446 [1954].)

bers) also would be expected to be more abundant on the basis of the second observation. These discrepancies may be due to the fact that large amounts of these materials (as sulfides, silicates, and other compounds) sank into the earth's interior while the earth was molten.

Availability Versus Abundance. It is perhaps surprising to note in Fig. 30-3 that many of the elements that we usually think of as scarce are relatively abundant. For example, platinum is more abundant than silver, and uranium is more abundant than bismuth, mercury, silver, or cadmium. The availability of an element is determined not so much by its relative abundance as by its distribution. If the elements were uniformly distributed in the earth's crust, there would be no search for valuable ores, for one piece of land would be as suitable as the next for ore extraction.

Although the elements may have been uniformly distributed at one time, many chemical and physical processes have caused the formation of concentrated deposits. It is these concentrated deposits that comprise our workable ores, deposits that are sparsely scattered in the earth's crust.

Igneous Rocks. Studies of the crystallization and solidification of molten minerals indicate that, as the molten crust cooled to form igneous rocks, the high-melting oxides of magnesium, iron, and chromium must have crystallized first, forming what is known as the spinel type of mineral; then followed the crystallization of magnesium-iron silicates (olivines) and calcium-sodium-aluminum silicates (feldspars). Possibly, as the first solid compounds formed in the liquid mixture, they sank and formed the deeper rocks, which are low in

Chemical Nature of Our World

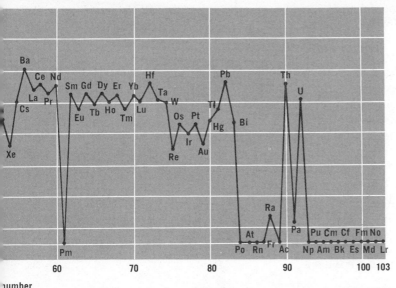

silicon and rich in magnesium. However, a good deal of the solid material remained suspended; hence when complete solidification occurred, large crystals of the solids formed earlier were embedded in the rock structure that resulted from the solidification of the last liquid portion (see Fig. 30-4).

Fig. 30–4. Three igneous rocks. From left to right: obsidian, granite, and basalt. Note the large crystals embedded in the fine-grained structure of granite and basalt.

The order of these crystallization processes depended not only on the melting points of the minerals, but also on such other factors as ionic radius, ionic charge, and the type of crystalline structures being formed (see Chap. 23). Any ion in a crystal can usually be replaced by another of about the same size without causing serious deformation of the structure. Thus, Fe^{3+}, Cr^{3+}, and Al^{3+} are mutually replaceable in many crystals, as are Mg^{2+} and Fe^{2+}. Even ions of the same size but with different charges may replace each other, provided electrical neutrality is maintained by other variations in the crystal. Thus, Na^+

may substitute for Ca^{2+} in feldspars, provided equivalent changes occur in the proportion of Al^{3+} and Si^{4+}. Compare $NaAlSi_3O_8$ and $CaAl_2Si_2O_8$ (Table 23-2).

Sedimentary Rocks. After the earth's crust solidified, other events brought about further changes in the distribution of the minerals in it. The action of water, wind, carbonic acid, and other agents produced sedimentary material that tended to collect in rivers, lakes, and oceans.

When the sediments were first deposited, they were soft and contained much water. As the amount of sedimentary matter increased, the lower layers were subjected to pressure that squeezed out most of the water. Minerals that were precipitated from both trapped and seepage water tended to cement the particles together, so that the deposit gradually took on the character of rock. In this manner, such sediments as clay, sand, and lime-mud became shale, sandstone, and limestone, respectively. Rocks of this type, called **sedimentary rocks**, constitute about 5 percent of the earth's crust.

THE HYDROSPHERE

The **hydrosphere** includes the oceans, seas, lakes, rivers, underground water, ice, and snow. The oceans and seas form by far the greatest portion of the hydrosphere, perhaps 98 percent. They cover about 71 percent of the earth's surface to an average depth of 2.2 miles. This quantity of water and dissolved materials has a volume of about 32 million cubic miles and a weight of about 1400 million billion tons (1.4×10^{18} tons).

Composition of Sea Water. The relative amounts of the major dissolved minerals in sea water are shown in Table 30-3. Although this table shows only five cations and six anions, ocean water contains traces of essentially all metallic and nonmetallic ions.

TABLE 30-3.

Major Ionic Components of Sea Water

CATION	PERCENT	ANION	PERCENT
Na^+	1.06	Cl^-	1.90
Mg^{2+}	0.13	SO_4^{2-}	0.26
Ca^{2+}	0.04	HCO_3^-	0.01
K^+	0.04	Br^-	0.0065
Sr^{2+}	0.01	F^-	0.0001
		I^-	0.000005

Vast amounts of minerals are held in solution in the oceans. If all the dissolved minerals in a cubic mile of ocean water were separated by evaporation of the water, the residue would weigh about 140 million (long) tons and would be composed mostly of such compounds as sodium chloride, magnesium sulfate, magnesium chloride, sodium bromide, calcium sulfate, potassium chloride, etc.

Chemical Nature of Our World

Complete recovery of certain of the elements in a cubic mile of sea water would yield:

81,000,000 tons of chlorine	200 tons of iodine
43,000,000 tons of sodium	2 tons of uranium
5,600,000 tons of magnesium	57 lb of gold
280,000 tons of bromine	0.5 g of radium

At present, only a few substances are produced commercially from sea water—magnesium and bromine in particular. Table salt is recovered in many parts of the world by solar evaporation. The present research on the desalination of sea water to produce potable water may also lead to the recovery of salts and concentrated salt solutions from which compounds may be produced.

THE ATMOSPHERE

Figure 30-5 shows the different zones of the atmosphere and the pressure at different heights. The outer boundary of the atmosphere is not known; possibly

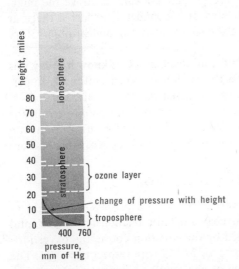

Fig. 30–5. The major layers of the atmosphere. There is no abrupt change from one layer to the next. As can be seen, the pressure is practically zero above an altitude of 20 miles.

even at a height of 35,000 miles the concentration of ions derived from our atmosphere is greater than the concentration in outer space. But more than 99 percent of the total mass of the atmosphere is below an altitude of 20 miles; indeed, as Fig. 30-5 shows, the atmospheric pressure approaches zero at a height of about 20 miles. The lowest region, the one we live in, is called the **troposphere**; it extends to a height ranging from about 5 miles over the poles to about 11 miles over the equator. It is in this region that common atmospheric phenomena, such as cloud formation, air movements, and wide temperature gradations, occur. The next layer is called the **stratosphere.** This region is characterized by the presence of several layers (hence the name) between which there is apparently no strong vertical circulation. Clouds and winds do not exist; it is very cold in the lower strata and warmer in the upper strata. There is a layer of warm ozone, between about 20 and 40 miles, that is due to the action of ultraviolet light on oxygen.

The **ionosphere** is an ionized layer at a fairly high temperature, presumably warmed and ionized by short-wave radiations from the sun. The ionizing reactions in the upper atmosphere absorb most of the high-energy portions of the sun's electromagnetic radiation; thus, the atmosphere protects the surface of the earth from radiation that would be damaging to living things. Only two narrow regions of the sun's spectrum are not efficiently absorbed by the atmosphere. One of these regions, 3000 to 25,000 A, stretches from just inside the ultraviolet, through the visible, and into the near infrared region; the other is in the radio-wave region from about 0.01 to 40 meters wavelength. Radio telescopes are therefore able to receive radiation not only from the sun, but from other stars, some of which have not been detected by optical telescopes but have only been "seen" with radio telescopes.

So much has been learned about distant stars and about our own sun through radio astronomy that astronomers are eagerly pressing forward with the construction of X-ray and ultraviolet telescopes that are carried above the interfering atmosphere by balloons or satellites. It is confidently expected that our atmosphereless moon will someday be the site of an astronomical observatory par excellence.

Origin. It seems highly probable that the atmosphere as we know it today was formed during an extensive period after the earth had cooled sufficiently to permit life to develop. The origin of oxygen has given rise to several controversial theories. According to one hypothesis, most of the oxygen now present in the air is due to *photosynthesis*, a reaction whereby chlorophyll-bearing plants are able to convert carbon dioxide and water into carbohydrates (organic carbon) and oxygen.

$$6CO_2 + 6H_2O \rightarrow C_6H_{12}O_6 + 6O_2$$

If the oxygen of the air was indeed formed by such a process, the weight of the total amount of this oxygen should bear a definite relationship to the total amount of organic carbon. As indicated by the equation above, the oxygen-to-carbon ratio is 2 : 1 with respect to atoms, or 32 : 12 with respect to weight. The organic carbon has largely accumulated as coal and to a lesser extent as petroleum. A small amount is present in the now existing living matter. Although it is difficult to approximate the weights of such huge quantities of matter, the best estimates indicate that the amounts of atmospheric oxygen and organic carbon are in about the ratio that would be expected on the assumption that both were formed by the photochemical process. Note that we need not consider the organic matter that decayed instead of becoming coal, because decay involves the removal of oxygen and the formation of carbon dioxide and water.

The presence of the other components can be accounted for in part as follows. *Gases* trapped in igneous rock were released later by volcanic activity. For example, igneous rocks are known to contain about fifty times as much *nitrogen* as is present in the atmosphere. *Helium* was produced by the radioactive decay of uranium and thorium minerals. *Argon* was produced by the decay of radioactive potassium.

It must be remembered that the theories regarding the earth's origin are quite speculative and that scientists are not in agreement on all the various points. Even though each point is based on observable evidence, in many cases the evidence is of such a nature that it can be interpreted equally well by different theories.

Composition of Dry Air. The composition of dry air is given in Table 30-4. Inspection of this table reveals that three gases—nitrogen, oxygen, and argon—make up 99.96 percent of dry air. Water vapor is not included, because the amount varies widely from day to day and season to season. If water vapor is neglected, the composition of the air is remarkably constant from place to place around the globe.

TABLE 30-4.

Average Composition of Dry Air

GAS	FORMULA	PERCENT BY VOLUME
Nitrogen	N_2	78.08
Oxygen	O_2	20.95
Argon	Ar	0.93
Carbon dioxide	CO_2	0.03
Neon	Ne	0.002
Helium	He	0.0005
Methane	CH_4	0.0002
Traces of krypton, nitrous oxide, hydrogen, xenon, and ozone		

The constancy of the composition of the air can be accounted for in two ways: (1) because of the large amount of air and its constant stirring by winds, localized chemical and physical processes that tend to use up one or more components have little effect on the over-all composition; (2) natural oxidation, a major process in the removal of oxygen, is balanced by photosynthesis, a process that restores oxygen.

Our way of life poses the serious problem of air pollution for large cities. It is estimated that the average automobile discharges about 1 lb of contaminants per day into the atmosphere. In a city with 2 million automobiles, this amounts to 1000 tons per day of such impurities as lead compounds, unburned hydrocarbons, carbon, and other substances (the CO_2 and H_2O formed are excluded from this calculation of "impurities"). Such contaminants, along with those reaching the air from industrial plants, are a serious menace to the health of the people living in areas thus affected. The most undesirable components of automobile exhaust are apparently oxides of nitrogen, unsaturated hydrocarbons, and carbon monoxide. From coal and fuel oil combustion a major offender is sulfur dioxide. The smogs afflicting cities also contain compounds formed when the initial combustion products are chemically activated by the sun's ultraviolet radiation.

Although we may regard the composition of "natural" dry air as approximately constant from one place to another, it apparently is changing slowly with time. At present, the amount of carbon dioxide in the atmosphere is increasing because of man's intense industrial and agricultural activity. The burning of fuels produces 6×10^9 tons of this gas per year; destruction of forests in favor of cultivated fields leaves another 2×10^9 tons in the atmosphere.[3] These huge amounts are sufficient to displace the equilibria involved in natural processes that maintain a rough balance between the earth and its atmosphere (see Fig. 30-6).

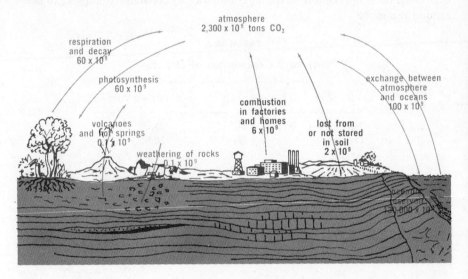

Fig. 30-6. Man's effect (black arrows) on the carbon dioxide cycle.

One possible result of this change in the atmosphere is a change in the temperature of the earth. According to the so-called greenhouse theory, carbon dioxide in the air acts to trap heat. The visible radiant energy of the sun comes through the atmosphere efficiently and warms the earth's surface. The earth reradiates much of this energy, but in the infrared rather than in the visible region of the spectrum. Carbon dioxide absorbs infrared radiation efficiently and the radiant heat thus trapped warms the atmosphere and thereby keeps the earth warmer.

Weather records indicate that the average temperature has increased 1°F over the past 100 years. This is thought to be due to the 360×10^9 tons of carbon dioxide poured into the atmosphere by man's activities in combustion and his destruction of forests. However, recent calculations indicate that the rate of warming by CO_2 decreases as the amount of CO_2 in the atmosphere increases.

---

[3] Forests are more efficient in using CO_2 from the atmosphere than cultivated crops are.

Another factor that is becoming increasingly important with the world's growing population is the amount of particulate matter (dust, smog, etc.) in the atmosphere. It has been calculated that if the equilibrium dust content of the air quadruples, the average global temperature will fall by 3.5°C. Such a decrease, if maintained for a few years, would probably start another ice age.

Noble Gases. The abundance of the noble gases—helium, neon, argon, krypton, and xenon—in the atmosphere is indicated in Table 30-4. (Radon, also one of the noble gases, is radioactive with a very short half-life; it does not occur in detectable amounts in the atmosphere.) The noble gases may be obtained by careful fractional distillation of liquid air. However, because the concentrations of helium, krypton, and xenon are so low, they cannot be obtained in appreciable amounts by this process.

THE BIOSPHERE

The **biosphere** includes all the living matter of the world. Because radiant energy from the sun is required for the growth of plant materials and because plants constitute the food of animals, life is largely restricted to areas where sunlight is available, including the oceans to a certain depth.

Of the three main areas where there is life—land, fresh water, and sea water—sea water is by far the most important from the standpoint of quantity. It provides approximately three hundred times as much inhabitable space as the other two combined. Estimates of the amount of organic carbon and oxygen produced annually by photosynthesis in different environments (ocean, cultivated land, forests, plains, deserts, etc.), result in the following averages per square mile: land, 62 tons; ocean, 132 tons. That is, per square mile, the ocean is over twice as productive as the land is. Because the oceans make up about three-fourths of the earth's surface, most of our planet's oxygen is produced in it, as well as most of the organic matter. The oceans are becoming more and more important as a source of food for the world's increasing population.

The great bulk of life in the oceans consists of small organisms, the phytoplankton and the zooplankton. These are found in all oceans, from the surface as far down as there is effective light penetration, to a depth of about 150 to 500 feet, depending on the turbidity.

The oceans, with their huge volumes of water, have long been considered a safe dumping ground for all kinds of waste. It is obvious now that this practice must be corrected. With over half of our population living near or in coastal areas, the large amounts of untreated human waste, industrial waste, and certain agricultural pesticides that flow daily into the oceans can cause irretrievable damage to marine ecosystems. Lifeless zones have already been created in certain coastal regions and large kills of fish have been brought about by industrial wastes. Shellfish have been found that contain polio virus, hepatitis virus, and other pathogens.

Tillable Soil. Life on the land is directly or indirectly dependent on the thin layer of small rock particles, sand, clay minerals, and decaying organic matter that we call the **soil**. Good soil must be of the proper texture so that roots, air, and water can penetrate it. It must retain water and plant nutrients, and must not harden excessively during dry seasons. The formation of such soil requires a very long time, a fact that every progressive farmer appreciates. However, in spite of our efforts at conservation, large quantities of cultivated soil are lost annually by the action of water and winds.

Soil is formed from rocks by the mechanical action of winds, water, and other agents, and the chemical action of such substances as oxygen, water, and carbonic acid. To illustrate the chemical action of water and carbonic acid, we shall use the chemical reaction that may occur when one of the feldspars (anorthite, $CaAl_2Si_2O_8$) is attacked by these two agents and forms water-soluble or colloidal substances that eventually may become limestone, clay, etc.:

$$\underset{\text{anorthite}}{CaO \cdot Al_2O_3 \cdot 2SiO_2} + 2H_2CO_3 + H_2O \rightarrow \underset{\substack{\text{limestone,} \\ \text{calcite, marble,} \\ \text{coral, etc.}}}{Ca(HCO_3)_2} + \underset{\text{clay, shale}}{Al_2O_3 \cdot 2SiO_2 \cdot 2H_2O}$$

Physical changes are important, too, in making rock surfaces available for chemical action. For example, mica is a potential source of potassium ions, one of the primary plant foods necessary in fertile soil. Normally tightly held between sheets of silicate tetrahedra in mica, potassium ions are exposed and made available for ion exchange when the mica sheets weather, develop cracks, then roll back and flake as the sheets separate. Reactions such as this produce water-soluble compounds of calcium, sodium, potassium, magnesium, and phosphorus, among others, that can be used by growing plants.

The accumulation of decaying organic matter (humus) in the soil not only returns to it the elements in plant and animal materials but also improves its texture and thus increases its value as a medium for plant growth.

THE MOON

Because it has no surface water or atmosphere, the moon's surface has not been subjected to erosion of a type similar to that taking place on the earth. Yet the moon's surface has obviously undergone constant change. Large and small meteoroids have bombarded it in great numbers, churning up its surface and causing lunar matter to melt and splash considerable distances. Seismic signals recorded from Apollo lunar module impacts on the moon indicate that the lunar structure to a depth of 6 to 12 miles is quite different from that typical of the earth's crust. One conclusion is that the moon's crust is not a more or less solid sheet of rock covered by a few feet of small rocks and rock powder, as is the case of the earth's crust; instead, a deep layer of rock powder is indicated.

With no atmosphere to shield it, the moon's surface is bombarded incessantly by cosmic rays.[4] As the rays move through rock, they produce permanent physical and chemical changes, thereby leaving tracks. Studies of these tracks

---

[4] Cosmic rays are mainly H nuclei and, to a much lesser extent, He, Li, and other small nuclei, that move about through interstellar space at velocities approaching the speed of light.

have led to some interesting conclusions about the history of the moon. For example, it has been concluded from such a study that one of the small rocks brought back by the Apollo 12 astronauts was pushed to within several feet of the surface about 70 million years ago. There it lay for about 68.3 million years until it was suddenly churned to the surface, probably by meteoroid impact. For 700,000 years it lay on one side and then was turned over by another meteoroid impact. Finally, about 9000 years ago, it was covered with molten glass, supposedly owing to the heat evolved by a nearby meteoroid impact.

Analyses of the first samples of moon rocks and soil brought back to earth in July 1969 yielded confirmation of some expectations and some surprises. Silicates similar to minerals on earth were found on the moon, with basaltic igneous minerals among the most common. Several silicates unlike any known terrestrial minerals were also brought back. Many small fragments of iron meteorites and carbonaceous meteorites were found mixed with the lunar soil. The soil is a mixture of crystalline and glassy fragments, many small and dustlike, that has evidently resulted from the incessant bombardment of the surface by meteorites, with the accompanying melting and splashing of molten rock, evaporation, and condensation. Single rocks tend to have pits which have been made by the impact of tiny meteorites.

The average age of the oldest rocks in these first samples from the lunar crust is 4.6×10^9 years. Some of the igneous rocks were found to be only 3.7×10^9 years old. It is speculated that these igneous rocks were formed in some intense meteoric bombardment or that they were expelled from a lower molten layer of the moon and solidified on the surface.

The average composition of the moon samples differs from that of the earth's crust. In general, volatile elements are less abundant on the moon, which is consistent with the thesis that these elements tended to escape from the weak gravitational field at a time when the moon was very hot. There is also evidence that water has played little or no part in the chemical history of the lunar surface. This, together with the absence of any evidence of complex carbon compounds such as amino acids, makes it appear that life has never existed there.

How the moon originated has been the subject of much speculation. According to modern views, the moon was formed by accretion of matter from a solar nebula. Recently, it has been postulated that the final accretion took place well outside the earth's gravitational field and that the resulting planet was captured later by the earth. Supposedly, during the accretion much or all of the moon became molten and the separation of chemical entities occurred.

EXERCISES

1. The wavelength of a certain spectral line from a distant star was determined as 4357 A. In the laboratory, this same spectral line has a wavelength of 4356 A. What can be concluded from this observation?
2. In general, how is the composition of a star determined?

3. Is the earth believed to be a typical sample of matter present in the universe? Why or why not?
4. What type of evidence is cited to support the concept of an expanding universe?
5. Some stars are described as white, others as red. Account for the differences in the nature of the light coming from these stars.
6. Even though the earth probably was molten at one time, why might some rather dense minerals have concentrated near the surface?
7. The planet Venus has a diameter of 7710 miles and a mass of 4.90×10^{24} kg. Jupiter has a diameter of 87,000 miles and a mass of 1.90×10^{27} kg. Calculate their densities in g/cc.
8. From the densities calculated in Exercise 7, what if any conclusions can be made concerning a common origin of Venus, Jupiter, and Earth? Compare your conclusion with the footnote of Table 30-1.
9. What happens to the energy of the ultraviolet light that is absorbed in the upper atmosphere?
10. What reasons are there for believing that the atmosphere as we know it today was formed over a rather long period after the earth had cooled considerably?
11. If we assume that the oxygen currently present in the atmosphere was formed by photosynthesis, we must also assume that the plant kingdom developed before the animal kingdom. Why?
12. What evidence leads to the belief that the earth consists of several layers, each composed of a different type of material?
13. By referring only to the periodic table, predict which element in each of the following pairs is the more abundant in the earth's crust: (a) Sr or Y; (b) Os or Ir; (c) P or Hg.
14. List as many processes as you can that produce carbon dioxide. The amount of carbon dioxide in the atmosphere is increasing, however, only slowly. Why is this?
15. In what type of climate would you expect chemical erosion to be very important?
16. What weight of seawater would have to be processed to obtain 1 ton of iodine (assume 80 percent recovery)? See Table 30-3.
17. A sample of dry air taken in the mountains of China has the same composition as a sample taken in Colorado. How is this possible?
18. Relative humidity is defined as follows:

$$\frac{water\ vapor\ present}{water\ vapor\ at\ saturation} \times 100$$

Calculate the relative humidity in a room at 20°C if the relative humidity of the same air at 5°C was 90. See Table 2 in the Appendix for the amount of water vapor (expressed in terms of pressure) at saturation at different temperatures.
19. Potassium-40 is radioactive and decays to the stable argon-40. The half-life is 1.28×10^{10} years. Suppose a rock sample from the moon was found to contain a $^{40}K : {}^{40}Ar$ ratio of 1:15 by atoms. What is the calculated age of the rock? What assumptions are made in this calculation? Would such a finding be in accord with the present belief about the age of the universe?
20. Explain how carbon dioxide in the atmosphere can allow radiant energy from the sun to reach the earth's surface, yet act so as to prevent the radiation of energy from the earth to outer space.
21. How could excessive particulate matter in the atmosphere cause a drop in the earth's temperature?

SUGGESTED READING

Anders, E., and G. G. Goles, "Theories on the Origin of Meteorites," *J. Chem. Educ.*, **38**, 58 (1961).

Asimov, I., "The Elementary Composition of the Earth's Crust," *J. Chem. Educ.*, **31**, 70 (1954).

Baldwin, R. B., "Summary of Arguments for a Hot Moon," *Science*, **170**, 1297 (1970).

Bassham, J. A., "Photosynthesis," *J. Chem. Educ.*, **36**, 548 (1959).

Dorf, E., "The Petrified Forests of Yellowstone Park," *Sci. Amer.*, **210**(4), 106 (1964).

Editorial Staff, "The Latest on Lunar Samples," *Chemistry*, **44**(3), 22 (1971).

Fleischer, M., "The Abundance and Distribution of the Chemical Elements in the Earth's Crust," *J. Chem. Educ.*, **31**, 446 (1954).

Othmer, D. F., "Water and Life," *Chemistry*, **43**(10), 13 (1970).

Rakestraw, N. W., "Controlling Breathing Atmospheres," *Chemistry*, **43**(9), 18 (1970).

Rasool, S. I., and S. H. Schneider, "Atmospheric Carbon Dioxide and Aerosols: Effects of Large Increases on Global Climate," *Science*, **173**, 138 (1971).

ANSWERS TO SOME EXERCISES

(*In working problems, atomic weights are rounded to the nearest tenth of a unit unless the data permit a more precise calculation. Most of these answers are rounded to show the proper number of significant figures.*)

CHAPTER 1 7. 9.7×10^8 g cm^2 sec^{-2}. 8. Pb, 0.67 cm; Cu, 0.85 cm; Fe, 0.95 cm. 9. (a) 8.1×10^3; (b) 1.6×10^2; (c) 2×10^7. 10. 11.3 g per ml. 11. (a) $-18°C$; (b) $177°C$; (c) $-115°C$. 12. 1.5×10^8 km. 14. (a) 2.9 g per ml; (b) 5.5 g per ml. 15. (b) 5000 cal. 16. 0.091 cal/(g $\times$ deg C). 17. (a) 1740 g; (b) 4.4×10^{25} atoms; (c) 4.0×10^{-23} g per atom. 18. (d) 6.1 miles per hr.

CHAPTER 2 10. 105 g. 12. Ratio of O that combines with 0.881 g of metal = $0.119/0.237 = 1:2$. 13. MO. 22. 4.2×10^{-22} g.

CHAPTER 3 12. (a) 6.02×10^{23} atoms; (c) 2.5×10^{21} atoms. 13. 24.3 awu. 14. 1.18×10^9 atoms. 18. 35.454. 19. 51.91 awu.

CHAPTER 4 7. A lower ionization energy for rubidium. 11. (a) Noble gas element; (b) Atomic number 113. 14. (a) 10; (b) 2; (c) 14; (d) 2; (e) 32.

CHAPTER 5 3. (a) H:S̈:H; (e) :S̈::Ö:, :Ö:S̈::Ö:, :Ö:S:Ö:. 4. Mainly ionic: K$_2$O, Al$_2$O$_3$, CaF$_2$, AlF$_3$. 5. NO$_2$, SF$_6$, CuF$_2$. 13. 0; No. 14. Mixture consists of MgS and unreacted sulfur. 15. 50 e^-. 18. (a) 1000; (b) 200; (c) 2.41×10^{24}.

CHAPTER 6 1. XeOF$_4$: +6, -2, -1; (NH$_4$)$_3$AsO$_4$: -3, $+1$, $+5$, -2. 5. $3A + B_3 \rightarrow 3AB$. 9. (a) Mg^{2+} + 2e^- $\rightarrow$ Mg; (b) 2H$^+$ + 2e^- $\rightarrow$ H$_2$; (a) and (b) 2Cl$^-$ $\rightarrow$ Cl$_2$ + 2e^-. 14. (b) HCO$_3^-$; (c) H$_2$O. 17. Only one reaction occurs: H$_3$O$^+$ + OH$^-$ $\rightarrow$ 2H$_2$O.

CHAPTER 7 1. Star: 2.5×10^{13} miles; atoms: 7×10^{10} miles. 2. S^{2-}, 32.1 g; CaCO$_3$, 100.0 g. 3. 2.7×10^{26} molecules. 4. (c) 1.2×10^{23} Ca^{2+} ions. 5. (a) 71.5 percent Ca; 28.5 percent O. (d) 21.2 percent N; 6.1 percent H; 24.3 percent S; 48.4 percent O. 6. -118.2 kcal. 7. -523 kcal. 8. CH$_4$: -133 kcal; C$_2$H$_4$: -118 kcal. 9. 0.10 cal/(g $\times$ deg C). 10. 18.2 g. 11. 5.7 g NH$_3$. 12. 1.8 lb. 13. CuI. 14. (a) C$_2$H$_5$O. 15. 82.9 percent. 16. (a) 60.0 percent C; 13.3 percent H; 26.7 percent O. 17. 232 lb. 21. -4350 cal.

CHAPTER 8 5. 450 g H$_2$O; 400 g O$_2$. 10. 110.6 kcal/mole of bonds. 12. (a) -13 kcal; (c) -16 kcal; (d) 4 g O$_3$. 14. BAEHDFC. 18. 34.0 percent Zn; 66.0 percent Cu.

CHAPTER 9 4. 266°K. 9. 267 ml; 534 ml. 11. 2.7 atm. 13. 516 ml. 14. 183 ml. 15. (b) $-73°C$. 18. 2.1 kg/cm^2. 19. 0.72 g/liter.

CHAPTER 10 3. (b) 710 ml; (c) 2900 ml. 5. (a) 1.78 g/liter; (b) 0.18 g/liter; (c) 2.86 g/liter. 6. (b) 760 ml. 7. 20.0 g/mole; 20.0 awu/molecule. 8. 95.1 awu. 9. 0.70 atm. 12. (a) $H_2/D_2 = 1.4/1.0$; (d) $SF_6/SO_2 = 1.5/1.0$. 13. 39 awu. 15. 2.48×10^{22} O_2 molecules. 16. 47.5 percent. 18. (a) No. 4; (b) No. 2. 19. 30 percent.

CHAPTER 11 6. About 83°C. 13. 4620 cal. 16. 27°C. 17. 60 lb. 19. 2.7 cal/g. 20. (a) 1630 cal/mole; (b) 7360 cal/mole.

CHAPTER 12 1. (c) contains smallest number of water molecules. 6. 9.8 g K; 17 g K. 7. No. 10. $MgSO_4 \cdot 7H_2O$. 21. O_2.

CHAPTER 13 1. 7.3 percent. 2. (a) 0.76 M; (b) 0.80 m; (c) 1.5 N. 3. 2.2 g. 5. 47 ml. 7. 3.5 percent. 8. 95 g; 2.1 M. 9. 100.85°C. 11. (a) $-1.2°C$; (b) 2.2°C; (c) 14°C. 12. 36 awu; $-2.83°C$. 15. NaCl. 16. 6 percent. 17. 100 awu.

CHAPTER 14 1. 4.5×10^8. 13. Condensation method. 20. 4.7×10^9 molecules/smoke particle. 23. 46 min.

CHAPTER 15 4. (b) 1.6. 15. 0.01 mole/liter. 16. 0.009 mole/liter. 20. At 300°C: 4.1×10^{-3} atm^{-2}; at 400°C: 1.7×10^{-4} atm^{-2}. 23. 0.8 mole/liter.

CHAPTER 16 7. 0.02 percent; 0.2 percent. 9. (a) 2; (b) 12; (c) 12.7. 10. 11. 11. 9.3. 12. 8×10^{-14} mole/liter. 16. (a) 3×10^{-2} mole/liter; (b) 8×10^{-9} mole/liter. 18. (b) 1×10^{-8}; (c) 5×10^{-7} mole/liter. 20. 4.1×10^{-5}. 22. 6.5×10^2. 23. 3×10^{-9} g. 25. No.

CHAPTER 17 5. 6; 4. 6. $^{55}_{22}Ti$, a beta particle; $^{48}_{24}Cr$, an alpha particle. 8. $^{216}_{84}Po$; $^{228}_{90}Th$. 10. 2.5 yr. 11. 8.0 g. 17. About 6.0×10^{10} cal/g; 2.4×10^8 kcal/mole. 21. Endothermic.

CHAPTER 18 6. Mg/Na = 1.0/1.8; Mg/Au = 11/1. 14. Average = 92.8 percent. 20. Mg:3.8 g per ml; Ca:3.0 g per ml.

CHAPTER 19 3. Cl_2. 4. Br^-. 10. (a) 0.5; (b) 2; (c) 1; (d) 3.1 ft^3. 13. (a) 2.0×10^7 kcal; 5.8×10^7 kcal. 18. HCl; $BaCl_2$; CsCl.

CHAPTER 20 10. (b) 1.64 volts; (e) 0.62 volts; (f) 0.003 volts. 12. 9.2×10^{21}. 14. 3.2×10^4 g. 15. (a) 1.53 volts; (b) 0.44 volts. 21. (a) 3.7 g. 22. Cl_2, 1.93×10^5 coulombs; Bi, 2.89×10^5 coulombs. 25. (a) 53.2; (b) 106.4.

CHAPTER 21 12. 32 lb. 15. 4.1. 16. 59.7 g; 48 g. 20. Hydrated Al^{3+} ions act as acids.

CHAPTER 22 17. 1.5×10^3 ft^3. 18. 3.2×10^2 kcal. 20. 0.84 ton. 24. 14 lb $(NH_4)_2SO_4$; 6.6 lb $Ca(H_2PO_4)_2$; 4.8 lb KCl.

CHAPTER 23 21. (a) -3.6×10^5 kcal; (b) -1.1×10^5 kcal.

CHAPTER 24 6. Most like Ra. 14. Ni^{2+}: $(Ar)3d^8$; Sn^{2+}: $(Kr)4d^{10}5p^2$.

CHAPTER 25 14. 0.019 ton; 0.038 ton. 15. 0.33 ton. 23. 11.1 in^3. 24. (a) 2.6×10^7 tons; 3.0×10^6 tons.

CHAPTER 26 5. No. 11. 5 isomers. 15. 8 isomers. 20. 2 isomers; 3 isomers.

CHAPTER 27 1. 8 isomers. 6. (a) $-7.5°C$; (b) 52 g. 12. $C_6H_5O^-$; greater than 7.

CHAPTER 30 7. 4.89 g/cm^3, Venus; 1.3 g/cm^3, Jupiter. 16. 3×10^7 tons. 18. 34 percent.

APPENDIX: TABLES

APPENDIX TABLES

TABLE 1.

Conversion Equivalents

1 lb (avoirdupois) = 453.59 g	1 in. = 2.54 cm
1 oz (avoirdupois) = 28.350 g	1 m = 39.370 in.
1 kg = 2.2046 lb (avoirdupois)	1 angstrom (A) = 1×10^{-8} cm
1 liter = 1.0567 qt (U.S.)	1 km = 0.62137 miles
1 gal (U.S.) = 3.7854 liters	1 atm = 760 mm Hg = 760 torr
1 cu ft = 28.3 liters	1 atm = 14.7 psi
1 ml = 0.06102 cu in.	$\Delta 1°C = \Delta 1.8°F$
1 ml = 1 cc	1 ev = 1.602×10^{-12} erg
1 cu in. = 16.39 ml	1 cal = 4.184×10^7 ergs
1 cu ft = 28,317 ml	1 Btu = 251.98 cal

Relations that are useful to remember:
1 lb = 454 g   1 in. = 2.54 cm   1 liter = 1.06 qt $\Delta 1°C = \Delta 1.8°F$
Common prefixes, abbreviations in parentheses:

centi- (c-)	milli- (m-)	micro- (μ)	kilo- (k-)
1/100	1/1,000	1/1,000,000	1,000

TABLE 2.

Vapor Pressure of Water at Different Temperatures

TEMPERATURE, °C	VAPOR PRESSURE, MM OF Hg	TEMPERATURE, °C	VAPOR PRESSURE, MM OF Hg
0	4.58	29	30.04
5	6.54	30	31.82
10	9.21	31	33.70
11	9.84	32	35.66
12	10.52	33	37.73
13	11.23	34	39.90
14	11.99	35	42.18
15	12.79	40	55.32
16	13.63	45	71.88
17	14.53	50	92.51
18	15.48	55	118.04
19	16.48	60	149.38
20	17.54	65	187.54
21	18.65	70	233.7
22	19.83	75	289.1
23	21.07	80	355.1
24	22.38	85	433.6
25	23.76	90	525.8
26	25.21	95	633.9
27	26.74	100	760.0
28	28.35	150	3570.5

TABLE 3.
Solubility Product Constants[a]

COMPOUND	SOLUBILITY PRODUCT CONSTANT
Arsenic(III) sulfide, As_2S_3	4×10^{-29}
Barium sulfate, $BaSO_4$	1×10^{-10}
Bismuth sulfide, Bi_2S_3	6.8×10^{-97}
Cadmium sulfide, CdS	7.8×10^{-27}
Calcium carbonate, $CaCO_3$	1×10^{-8}
Calcium oxalate, CaC_2O_4	1.78×10^{-9}
Calcium sulfate, $CaSO_4$	2×10^{-4}
Cobalt sulfide, CoS	3×10^{-26}
Copper(II) sulfide, CuS	8.7×10^{-36}
Iron(III) hydroxide, $Fe(OH)_3$	1×10^{-36}
Lead chloride, $PbCl_2$	1×10^{-4}
Lead sulfate, $PbSO_4$	1×10^{-8}
Lead sulfide, PbS	8.4×10^{-28}
Magnesium hydroxide, $Mg(OH)_2$	1.2×10^{-11}
Manganese sulfide, MnS	1.4×10^{-11}
Mercury(II) sulfide, HgS	3.5×10^{-52}
Nickel sulfide, NiS	1.8×10^{-21}
Silver chloride, AgCl	1×10^{-10}

[a] K_{sp} varies with temperature. The constants listed were determined at temperatures ranging from 18 to 25°C. For sulfides, see W. H. Waggoner, *J. Chem. Educ.*, 35, 339 (1958).

TABLE 4.
Solubilities of Common Metal Compounds^a in Water

COMPOUND	SOLUBILITY
Acetates	All soluble except Ag^+, Hg^+, Bi^{3+}
Nitrates	All soluble
Nitrites	All soluble except Ag^+
Chlorides	All soluble except Ag^+, Hg^+, Pb^{2+}, Cu^+
Bromides	All soluble except Ag^+, Hg^+, Pb^{2+}
Iodides	All soluble except Ag^+, Hg^+, Pb^{2+}, Bi^{3+}
Sulfates	All soluble except Pb^{2+}, Ba^{2+}, Sr^{2+}
Sulfites	All insoluble except Na^+, K^+, NH_4^+
Sulfides	All insoluble except Na^+, K^+, NH_4^+, Ba^{2+}, Sr^{2+}, Ca^{2+}
Phosphates	All insoluble except Na^+, K^+, NH_4^+
Carbonates	All insoluble except Na^+, K^+, NH_4^+
Oxalates	All insoluble except Na^+, K^+, NH_4^+
Oxides	All insoluble except Na^+, K^+, Ba^{2+}, Sr^{2+}, Ca^{2+}
Hydroxides	All insoluble except Na^+, K^+, NH_4^+, Ba^{2+}, Sr^{2+}, Ca^{2+}

^{a} The compounds listed here include only those of the common metals of groups IA, IB, IIA, and IIB, and Mn, Fe, Co, Ni, Al, Sn, Pb, Sb, and Bi. The polyatomic ion NH_4^+ is included because of its importance.

TABLE 5.

Four-place Logarithms

NATURAL NUMBERS	0	1	2	3	4	5	6	7	8	9	\multicolumn{9}{c}{PROPORTIONAL PARTS}
											1	2	3	4	5	6	7	8	9
10	0000	0043	0086	0128	0170	0212	0253	0294	0334	0374	4	8	12	17	21	25	29	33	37
11	0414	0453	0492	0531	0569	0607	0645	0682	0719	0755	4	8	11	15	19	23	26	30	34
12	0792	0828	0864	0899	0934	0969	1004	1038	1072	1106	3	7	10	14	17	21	24	28	31
13	1139	1173	1206	1239	1271	1303	1335	1367	1399	1430	3	6	10	13	16	19	23	26	29
14	1461	1492	1523	1553	1584	1614	1644	1673	1703	1732	3	6	9	12	15	18	21	24	27
15	1761	1790	1818	1847	1875	1903	1931	1959	1987	2014	3	6	8	11	14	17	20	22	25
16	2041	2068	2095	2122	2148	2175	2201	2227	2253	2279	3	5	8	11	13	16	18	21	24
17	2304	2330	2355	2380	2405	2430	2455	2480	2504	2529	2	5	7	10	12	15	17	20	22
18	2553	2577	2601	2625	2648	2672	2695	2718	2742	2765	2	5	7	9	12	14	16	19	21
19	2788	2810	2833	2856	2878	2900	2923	2945	2967	2989	2	4	7	9	11	13	16	18	20
20	3010	3032	3054	3075	3096	3118	3139	3160	3181	3201	2	4	6	8	11	13	15	17	19
21	3222	3243	3263	3284	3304	3324	3345	3365	3385	3404	2	4	6	8	10	12	14	16	18
22	3424	3444	3464	3483	3502	3522	3541	3560	3579	3598	2	4	6	8	10	12	14	15	17
23	3617	3636	3655	3674	3692	3711	3729	3747	3766	3784	2	4	6	7	9	11	13	15	17
24	3802	3820	3838	3856	3874	3892	3909	3927	3945	3962	2	4	5	7	9	11	12	14	16
25	3979	3997	4014	4031	4048	4065	4082	4099	4116	4133	2	3	5	7	9	10	12	14	15
26	4150	4166	4183	4200	4216	4232	4249	4265	4281	4298	2	3	5	7	8	10	11	13	15
27	4314	4330	4346	4362	4378	4393	4409	4425	4440	4456	2	3	5	6	8	9	11	13	14
28	4472	4487	4502	4518	4533	4548	4564	4579	4594	4609	2	3	5	6	8	9	11	12	14
29	4624	4639	4654	4669	4683	4698	4713	4728	4742	4757	1	3	4	6	7	9	10	12	13
30	4771	4786	4800	4814	4829	4843	4857	4871	4886	4900	1	3	4	6	7	9	10	11	13
31	4914	4928	4942	4955	4969	4983	4997	5011	5024	5038	1	3	4	6	7	8	10	11	12
32	5051	5065	5079	5092	5105	5119	5132	5145	5159	5172	1	3	4	5	7	8	9	11	12
33	5185	5198	5211	5224	5237	5250	5263	5276	5289	5302	1	3	4	5	6	8	9	10	12
34	5315	5328	5340	5353	5366	5378	5391	5403	5416	5428	1	3	4	5	6	8	9	10	11
35	5441	5453	5465	5478	5490	5502	5514	5527	5539	5551	1	2	4	5	6	7	9	10	11
36	5563	5575	5587	5599	5611	5623	5635	5647	5658	5670	1	2	4	5	6	7	8	10	11
37	5682	5694	5705	5717	5729	5740	5752	5763	5775	5786	1	2	3	5	6	7	8	9	10
38	5798	5809	5821	5832	5843	5855	5866	5877	5888	5899	1	2	3	5	6	7	8	9	10
39	5911	5922	5933	5944	5955	5966	5977	5988	5999	6010	1	2	3	4	5	7	8	9	10
40	6021	6031	6042	6053	6064	6075	6085	6096	6107	6117	1	2	3	4	5	6	8	9	10
41	6128	6138	6149	6160	6170	6180	6191	6201	6212	6222	1	2	3	4	5	6	7	8	9
42	6232	6243	6253	6263	6274	6284	6294	6304	6314	6325	1	2	3	4	5	6	7	8	9
43	6335	6345	6355	6365	6375	6385	6395	6405	6415	6425	1	2	3	4	5	6	7	8	9
44	6435	6444	6454	6464	6474	6484	6493	6503	6513	6522	1	2	3	4	5	6	7	8	9
45	6532	6542	6551	6561	6571	6580	6590	6599	6609	6618	1	2	3	4	5	6	7	8	9
46	6628	6637	6646	6656	6665	6675	6684	6693	6702	6712	1	2	3	4	5	6	7	7	8
47	6721	6730	6739	6749	6758	6767	6776	6785	6794	6803	1	2	3	4	5	5	6	7	8
48	6812	6821	6830	6839	6848	6857	6866	6875	6884	6893	1	2	3	4	4	5	6	7	8
49	6902	6911	6920	6928	6937	6946	6955	6964	6972	6981	1	2	3	4	4	5	6	7	8
50	6990	6998	7007	7016	7024	7033	7042	7050	7059	7067	1	2	3	3	4	5	6	7	8
51	7076	7084	7093	7101	7110	7118	7126	7135	7143	7152	1	2	3	3	4	5	6	7	8
52	7160	7168	7177	7185	7193	7202	7210	7218	7226	7235	1	2	2	3	4	5	6	7	7
53	7243	7251	7259	7267	7275	7284	7292	7300	7308	7316	1	2	2	3	4	5	6	6	7
54	7324	7332	7340	7348	7356	7364	7372	7380	7388	7396	1	2	2	3	4	5	6	6	7

NATURAL NUMBERS	0	1	2	3	4	5	6	7	8	9	PROPORTIONAL PARTS								
											1	2	3	4	5	6	7	8	9
55	7404	7412	7419	7427	7435	7443	7451	7459	7466	7474	1	2	2	3	4	5	5	6	7
56	7482	7490	7497	7505	7513	7520	7528	7536	7543	7551	1	2	2	3	4	5	5	6	7
57	7559	7566	7574	7582	7589	7597	7604	7612	7619	7627	1	2	2	3	4	5	5	6	7
58	7634	7642	7649	7657	7664	7672	7679	7686	7694	7701	1	1	2	3	4	4	5	6	7
59	7709	7716	7723	7731	7738	7745	7752	7760	7767	7774	1	1	2	3	4	4	5	6	7
60	7782	7789	7796	7803	7810	7818	7825	7832	7839	7846	1	1	2	3	4	4	5	6	6
61	7853	7860	7868	7875	7882	7889	7896	7903	7910	7917	1	1	2	3	4	4	5	6	6
62	7924	7931	7938	7945	7952	7959	7966	7973	7980	7987	1	1	2	3	3	4	5	6	6
63	7993	8000	8007	8014	8021	8028	8035	8041	8048	8055	1	1	2	3	3	4	5	5	6
64	8062	8069	8075	8082	8089	8096	8102	8109	8116	8122	1	1	2	3	3	4	5	5	6
65	8129	8136	8142	8149	8156	8162	8169	8176	8182	8189	1	1	2	3	3	4	5	5	6
66	8195	8202	8209	8215	8222	8228	8235	8241	8248	8254	1	1	2	3	3	4	5	5	6
67	8261	8267	8274	8280	8287	8293	8299	8306	8312	8319	1	1	2	3	3	4	5	5	6
68	8325	8331	8338	8344	8351	8357	8363	8370	8376	8382	1	1	2	3	3	4	4	5	6
69	8388	8395	8401	8407	8414	8420	8426	8432	8439	8445	1	1	2	2	3	4	4	5	6
70	8451	8457	8463	8470	8476	8482	8488	8494	8500	8506	1	1	2	2	3	4	4	5	6
71	8513	8519	8525	8531	8537	8543	8549	8555	8561	8567	1	1	2	2	3	4	4	5	5
72	8573	8579	8585	8591	8597	8603	8609	8615	8621	8627	1	1	2	2	3	4	4	5	5
73	8633	8639	8645	8651	8657	8663	8669	8675	8681	8686	1	1	2	2	3	4	4	5	5
74	8692	8698	8704	8710	8716	8722	8727	8733	8739	8745	1	1	2	2	3	4	4	5	5
75	8751	8756	8762	8768	8774	8779	8785	8791	8797	8802	1	1	2	2	3	3	4	5	5
76	8808	8814	8820	8825	8831	8837	8842	8848	8854	8859	1	1	2	2	3	3	4	5	5
77	8865	8871	8876	8882	8887	8893	8899	8904	8910	8915	1	1	2	2	3	3	4	4	5
78	8921	8927	8932	8938	8943	8949	8954	8960	8965	8971	1	1	2	2	3	3	4	4	5
79	8976	8982	8987	8993	8998	9004	9009	9015	9020	9025	1	1	2	2	3	3	4	4	5
80	9031	9036	9042	9047	9053	9058	9063	9069	9074	9079	1	1	2	2	3	3	4	4	5
81	9085	9090	9096	9101	9106	9112	9117	9122	9128	9133	1	1	2	2	3	3	4	4	5
82	9138	9143	9149	9154	9159	9165	9170	9175	9180	9186	1	1	2	2	3	3	4	4	5
83	9191	9196	9201	9206	9212	9217	9222	9227	9232	9238	1	1	2	2	3	3	4	4	5
84	9243	9248	9253	9258	9263	9269	9274	9279	9284	9289	1	1	2	2	3	3	4	4	5
85	9294	9299	9304	9309	9315	9320	9325	9330	9335	9340	1	1	2	2	3	3	4	4	5
86	9345	9350	9355	9360	9365	9370	9375	9380	9385	9390	1	1	2	2	3	3	4	4	5
87	9395	9400	9405	9410	9415	9420	9425	9430	9435	9440	0	1	1	2	2	3	3	4	4
88	9445	9450	9455	9460	9465	9469	9474	9479	9484	9489	0	1	1	2	2	3	3	4	4
89	9494	9499	9504	9509	9513	9518	9523	9528	9533	9538	0	1	1	2	2	3	3	4	4
90	9542	9547	9552	9557	9562	9566	9571	9576	9581	9586	0	1	1	2	2	3	3	4	4
91	9590	9595	9600	9605	9609	9614	9619	9624	9628	9633	0	1	1	2	2	3	3	4	4
92	9638	9643	9647	9652	9657	9661	9666	9671	9675	9680	0	1	1	2	2	3	3	4	4
93	9685	9689	9694	9699	9703	9708	9713	9717	9722	9727	0	1	1	2	2	3	3	4	4
94	9731	9736	9741	9745	9750	9754	9759	9763	9768	9773	0	1	1	2	2	3	3	4	4
95	9777	9782	9786	9791	9795	9800	9805	9809	9814	9818	0	1	1	2	2	3	3	4	4
96	9823	9827	9832	9836	9841	9845	9850	9854	9859	9863	0	1	1	2	2	3	3	4	4
97	9868	9872	9877	9881	9886	9890	9894	9899	9903	9908	0	1	1	2	2	3	3	4	4
98	9912	9917	9921	9926	9930	9934	9939	9943	9948	9952	0	1	1	2	2	3	3	4	4
99	9956	9961	9965	9969	9974	9978	9983	9987	9991	9996	0	1	1	2	2	3	3	3	4

INDEX

Absolute temperature, 133
Absolute zero, 134
Absorption spectra, 44
Accelerators, 280–281
Acetaldehyde, 471
Acetate rayon, 510
Acetic acid, 473, 474
  glacial, 474
Acetone, 473
Acetylene, see Ethyne
Acetylsalicylic acid, 479
Acid anhydrides, 300
Acid salts, see Hydrogen salts
Acidic oxides, 300
Acidity, 260
Acids, carboxylic, 473–475
  Brønsted-Lowry definition, 89
  conjugate, 252, 253
  diprotic, 253
  fatty, 474, 493
  ionization of, 89, 253
  Lewis definition, 252
  monoprotic, 253
  naming, 80
  organic, 473
  properties, 89
  relative strengths, 89, 252
  triprotic, 253
Acrilan, 507
Acrylonitrile, 507
Actinide series, 304, 416
Actinium, 416
Activation energy, 231
Activity series, 121
Addition, to double bond, 460–461
  to triple bond, 461
Addition polymers, 504
Addition reaction, 460
Adipic acid, 507, 514
Adrenalin, 497
Adsorption, 216, 220, 232
  of ions by colloids, 216
Aerosol, 209
Agate, 409
Age, of earth, 277
  of elements, 518
  of organic material, 277
  of stars, 517–518
Air, see Atmosphere
Albite, 403
Alcohols, 466–469
  primary, 471
  secondary, 473
Aldehydes, 471–472
Aliphatic hydrocarbons, 446
Alizarin, 262
Alkali family, 304–311
  atomic radii, 305
  chemical properties, 306
  compounds, 308–311
  flame spectra, 42, 306

occurrence, 307
physical properties, 305
preparation, 435
uses, 435
Alkaline earth family, 304–311
  chemical properties, 306
  compounds, 308–311
  flame spectra, 42, 306
  occurrence, 307
  physical properties, 305
  preparation, 435
  uses, 435
Alkanes, 446
  structure, 448
Alkenes, 447
  structure, 451
Alkyl radicals, 458
Alkynes, 447
  structure, 461
Allotropic forms, 167
Alloy steels, 442
Alloys, 441–442
  nonferrous, 442
  of As, Sb, and Bi, 388, 389
Alpha emission, 273
Alpha helix, 481
Alpha particle, 272
Alternating current, 84
Aluminum, 417, 419–424
  production, 437
Aluminum family, 417, 419–424
Amethyst, 409
Aminoacetic acid, 483
Amino acids, 483
p-Aminobenzoic acid, 496
Ammonia, 385, 389–391
  molecule, 390
  production, 390
  reaction with water, 256, 389
Ammonia complexes, 427
Ammonia fountain, 389–390
Ammonia water, 389
Ammoniates, 427
Ammonium carbonate, 391
Ammonium chloride, 391
Ammonium compounds, 391–392
Ammonium fluoride, 391
Ammonium hydroxide, 389
Ammonium ion, 391
Ammonium nitrate, 391, 392
Ammonium nitrite, 391
Ammonium polysulfide, 372
Ammonium sulfate, 391
Ammonium sulfide, 391
Amorphous substances, 165
Ampere, 337
Amphiboles, 403
Amphoteric hydroxides, 264, 386
Analytical methods, 105
Angstrom unit, 4
Anhydrous compound, 177

Aniline, 478
Anions, see Negative ions
Anode, 30, 84, 342
Anorthite, 403
Answers to exercises, 539–541
Antibonding electrons, 67
Antimetabolites, 496
Antimony, properties, 380–386
  uses, 387–389
    See also Nitrogen family
Antimony oxide, 386
Antimony sulfide, 386
Antioxidants, 514
Apatite, 386
Aqua regia, 393, 428
Aquated ion, 178
Aragonite, 408
Arc process for nitric acid, 384
Argon, 330–333
Arithmetical procedures, 9–13
Aromatic hydrocarbons, 446
  structure, 453
Arsenic, properties, 380–386
  uses, 387–389
    See also Nitrogen family
Arsenic sulfide, 386
Arsenic trioxide, 385
Arsenical pyrites, 386
Arsine, 385
Asbestos, 403
Aspirin, 479
Association colloids, 212
Astatine, 314
Aston, F. W., 35
Asymmetric carbon atom, 485
Atmosphere, 527–531
  composition, 529
  origin, 528
  zones, 527
Atmospheric pressure, 132
Atomic bomb, 286, 294
Atomic chlorine, 317
Atomic diameter, 39, 302
Atomic energy, 3
    See also Nuclear reactions
Atomic mass, see Atomic weight
Atomic number, 34
Atomic orbitals, see Orbitals
Atomic pile, 290
Atomic structure, 29–39, 58
Atomic theory of Dalton, 21–23
Atomic weight, 22, 38
  determination, 36, 102
  standard of, 36
Atomic weight unit, 36
Atomos, 21
Atoms, 21
  diagrams, 48
  how they combine, 58–73
  size, 39, 302
  structure, 29–39
  weights, 22, 38

553

554

Index

Attraction, molecular, *see* van der Waals forces
Avogadro, Amadeo, 147
Avogadro's law, 147
  and molecular weights, 147
Avogadro's number, 95
Awu, 36

Babbitt, 389, 442
Bakelite, 507
Baking powder, 407
Baking soda, 407
Barite, 369
Barium, physical constants, 305
  See also Alkaline earth family
Barium carbonate, 407
Barium oxide, 309
Barium peroxide, 309
Barium sulfate, 311
Barometers, 131
Basalt, 525
Basaltic rocks, 522
Bases, 89
  Brønsted-Lowry definition, 89
  conjugate, 252, 253
  Lewis definition, 252
  properties, 89
  strength, 89, 252
Basic oxides, 300
Battery, 343
  See also Voltaic cell
Becker, H., 279
Becquerel, Henri, 270
Benzaldehyde, 478, 513
Benzene, 447
  addition reactions of, 461
  structure, 453
Benzene ring, 453–454
Benzoic acid, 478
Beryllium, 304
  See also Alkaline earth family
Berzelius, J. J., 23
Bessemer process, 441
Beta emission, 275
Beta particle, 272
Betatron, 281
Bicarbonates, 407–409
Bidentate, 427
Big-bang theory, 517
Binary compounds, 79
Binding energy, 285
Biocatalysts, 494–497
Biochemistry, 481–500
Biodegradable, 512
Biosphere, 531
Bismite, 386
Bismuth, properties, 380–386
  uses, 387–389
  See also Nitrogen family
Bismuth oxide, 386
Bismuth sulfide, 386
Bismuthine, 385
Bismuthite, 386
Bismuthyl ion, 386
Blast furnace, 438
Bleaching, 126
Blood sugar, 489

Bohr, N., 41
Bohr atom, 41
Boiler scale, 409
Boiling point, 161
  effect of structure, 324
  normal, 162
  of solutions, 198
  variation with pressure, 162
Bombardment reactions, 279
Bond energy, 117, 126
Bonding electrons, 67
Bonds, covalent, 63, 70, 168
  electrovalent, 62, 70
  hydrogen, 175, 324
  metallic, 168
  nonpolar, 465, 466
  pi, 66, 451, 452
  polar, 71
  polar covalent, 71
  sigma, 64, 448, 454
  sp, 452
  sp^2, 451, 453
  sp^3, 65–66, 448
Boneblack, 402
Borates, 412
Borderline elements, 300
Boron, 396
Boron trichloride, 252
Bothe, W., 279
Boyle's law, 138
Brasses, 442
Brimstone, 368
British thermal unit (Btu), 538
Bromic acid, 327
Bromine, compounds of, 323–330
  occurrence, 318
  preparation, 322
  properties, 314–318
  uses, 323
Bromo-carnallite, 318
Brønsted, J. N., 89
Bronze, 442
Brown, Robert, 218
Brownian movement, 218
Btu, 538
Buffered solution, 264
Buna rubber, 507
Butadiene, 507
Butane, 447
Butanols, 467
Butene, 447
Butyne, 447
Butyraldehyde, 503
Butyric acid, 474, 504

Cadmium, *see* Zinc family
Calcium, physical constants, 304
  See also Alkaline earth family
Calcium bicarbonate, 408
Calcium carbide, 477
Calcium carbonate, 311, 408
Calcium chloride, 310
Calcium fluoride, 318
Calcium hydroxide, 309
Calcium oxide, 309
Calcium phosphate, 386, 407

Calcium silicate, 412, 440
Calcium sulfate, 377
Calorie, 8
Calorimeter, 99
Carat, 182
Carbohydrates, 488–492
  metabolism, 498
Carbon, 396–413
  amorphous, 397
  chemical properties, 399–401
  crystalline, 397
  occurrence, 401
  oxides, 404–407
  physical properties, 397
  preparation, 402
  uses, 402
Carbon-14, 277
Carbon black, 402
Carbon dioxide, 404–407
  in photosynthesis, 528
  weathering of rocks, 531
Carbon dioxide cycle, 530
Carbon dioxide fire extinguishers, 405
Carbon monoxide, 404
  heating value, 404
  physiological action, 404
  reduction, with, 404
Carbonates, 407–409
Carbonic acid, 405, 407
  ionization, 254
Carboxyl group, 473
Carboxylic acid, 473–476
Cast iron, 440
Catalysis, 231
Catalysts, 115, 231
  mechanism of, 231–233
Cathode, 30, 84, 342
Cathode rays, 31
Cathodic protection, 422
Cation hydration, 176
Cations, 86
Caustic soda, 310
Cave formation, 408
Cavendish, Henry, 112
Cell, 481
Cellophane, 510
Cellulose, 491, 509
Cellulose acetate, 510
Celsius, Andrew, 13
Celsius scale, 13
Cement, 411
Centigrade scale, 7, 13
Centimeter, 4
Ceramics, 411
Cesium, physical constants, 305
  See also Alkali family
Chadwick, J., 37, 279
Chain reaction, 317
Chalcopyrite, 369
Chalk, 408
Changes, chemical, 16
  physical, 16
Charcoal, 402
  as adsorbent, 288
Charge density, 177, 185
Charged bodies, 30–32
Charles, J., 139
Charles' law, 139

Index

Chelates, 427
Chemical bonds, see Bonds
Chemical change, 16
Chemical combination, see Compounds
Chemical energy, 3
Chemical equations, see Equations
Chemical equilibria, see Equilibrium
Chemical formulas, see Formulas
Chemical reactions, 16
 endothermic, 17
 exothermic, 16
 neutralization, 91
 oxidization-reduction, 120, 356–359
 spontaneous, 340
 volume relationships, 146, 151–153
 weight relationships, 103–105
Chemistry, definition, 3
Chemotherapy, 510
Chilean nitrate, 386
Chloric acid, 327
Chlorine, compounds of, 323–330
 occurrence, 318
 preparation, 321–322
 properties, 314–318
 uses, 321
Chlorine electrode, 345
Chloroprene, 507
Chlorous acid, 327
Chromatography, 221–223
Chromic acid, 417
Chromic oxide, 417
Chromium, 417
Chromium family, 417–423
Cinnabar, 369
Cis isomer, 505
Citric acid, 475
Classification of compounds, 83–91
Clay, 411
Clorox, 329
Cloud chamber, 272
Coal gas, 461
Coal tar, 461
Cobalt, see Iron family
Coke, 402
Colligative properties, 196
Colligative properties law, 196, 213
Colloid mill, 210
Colloidal dispersions, 207
 coagulation, 223
 color, 217
 formation, 210
 hydrophilic, 211
 hydrophobic, 211, 214
 lyophilic, 211
 lyophobic, 211, 214
 properties, 217–223
 rate of settling, 219
 stabilization, 211–216
 types, 209
Colloidal particles, 208
Colloidal solutions, 211

Colloidal, state, 207–226
Color, and chemical constitution, 423
 of colloids, 217
 of ions, 423
Combustion, 119
 of common fuels, 119
Common ion effect, 263
Complex ions, 425, 428
Composition of compounds, 102
Compounds, 18, 60
 classification, 83
 composition, 102
 intermediate, 231–232
 naming, 79, 80, 328
Concentration, effect on reaction rate, 233
 of solutions, 191
Condensation colloids, 210
Condensation polymers, 504
Conjugate base, 252–253
Conservation of energy, 17, 283
 of mass, 19, 283
Constant-boiling mixture, 203
Contact process, 375
Continuous spectrum, 41, 44
Conversion factors, 12, 139
Coordination, octahedral, planar, and tetrahedral, 426
Coordination compounds, 424–428
 uses, 428
Copper, 417, 419–424
Copper family, 417, 419–424
Copper sulfate, 377
Corrosion, 422
Cosmic rays, 533
Cottrell precipitator, 224
Coulomb, 337
Covalent bond, 63, 70
Covalent compounds, 63
Covalent molecules, 71
Covalent radius, 418
Creosote oil, 479
Cresols, 479
Crick, F. H. C., 487
Critical mass, 287
Critical pressure, 154–155
Critical temperature, 154–155
Crust, earth's, 521, 523
Cryolite, 318, 437
Crystalline solids, 166, 170
 ionic substances, 168
 isomorphous, 166
 polymorphous, 167
 structure of metals, 168
 types, 166
Cubic centimeter, 5
Cumene, 504
Curie, Pierre and Marie, 271
Current of electricity, 84
Cyanides, 426
Cyclamate, 514
Cyclohexane, 461
Cyclotron, 280
Cytoplasm, 481

Dacron, 507
Dalton, John, 21

Dalton's atomic theory, 21–23
Dalton's law, 141
Daniell cell, 342
Dating by radioactivity, 277
Decanes, 447
Dees, 280
Degree, of ionization, 252–258
 of rotation, 485
Dehydrating agent, 376
Deliquescence, 310
Denatured alcohol, 469
Density, 6
 of common substances, 7
 of gases, 152
 of metals, 305
 of water, 174, 175
Dentate, 427
Deoxyribonucleic acid, 486
Derivatives of hydrocarbons, 465–479
Destructive distillation, of coal, 402
 of wood, 402
Detergents, 511, 512
Deuterium, 123, 179
Deuterium oxide, 178–180
Dextrose, see Glucose
Dialysis, 224
Diamagnetic, 127
Diamond, 397, 401, 402
Diesel oil, 463
Diethyl ether, 470
Diffusion, of gases, 134, 149
 of liquids, 159
 of solids, 163
Digestion of foods, 497
Dimethyl ether, 470
Dimethyl ketone, 472
Diphenyl ether, 478
Dipole, 71
Diprotic acid, 253
Direct current, 84
Disaccharides, 488, 490
Dispersed phase, 209
Dispersing phase, 209
Dispersion, colloidal, 207
Dispersion medium, 209
Distillation, 18, 202–204
 of petroleum, 462
Distribution curve, 231
DNA, 486
Döbereiner, J., 24
Doppler effect, 517
Double bond, 451
Drugs, 500
Dry cell, 343
Dry ice, 406
Dulong and Petit law, 101

Earth, density, 519
 formation, 520–522
 major zones, 521
Earthquake waves, 520
Earth's core, 521
Earth's crust, 523
 composition, 523
 formation, 520–522
Effervescence, 405

Effusion, 134, 149
Einstein, Albert, 283
    mass-energy law, 283
Electric current, 84
Electrical charges, 30
Electrical conductivity, 84
Electrical energy, 2
Electrical units, 337
Electrochemical series, 121
    See also Oxidation potentials
Electrochemistry, 336–356
Electrode, 84
    hydrogen, 344
    potential, 345–348
    products, 338
    standard, 345
Electrolysis, 85, 338
    of water, 116
Electrolyte, 83–89
    solutions of, 200
    strong, 88
    weak, 88
Electromagnetic radiation, 3
Electron-dot formulas, 64
Electron levels, *see* Energy levels
Electron shells, *see* Energy levels
Electron structure, 29–39, 48, 49–55
    of ions, 61
    and the periodic law, 48
Electron volt, 45
Electronegativity, 71–73, 418
Electrons, 31
    arrangement in atoms, 48–54
    evidence for, 31, 43
    excited state, 43
    ground state, 43
    sharing, 62
    transfer, 60
Electrophoresis, 219
Electroplating, 339
Electrostatic forces, 30
Electrovalent bonds, 62, 70
Electrovalent compounds, 62, 86
Elements, 18, 34
    age of, 277, 518
    classification, 23–26
    natural abundance, 523–525
    synthesis, 282
Emission spectra, 41
Emissions, nuclear, 271
Empirical formula, 105
Emulsifying agents, 215
Emulsion, 209
Endpoint of titration, 194
Endothermic reactions, 17
Energy, 2
    activation, 231
    binding, 284
    in chemical changes, 16, 17
    conservation, 17, 283
    ionization, 45–46
    kinds, 2
    and mass, 283
Energy levels, 44, 48, 68

Energy levels, MO, 68
Energy sublevels, 49–54, 68
Enthalpy, 97–102, 245
    of fusion, 170
    of reactions, 245
    of vaporization, 162
Entropy, 185, 245
Enzymes, 494–496
Epinephrine, 497
Epsom salts, 377
Equanil, 501
Equations, 60, 81, 82
    oxidation-reduction, 356–359
Equilibrium, 87
    chemical, 87, 228–268
    factors that influence, 237–247
    ionic, 251–268
    physical, 186
Equilibrium constant, 239, 254
    and free energy, enthalpy, and entropy, 245–246
    and temperature, 245
Equivalent weight, 193
Erg, 284
Esters, 475–476
Ethanal, 471
Ethane, 447, 448
Ethanoic acid, 473–474
Ethanol, 466–468
Ethene, 447, 451
Ether (common), 470
Ethers, 469–470
Ethyl acetate, 476, 513
Ethyl alcohol, 466–468
Ethyl butyrate, 513
Ethyl formate, 475
Ethyl radical, 458
Ethylene, 447
Ethylene glycol, 469
Ethyne, 447, 452
Eutrophication, 512
Excited states, 43
Exothermic reactions, 16
Exponents, 10
Extrinsic properties, 15

Factor-units method, 12
Fahrenheit scale, 7
Fallout, 288
Families of elements, 47
Faraday, 337
Faraday, Michael, 454
Faraday's law, 355
Fats, 492–494, 511
    metabolism, 498
Fatty acids, 474, 493
Fayalite, 522
Feldspars, 403
Ferric, 80
Ferromagnetic, 140
Ferrous, 80
Fertilizers, 391
Fire extinguisher, 405
Fission, nuclear, 294
Fission bombs, 286
Fixation of nitrogen, 384
Flame spectra, 42, 306
Flavors, 512

Flotation of ores, 433
Fluorescence, 31, 272
Fluorine, compounds, 323–330
    occurrence, 318
    preparation, 319
    properties, 314–318
    uses, 319
Fluorspar, 318
Flux, 434
Foam, 209
Fog, 209
Food additives, 512–515
Fool's gold, 372
Formaldehyde, 471, 472
Formalin, 472
Formic acid, 473, 474
Formula weight, *see* Molar weight
Formulas, 76
    dot, 64
    empirical, 105
    molecular, 106
    and molecular weights, 107
    from oxidation numbers, 78
    and percentage composition, 102
    structure, 449
Forsterite, 522
Fractional distillation, 202, 462
Fractionating columns, 203, 462
Francium, *see* Alkali family
Frasch, Herman, 369
Frasch process, 369
Fraunhofer lines, 44
Free energy, 246
Freezing point, effect of solute, 254
    See also Melting point
Frequency of light, 3
Fructose, 490
Fuel cells, 353
Fuel oil, 463
Functional groups, 465
Fusion, nuclear, 286

Galena, 369
Gallium, *see* Aluminum family
Galvanic cell, *see* Voltaic cell
Galvanized iron, 422
Gamma rays, 272
Gamow, George, 517
Gangue, 433
Gas chromatography, 222
Gas laws, 138–144
    deviations from, 144
    ideal gas, 153
Gas oil, 463
Gaseous diffusion, 134
Gases, 130
    behavior, 130–144, 153
    densities, 152
    and the kinetic theory, 134–144
    volume-weight relationship 151–153
Gasoline, 463
Gay-Lussac, J. L., 146
Gay-Lussac's laws, 140, 146
Geiger, H., 33

Index

Geiger counter, 273
Gels, 225
Geneva System, 457
Geometrical isomers, 505
Germanium family, 417, 419–421
Gibbs free energy, 246
Glacial acetic acid, 474
Glass, 411–412
Glassy structure, 412
Glucose, 489–490
Glycerin (glycerol), 469, 493
Glycine, *see* Aminoacetic acid
Glycol, *see* Ethylene glycol
Gold, *see* Copper family
  white, 442
Goldstein, E., 32
Gouy balance, 127
Graham, Thomas, 149
Graham's law, 149
Grain alcohol, 468
Gram-equivalent weight, 193
Gram-molecular volume, *see* Molar gas volume
Gram-molecular weight, *see* Molar weight
Granite, 522, 525
Granitic rocks, 522
Graphite, 397, 402
Gravity cell, 343
Ground state, 43
Group IA, *see* Alkali family
Group IB, *see* Copper family
Group IIA, *see* Alkaline earth family
Group IIB, *see* Zinc family
Group IIIA, *see* Aluminum family
Group IIIB, *see* Scandium family
Group IVA, *see* Germanium family
  See also Carbon, Silicon
Group IVB, *see* Titanium family
Group VA, *see* Nitrogen family
Group VB, *see* Vanadium family
Group VIA, *see* Sulfur family
Group VIB, *see* Chromium family
Group VIIA, *see* Halogen family
Group VIIB, *see* Manganese family
Group VIIIA, *see* Noble gases
Group VIIIB, *see* Iron family; Platinum family
Groups, of elements, 47
Gutta-percha, 506
Gypsum, 369, 377

Haber, Fritz, 390
Haber process, 390
Habits of crystals, 167
Hafnium, *see* Titanium family
Half-life, 276
Halides, 323–327
Halite, 318
Hall, Charles M., 437
Hall process, 437
Halogen family, 314–330
  chemical properties, 315–318
  compounds of, 323–330
  occurrence, 318
  oxyhalogens, 327–330
  physical properties, 314–315
  preparation, 319–323
  uses, 319–323
Hardness of water, 413
Heat, 8
  of fusion, 163, 169
  of reaction, 245
  of solidification, 169
  of vaporization, 162, 163
  See also Enthalpy
Heat energy, 2, 8
Heavy hydrogen, 123, 179
Heavy water, 178–180
Heisenberg uncertainty principle, 50
Helium, 330–333
  discovery, 43
Hematite, 438
Henry's law, 188
Heptanes, 447
Héroult, Paul, 437
Hess, G. H., 100
Hess's law, 100
Heterogeneous mixture, 182
Hexagonal crystals, 166
Hexanes, 447
Homogeneous mixtures, 182
Homolog, 447
Homologous series, 447
Hormones, 497
Hund's principle, 53
Hybridized orbitals, *see* Orbitals
Hydrated ions, 176
Hydrates, 176–178, 427
Hydration, 176
Hydrazine, 389
Hydriodic acid, 323–327
Hydrobromic acid, 323–327
Hydrocarbons, 446–463
  aliphatic, 446
  alkanes, 446
  alkenes, 446
  alkynes, 446
  aromatic, 446
  chemical properties, 460
  derivatives, 465
  nomenclature, 457–459
  physical properties, 459
  source, 461–463
  structure, 447–454
Hydrochloric acid, 323–327
  ionization of, 257
Hydrocyanic acid, 257
Hydrofluoric acid, 323–327
Hydrogen, atomic, 117, 317
  chemical properties, 117–120
  commercial preparation, 116
  determination of, 105
  discovery of, 111–113
  isotopes, 123
  laboratory preparation, 114–116
  occurrence, 113–114
  physical properties, 114
  uses, 124
Hydrogen acetate, *see* Acetic acid
Hydrogen bomb, 294
Hydrogen bond, 175, 324
Hydrogen bromide, 323
Hydrogen chloride, 323
Hydrogen chloride fountain, 325
Hydrogen electrode, 344
Hydrogen fluoride, 323
Hydrogen halides, 323–327
Hydrogen iodide, 323
Hydrogen ion, 91
Hydrogen ion concentration, 260–262
Hydrogen nitrate, 392
Hydrogen peroxide, 180–181
Hydrogen phosphate, 393
Hydrogen polonide, 370
Hydrogen selenide, 370
Hydrogen sulfide, 370
Hydrogen telluride, 370
Hydrolysis, 262
  of fats, 494
  of polysaccharides, 492
  of proteins, 498
Hydronium ion, 86, 90
Hydronium ion concentratiion, 260–262
Hydrophilic colloid, 211
Hydrophobic colloid, 211, 214
Hydrosphere, 521, 526
Hydrosulfuric acid, 371
Hydroxide bases, 89
Hydroxide ion, 91
Hydroxylamine, 389
Hypertonic solutions, 205
Hypobromous acid, 327
Hypochlorous acid, 327
Hypoiodous acid, 327
Hypothesis, 2
Hypotonic solutions, 205

Ice, 175
Ideal density of a gas, 154
Ideal gas, 153
Ideal solutions, 201
Igneous rock, 524
Immiscible liquids, 186
Indicators, 195
  and *p*H, 262
  table of, 262
Indium, *see* Aluminum family
Inert gases, *see* Noble gases
Inhibitor, 231
Inner transition series, 304
Inorganic compounds, 83
Insoluble substances, 186
Interconversion of matter and energy, 283
Interferon, 499
Intermediate compound, 232
Intermetallic compounds, 441
Intrinsic properties, 15
Iodic acid, 327

Iodine, compounds of, 323–330
  occurrence, 318
  preparation, 323
Ion, 35
  electronic structure, 61
  hydrated, 176
  polyatomic, 71
  size, 302
    See also Complex ions; Negative ions; Positive ions
Ion exchange, 413
Ion product of water, 259
Ionic bond, see Electrovalent bond
Ionic compounds, 60, 168
Ionic equation, 82
Ionic equilibria, see Equilibrium
Ionization, 87, 253
  of water, 258
  of weak acids, 253, 257
  of weak bases, 257
Ionization constants, 254, 267
Ionization energies, 45–46
  IA elements, 305
  IIA elements, 305
  VA elements, 382
  VIA elements, 365
  VIIA elements, 315
  and orbitals, 55
  transition elements, 418
Ionizing radiations, 271
Ionosphere, 528
Iridium, see Platinum family
Iron, 417–424
  physical constants, 418–420
  production, 438, 441
Iron alloys, 441
Iron family, 417–424
Iron pyrites, 372
Isoamyl acetate, 476, 513
Isobutane, 455
Isomers, 455
  optical, 485
  ortho, meta, para, 457
Isometric crystals, 166
Isomorphous crystals, 166
Isoprene, 505
Isopropyl alcohol, 466, 504
Isotonic solutions, 205
Isotopes, 36, 276

Jadeite, 403
Janssen, P., 43
Jellied alcohol, 226
Jelly, 225

K, 239
K_i, 254
K level, see Energy levels
K_p, 249
K_{sp}, 265
K_w, 259
Kaolinite, 411
Kekulé, F. A., 453
Kelvin scale, 134
Kelvin temperature, 134
Kerosene, 463
Ketones, 472

Kilogram, 5
Kilowatt, 337
Kilowatt-hour, 337
Kinetic energy, 2
  distribution curve, 231
Kinetic-molecular theory, of gases, 134–138
  of liquids, 158–162
  of solids, 163–165
Kinetics, 228
Krypton, 330–333
Kurchatovium, see table inside front cover, 283

Labile, 426
Lactic acid, 475, 514
Lactose, 490
Lambda (λ), wavelength, 3
Lampblack, see Carbon black
Lanthanide series, 304, 416
Lanthanum, see Scandium family
Larsenite, 522
Lauric acid, 493
Lavoisier, Antoine L., 112
Law, 2
  Amonton's, 140
  Avogadro's, 147
  Boyle's, 138
  Charles', 139
  colligative properties, 196
  combining volumes, 146
  conservation of energy, 17
  conservation of mass, 19
  Dalton's, 141
  definite composition, 20
  of Dulong-Petit, 101
  Faraday's, 355
  first law of thermodynamics, 94
  Gay-Lussac's, 140, 146
  Graham's, 149
  Henry's, 188
  Hess's, 100
  multiple proportions, 20
  periodic, 24
  Raoult's, 196
  second law of thermodynamics, 185
Lawrence, E. O., 281
Lead, see Germanium family
  as end product of radioactivity, 275
Lead glass, 412
Lead storage cell, 352
Lead sulfide, 369
Le Chatelier's principle, 242, 244
Lepidolite, 403
Levulose, 490
Lewis acids and bases, 250, 425
Ligands, 425
Light, absorption, 44
  diffusion by colloidal system, 217
  plane polarized, 486
    See also Radiant energy
Lignin, 509
Lime, 309

Lime glass, 412
Limestone, 408
Limonite, 438
Line spectra of elements, 42
Linoleic acid, 493
Linolenic acid, 493
Linotype metal, 442
Liquefaction of gases, 154
Liquid air, 116
Liquid solutions, 183
Liquids, 130
  behavior, 158–163
  kinetic-molecular theory, 158–162
Liter, 5
Lithium, 304
    See also Alkali family
Lithosphere, 521, 523
Litmus, 262
Logarithm table, 548
LSD, 501
Lubricating oils, 463
Lucretius, 367
Lye, 310
Lyophilic colloids, 211
Lyophobic colloids, 211, 213
Lysergic acid diethylamide, 501

Macromolecules, 211
Magnesia, 309
Magnesium, 304
  production, 435
    See also Alkaline earth family
Magnesium bicarbonate, 407
Magnesium carbonate, 407–409
Magnesium silicate, 407
Magnesium sulfate, 377
Magnetic fields, charged particles in, 30–31
Magnetite, 438
Malic acid, 475
Maltol, 513
Maltose, 490
Manganese, 417–424
Manganese family, 417–424
Mantle, 521
Marble, 408
Marsden, E., 33
Mass, 5, 19
  and energy, 283
Mass loss, 284
Mass number, 37
Mass spectrograph, 35–36
Matches, 388
Matter, 2
  classes of, 18
  states, 130
Mayonnaise, 215
Mechanism of reaction, 228
Melting point, 169
  effect of structure, 410
Mendeleev, D. I., 24
Meprobamate, 501
Mercury, see Zinc family
Mercury oxide, 112
Mercury sulfide, 369
Mesons, 29
Metabolism, 498–499

Index

Meta isomers, 457
Metallic bond, 168
Metallic properties, 418
Metallo-acid elements, 300, 417
Metalloids, 59, 300
Metallurgy, 432
Metals, 59, 300
  activity series, 122
  alloys of, 441
  natural sources, 431–432
  production, 432–441
  properties, 304, 418
  structure of, 168, 418
Metasilicic acid, 400
Meteorites, 519
Meter, 4
Methanal, see Formaldehyde
Methane, 447, 448
Methanoic acid, see Formic acid
Methanol, 466, 468
Methyl acetate, 476
Methyl alcohol, see Methanol
Methyl benzoate, 478
Methyl formate, 475
Methyl isatin-β-thiosemicarbazone, 500
Methyl orange, 262
Methyl radical, 458
Methyl salicylate, 479, 513
Methyl silicone, 412
Methyl yellow, 262
Metric system, 4–10
Meyer, Lothar, 24
Micas, 403
Micelles, 213
Milk of magnesia, 310
Milliliter, 5
Minerals, 431
Mixtures, 18
MO theory, 67–69
Moderator, 291
Molal boiling-point constant, 199
Molal freezing-point constant, 197
Molal solutions, 192
Molality, 192
Molar gas volume, 147
Molar heat capacity, 170
Molar solutions, 191
Molar weight, 95
Mole, 94–96
Molecular association, 175
Molecular attraction, 153
Molecular bombardment, see Brownian movement
Molecular motion, see Kinetic-molecular theory
Molecular orbitals, 63–69
Molecular weight, from bp and fp change, 196–200
  from gas densities, 147
  from Graham's law, 149
  precise, 108
Molecules, defined, 63
  diatomic, 75, 135
  distribution of energy, 231
  large, 397
    See also Polymers
  monatomic, 81, 135
  shape, 64
  structure, 64
  triatomic, 135
Molybdenum, see Chromium family
Monoclinic crystals, 166
Monoprotic acid, 253
Monosaccharides, 488–490
Monosodium glutamate, 512
Monticellite, 522
Moon, 532–533
  age of rocks, 533
  composition, 533
  origin, 533
Moseley, H. G. J., 34
Muriatic acid, 326
Muscovite, 403

Natrolite, 413
Natural gas, 462, 463
Natural radioactive families, 273
Negative catalyst, 231
Negative ions, 86
Neighbors of transition elements, 417
  physical properties, 419
Neon, 330–333
Neptunium, 283
Neutral solution, 260
Neutralization, 91
Neutrino, 29
Neutron-proton ratio, 275
Neutrons, 36–37, 279
  discovery, 36
Newlands, J., 24
Nickel, see Iron family
  as catalyst, 232
Nickel silver, 442
Niobium, see Vanadium family
Nitrate ion, 71
  See also Nitrogen family
Nitric acid, 392
Nitrides, 384
Nitrogen, 380–394
  chemical properties, 383–386
  occurrence, 387
  origin in air, 528
  physical properties, 380–383
  production, 387–389
  uses, 387–389
Nitrogen base, 487
Nitrogen cycle, 384
Nitrogen dioxide, 389
Nitrogen family, 380–394
  chemical properties, 383–386
  compounds of, 385–394
  occurrence, 387
  physical properties, 380–383
  preparation of, 387–389
  uses of, 387–389
Nitrogen fixation, 384
Nitrogen molecule structure, 66, 67
Nitrogen oxides, 384, 389
Nitrogen pentoxide, 389
Nitrogen trioxide, 389
Nitroglycerin, 476, 504
Nitrous acid, 386
Nitrous oxide, 389
Noble gases, 59, 300, 330–333
  compounds of, 332
  discovery of, 330
  electron structure, 59
  uses, 332–333
Noble metals, 420
Nomenclature, of inorganic compounds, 79, 328
  of organic compounds, 457
Nonanes, 447
Nonelectrolyte, 83
Nonferrous alloys, 442
Nonmetals, 59, 300
Normal alkanes, 455
Normal solutions, 191
Normality, 191
Nuclear bomb, 286, 294
Nuclear chain reaction, 288
Nuclear chemical equations, 279–283
Nuclear energy, 3
Nuclear fission, 285
Nuclear fission bombs, 286
Nuclear fissionable materials, 290
Nuclear fusion, 286
Nuclear fusion bombs, 294
Nuclear reactions, 279–283
  applications, 283
  fission bombs, 286
  fusion bombs, 286
  industry, 295
  medicine, 296
  production of energy, 290, 294
  research, 296
Nuclear reactors, 290, 294
Nucleus of the atom, 32–38
Nylon, 507

Obsidian, 525
Ocean water, 526
Octahedral, 426
Octanes, 447
Olefins, see Alkenes
Oleic acid, 493
Oleum, 375
Olivine minerals, 522
Onyx, 409
Opal, 409
Open-hearth process, 441
Optical activity, 485–486
Optical isomers, 485
Orbitals, 49–55
  in coordination compounds, 426
  hybrid, 65
  and ionization energies, 55
  order of filling, 52, 68
  p, 50
  pi, 66, 451, 452
  s, 50
  sigma, 64
  sp, 452
  sp^2, 451, 453
  sp^3, 65–66, 448

Ores, 431
    concentration, 433
    reduction, 434
    roasting, 434
Organic acids, 473, 493
Organic chemistry, 445
Organic compounds, 83, 446
    nomenclature, 457–459
Orlon, 507
Orpiment, 386
Orthoclase, 403
Ortho isomers, 458
Orthorhombic crystals, 166
Orthosilicic acid, 400
Osmium, see Platinum family
Osmosis, 204
Oxalic acid, 475
Oxidation, 119–121
    of alcohols, 471, 472
    of aldehydes, 472
    of hydrocarbons, 460
Oxidation numbers, 76–78, 356
    in naming, 80
    from periodic table, 76
    transition elements, 420
    transition elements, neighbors, 421
    VA elements, 385
    VIA elements, 367
    VIIA elements, 327
Oxidation potentials, 345–349
Oxidation-reduction equations, 356–359
Oxidation state, see Oxidation numbers
Oxides, acidic and basic, 300
    metallic, 300
    nonmetallic, 300
Oxidizing agents, 120
Oxy-acids, of halogens, 327
    of sulfur family, 373–377
Oxy-halogen compounds, 327–330
Oxy-hemoglobin, 404
Oxy-ions, 327
Oxygen, atomic, 117
    chemical properties, 117–120
    discovery of, 111–113
    isotopes of, 123
    occurrence, 113–114
    origin of, 528
    physical properties, 114
    preparation, 114–116
    uses, 124
Ozone, 125–127

Palladium, see Platinum family
Palmitic acid, 474, 493
Paper, 509
Paraffin hydrocarbons, 446
Paraffin wax, 463
Para isomers, 458
Paramagnetic, 127
Parent compound, 458
Parent element, 273
Partial pressures, 247
Pauling, Linus, 72
Pectin, 225
Pentane, 447

Pentene, 447
Pentynes, 447
Peptide bond, 485
Peptization, 210
Peptizing agent, 210
Percentage composition, 102–103
Perchloric acid, 327
Period, 47
Periodic acid, 327
Periodic law, 24
Periodic table, 25–27, 47
    and filling orbitals, 54
    general divisions, 299
    Mendeleev's, 24–27
    trends in, 300–304
Peroxides, 180–181, 309–310
Perrin, Jean, 218
Petroleum, 462
    hydrocarbon fractions, 463
Petroleum ether, 463
Petroleum jelly, 463
Pewter, 442
pH, 260–262
    of some common solutions, 261
Phenol, 478
Phenolphthalein, 262
Phenols, 478–479
Phlogiston, 111
Phosphate ion, 394
Phosphine, 385
Phosphoric acid, 257, 393
Phosphoric oxide, 384, 394
Phosphorite, 386
Phosphorus, properties, 380–386
    red, 381
    uses, 387–389
    white, 381
    See also Nitrogen family
Phosphorus halides, 385
Phosphorus (III) oxide, 384
Phosphorus (V) oxide, 384
Phosphorus pentoxide, 384
Phosphorus trioxide, 384
Photochemical reaction, 317
Photoconductivity, 364
Photosynthesis, 528
Physical change, 16
Pi bond, 66, 451, 452
Pig iron, 440
Pile, 290
Planar structure, 426
Plane-polarized light, 486
Plankton, 531
Plasma, 293
Plaster of paris, 369
Plastics, 507–508
    addition, 504, 508
    condensation, 504, 508
Platinum, 417–424
Platinum family, 417–424
Plücker, J., 31
Plutonium, 283, 286
Polar covalent bonds, 71
Polar covalent compounds, 86
Polar molecule, 71
Polar solvent, 184

Polarized light, 486
Polonium, 363
Polyatomic ions, 71
Polychloroprene, 507
Polyethylene plastics, 507
Polymers, 504–508
Polymorphous crystals, 167
Polynucleotide, 487
Polypropylene, 508
Polysaccharides, 488, 491
Polysulfides, 372
Poly(vinyl chlorides), 507
Portland cement, 411
Positive ions, 86
Positron, 29
Potassium physical constants, 305
    See also Alkali family
Potassium acid tartrate, 407
Potassium chlorate, 329
Potassium chloride, 310, 318
Potassium cyanide, 426
Potassium dichromate, 471
Potassium ozonide, 310
Potassium perchlorate, 330
Potassium permanganate, 322
Potassium peroxide, 310
Potassium sulfate, 311
Potassium superoxide, 310
Potential energy, 2
Potentials, 345–349
Pound moles, 95
Precipitates, formation, 266
Prefixes, 538
Pressure of gases, 131–132
Priestley, Joseph, 112
Primary alcohols, 471
Principle of Le Chatelier, 242, 244
Prism spectroscope, 42
Propane, 447
Propanols, 466
Propene, 447
Properties, 15
    chemical, 16
    extrinsic, 15
    intrinsic, 15
    magnetic, 127
    physical, 15
Propionic acid, 472
Propyl alcohol, 467
Propyl radical, 458
Propylene, 447, 504
Propyne, 447
Protective colloids, 214
Proteins, 482–488
    metabolism, 499
    synthesis, 486
Protium, 123
Proton, 32
Pyrex glass, 412
Pyrite, 372
Pyrosulfuric acid, 375
Pyroxenes, 403

Quadridentate, 427
Quartz, 403
Quicklime, see Lime
Quinine, 510

Index

Radiant energy, 3
Radiation, detection, 272
   natural, 271, 272
Radical, 458
Radioactive decay, 274
Radioactive elements, 271
Radioactive families, 273
Radioactive isotopes, common
   elements, 273–275
   production, 282, 294
   rays, 271
   uses, 277, 290, 295
Radioactivity, 270
Radium, 271
   See also Alkaline earth
     family
Radon, 332
Ramsay, Sir William, 330
Raoult's law, 196
Rare earth metals, *see*
   Lanthanide series
Rate, of diffusion, 149
   of reaction, *see* Reaction rate
Rate constants, 233
Rayon, 509
Reaction, *see* Chemical
   Reaction
Reaction mechanism, *see*
   Mechanism of reaction
Reaction rate, 229, 234
   constant, 233
   determining step, 234
   factors affecting, 229–234
Reactor, nuclear, 290, 294
Realgar, 386
Reducing agent, 120
Reduction, 119–121
Reduction potentials, 345–348
Refining of metals, 435
Relative atomic weights, 22
René, 41, 442
Reversible reactions, 87
Rhenium, *see* Manganese
   family
Rhodium, *see* Platinum family
Ribonucleic acid, 488
RNA, 488
Roasting of ores, 434
Rocket propulsion, 124
Rocks, igneous, 524
   sedimentary, 526
Rubber, 505
Rubidium, 304
   See also Alkaline earth
     family
Rule of eight, 61
Ruthenium, *see* Platinum family
Rutherford, Ernest, 33
Rutherford's experiment, 33

Saccharin, 514
Salicylic acid, 479
Salt bridge, 341
Salts, 91
   naming, 79–80
   reaction with water, 177
   solubilities, 539
Sand, 409
Sandstone, 526

Sanger, F., 484
Saponification, 511
Saturated hydrocarbons, 454
Saturated solutions, 186
Scandium, 416–424
Scandium family, 416–424
Science, 1
Seawater composition, 526
Seaborg, G. T., 283
Secondary alcohols, 473
Sedimentary rock, 526
Seismograph, 520
Selenates, 377
Selenides, 370
Selenium, compounds of,
   370–377
   occurrence, 368
   properties, 363–368
   uses, 370
Semipermeable membranes, 204
Senarmonite, 386
Sexadentate, 427
Shale, 526
Sharing of electrons, 62
Shells of electrons, *see* Energy
   levels
Sigma bond, 64, 448, 454
Sigma orbital, 64
Significant figures, 10
Silica, 403, 409
Silica gel, 399
Silicates, 400, 403, 411
Silicic acid, 400
Silicon, 396–413
   chemical properties, 399–401
   occurrence, 403
   physical properties, 397
   preparation, 403
   uses, 403
Silicon dioxide, 403, 409
Silicon tetrafluoride, 326
Silicones, 412
Silver, *see* Copper family
   sterling, 442
Single bond, *see* Covalent bond
Slag, 434
Slaked lime, 309
Slaking, 309
Smokes, 209
Soap film, 213
Soaps, 511
Soda, *see* Sodium bicarbonate
Soda ash, *see* Sodium
   carbonate
Sodium, 304
   production, 435
   See also Alkali family
Sodium aluminum fluoride, 318
Sodium aluminum sulfate, 407
Sodium bicarbonate, 407
Sodium carbonate, 311, 407
Sodium chloride, 310, 318
Sodium dichromate, 322
Sodium glutamate, 512
Sodium hydroxide, 310
Sodium hypochlorite, 329
Sodium iodate, 318
Sodium nitrate, 386
Sodium palmitate, 212

Sodium periodate, 318
Sodium peroxide, 310
Sodium phosphate, 394
Sodium silicate, 400
Sodium sulfate, 311
Soil, 531
Solder, 442
Solid alcohol, 226
Solid solutions, 182
Solids, 130
   behavior of, 163–170
   crystalline, 165–169
   kinetic theory of, 163–164
   particles in, 164
Sols, 209
Solubility, of certain solids, 188
   factors influencing, 187–188
   of salts, 547
Solubility product constant, 265
   table of, 546
Solute, 182
Solutions, 181–205
   boiling point, 198
   buffered, 264
   concentration, 191
   defined, 182
   of electrolytes, 200
   freezing point, 197
   gaseous, 183
   ideal, 201
   liquid, 183
   molal, 193
   molar, 191
   normal, 193
   saturated, 186
   seeding of, 187
   solid, 182, 183
   standard, 194
   supersaturated, 187
   types of, 183
   unsaturated, 187
   vapor pressure, 198
   why substances dissolve,
     183–186
Solvay process, 310
Solvent, 182
Specific heat, 8, 102
Spectator ion, 92
Spectra, 41–44
   absorption, 44
   bright line, 42
   continuous, 41
   discontinuous, 41
   emission, 41
   flame, 42, 306
   theory of origin, 43
   visible, 41–44
   X-ray, 34
Spectrograph, mass, 35
Spectroscope, 42
Spectrum of the sun, 43
Speed of reaction, *see* Reaction
   rate
Sphalerite, 369
sp orbitals, 452
sp^2 orbitals, 451, 453
sp^3 orbitals, 65–66, 448
Spontaneous combustion, 119
Spontaneous reaction, 340, 351

Stalactites, 409
Stalagmites, 409
Standard conditions, 141
Standard electrode potentials, 345
Standard oxidation potentials, table of, 348
Standard pressure, 132
Standard temperature, 141
Starch, 491
States of matter, 130
Stearic acid, 474, 493
Steel, 440–441
　alloys of, 442
Stellar energy, 293
Sterling silver, 442
Stibine, 385
Stibnite, 386
Storage battery, 352
STP, 141
Stratosphere, 527
Strong acid, 89
Strong base, 89
Strong electrolyte, 88
Strontium, physical constants, 305
　　See also Alkaline earth family
Structural formula, 449
Styrene, 507
Subcritical mass, 288
Subgroups, see Families
Sublevels, see Energy levels
Sublimation, 164
Substitution reactions, 461
Sucaryl, 514
Sucrose, 490
Sugar, see Sucrose
Sugar, threshold, 498
Sulfa drugs, 510
Sulfanilamide, 496, 510
Sulfapyridine, 510
Sulfates, 377
Sulfathiazole, 510
Sulfides, metallic, 370
　nonmetal, 368
　poly-, 372
Sulfites, 374
Sulfur, amorphous, 365
　chemical properties, 367–368
　compounds of, 370–377
　Frasch process, 368
　monoclinic, 365
　occurrence, 368
　physical properties, 363–367
　polymorphism, 365
　production, 368
　rhombic, 365
　uses, 370
Sulfur dioxide, 373–374
Sulfur family, 363–378
Sulfur hexafluoride, 368
Sulfur monochloride, 368
Sulfur trioxide, 374
Sulfuric acid, 375–377
　contact process, 375
　ionization, 331
Sulfurite, 369

Sulfurous acid, 374
Supercritical mass, 288
Suspension, 208
Sylvinite, 318
Sylvite, 318
Symbols, 23–24
Synthetic rubber, 507
Synthetic sweeteners, 514

Tantalum, see Vanadium family
Tartaric acid, 475, 514
Technetium, 282, see Manganese family
Teflon, 507
Tellurates, 377
Tellurides, 370
Tellurium, compounds of, 370–377
　occurrence, 368
　properties, 363–368
　uses, 370
Temperature, 7, 136
　absolute, 134
　and equilibrium concentrations, 243
　and pressure, 136
　and reaction rates, 230
　and solubility, 187
　and vapor pressure, 159–161
　and volume, 132, 139
Temperature scales, 7
Terdentate, 427
Terephthalic acid, 507
Tetragonal crystals, 166
Tetrahedral, 426
Thallium, see Aluminum family
Theory, 2
Thermochemical equation, 97
Thermochemistry, 97
Thermodynamics, first law of, 94
　second law of, 185
Thermonuclear bombs, 294
Thermonuclear reactions, 293
Thomson, J. J., 35
Thorium, radioactive, 273
Thymolphthalein, 262
Tin, see Germanium family
Titanium, 417–424
Titanium family, 417–424
Titration, 194
TNT, 392
Toluene, 447, 453
Torricelli, E., 132
Tracers, radioactive, 295
Tranquilizers, 501
Trans isomer, 505
Transfer of electrons, 60
Transition elements, 301
　classification, 416
　color of ions, 423
　complex ions, 424–428
　oxidation states, 420
　properties, 418–428
Transition temperature, 365
Transmutation, 270
Transuranium elements, 282–283
Travertine, 408

Tremolite, 403
Trends in periodic table, 300–304
Triads, 24
Triclinic crystals, 166
Trinitrotoluene, 392
Triple bond, 383, 453
Triprotic acid, 253
Tritium, 123
Troposphere, 527
Tungsten, see Chromium family
Tyndall effect, 217
Type metal, 442

Ultramicroscope, 218
Ultraviolet radiation, 3
Units of measure, 545
Universe, composition, 518
　theory of origin, 517
Unsaturated hydrocarbons, 454
Uranium-235, 286
Uranium-238, 273
Uranium hexafluoride, 319
Uranium radioactivity series, 274

Valence, 70
Valence bonds, 71
Valence electrons, 59
Vanadium, 417–424
Vandium family, 417–424
Vanadium oxide, 375
van der Waals forces, 153
Vanillin, 513
Vapor pressure, 159
　and distillation, 202
　effect of solute, 198–201
　effect of temperature, 160
　measurement of, 160
　of solutions, 198–201
　of water, 161
Vaporization, 159
Vaseline, 463
Vinegar, 474
Vinyl alcohol, 477
Vinyl chloride, 507
Viruses, 499
Viscose, 509
Viscosity, 366
Visible spectra, 41–42
Vital force theory, 445
Vitamins, 496
Volatile, 159
Volt, 337
Voltaic cell, 341
　potential of, 345–349
　representation of, 349
Volume of gases, and pressure, 138
　and temperature, 139
　and weight, 151
Volumetric flask, 192
Vulcanization, 505

Washing soda, see Sodium carbonate
Water, association of, 174
　chemical properties, 175–179

Water (continued)
　density, 173, 174
　electrolysis, 116
　hard and soft, 409, 413
　heavy, 178
　hydrogen bonds in, 175
　ion product, 259
　ionization, 258
　pH of, 260
　physical properties, 172
　softening, 413
　structure, 174–175
　vapor pressure, 162, 545
Water-attracting group, 212
Water of crystallization, 177
Watson, J. D., 487

Watt, 337
Wavelength, of electromagnetic
　　radiation, 3
　of light, 3, 42
Weak acid, 89
Weak base, 89
Weak electrolyte, 88
Weathering of rocks, 531–532
Weight, 5
Wilson, C. T. R., 272
Wintergreen, 479
Wöhler, Frederick, 445
Wolfram, *see* Tungsten
Wood, 508
Wood alcohol, *see* Methanol
Wood's metal, 442

X-ray tube, 34
X rays, 3, 34
　and atomic number, 34
Xenon, 330–333
Xylenes, 447, 456, 457

Yttrium, 416

Zeolite, 413
Zinc, 417
Zinc family, 417, 419–423
Zinc sulfide, 272
Zirconium, *see* Titanium family
Zooplankton, 531

72 73 74 75 76 9 8 7 6 5 4 3 2 1

INTERNATIONAL ATOMIC WEIGHTS 1969

(Based on carbon − 12 = 12)

[a] A value in brackets denotes the mass number of the most stable known isotope.

[b] Names and symbols for elements 104 and 105 have not been officially accepted. See page 283.

	SYMBOL	ATOMIC NUMBER	ATOMIC WEIGHT[a]
Actinium	Ac	89	[227]
Aluminum	Al	13	26.9815
Americium	Am	95	[243]
Antimony	Sb	51	121.75
Argon	Ar	18	39.948
Arsenic	As	33	74.9216
Astatine	At	85	[210]
Barium	Ba	56	137.34
Berkelium	Bk	97	[247]
Beryllium	Be	4	9.01218
Bismuth	Bi	83	208.980
Boron	B	5	10.81
Bromine	Br	35	79.904
Cadmium	Cd	48	112.40
Calcium	Ca	20	40.08
Californium	Cf	98	[249]
Carbon	C	6	12.011
Cerium	Ce	58	140.12
Cesium	Cs	55	132.9055
Chlorine	Cl	17	35.453
Chromium	Cr	24	51.996
Cobalt	Co	27	58.9332
Copper	Cu	29	63.546
Curium	Cm	96	[245]
Dysprosium	Dy	66	162.50
Einsteinium	Es	99	[254]
Erbium	Er	68	167.26
Europium	Eu	63	151.96
Fermium	Fm	100	[253]
Fluorine	F	9	18.9984
Francium	Fr	87	[223]
Gadolinium	Gd	64	157.25
Gallium	Ga	31	69.72
Germanium	Ge	32	72.59
Gold	Au	79	196.9665
Hafnium	Hf	72	178.49